NONINDIGENOUS FRESHWATER ORGANISMS

Vectors, Biology, and Impacts

NONINDIGENOUS FRESHWATER ORGANISMS

Vectors, Biology, and Impacts

EDITED BY

Renata Claudi

and

Joseph H. Leach

LEWIS PUBLISHERS

Boca Raton London New York Washington, D.C.

Library of Congress Cataloging-in-Publication Data

Claudi, Renata
 Nonindigenous freshwater organisms: vectors, biology, and impacts
  p. cm.
 Includes bibliographical references.
 ISBN 1-5667-0449-9
 1. Freshwater ecology--North America. 2. Animal introduction-
-North America. 3. Nonindegenous aquatic pests--North America.
 I. Leach, J. (Joseph) II. Title.
 QH102.C56 +999 2000
 577.6'097—dc21
                    99-28607
                    CIP

© 1999 by CRC Press LLC
Lewis Publishers is an imprint of CRC Press LLC

No claim to original U.S. Government works
International Standard Book Number 1-5667-0449-9
Library of Congress Card Number 99-28607
Printed in the United States of America 3 4 5 6 7 8 9 0
Printed on acid-free paper

Preface

R. Claudi and J.H. Leach

Invasions of North American aquatic ecosystems by nonindigenous species have been ongoing for centuries. Invasions occurred through intentional introductions of various organisms as well as accidents that were by-products of other human activities. Concern in the scientific community about ecosystem damage from such invasions was insufficient to provoke widespread interest in the general public or legislative bodies. The arrival of the zebra mussel (*Dreissena polymorpha*) to the Laurentian Great Lakes in the 1980s changed all that, swiftly and dramatically. The actual and potential economic impacts of the zebra mussel as a biofouler immediately attracted the interest of the media, the public, and politicians. Probably never before in the history of invasions has so much attention been paid by so many to so small an organism. But then again, when has an invader cost as much hard cash as the various users of the Great Lakes water had to spend to keep operating? While the impact of the zebra mussels on the industry is by now well quantified, it will be a while before we can say with any certainty what the total impact of the zebra mussel has been on the Great Lakes ecosystem.

Although the various invasions of the aquatic ecosystems have been accomplished through a number of different pathways, the zebra mussel appears to have arrived in the ballast water of a foreign ship. This helped focus public concern and later legislative effort on the need to control ballast water discharge as the means to prevent further introductions of nonindigenous species. While partial ballast water control is now in place, other pathways of introduction still pose considerable risk of further invasions. Nonindigenous aquatic organisms continue to spread into new ecosystems throughout North America.

This book is a collection of chapters dealing with various aspects of nonindigenous aquatic organisms including their biology, as well as their ecological, and sometimes economic, impact. The chapters are organized into sections based on the primary pathway, or vector, of introduction. The final section of the book is devoted to how to predict the success or failure of an aquatic invader. Each pathway or vector is described at the beginning of the section, giving an overview of what potential exists for further introductions and some ideas on how to minimize this risk. The content of the volume appears to be centered on two main groups of organisms—fishes and molluscs. This was not intentional, but occurred because these organisms have caused most of the recognized ecosystem damage to date, and hence, more is known about them. Serious damage from other invasive organisms, including bacteria, viruses, and parasites, is possible and likely but largely undocumented at this time.

As this volume is multi-authored, some overlap between topics is inevitable. This will allow the reader to obtain information on a specific vector or organism without reading the entire book. For those who do read the entire book, it quickly becomes obvious that there are no easy answers. Still, unless we accept globalization of our ecosystems as inevitable, we need to devote more attention to preventing entry of nonindigenous organisms by any pathway.

Contributors

A.J. Benson
U.S.Geological Survey
Biological Resources
Florida Caribbean Science Center
7920 NW 71st Street
Gainesville, FL 32653-3071

C.P. Boydstun
U.S. Geological Survey
Biological Resources
Florida Caribbean Science Center
7920 NW 71st Street
Gainesville, FL 32653

Dr. A.M. Brasher
75-5776 Iuna Place
Kailua-kona, HI 96740

Dr. L.R. Brown
5083 Veranda Terrace
Davis, CA 95616

G. Castillo
Oregon Cooperative Fish and Wildlife
 Research Unit (USGS-BRD)
Department of Fisheries
 and Wildlife
Oregon State University
Corvallis, OR 97331

R. Claudi
1605 Bramsey Dr.
Mississauga, Ontario, L5J 2H8

A. Contreras-Arquieta
Cuemanco No. 617
Lomas de Anahuac
64260 Monterrey
Nuevo Leon, Mexico

S. Contreras-Balderas
Bioconservacion, A.C.
Apartado Postal 504,
66450 San Nicolas de los Garza
Nuevo Leon, Mexico

J.R. Chrisman
Cornell Biological Field Station
900 Shackelton Point Rd.
Bridgeport, NY 13030

Dr. J.M. Cooley
Great Lakes Laboratory of Fisheries and
 Aquatic Sciences
Fisheries and Oceans Canada
Bayfield Institute, P.O. Box 5050, 867
 Lakeshore Rd.
Burlington, Ontario L7R 4A6

M.A. Coscarelli
Michigan Department of Environmental Quality
P.O. Box 30473
Lansing, MI 48909-7973

Dr. W.R. Courtenay, Jr.
Dept. of Biological Sciences
Florida Atlantic University
Boca Raton, FL 33431

Dr. E.J. Crossman
Royal Ontario Museum
Ichthyology Section, CBCB
100 Queen's Park
Toronto, Ontario M5S 2C6

B.C. Cudmore
Royal Ontario Museum
Ichthyology Section, CBCB
100 Queen's Park
Toronto, Ontario M5S 2C6

A.J. Dextrase
Ontario Ministry of Natural Resources
Box 7000
Peterborough, Ontario K9J 8M5

J.A. Drake
Dept. of Zoology
University of Oklahoma
730 Van Vleet Oval, Room 314
Norman, OK 73019

G.L. Fahnenstiel
Great Lakes Environmental Laboratory, NOOA
Lake Michigan Field Station
1431 Beach Street
Muskegon, MI 49441

Dr. W.G. Franzin
Environmental Science Division
Freshwater Institute, Central and Arctic Region
Fisheries and Oceans Canada
510 University Crescent
Winnipeg, Manitoba R3T 2N6

M.M. Fuller
Dept. of Zoology
University of Oklahoma
730 Van Vleet Oval, Room 314
Norman, OK 73019

C.D. Goodchild
2064 Esson Line, RR 1,
Indian River, Ontario K0L 2B0

B.C. Harvey
U.S. Forest Service
Redwood Sciences Laboratory
1700 Bayview Drive, Arcata, CA 95521

K.T. Holeck
Cornell Biological Field Station
900 Shackelton Point Rd.
Bridgeport, NY 13030

T.H. Johengen
Cooperative Institute for Limnology and
  Ecosystem Research
University of Michigan
2200 Bonisteel Blvd.
Ann Arbor, MI 48909

Dr. J.H. Leach
Limnotechnics Environmental Consulting
111 Division St. S.
Kingsville, Ontario N9Y 1P5

D.P. Lewis
Ecological Services for Planning
361 Southgate Drive, Guelph
Ontario M1G 3M5

Dr. H.W. Li
Oregon Cooperative Fish and Wildlife
  Research Unit (USGS-BRD)
Department of Fisheries
  and Wildlife
Oregon State University
Corvallis, OR 97331

Dr. M.K. Litvak
Associate Professor of Biology
Centre for Coastal Studies
  and Aquaculture
University of New Brunswick
P.O. Box 5050
Saint John, New Brunswick E2L 4L5

Dr. C.Macias-Garcia
Instituto de Ecologia
Apartado Postal 70-275
Ciudad Universitaria
UNAM 04510, Mexico

Dr. G.L. Mackie
Dept. of Zoology
University of Guelph
Guelph, Ontario N1G 2W1

Dr. N.E. Mandrak
Dept. of Biological Sciences
One University Plaza
Youngstown, OH 44555-0001

M. Matthews
519 East 8th St. #2
Davis, CA 95616

Dr. R.F. McMahon
Center for Biological Macrofouling Research
Dept. of Biology
Box 19498
University of Texas at Arlington
Arlington, TX 76019

Dr. E.L. Mills
Cornell Biological Field Station
900 Shackelton Point Rd.
Bridgeport, NY 13030

Dr. C.K. Minns
Systems Ecologist
Great Lakes Laboratory of Fisheries and Aquatic
  Sciences
Fisheries and Oceans Canada
P.O. Box 5050, 867 Lakeshore Rd.
Burlington, Ontario L7R 4A6

Dr. T.F. Nalepa
Great Lakes Environmental Laboratory, NOOA
2205 Commonwealth Blvd.
Ann Arbor, MI 48105

Stan A. Orchard
Canadian Amphibian and Reptile
  Conservation Network
1745 Bank Street
Victoria, British Columbia V8R 4V7

Dr. P.H. Patrick
6345 Bell School Line
Milton, Ontario L9T 2T1

P.A. Rossignol
Department of Entomology
Oregon State University
Corvallis, OR 97331

M. Shafer
Elsag Bailey (Canada) Inc.
134 Norfinch Drive
Toronto, Ontario M3N 1X7

Dr. G.E. Whitby
Suntec Environmental
4699 Keele St., Unit 4
Toronto, Ontario M3J 2N8

C.J. Wiley
Transport Canada,
Ship Safety, Ontario Region
201 Front Street
Sarnia, Ontario N7T 8B1

D.G. Wright
Environmental Sciences Division
Freshwater Institute
Fisheries and Oceans Canada
501 University Crescent
Winnipeg, Manitoba R3T 2N6

Dr. L. Zambrano
Wolfson College
Oxford
OX2 6UD, U.K.

Table of Contents

Section III — Bait

Section IV — Ballast Water Vector

Section V — Aquaculture Vector

1 Documenting Over a Century of Aquatic Introductions in the United States

A.J. Benson

INTRODUCTION

Nonindigenous species, exotics, what is it all about? What do they mean for us and the environment? After thousands of introductions over the past several hundred years, we are finally taking notice of nonindigenous species and attempting to learn and to understand the effects and impacts they can have on our lives and our environment. Nonindigenous species deserve special attention if we are to preserve our natural resources and biodiversity. Not only have we seen the environment pay a price, but monitoring, eradication, and control of nonindigenous species have been very costly in the economic sense.

What we recognize as nonindigenous species first arrived in North America with the early European settlers. Even then, plants and animals were introduced intentionally and unintentionally as they continue to be today. We have defined nonindigenous aquatic species as any individual, group, subspecies, or population that enters a body of water or aquatic ecosystem outside of its historic native range. Most nonindigenous introductions are a result of human activities since the European colonization of North America. These include not only species that arrived from outside of North America, commonly referred to as "exotics," but, also species native to North America but introduced to waters outside their historic North American native ranges. An example of the former is the brown trout (*Salmo trutta*), a native of Europe first imported to the U.S. in 1883 from Germany as a food fish. An example of the latter would be the coho salmon (*Oncorhynchus kisutch*), a native of the Pacific coast from northern California to Alaska, which was introduced into the Great Lakes as early as the 1930s. The importance of studying nonindigenous organisms is in understanding the effects they may have on native organisms and the physical environment. In response to this, a large effort is underway to document the history and monitor the distribution of aquatic introductions in North America.

The Florida Caribbean Science Center (FCSC), in Gainesville, Florida, is part of the Biological Resources Division of the U.S. Geological Survey, an agency within the U.S. Department of the Interior. Since 1978, researchers at the FCSC have monitored the status and distribution of nonindigenous fish species in open waters in the U.S. Williams and Jennings (1991) gave the early history of this database. In 1995, FCSC biologists also began gathering information on introductions of other aquatic vertebrates, invertebrates, and plants. Data are compiled by searching publications and reports; by contacting state and federal agencies, land managers, researchers, taxonomists, and museum curators; and by searching laboratory and museum collections and computer databases. An informal network of biologists from many federal, state, and local governments, as well as from universities and private companies, was established to exchange information and openly discuss nonindigenous species issues. To facilitate information transfer to the user, these data are continually being compiled, verified, and stored in a geographic information system (GIS).

The GIS for nonindigenous aquatic species includes spatially referenced data sets of information on taxonomy, locality, date of collection, method of collection, disposition of specimens, origin, and status. The GIS currently contains more than 34,000 records of over 620 species of fishes, mammals, reptiles, amphibians, molluscs, crustaceans, bryozoans, coelenterates, plants, diseases, and parasites. Fuller et al. (1999) provide detailed descriptions of all nonindigenous freshwater fishes for which there is information in the database. This GIS is used as an analytical tool, providing valuable data and analyses to educators, researchers, natural resource managers, and policymakers.

LEGISLATION

UNITED STATES

The first legislation aimed at controlling unwanted introductions was the Lacey Act, passed in 1900. This act regulated the importation and interstate transfer of wildlife and allowed the Secretary of the Interior to designate injurious wildlife. The Federal Noxious Weed Act of 1974 was designated to prevent the further spread and introduction of noxious weeds in the U.S. However, it appears to be inadequately written, providing only partial protection. There is no regulation on interstate transport of noxious weeds once a weed enters the country. More recently, the harm caused by the introduction of the zebra mussel (*Dreissena polymorpha*) and the concern over a possible increase in the number of unintentional introductions resulted in the passage of a substantial piece of legislation, the Nonindigenous Aquatic Nuisance Prevention and Control Act of 1990 (NANPCA). This statute mandated development and implementation of a comprehensive national program to prevent and respond to problems caused by unintentional introductions of nonindigenous aquatic species into waters of the U.S. NANPCA called for the coordination of federally conducted, funded, or authorized research, prevention, control, information dissemination, and other activities regarding the zebra mussel and other aquatic nuisance species. An aquatic nuisance species is defined in the Act as any nonindigenous species that threatens the diversity or abundance of native species or the ecological stability of infested waters, or commercial, agricultural, aquacultural or recreational activities dependent on such waters. NANCPA addressed all nonindigenous aquatic species, not just extracontinental "exotics," and dealt primarily with unintentional introductions. It established a nationwide program to reduce the risk of more introductions and to understand and minimize economic and ecological impacts of nonindigenous aquatic nuisance species that become established. This meant developing and carrying out environmentally sound control methods to prevent, monitor, and control unintentional introductions of nonindigenous species from pathways other than ballast water exchange.

Special attention was given to preventing more ballast water introductions in the Great Lakes and other vectors associated with transcontinental shipping. A National Ballast Water Control Program was established, part of which consisted of a ballast exchange study. The study would assess the environmental effects of ballast water exchange on the diversity and abundance of native species in receiving estuarine, marine, and fresh waters of the U.S. It would also identify areas within the waters of the U.S. and the exclusive economic zone, if any, where the exchange of ballast water does not pose a threat of infestation or spread of aquatic nuisance species in the Great Lakes and elsewhere.

A task force made up of delegates from eight federal bureaus was established to develop and implement a program for the waters of the U.S. to prevent introductions and dispersal of aquatic nuisance species. The program was to provide technical assistance to state and local government by assisting states with nonindigenous management plans, by providing grants, and educating the affected entities and public on how to identify and prevent spread. Research was to be conducted concerning the environmental and economic risks associated with the introduction of aquatic nuisance species into U.S. waters; the principal pathways by which aquatic nuisance species are introduced and dispersed; possible methods for the prevention, detection, monitoring, and control of aquatic nuisance species; and the assessment of the effectiveness of prevention, monitoring, and control methods. The task force also conducted an intentional introductions policy review in 1994 in which their recommendations promoted education, cooperation, and accountability. Because prevention is the best way to lower risk of a potentially harmful introduction, recommendations centered around the decision-making process whether to introduce, or not to introduce (Aquatic Nuisance Species Task Force, 1994). On October 3, 1996, the NANPCA was reauthorized and renamed the National Invasive Species Act of 1996. This new legislation focused on ballast water management and is aimed at preventing the introduction of more nonindigenous aquatic species.

CANADA

In Canada, there is a host of regulations and policies to prevent unwanted aquatic species from entering the country and to control intentional introductions. Intentional introductions of aquatic organisms are governed by the Federal Fisheries Act. The use of bait is controlled by the provinces by regulations made under the Federal Fisheries Act. These regulations also prohibit the importation and possession of certain undesirable species in some provinces. Aquaculture is controlled through provincial fish and wildlife statutes. Ballast water in the Great Lakes is dealt with through voluntary guidelines and pollution prevention provisions under the Canada Shipping Act. The aquarium trade and the live food fish industry are largely unregulated with respect to import. The import and control of some plant species is regulated under the Federal Seeds Act and Provincial Weed Acts. A new piece of federal legislation has a "dirty list" provision similar to the Lacey Act in the U.S., but no aquatic species have been listed to date. With the steady rate of increase of aquatic introductions, the current regulations and policies appear inadequate. Additional legislation and a better effort toward enforcement and compliance may help eliminate unwanted introductions.

TYPES OF ORGANISMS

As previously stated, nonindigenous aquatic species are not confined to one taxa of organisms (Table 1.1). For ease of classification and discussion, information on freshwater nonindigenous species has been organized in the GIS by the following taxa: amphibians, bryozoans, coelenterates, crustaceans, fishes, mammals, molluscs, plants, and reptiles. We have excluded nonindigenous

TABLE 1.1
Numbers of Nonindigenous Aquatic Species Introduced into the United States

	Foreign to United States	Native to United States
Amphibians	13	24
Bryozoans	1	2
Coelenterates	4	1
Crustaceans	15	15
Fishes	176	331
Mammals	1	0
molluscs	29	7
Plants	74	15
Reptiles	21	33

diseases and parasites from some of the discussion because our efforts until recently have focused mostly on fishes; therefore, the database has few entries for diseases and parasites. However, we recognize the extreme importance of nonindigenous diseases and parasites. A few examples of their impacts are mentioned later.

Because of the recent onslaught of exotic aquatic introductions, it was determined to be equally as important to document not only species of fish, but other vertebrates and invertebrates as well (Table 1.2). Therefore, the objective became clear to document occurrences of all nonindigenous aquatic species. We consider species as "aquatic" if they are obligated to live in a water body or wetland during all or part of their lives. Amphibians are included because all members of this group use aquatic habitats at the larval stage and most as adults. A few examples of amphibians reported outside their native range are the marine toad (*Bufo marinus*), the Cuban treefrog (*Osteopilus septentrionalis*), and the African clawed frog (*Xenopus laevis*). Many reptiles are considered to be aquatic because of the amount of time spent using aquatic habitats. Examples of reptiles introduced outside their native ranges are the red-eared slider (*Trachemys scripta elegans*), the spectacled caiman (*Caiman crocodilis*), and the Rio Magdalena river turtle (*Podocnemis lewyana*). The group that has the fewest species considered to be aquatic and that spends a minimal time in the water relative to the other groups is the mammals. The nutria (*Myocastor coypus*), a rodent similar in appearance to the muskrat but larger in size, spends much of its time in the water. It feeds on aquatic and semiaquatic vegetation and can breed all year round (Davis, 1974). It is a native of South America, but can now be found in the U.S. from Delaware to Texas, Oregon, and Washington Nutria have also been known to occur in Indiana, Michigan, Ohio, Idaho, and Montana.

The true aquatic organisms have evolved to live their entire lives in the water. Organisms such as bryozoans are completely dependent on aquatic habitats. They are sessile organisms with characteristic ciliated tentacles surrounding the mouth and live in large colonies attached to submerged surfaces. Individual bryozoans called zooids, identical in structure and only several millimeters in length, seamlessly fuse together to form a single structure and can be found in almost any freshwater lake or stream. Bryozoans have the reputation as biofoulers in water intake pipes (Wood 1991). An example of a bryozoan reported outside its native range is the freshwater bryozoan (*Lophopodella carteri*) found in the Great Lakes and in several Hawaiian reservoirs. Coelenterates consist of animals we know as jellyfish, anemones, corals, and polyps. Most coelenterates are sessile; however, jellyfish are free-swimming, making them more susceptible to being relocated. Coelenterates are characterized by structures called cnidaria or nematocysts. Nematocysts are long and needle-like and are fired into the coelenterate's prey (Slobodkin and Bossert 1991). Most coelenterates are marine inhabitants, but some such as the Black Sea jellyfish (*Maeotias inexspectata*) are found in fresh water. It is one of the few coelenterates found outside its native European range in North America. Crustaceans, which includes crayfish, crabs, shrimps, amphipods, cladocerans, and copepods, inhabit most every aquatic environment, but only a small portion of this group inhabits fresh water. Examples of nonindigenous crustaceans in North America are the spiny water flea (*Bythotrephes cederstroemi*) in the Great Lakes and the Chinese mitten crab (*Eriocheir sinensis*) found in the San Francisco Bay area. Both are assumed to be ballast water introductions from Europe and Asia, respectively. Not unlike the crustaceans, fishes are also found in nearly every aquatic environment. They have been introduced for a wide variety of reasons both intentionally and unintentionaly. Several examples of nonindigenous fishes found in the U.S. are the common carp (*Cyprinus carpio*), brown trout (*Salmo trutta*), and Mozambique tilapia (*Tilapia mossambicus*), each originating from a different continent. Within the molluscs, there are two large groups, bivalves and gastropods. Both groups inhabit freshwater lakes, streams, rivers, and terrestrial habitats all across North America. Some examples of several nonindigenous molluscs are the Asian clam (*Corbicula fluminea*) and zebra mussel representing the bivalves, and the New Zealand mudsnail (*Potamopyrgus antipodarum*) representing the gastropods.

Nonindigenous aquatic plants are also common to aquatic ecosystems. In the database, plants are categorized by natural divisions that include ferns, dicots, and monocots, as well as a variety

TABLE 1.2
Nonindigenous Aquatic Species Introduced in the United States

Name	Nonindigenous Distribution	Pathway	Origin
Amphibians			
Acris crepitans (northern cricket frog)	CO	stocked	North America
Ambystoma tigrinum (tiger salamander)	AZ, CA, CO, LA, NV, TX	bait release	North America
Atelopus zeteki (Panamanian golden frog)	FL	released pets	unknown
Bufo americanus (American toad)	MA	released pets	North America
Bufo blombergi (Colombian giant toad)	FL	released pets	unknown
Bufo boreas (California toad)	HI	biocontrol	North America
Bufo marinus (marine toad)	FL, HI, MA	released pets / biocontrol	South and Central America
Cynops pyrrhogaster (Japanese fire-bellied salamander)	FL, MA	released pets	Asia
Dendrobates auratus (dart poison frog)	HI	biocontrol	Central America
Dendrobates leucomelas (yellow banded poison arrow frog)	HI	unknown	South and Central America
Desmognathus quadramaculatus (blackbelly salamander)	GA	bait release	North America
Desmognathus santeetlah (Santeetlah dusky salamander)	GA	unknown	North America
Eleutherodactylus coqui (coqui)	FL, LA	in plant shipments	Puerto Rico
Eleutherodactylus planirostris planirostris (greenhouse frog)	FL, LA	released pets	North America
Eleutherodactylus portoricensis (frog)	FL	released pets	unknown
Hyla cinerea (green treefrog)	MA, TX	in plant shipments	North America
Hyla eximia (mountain treefrog)	AZ	bait release	North America
Hyla gratiosa (barking treefrog)	NJ	unknown	North America
Hymenochirus boettgeri (Zaire dwarf clawed frog)	FL	released pets	Africa
Litoria aurea (green and golden bell frog)	HI	stocked	Africa
Necturus maculosus (mudpuppy)	CT, MA, ME, NH, RI, VT	aquarium release	North America
Notophthalmus viridescens (red spotted newt)	FL	released pets	North America
Osteopilus septentrionalis (Cuban treefrog)	CO, FL, HI, MD	accidentally in shipping crates	North America
Pachymedusa dacnicolor (red-eyed treefrog)	FL	released pets	Central America
Pseudacris regilla (Pacific chorus frog)	AK, AZ, CO	released pets	North America
Rana aurora (red-legged frog)	NV	food	North America
Rana berlandieri (Rio Grande leopard frog)	AZ, CA	stock contamination	North America

TABLE 1.2 (CONTINUED)
Nonindigenous Aquatic Species Introduced in the United States

Name	Nonindigenous Distribution	Pathway	Origin
Rana blairi (plains leopard frog)	CO	accidental	North America
Rana catesbeiana (bullfrog)	AZ, CA, CO, HI, ID, MA NM, NV, OR, UT, WA, WY	stocked	North America
Rana clamitans (green frog)	AZ, UT, WA	stocked	North America
Rana nigromaculata (black spotted frog)	HI	biocontrol	Asia
Rana pipiens (Northern leopard frog)	CA, HI, MA, NV	unknown	North America
Rana rugosa (wrinkled frog)	HI	biocontrol	Asia
Rana utricularia (Southern leopard frog)	MA	released pets	North America
Syrrhophus cystignathoides (Rio Grande chirping frog)	TX	in plant shipments	North America
Taricha granulosa (rough-skinned newt)	ID	released pets	North America
Xenopus laevis (African clawed frog)	AZ, CA, CO, FL, MA, VA	released pets	Africa
Bryozoans			
Lophopodella carteri (freshwater bryozoan)	HI, Lake Erie, MI	unknown	Asia
Plumatella repens (freshwater bryozoan)	HI, Lake Erie	unknown	North America
Urnatella gracilis (ectoproct)	CA	aquarium release	North America
Coelenterates			
Blackfordia virginica (Black Sea jellyfish)	CA, VA	ballast water	Europe
Cordylophora caspia (hydroid)	CA, OH	ballast water	North America
Craspedacusta sowerbyi (freshwater jellyfish)	HI, KY, MI, NY, RI	aquarium release	Asia
Maeotias inexspectata (Black Sea jellyfish)	CA, SC, VA	ballast water	Europe
Moerisia lyonsi (jellyfish)	VA	ballast water	Africa
Crustaceans			
Argulus japonica (parasitic copepod)	widely in the U.S.	aquarium trade	Asia
Bythotrephes cederstroemi (spiny water flea)	All Great Lakes, MN	ballast water	Europe
Cercopagis pengoi (water flea)	Lake Ontario	ballast water	Asia
Daphnia lumholtzi (daphnia)	AL, AR, FL, IL, KS, KY, LA, MO, MS, NC, OH, OK, SC, TN, TX	unknown	Africa and Australia
Echinogammarus ischnus (an amphipod)	Lakes Erie, Huron, Ontario	ballast water	Europe
Eriocheir sinensis (Chinese mitten crab)	CA, LA, WA, Lakes St. Clair, Erie	ballast water	Asia
Eubosmina coregoni (water flea)	All Great Lakes, NY, PA, VT	ballast water	Europe

TABLE 1.2 (CONTINUED)
Nonindigenous Aquatic Species Introduced in the United States

Name	Nonindigenous Distribution	Pathway	Origin
Eurytemora affinis (calanoid copepod)	All 5 Great Lakes	ballast water	Atlantic Ocean
Macrobrachium lar (Tahitian prawn)	HI	stocked	South Pacific
Macrobrachium olfersi (bristled river shrimp)	FL	released	North America
Macrobrachium rosenbergii (Malaysian prawn)	HI	aquaculture escape	Asia
Mysis relicta (opossum shrimp)	CA, CO, ID, MT, NV, WY	stocked as forage	North America
Orconectes causeyi (crayfish)	NM	bait bucket release	North America
Orconectes immunis (calico crayfish)	CT, ME, MA, NH, RI, VT	bait bucket release	North America
Orconectes rusticus (rusty crayfish)	CT, IL, IA, ME, MA, MN, NH, NJ, NM, NY, PA, TN, VT, WV, WI	bait bucket release	North America
Orconectes virilis (virile crayfish)	AL, AZ, CA, CT, MA, ME, MD, MS, NH, NJ, NM, PA, RI, TN, VT, VA, WV	bait bucket release	North America
Pacifastacus leniusculus (signal crayfish)	CA, NV, UT	bait bucket release, stocked	North America
Palaemonetes plaudosus (riverine grass shrimp)	AZ, CA	stocked	North America
Platychirograpsus spectabilis (saber crab)	FL	on imported logs	Central America
Procambarus acutus (White River crayfish)	CA, CT, GA, ME, MA, RI	bait bucket release	North America
Procambarus clarkii (red swamp crawfish)	AZ, CA, GA, HI, ID, MD, NV, NM, NC, OH, OR, SC, UT, VA	stocked	North America
Rhithropanopeus harrisii (Harris mud crab)	CA	ballast water	North America
Skistodiaptomus pallidus (calanoid copepod)	Lakes Erie, Ontario, St. Clair	bait bucket, stock contamination	North America

Fishes (Introduced into 10 or More U.S. States) (Modified from Fuller et al., 1999)

Name	Nonindigenous Distribution	Pathway	Origin
Cyprinus carpio (common carp)	all 50 U.S. states	stocked for food	Asia
Carassius auratus (goldfish)	all U.S. states except AK	ornamental stocking, bait release	Asia
Oncorhynchus mykiss (rainbow trout)	all U.S. states	stocked for sport	North America
Salmo trutta (brown trout)	all U.S. states except AL and MS	stocked for food and sport	Europe
Ctenopharyngodon idella (grass carp)	all U.S. states except CT, ME, MT, RI, and VT	biocontrol	Asia
Stizostedion vitreum (walleye)	All U.S. states except FL, IL, MO, ND, and RI	stocked for sport	North America
Micropterus salmoides (largemouth bass)	All U.S. states except AL, FL, IN, LA, SC, TN	stocked for sport	North America
Micropterus dolomieu (smallmouth bass)	All U.S. states except FL, IA, IN, LA, OH, TN, VT	stocked for sport	North America

TABLE 1.2 (CONTINUED)
Nonindigenous Aquatic Species Introduced in the United States

Name	Nonindigenous Distribution	Pathway	Origin
Salmo salar (Atlantic salmon) 40 states	AK, CA, CO, CT, IA, ID, IL, IN, KS, KY, MA, MD, ME, MI, MN, MO, MS, MT, NC, ND, NE, NH, NJ, NM, NV, NY, OH, OR, PA, RI, SC, TN, TX, UT, VA, VT, WA, WI, WV, WY	stocked for sport	North America
Salvelinus fontinalis (brook trout) 39 states	AK, AR, AZ, CA, CO, CT, DE, GA, HI, ID, IL, IN, KS, KY, MD, ME, MI, MN, MT, NC, ND, NE, NH, NJ, NM, NV, NY, OH, OR, PA, SC, SD, TN, TX, UT, VA, WA, WI, WY	stocked for sport	North America
Pomoxis annularis (white crappie) 38 states	AL, AR, AZ, CA, CO, CT, DE, FL, GA, ID, IN, KS, KY, MA, MD, MN, MT, NC, ND, NE, NJ, NM, NV, NY, OH, OK, OR, PA, SC, SD, TX, UT, VA, VT, WA, WI, WV, WY	stocked for sport	North America
Esox lucius (northern pike) 37 states	AK, AR, AZ, CA, CO, CT, GA, IA, ID, KS, KY, MA, MD, ME, MN, MO, MT, NC, NE, NH, NJ, NM, NV, OH, OK, OR, PA, RI, SD, TN, TX, UT, VA, WA, WI, WV, WY	stocked for sport	North America
Lepomis macrochirus (bluegill) 36 states	AR, AZ, CA, CO, CT, DE, GA, HI, ID, IL, IN, KS, KY, MA, MD, MI, MN, MT, NC, NE, NH, NJ, NM, NV, NY, OH, OK, OR, RI, SC, SD, TX, UT, WA, WV, WY	stocked for sport	North America
Morone saxatilis (striped bass) 36 states	AL, AR, AZ, CA, CO, FL, GA, HI, IA, IL, IN, KS, KY, LA, MD, MN, MO, MS, NC, ND, NE, NJ, NM, NV, NY, OH, OK, OR, PA, SC, TN, TX, UT, VA, WA, WV	stocked for sport	North America
Pimephales promelas (fathead minnow) 35 states	AL, AR, AZ, CA, CO, CT, DE, FL, GA, ID, KY, LA, MA, MD, ME, MN, MO, MS, MT, NC, NE, NH, NV, OH, OK, OR, PA, TN, TX, UT, VA, WA, WI, WV, WY	stocked as forage, bait bucket release	North America
Gambusia affinis (mosquitofish) 34 states	AK, AL, AZ, CA, CO, FL, HI, IA, ID, IL, IN, KS, KY, MA, MI, MN, MO, MS, MT, NC, NE, NJ, NM, NV, NY, OH, OR, PA, TN, TX, UT, WA, WI, WY	stocked for biocontrol	North America

TABLE 1.2 (CONTINUED)
Nonindigenous Aquatic Species Introduced in the United States

Name	Nonindigenous Distribution	Pathway	Origin
Ictalurus punctatus (channel catfish) 34 states	AR, AZ, CA, CO, CT, DE, GA, HI, IA, ID, IL, KS, MA, MD, MN, MO, MC, NE, NH, NJ, NM, NV, OH, OK, OR, SC, TX, UT, VA, VT, WA, WI, WV, WY	stocked for sport	North America
Lepomis cyanellus (green sunfish) 33 states	AZ, CA, CO, CT, DE, FL, GA, HI, ID, MA, MD, ME, MI, MN, MT, NC, ND, NJ, NM, NV, NY, OH, OR, PA, SC, SD, TX, UT, VA, WA, WI, WV, WY	stock contaminant, stock misidentified	North America
Pomoxis nigromaculatus (black crappie) 33 states	AZ, CA, CO, CT, DE, ID, IN, KS, KY, MA, MD, ME, MT, ND, NE, NH, NJ, NM, NV, NY, OK, OR, PA, RI, SD, TX, UT, VA, VT, WA, WI, WV, WY	stocked for sport	North America
Oncorhynchus kisutch (coho salmon) 32 states	AK, AZ, CA, CO, IA, ID, IL, IN, KY, MA, MD, ME, MI, MN, MT, ND, NE, NH, NM, NV, NY, OH, OR, PA, SD, TN, TX, UT, VA, WA, WI, WY	stocked for sport	North America
Perca flavescens (yellow perch) 31 states	AL, AZ, CA, CO, FL, GA, ID, IL, IN, KS, KY, MD, MO, MT, NC, NE, NJ, NM, NV, NY, OH, OK, OR, SC, SD, TN, TX, UT, WA, WV, WY	stocked for sport	North America
Tinca tinca (tench) 31 states	AL, AR, AZ, CA, CO, CT, DE, FL, GA, IA, ID, IL, IN, KS, KY, LA, MA, MD, MI, MO, NC, NE, NM, NV, NY, OH, OK, OR, PA, SC, TN, TX, UT, VA, WA, WI, WV	stocked for sport	Europe
Osmerus mordax (rainbow smelt) 29 states	AR, CO, CT, GA, IA, ID, IL, IN, KS, KY, LA, MA, MD, MI, MN, MO, MT, NC, ND, NE, NH, NY, OH, PA, SD, TN, VA, VT, WI	stocked for forage	North America
Salvelinus namaycush (lake trout) 29 states	CA, CO, CT, IA, ID, IL, IN, KS, KY, MA, MD, MI, MN, MT, ND, NE, NH, NJ, NM, NV, NY, OR, PA, SD, TN, UT, VA, WA, WI	stocked for sport	North America
Ambloplites rupestris (rock bass) 28 states	AR, AZ, CA, CO, CT, DE, ID, IN, KS, MA, MD, MO, MT, NC, NE, NH, NJ, NM, OK, OR, PA, SD, TX, UT, VA, WA, WV, WY	stocked for sport	North America

TABLE 1.2 (CONTINUED)
Nonindigenous Aquatic Species Introduced in the United States

Name	Nonindigenous Distribution	Pathway	Origin
Dorosoma petenense (threadfin shad) 26 states	AL, AR, AZ, CA, CO, DE, GA, HI, IL, KS, KY, MD, MO, NC, ME, NM, NV, OH, OK, OR, PA, SC, TN, UT, VA, WV	stocked for forage	North America
Esox masquinongy (muskellunge) 26 states	AR, AZ, CA, CO, DE, GA, IA, IL, KY, MA, MD, ME, MN, MO, NC, ND, NE, NJ, NY, OH, PA, SD, TN, TX, VA, WV	stocked for sport	North America
Lepomis microlophus (redear sunfish) 26 states	AR, AZ, CA, CO, DE, IA, IL, IN, KS, KY, MI, MO, NC, NE, NM, NV, OH, OK, OR, PA, SC, TN, TX, UT, VA, WV	stocked for sport	North America
Oncorhynchus tshawytscha (chinook salmon) 26 states	CO, HI, ID, IL, IN, KS, MA, MD, ME, MI, MN, MT, ND, NE, NH, NJ, NV, NY, OH, PA, SD, TX, UT, VA, VT, WI	stocked for sport	North America
Morone chrysops (white bass) 24 states	AL, AR, AZ, CA, CO, DE, FL, GA, KS, KY, MO, MS, MT, NC, ND, NE, NJ, NM, NV, OH, OK, PA, SC, SD, TN, TX, UT, VA, WA WV	stocked for sport	North America
Lepomis gibbosus (pumpkinseed) 23 states	AZ, CA, CO, IA, ID, IN, KY, MD, MN, MO, MT, NC, ND, NE, NV, OH, OR, PA, SD, TN, WA, WV, WY	stocked for sport	North America
Notemigonus crysoleucas (golden shiner) 23 states	AR, AZ, CA, CO, FL, IA, KY, MN, MT, NC, NE, NM, NV, NY, OH, OR, SD, TN, TX, UT, VA, WV, WY	stocked as forage, bait bucket release	North America
Oncorhynchus nerka (sockeye salmon) 22 states	AZ, CA, CO, CT, ID, MA, MN, MT, NC, ND, NE, NM, NV, NY, PA, SD, TN, UT, VT, WA, WI, WY	stocked for sport	North America
Esox lucius x masquinongy (tiger muskellunge) 21 states	AR, CO, DE, IA, ID, IL, KS, MA, MI, MN, MO, MT, ND, NE, NY, PA, SD, TX, VA, WA, WI	stocked for sport	North America
Scardinius erythrophthalmus (rudd) 21 states	AL, AR, CO, CT, IL, KS, MA, ME, MO, MS, NE, NJ, NY, OK, PA, SD, TX, VA, VT, WI, WV	stocked for sport, bait bucket release	Europe
Alosa sapidissima (American shad) 20 states	AK, AZ, CA, CO, ID, IL, MA, MD, NC, NE, NJ, NV, NY, OH, OR, TX, UT, VA, WA, WI	stocked for sport, food, forage	North America
Ameiurus melas (black bullhead) 20 states	AZ, CA, CO, CT, FL, GA, ID, KS, MA, MD, MT, NC, NM, NV, OR, TX, UT, VA, WA, WI	stocked for sport	North America
Esox niger (chain pickerel) 20 states	AR, CO, CT, FL, IA, IN, KY, MA, MD, MN, NE, NJ, NY, OH, OK, PA, TX, VA, VT, WV	stocked for sport	North America

TABLE 1.2 (CONTINUED)
Nonindigenous Aquatic Species Introduced in the United States

Name	Nonindigenous Distribution	Pathway	Origin
Ictalurus furcatus (blue catfish) 20 states	AL, AR, CA, CO, FL, GA, ID, MD, MN, MO, NC, NJ, NM, OH, OK, OR, PA, SC, TX, VA	stocked for sport	North America
Oncorhynchus clarki (cutthroat trout) 20 states	AR, AZ, CA, CO, ID, MD, MI, MN, ND, NE, NJ, NM, NV, SD, TN, UT, VA, VT, WI, WY	stocked for sport, restoration	North America
Serrasalmus or Piaractus sp. (piranha sp.) 20 states	CA, FL, HI, IA, ID, IL, KY, MA, MI, MN, MO, NH, NY, OH, OK, PA, TX, UT, VA, WA	aquarium release	South America
Ameiurus nebulosus (brown bullhead) 19 states	AR, AZ, CA, CO, HI, ID, KS, KY, LA, MO, NE, NH, NM, NV, OH, OR, VA, WA, WV	stocked for sport	North America
Pylodictis olivaris (flathead catfish) 19 states	AR, AZ, CA, CO, FL, GA, ID, KS, NC, NE, NM, OK, OR, SC, TX, VA, WA, WI, WY	stocked for sport	North America
Thymallus articus (Arctic grayling) 19 states	AK, AZ, CA, CO, ID, ME, MI, MN, MT, NE, NM, NV, OR, UT, VA, VT, WA, WI, WY	stocked for sport	North America
Coregonus clupeaformis (lake whitefish) 18 states	CA, CO, IA, ID, IN, ME, MN, MT, ND, NE, NH, NV, NY, OR, PA, SD, UT, WA	stocked for sport, commercial fishing	North America
Lepomis gulosus (warmouth) 18 states	AZ, CA, CO, DE, ID, IN, KS, KY, MD, NC, NM, NV, NY, OH, OR, TX, WA, WI	stocked for sport	North America
Ameiurus catus (white catfish) 17 states	AL, AR, CA, CT, FL, IL, IN, KY, MA, MS, NC, NV, OH, OR, PA, RI, WA	stocked for sport	North America
Micropterus punctulatus (spotted bass) 17 states	AL, AZ, CA, CO, FL, GA, IA, KS, KY, MO, NC, NE, NM, NV, TN, VA, WV	stocked for sport	North America
Hypophthalmichthys nobilis (bighead carp) 16 states	AL, AR, AZ, CO, FL, IA, IL, IN, KS, KY, LA, MO, MS, OH, TN, TX	aquaculture escapes, stocked for biocontrol	Asia
Alosa pseudoharengus (alewife) 15 states	CO, GA, IL, IN, KY, MN, NE, NH, NY, OH, PA, TN, VA, WI, WV	via canals	North America
Stizostedion canadense (sauger) 15 states	AL, AR, CO, FL, GA, ID, LA, MS, NC, NE, OK, PA, SC, TX, WI	stocked for sport	North America
Poecilia reticulata (guppy) 14 states	AZ, CA, CO, FL, HI, ID, MO, MT, NM, NV, TX, UT, WI, WY	aquarium release	North America
Lepomis auritus (redbreast sunfish) 13 states	AL, AR, GA, KY, LA, NC, NY, OK, PA, TN, TX, VA, WV	stocked for sport	North America
Morone americana 13 states	CO, IL, IN, MA, MI, MN, NE, NH, NY, OH, PA, VA, WI	via canals	North America
Tilapia mossambicus (Mozambique tilapia) 13 states	AL, AZ, CA, CO, FL, GA, HI, ID, IL, MT, NC, NY, TX	stocked for sport, biocontrol, food	Africa

TABLE 1.2 (CONTINUED)
Nonindigenous Aquatic Species Introduced in the United States

Name	Nonindigenous Distribution	Pathway	Origin
Oreochromis aureus (blue tilapia) 12 states	AL, AZ, CA, CO, FL, GA, ID, KS, NC, OK, PA, TX	stocked for sport, biocontrol, food	Africa
Oncorhynchus aguabonita (golden trout) 11 states	AZ, CA, CO, ID, MT, NM, NV, OR, UT, WA, WY	stocked for sport	North America
Tilapia zillii (redbelly tilapia) 11 states	AL, AR, AZ, CA, FL, HI, ID, NC, NV, SC, TX	aquaculture escapes, biocontrol, food	Africa
Xiphophorus helleri (green swordtail) 11 states	AZ, CA, CO, FL, HI, ID, MT, NV, OK, TX, WY	aquarium release	North America
Mammals			
Myocastor coypus (nutria)	AL, DE, FL, GA, ID, IN, LA, MD, MI, MS, MT, NC, NY, OH, OR, SC, TX, VA, WA	escaped farms, farm releases	South America
molluscs			
Biomphalaria glabrata (bloodfluke planorb)	FL	unknown	South America
Bithynia tentaculata (mud bithynia)	IL, MI, MT, NY, OH, PA, VA, VT, WI	in packaging material or ballast water	Europe
Cipangopaludina chinensis malleata (Chinese mysterysnail)	AZ, CA, DE, FL, HI, MA, MD, ME, MI, MN, NC, NE, NH, NJ, NY, OH, OK, PA, RI, TX, VT, WI	imported as live food, aquarium releases	Asia
Cipangopaludina japonica (Japanese mysterysnail)	IN, MI, OH	imported as live food	Asia
Corbicula fluminea (Asian clam)	AL, AR, AZ, CA, CT, DE, FL, GA, HI, IA, ID, IL, IN, KS, KY, LA, MD, MI, MN, MO, MS, NC, NE, NJ, NM, NV, NY, OH, OK, OR, PA, SC, TN, TX, UT, VA, WA, WI, WV	imported as live food	Asia
Dreissena polymorpha (zebra mussel)	AL, AR, CT, IL, IN, IA, KY, LA, MI, MN, MO, MS, NY, OH, OK, PA, TN, VT, WI, WV	ballast water	Eurasia
Dreissena bugensis (quagga mussel)	Lakes Erie, Huron, Ontario	ballast water	Eurasia
Drepanotrema aeruginosum (rusty rams-horn)	unspecified	unknown	unknown
Drepanotrema cimex (ridged rams-horn)	unspecified	unknown	unknown
Drepanotrema kermatoides (crested rams-horn)	unspecified	unknown	unknown
Elimia virginica (Piedmont elimia snail)	NY	canals	Atlantic Ocean
Gillia altilis (buffalo pebblesnail)	NY	canals	Atlantic Ocean
Helisoma sp. (unidentified rams-horn)	HI	unknown	North America
Lasmigona subviridis (green floater mussel)	NY, VA, WV	canals	Atlantic Ocean

TABLE 1.2 (CONTINUED)
Nonindigenous Aquatic Species Introduced in the United States

Name	Nonindigenous Distribution	Pathway	Origin
Marisa cornuarietis (giant rams-horn snail)	FL, TX	released pets	South America
Melanoides tuberculatus (red-rimmed melania)	AZ, CA, FL, MT, TX	released pets	Africa
Melanoides turricula (faune melania)	FL	unknown	South Pacific
Mytilopsis leucophaeta (dark falsemussel)	AL, KY, TN, WV	shipping activities	Atlantic Ocean
Physa skinneri (glass physa)	unspecified	unknown	unknown
Physella acuta (European physa)	HI	unknown	Europe
Pisidium amnicum (greater European peaclam)	NJ, NY, PA	shipping activities	Europe
Pisidium henslowanum (Henslow peaclam)	Great Lakes	unknown	Europe
Pisidium supinum (humpback peaclam)	unspecified	unknown	unknown
Pomacea bridgesi (spiketop applesnail)	FL	aquarium release	S. America
Pomacea canaliculata (applesnail)	HI	aquarium release	unknown
Pomacea haustrum (titan applesnail)	unspecified	unknown	unknown
Pomacea paludosa (Florida applesnail)	FL, LA, OK	unknown	West Indies
Potamopyrgus antipodarum (New Zealand mudsnail)	ID, MT, NY	stock contamination	New Zealand
Pseudosuccinea columella (mimic lymnae)	CA	unknown	North America
Radix auricularia (big-eared radix)	CO, CA, IL, KY, MA, MI, MT, NJ, NM, NY, OH, PA, VT	ornamental plants	Europe
Sphaerium corneum (European fingernailclam)	NY	unknown	Eurasia
Stenophysa marmorata (marbled aplexa)	TX	released pet	unknown
Stenophysa maugeriae (tawny aplexa)	TX	unknown	North America
Tarebia granifera (quilted melania)	FL, TX	released pet	unknown
Valvata piscinalis (European stream valvata)	NY, MI	in packaging material	Europe
Viviparus georgianus (banded mysterysnail)	MI, NY	aquarium release	North America
Viviparus viviparus (mysterysnail)	MD	unknown	Europe

Reptiles

Name	Nonindigenous Distribution	Pathway	Origin
Agkistrodon piscivorus (cottonmouth)	MA	released pet	North America
Alligator mississippiensis (American alligator)	AZ, CO, VA	pet release	North America

TABLE 1.2 (CONTINUED)
Nonindigenous Aquatic Species Introduced in the United States

Name	Nonindigenous Distribution	Pathway	Origin
Apalone ferox (Florida softshell)	FL	released pet	North America
Apalone spinifera emoryi (Texas spiny softshell)	AZ, CA, NM, NV	released pet	North America
Apalone spinifera spinifera (eastern spiny softshell)	MA, NJ	released pet	North America
Caiman crocodilus (spectacled caiman)	FL, MA, TX	released pet	South America
Chelus fimbriatus (matamata)	FL	released pet	South America
Chelydra serpentina (snapping turtle)	CA, OR	released pet	North America
Chinemys reevesii (Chinese 3-keeled pond turtle)	CA, MA	released pet	Asia
Chrysemys picta (painted turtle)	CA	released pet	North America
Chrysemys picta dorsalis (southern painted turtle)	FL	released pet	North America
Chrysemys picta belli (western painted turtle)	FL	released pet	North America
Clemmys insculpta (wood turtle)	MA	released pet	North America
Clemmys muhlenbergii (bog turtle)	MA	released pet	North America
Cuora flavomarginatus (yellow-margined box turtle)	MA	released pet	Asia
Deirochelys reticularia (Florida chicken turtle)	FL	pet release	North America
Elaphe obsoleta (yellow rat snake)	MA	unknown	North America
Emydoidea blandingii (Blandking's turtle)	OH	pet release	North America
Graptemys barbouri (Barbour's map turtle)	FL	unknown	North America
Graptemys geographica (common map turtle)	MA	released pet	North America
Graptemys kohnii (Mississippi map turtle	FL, VA	pet release	North America
Graptemys ouachitensis (Ouachita map turtle)	AL	unknown	North America
Graptemys pseudogeographica (false map turtle)	FL	released pet	North America
Kinosternon flavescens (yellow mud turtle)	AZ	pet release	North America
Kinosternon scorpiodes (scorpion mud turtle)	FL	released pet	unknown
Kinosternon subrubrum hippocrepis (Mississippi mud turtle)	MA	released pet	North America
Kinosternon subrubrum (eastern mud turtle)	CA, MA	released pet	North America
Macroclemys temminckii (alligator snapping turtle)	AZ	pet release	North America

TABLE 1.2 (CONTINUED)
Nonindigenous Aquatic Species Introduced in the United States

Name	Nonindigenous Distribution	Pathway	Origin
Malaclemmys terrapin (diamondback terrapin)	MA	stocked	North America
Malayemys subtrijuga (Malayan snail-eating turtle)	NM	pet release	Asia
Mauremys caspica (Caspian turtle)	MA	released pet	Asia
Natrix tesselata (tesselated water snake)	VA	accidental	Eurasia
Nerodia fasciata pictiventris (Florida water snake)	TX	unknown	North America
Nerodia taxispilota (brown water snake)	CO	unknown	North America
Palea steindachneri (wattleneck softshell turtle)	HI	food	Asia
Pelodiscus sinensis (Chinese softshell turtle)	HI	food	Asia
Platemys platycephala (twist-necked turtle)	FL	pet release	South America
Podocnemis lewyana (Rio Magdalena river turtle)	FL	released pet	South America
Podocnemis sexituberculata (six tubercled Amazon River turtle)	FL	released pet	South America
Podocnemis unifilis (yellow spotted Amazon River turtle)	FL	released pet	South America
Pseudemys concinna (river cooter)	MA	released pet	North America
Pseudemys dorbigni (Brazilian slider)	FL	released pet	unknown
Pseudemys floridana (Florida cooter)	MA	released pet	North America
Pseudemys nelsoni (Florida redbelly turtle)	TX	pet release	North America
Pseudemys rubriventris (redbelly turtle)	CA	pet release	North America
Rhinoclemmys pulcherrima (painted wood turtle)	MA	released pet	Central America
Sternotherus odoratus (common musk turtle)	CA	pet release	North America
Trachemys dorbigni (Brazilian slider)	FL	pet release	South America
Trachemys scripta elegans (red-eared slider)	CA, FL, HI, MA, MD, OR	released pet	North America
Trachemys scripta callirostris (slider)	FL	released pet	South America
Trachemys stenjnegeri malonei (Central Antillean slider)	FL	released pet	Caribbean
Varanus niloticus (Nile monitor)	FL	pet release	Africa
Varanus salvator (water monitor)	AZ	pet release	Asia

TABLE 1.3
Selected Invasive Nonindigenous Aquatic Plants in the United States

Name	No. of States Reported	Origin	Pathway
Alternanthera philoxeroides (alligatorweed)	14	South America	ballast water
Butomus umbellatus (flowering rush)	14	Europe	dry ballast
Egeria densa (Brazilian waterweed)	30	South America	aquarium release
Eichhornia crassipes (water hyacinth)	18	South America	cultivation
Hydrilla verticillata (hydrilla)	16	Asia	aquarium release
Hydrocharis morsus-ranae (European frog-bit)	2	Europe	cultivation
Hygrophila polysperma (Indian hygrophila)	2	Asia	aquarium release
Iris pseudacorus (yellow iris)	37	Europe, Africa	cultivation
Limnophila sessiliflora (Asian marshweed)	4	Asia	aquarium release
Lythrum salicaria (purple loosestrife)	42	Eurasia	cultivation, ballast
Myriophyllum aquaticum (parrot-feather)	29	South America	cultivation
Myriophyllum spicatum (Eurasian water-milfoil)	44	Eurasia	cultivation
Najas minor (brittle naiad)	24	Europe	aquarium release
Nasturtium officinale (water-cress)	47	Europe	cultivation
Nymphoides peltata (yellow floating-heart)	20	Eurasia	cultivation
Panicum repens (torpedo grass)	6	Australia	cultivation
Pistia stratiotes (water-lettuce)	7	South America	ballast water
Potamogeton crispus (curly pondweed)	46	Europe	unknown
Salvinia molesta (giant salvinia)	6	South America	water garden trade
Spirodela punctata (dotted duckweed)	18	Asia	aquarium release
Trapa natans (water-chestnut)	8	Eurasia	cultivation

of algae, including both marine and freshwater species. Many nonindigenous members of these groups effectively colonize aquatic ecosystems, competing with and often displacing native plants. Documented invasions by nonindigenous plants include true aquatic and wetland types. Species such as purple loosestrife (*Lythrum salicaria*), a wetland type, and alligatorweed (*Alternanthera philoxeroides*), a true aquatic type, are extremely invasive. Purple loosestrife was brought to North America from Eurasia to be cultivated in household gardens. Alligatorweed was probably a ballast water introduction from South America and has spread throughout the southern U.S. A selected list of invasive plants is given in Table 1.3.

PATHWAYS OF INTRODUCTION

Aquatic organisms have been introduced intentionally as well as unintentionally (Table 1.4) for reasons ranging from aquaculture to fisheries enhancement (Carlton, 1992). Most common has been the intentional introduction of nonindigenous fishes for sportfishing. There have been more than 200 different species of fishes stocked in nonindigenous waters of North America for this purpose. Many of the introductions have provided excellent fishing opportunities with few consequences. However, some fish introductions have proved disastrous to the native fisheries through competition, hybridization, and diseases. Negative impacts also result from the introductions of other types of vertebrates, invertebrates, and plants.

The intentional introductions are dealt with in detail by Dextrase and Coscarelli in Chapter 4 of this volume. Some of the earliest documented fish introductions began in the late 1800s when East Coast fishes were transported to the West Coast and vice versa. The American shad (*Alosa sapidissima*) was introduced in the Great Lakes in an experimental effort to solve the problems of declining East Coast commercial fisheries. The Atlantic salmon (*Salmo salar*), endemic to the

TABLE 1.4
Numbers of Nonindigenous Aquatic Species in the United States by Pathway[a]

Group	Amphib.	Bryozoa	Coelent.	Crustac.	Fish	Mammal	mollusc	Plant	Rept.
Intentional Releases									
Bait bucket	3	0	0	7	84	0	0	—	0
Aquarium/Pets	22	2	2	1	81	0	9	—	44
Aquaculture/Farm	0	0	0	0	4	1	1	—	0
Biocontrol	6	0	0	0	14	0	0	—	0
Ornamental	0	0	0	0	18	1	0	—	0
Sportfishing	0	0	0	0	114	0	0	—	0
Forage	0	0	0	1	47	0	0	—	0
For food	1	0	0	0	19	0	0	—	2
Stocked	7	0	0	4	21	0	2	—	6
Conservation	0	0	0	0	21	0	0	—	0
Unknown	2	0	0	1	0	0	1	—	0
Unintentional Releases									
Aquaculture escapes	0	0	0	2	50	1	0	—	0
Ballast water	0	0	4	5	7	0	3	—	0
Stock contamination	2	0	0	1	31	0	1	—	0
Canals	0	0	0	0	41	0	3	—	0
Shipping (overland)	4	0	0	1	0	0	6	—	0
Imported live food	0	0	0	0	3	0	3	—	0
Unknown	12	1	0	1	15	0	15	—	12

[a] Determining the pathways for the introduction of all aquatic plants is not completed. However, pathways of introductions for some plant species include ballast water, dry ballast, aquarium release, contamination, water-garden nurseries, and cultivation.

Atlantic coast and Lake Ontario, was stocked in Lake Ontario in the late 1800s to bolster a declining population and in the other Great Lakes to provide fishing opportunities, which later proved unsuccessful (Parsons, 1973). This practice of stocking fish continued for many years, and by 1974, 34 fish species had been introduced into the Great Lakes, of which 17 became established (Emery, 1985).

Another common pathway for the large number of species introduced has been from aquarium and pet releases. This includes hobbyists and pet owners who release their pets, such as goldfish, piranhas, frogs, or turtles, as a humane method of disposal when they have outgrown their aquarium or owner's fondness, but with no intention of starting a self-sustaining population. Aquarium releases are also responsible for the introduction of invasive plants such as hydrilla (*Hydrilla verticillata*) and parrot feather (*Myriophyllum aquaticum*) (Couch and Nelson, 1991). Another pathway of introduction related to the aquarium pet industry includes escapes from the hundreds of fish farms in the U.S. (Courtenay and Williams, 1992).

Other types of introductions that are a direct result of the enthusiasm for sportfishing are bait bucket releases and escapes (Ludwig and Leitch, 1996; Litvak and Mandrak in Chapter 11 of this volume). Often, after an angler is through fishing for the day, the bucket of baitfishes is dumped into the water as a way of disposing of the bait or with the assumption that the bait will become food for the more desirable sport fish. Examples of common bait bucket releases are the rudd (*Scardinius erythropthalmus*), a minnow from Europe, and the native rusty crayfish (*Orconectes rusticus*), which have been introduced into 21 and 15 states in the U.S. respectively. In addition to bait bucket releases, the intentional stocking of forage fishes can also be associated with sportfishing. In some cases where sportfishes are introduced, there is no known forage base for them, so one must be provided. For this reason, the fathead minnow (*Pimephales promelas*) has been introduced beyond its native range in many states. Stock contamination provides another means for introducing a nonindigenous species. Nontarget species of fish, such as the tadpole madtom (*Noturus gyrinus*), have been inadvertently mixed in with the intended species to be stocked, channel catfish (*Ictalurus punctatus*) (Simpson and Wallace, 1978), just as white suckers (*Catostomus commersoni*) have been known to contaminate stockings of trout (Fuller et al., In prep.). Fish farming for human consumption also has provided an avenue for the introduction of escapes. Several species of tilapia and carp have found their way into open waters from aquaculture facilities (Courtenay and Williams, 1992).

In addition to the stocking of sportfishes and associated introductions, there are other explanations for the intentional release of nonindigenous organisms. Conservation has become an important issue as the development of natural areas has caused significant habitat loss for many species. Endangered species such as the Devils Hole pupfish (*Cyprinodon diabolis*) have been released outside their native range in hopes of conserving the species (Fuller et al., 1999). Biological control is another reason for the intentional release of nonindigenous organisms. The grass carp (*Ctenopharyngodon idella*), known to be a hearty vegetarian in its native Asia, has been released to control aquatic weeds in North America. However, as a result of stockings, it was discovered that reproducing grass carp populations can become a nuisance by becoming too numerous and eating all the vegetation present, including that which may be considered desirable. To keep grass carp confined and from reproducing, many states permit only sterile triploid fish to be released in waters where they can be readily contained. Insects such as the mosquito, regardless of whether they are native or not, can be a nuisance. Mosquitofish (*Gambusia affinis*) have been introduced in 34 states in the U.S. to control mosquito larvae in their aquatic stage. Unfortunately, the mosquitofish have replaced some of the native species that may actually have been better at controlling mosquito larvae (Courtenay and Meffe, 1989).

Boating and shipping activities, past and present, have had a major role in the introduction and spread of nonindigenous aquatic species in North America. Unintentionally, ballast water has been a significant contributor of nonindigenous aquatic organisms to North American coastal and inland freshwater ports. A few of the more notable ballast water introductions of recent years in fresh water include the ruffe (*Gymnocephalus cernuus*), round goby (*Neogobius melanostomus*), zebra mussel, quagga mussel (*Dreissena bugensis*), and spiny water flea. In the Great Lakes, 21 species of algae have been introduced through ballast water (Mills et al., 1993). Other types of boat movements are likely responsible for the spread of nonindigenous species once the organisms arrive here from overseas. Barge traffic in our navigable waterways has been known to carry zebra mussels for thousands of miles on their hulls up and down the Mississippi River and several of its tributaries (Keevin, 1992). On a smaller scale, recreational boats in the Great Lakes region are probably responsible for spreading nonindigenous organisms between neighboring lakes, including zebra mussels and undesirable plants such as Eurasian water-milfoil when left on hulls, propellers, and in live wells. From 1993 to mid-1999, 11 boats being hauled overland on trailers, when stopped at major highway agricultural inspection stations in California, had zebra mussels attached to their hulls (Janik, pers. comm., 1999). In 1993, 36% of over 1200 boats and trailers surveyed exited a county in Minnesota carrying fragments of Eurasian water-milfoil (Exotic Species Programs, 1993).

Manmade canals and diversions are responsible for the introduction of several fish species and probably some plants (Mills et al., this volume). The Welland Canal, which connects Lake Ontario to the rest of the Great Lakes, allowed the entry of two very harmful species, the sea lamprey (*Petromyzon marinus*) and alewife (*Alosa pseudoharengus*). The Chicago Sanitary and Shipping Canal, connecting Lake Michigan with the Illinois River, was responsible for allowing the zebra mussel to escape from the Great Lakes into the Mississippi River drainage. The importation of live food can also be a source of introductions. The Asian clam and other live food items have been introduced by escaping or by being intentionally released by individuals who hope that the organism will reproduce locally. Chinese immigrants, hungry for their native foods, may have brought the

Asian clam to North America. The first North American record came from Vancouver Island, British Columbia, in 1924 (Counts, 1981), while the first record in the U.S. was found in the state of Washington in 1937 (Counts, 1985). Prior to the more recently publicized zebra mussel, the Asian clam found its way into water-intake pipes, thus acquiring a reputation as an industrial biofouler. The range of the Asian clam is still expanding and currently extends to 38 U.S. states. Intentional introductions have also been made for ornamental purposes such as stocking a pond with goldfish (*Carassius auratus*) or koi and water hyacinths (*Eichhornia crassipes*).

Compounding the problems that the introduction of a new fish or plant may bring are the exotic diseases and parasites that can be introduced at the same time on the organism itself or in the water used for transporting. White spot disease has been imported into the U.S. along with Pacific shrimp. This disease has been known to cause 100% mortalities of shrimp in aquaculture growout ponds and was diagnosed at one aquaculture facility in Texas in 1995. The disease can be easily transmitted in contaminated water and in decomposing tissue and fecal matter (Johnson, 1996). The danger to native crustaceans from white spot disease is not yet known. In addition to harboring diseases, water that is used to ship any aquatic organisms can also be a source of other nonindigenous organisms invisible to the naked eye such as zooplankton. It is thought that an exotic daphnia, *Daphnia lumholtzi*, may have come into the U.S. in this manner (Havel, pers. comm., 1994).

Other less common pathways of introduction have been identified exhibiting the wide range of seemingly endless possibilities. These include the hybridization of two introduced species, creating a third introduced species. An example of this kind of hybridization has been known to occur between common carp and goldfish (Trautman, 1981). The black carp was introduced in 1994 when flood waters overcame a dike at a hatchery in Missouri (Fuller et al., 1999). Because of its appetite for molluscs, the black carp (*Mylopharygodon piceus*) was brought to the U.S. to forage on snails which serve as intermediate hosts for a catfish parasite. This has raised some questions about its possible use in biocontrol of zebra mussels. A more unusual pathway of introduction involves overland shipping. Snails, such as the European valve snail (*Valvata piscinalis*), have entered North America on plants used as packaging material of which the plants themselves are most likely also nonindigenous (Mills et al., 1993). Mistaken identity has been the cause of some introductions. Several species of fish have been stocked by accident because they were misidentified, such as green sunfish (*Lepomis cyanellus*) for bluegills (*Lepomis macrochirus*) (Fuller et al., In prep.).

PLACES OF ORIGIN

Nearly one half of all nonindigenous aquatic species introduced into waters in the U.S., including amphibians, crustaceans, fishes, molluscs, and reptiles, and more than 80% of the introduced plant species have origins outside of North America. Species from all taxonomic groups have come from every corner of the globe: South America, Central America, Asia, Europe, Africa, Australia, the Caribbean, the South Pacific, and Pacific and Atlantic Oceans (Table 1.5). South America is the largest contributor of nonindigenous aquatic species with 22%, followed closely by Asia with 20%, Eurasia with 16%, Europe with 13%, and Africa with 12% (Table 1.6). At least 176 species of fishes of foreign origin have been introduced into U.S. waters (Table 1.7). Aquarium releases are the largest contributor of the number of fish species of foreign origin introduced into North America. The majority of the ornamental fish imported for aquarium use into the U.S. come from Southeast Asia. The rest come from South American, African, and domestic (mostly Florida) origins. The cyprinids make up a large portion of earliest known imported fish species. Some of the earliest introductions in North America were goldfish and common carp from Asia. The only aquatic mammal whose distribution is well documented is the nutria, a large aquatic rodent native to South America and raised on farms in the U.S. for its fur. The nutria's introduction from escapes and intentional releases dates back to the 1930s (O'Neil, 1949).

TABLE 1.5
Numbers of Nonindigenous Aquatic Species in the United States by Origin[a]

Group	Amphib.	Bryozoa	Coelent.	Crustace	Fish	Mammal	molluscs	Plant[a]	Rept
North America	23	2	1	12	302	0	4	11	32
South America	0	0	0	0	66	1	3	16	9
Central America	2	0	0	1	28	0	0	4	1
South/Central Am.	2	0	0	0	0	0	0	4	0
West Indies	1	0	0	0	0	0	1	0	1
Europe	0	0	2	3	20	0	7	24	0
Asia	3	1	1	4	44	0	3	23	8
Eurasia	0	0	0	0	7	0	3	55	1
Africa	2	0	1	1	31	0	1	12	1
Australia	1	0	0	0	2	0	0	6	0
South Pacific	0	0	0	1	1	0	1	0	0
Pacific Ocean	0	0	0	0	2	0	0	0	0
Atlantic Ocean	0	0	0	1	0	0	4	7	0
Unknown	3	0	0	0	0	0	9	0	1

[a] Included in the plants are wetland species.

TABLE 1.6
Percentages of All Aquatic Species Introduced into the United States from Other Areas

South America	22%
Asia	20%
Eurasia	16%
Europe	13%
Africa	12%
Central America	6%
Australia	2%
South and Central America	1%
South Pacific	1%
Atlantic Ocean	1%
Pacific Ocean	>1%
Unknown	4%

IMPACTS

ECOLOGICAL

Nonindigenous species can impact industry, human health, and the conservation of natural areas. Impacts on industry and human health issues are first to receive political support, while many of the effects of nonindigenous species on natural areas go unaddressed due to a lack of financial incentives (U.S. Congress, 1993). Over the past 100 years, the extinction of 27 species and 13 subspecies of North American fishes has been attributed to nonindigenous species (Miller et al., 1989). In terms of the impacts nonindigenous species have had to their new environments, we have only begun to learn about and understand some of the effects and potential impacts. The most obvious potential impacts are competition with native species for food and space, predation on

TABLE 1.7
Nonindigenous Fish Species of Foreign Origin Introduced into the United States Since 1850

Species	Origin	Pathway
Acanthogobius flavimanus (yellowfin goby)	Asia	ballast water
Aequidens pulcher (blue acara)	South America	fish farm escape
Ameca splendens (butterfly splitfin)	North America (Mexico)	unknown
Anabas testudineus (climbing perch)	Asia	fish farm escape
Anchoa mundeoloides (anchovy)	South/Central America	stocked for forage
Ancistrus sp. (bristle-nosed catfish)	South/Central America	aquarium release
Anguilla anguilla (European eel)	Europe/Africa	aquaculture escape
Anguilla australis (shortfin eel)	Australia	fish farm escape
Aphyocharax anisitsi (bloodfin tetra)	South America	fish farm escape
Aplocheilus lineatus (striped panchax)	Asia	unknown
Astronotus ocellatus (oscar)	South America	fish farm release
Atherinops regius (silverside)	Central America	stocked for forage
Bairdiella icistia (bairdiella)	North America (Mexico)	stocked for sportfishing
Belonesox belizanus (pike killifish)	Mexico/Central America	research lab release
Betta splendens (Siamese fighting fish)	Asia	fish farms escapes
Brachydanio rerio (zebra danio)	Asia	fish farms escapes
Callichthys callichthys (cascarudo)	South America	aquarium release
Carassius auratus (goldfish)	Asia	imported as ornamentals
Carassius carassius (Curcian carp)	Eurasia	bait bucket release
Channa micropeltes (giant snakehead)	Asia	aquarium release
Channa striata (chevron snakehead)	Asia	imported as food
Chirostoma jordani (charal)	North America (Mexico)	stocked as food fish
Chitala ornata (clown knifefish)	Asia	aquarium release
Cichla ocellaris (peacock cichlid)	South America	stocked for sportfishing
Cichla temensis (speckled pavon)	South America	stocked for sportfishing
Cichlasoma beani (green guapote)	North America (Mexico)	aquarium release
Cichlasoma bimaculatum (black acara)	South America	fish farm escapes
Cichlasoma citrinellum (Midas cichlid)	Central/South America	aquarium release
Cichlasoma labiatum (red devil)	Central America	aquarium release
Cichlasoma managuense (jaguar guapote)	Central America	aquarium release
Cichlasoma meeki (firemouth cichlid)	Mexico/Central America	aquarium/fish farm release
Cichlasoma nigrofasciatum (convict cichlid)	Central America	aquarium release
Cichlasoma octofasciatum (Jack Dempsey)	Central America	aquarium release
Cichlasoma salvini (yellowbelly cichlid)	Central America	aquarium release
Cichlasoma spilurum (blue-eyed cichlid)	Central America	aquarium escape
Cichlasoma synspilum (redhead cichlid)	Central America	aquarium release
Cichlasoma trimaculatum (threespot cichlid)	Central America	aquarium release
Cichlasoma urophthalmus (Mayan cichlid)	Central America	aquarium release/farm escape
Clarias batrachus (walking catfish)	Asia	fish farm escapes
Clarias fuscus (whitespotted clarias)	Asia	imported as food
Colisa fasciata (banded gourami)	Asia	aquarium release
Colisa labiosa (thicklip gourami)	Asia	aquarium release/farm escape
Colisa lalia (dwarf gourami)	Asia	aquarium release/farm escape
Colossoma macropomum (black pacu)	South America	aquarium release
Coregonus albula (vendace)	Europe	stocked as food
Coregonus lavaretus (powan)	Eurasia	stocked as food
Corydoras aeneus (green corydoras)	South America	aquarium release
Corydoras sp. (corydoras)	South America	aquarium/fish farm release
Ctenopharyngodon idella (grass carp)	Asia	biocontrol of vegetation

TABLE 1.7 (CONTINUED)
Nonindigenous Fish Species of Foreign Origin Introduced into the United States Since 1850

Species	Origin	Pathway
Ctenopoma nigropannosum (twospot climbing perch)	Africa	fish farm escapes
Cynolebias bellottii (Argentine pearlfish)	South America	biocontrol of mosquitoes
Cynolebias nigripinnis (blackfin pearlfish)	South America	biocontrol of mosquitoes
Cynoscion othonopterus (scaly corvina)	Central America	stocked for sport
Cynoscion parvipinnis (shortfin corvina)	North America (Mexico)	stocked for sportfishing
Cynoscion xanthulus (orangemouth corvina)	North America (Mexico)	stocked for sportfishing
Cyprinus carpio (common carp)	Asia	stocked as food
Cyprinus carpio (koi)	Asia	stocked as ornamentals
Danio malabaricus (Malabar danio)	Asia	aquarium release/farm escape
Esox reicherti (Amur pike)	Asia	stocked for sportfishing
Geophagus brasiliensis (pearl eartheater)	South America	fish farm escape
Geophagus surinamensis (redstriped eartheater)	South America	aquarium release/farm escape
Gillichthys seta (mudsucker)	Central America	stocked for sport
Girella simplicidens (opaleye)	Central America	stocked for sport
Gymnocephalus cernuus (ruffe)	Eurasia	ballast water
Gymnocorymbus ternetzi (black tetra)	South America	aquarium release/farm escape
Helostoma temminckii (kissing gourami)	Asia	aquarium release/farm escape
Hemichromis elongatus (banded jewelfish)	Africa	aquarium release
Hemichromis letourneauxi (African jewelfish)	Africa	aquarium release
Hemigrammus ocellifer (head-and-taillight tetra)	South America	fish farm escape
Heros severus (banded cichlid)	South America	aquarium release
Hoplias malabaricus (trahira)	Central/South America	fish farm escape
Hoplosternum littorale (brown hoplo)	South America	unknown
Hyphessobrycon serpae (serpae tetra)	South America	aquarium release/farm escape
Hypomesus nipponensis (wagasaki)	Asia (Japan)	stocked as forage for trout
Hypophthalmichthys molitrix (silver carp)	Asia	biocontrol of phytoplankton
Hypophthalmichthys nobilis (bighead carp)	Asia	biocontrol of phytoplankton
Hypostomus plecostomus (suckermouth catfish)	Central/South America	aquarium release/farm escape
Kuhlia rupestris (nato)	Africa/Australia	stocked for sportfishing
Labeotropheus sp. (scrapermouth cichlid)	Africa	aquarium release
Lates angustifrons (Tanganyika lates)	Africa	stocked for sportfishing
Lates mariae (bigeye lates)	Africa	stocked for sportfishing
Lates niloticus (Nile perch)	Africa	stocked for sportfishing
Leuresthes sardina (Gulf grunion)	Central America	stocked for forage
Leporinus fasciatus (banded leporinus)	South America	aquarium release
Leuciscus idus (ide)	Eurasia	stocked as food
Limia vittata (Cuban limia)	Cuba	aquarium release
Macropodus opercularis (paradisefish)	Asia	fish farm escape
Melanochromis auratus (golden mbuna)	Africa	aquarium release
Melanochromis johanni (bluegray mbuna)	Africa	aquarium release
Melanotaenia nigrans (black-banded rainbowfish)	Australia	aquarium release
Menticirrhus nasus (corbina)	Central America	stocked for sport
Metynnis sp. (metynnis)	South America	aquarium release
Misgurnus anguillicaudatus (Oriental weatherfish)	Asia	aquarium release
Misgurnus mizolepis (Chinese fine-scaled loach)	Asia	aquarium release
Moenkhausia sanctaefilomenae (redeye tetra)	South America	aquarium release/farm escape
Monopterus albus (swamp eel)	Asia	aquarium release
Morulius chrysophekadion (black sharkminnow)	Asia	aquarium release/farm escape
Myleus rubripinnis (redhook pacu)	South America	aquarium release

TABLE 1.7 (CONTINUED)
Nonindigenous Fish Species of Foreign Origin Introduced into the United States Since 1850

Species	Origin	Pathway
Mylopharyngodon piceus (black carp)	Asia	aquaculture escape
Neogobius melanostomus (round goby)	Eurasia	ballast water
Nothobranchius guentheri (redtail notho)	Africa	biocontrol of mosquitoes
Oncorhynchus masou (cherry salmon)	Asia	stocked for sportfishing
Oreochromis aureus (blue tilapia)	Africa	sportfishing/aquatic plant control
Oreochromis macrochir (longfin tilapia)	Africa	unknown
Oreochromis mossambicus (Mozambique tilapia)	Africa	sportfishing/aquatic plant control
Oreochromis niloticus (Nile tilapia)	Africa	aquaculture escape
Oreochromis urolepis (Wami tilapia)	Africa	biocontrol of plants, insects
Oryzias latipes (Japanese medaka)	Asia	biocontrol of insects
Osphronemus goramy (giant gourami)	Asia	aquarium release
Osteoglossum bicirrhosum (arawana)	South America	aquarium release/farm escape
Otocinclus sp. (suckermouth catfish)	South America	fish farm escape
Oxydorus Niger (ripsaw catfish)	South America	aquarium release
Pangio kuhlii (coolie loach)	Asia	fish farm escape
Paralichthys aestuarius (halibut)	Central America	stocked for sport
Paralichthys woolmani (halibut)	Central America	stocked for sport
Paracheirodon innesi (neon tetra)	South America	fish farm escape
Parauchenipterus galeatus (driftwood catfish)	South America	aquarium release
Peckoltia sp. (clown pleco)	South America	aquarium release
Pelvicachromis pulcher (rainbow krib)	Africa	aquarium escape
Peprilus ovatus (Gulf butterfish)	Central America	stocked for sport
Perrunichthys perruno (leopard catfish)	South America	aquarium release
Phractocephalus hemioliopterus (redtail catfish)	South America	aquarium release
Piaractus brachypomus (redbellied pacu)	South America	aquarium release
Piaractus mesopotamicus (small-scaled pacu)	South America	aquarium release
Platichthys flesus (European flounder)	Atlantic Ocean	ballast water
Platydoras costatus (Raphael catfish)	South America	aquarium release
Plecoglossus altivelis (ayu)	Asia	imported as live food
Poecilia latipunctata (broadspotted molly)	North America (Mexico)	aquarium/fish farm release
Poecilia mexicana (shortfin molly)	Central America	aquarium/fish farm release
Poecilia petenensis (swordtail molly)	Central America	aquarium release
Poecilia reticulata (guppy)	South America	aquarium/fish farm release
Poecilia sphenops (liberty molly)	Central America	aquarium/fish farm release
Poeciliopsis gracilis (porthole livebearer)	Central America	aquarium release/farm escape
Proterorhinus marmoratus (tubenose goby)	Eurasia	ballast water
Pseudotropheus zebra (zebra mbuna)	Africa	aquarium release
Pterodoras granulosus (granulated catfish)	South America	aquarium release
Pterophyllum scalare (freshwater angelfish)	South America	aquarium release
Pterygoplichthys disjunctivus (vermiculated sailfin catfish)	South America	aquarium release/farm escape
Pterygoplichthys multiradiatus (Orinoco sailfin catfish)	South America	aquarium release/farm escape
Pterygoplichthys pardalis (Amazon sailfin catfish)	South America	aquarium release
Puntius conchonius (rosy barb)	Asia	fish farm escape
Puntius filamentosus (blackspot barb)	Asia	aquarium escape
Puntius gelius (dwarf barb)	Asia	fish farm escape
Puntius schwanenfeldii (tinfoil barb)	Asia	fish farm escape
Puntius semifasciolatus (green barb)	Asia	imported as an ornamental

TABLE 1.7 (CONTINUED)
Nonindigenous Fish Species of Foreign Origin Introduced into the United States Since 1850

Species	Origin	Pathway
Puntius tetrazona (tiger barb)	Asia	fish farm escape
Pygocentrus nattereri (red piranha)	South America	aquarium release
Rhamdia quelen (bagre)	South America	aquarium release
Rhodeus sericeus (bitterling)	Eurasia	aquarium release
Rivulus harti (giant rivulus)	South America	fish farm escape
Salmo letnica (Ohrid trout)	Europe	stocked for sportfishing
Salmo trutta (brown trout)	Eurasia/Africa	stocked for sportfishing
Sarotherodon melanotheron (blackchin tilapia)	Africa	fish farm escape
Scardinius erythrophthalmus (rudd)	Eurasia	imported as an ornamental
Serrasalmus rhombeus (white piranha)	South America	aquarium release
Simpsonichthys whitei (Rio pearlfish)	South America	biocontrol of mosquitoes
Stizostedion lucioperca (zander)	Eurasia	stocked for sportfishing
Symphysodon discus (red discus)	South America	fish farm escape
Telmatochromis bifrenatus (Lake Tanganyika dwarf cichlid)	Africa	aquarium release
Tilapia mariae (spotted tilapia)	Africa	fish farm escapes/releases
Tilapia rendalli (redbreast tilapia)	Africa	biocontrol of aquatic plants
Tilapia sparrmannii (banded tilapia)	Africa	fish farm escapes/releases
Tilapia zillii (redbelly tilapia)	Africa	biocontrol of aquatic plants/ insects
Tinca tinca (tench)	Eurasia	stocked for sportfishing
Totoaba macdonaldi (totuava)	Central America	stocked for sport
Trachinotus paitensis (Paloma pompano)	Central America	stocked for sport
Trichogaster leerii (pearl gourami)	Asia	fish farm escape
Trichogaster trichopterus (blue gourami)	Asia	fish farm escapes/releases
Trichopsis vittata (croaking gourami)	Asia	fish farm escape
Tridentiger bifasciatus (Shimofuri goby)	Asia	ballast water
Tridentiger trigonocephalus (chameleon goby)	Asia	imported on oysters/ballast water
Xenentodon cancila (Asian needlefish)	Asia	aquarium release
Xiphophorus helleri (green swordtail)	Central America	aquarium release/farm escape
Xiphophorus maculatus (southern platyfish)	Central America	aquarium/fish farm release
Xiphophorus variatus (variable platyfish)	North America (Mexico)	aquarium release

* established in the United States.

native species, habitat alteration, and the introduction of diseases and parasites (Krueger and May, 1991; Taylor et al., 1984). Nonindigenous species can have genetic effects that occur through hybridization and interbreeding with native species (Mills et al., 1994).

The sea lamprey was probably one of the first to be recognized as having a significant impact after it invaded Lake Erie and the upper Great Lakes via the Welland Canal. It caused the decline and near extinction of native lake trout in the Great Lakes. The alewife, a fish nonindigenous to the Great Lakes, entered the Great Lakes after completion of the Welland and Erie canals. This species accelerated the collapse of coregonid stocks in some of the Great Lakes due to competition and predation. Because of the depletion of the lake trout by the lamprey, the alewife had no natural predators during the 1960s. Alewife abundance fluctuated widely due to natural causes. In some years, massive die-offs fouled beaches and clogged water intakes. Currently, several introduced species of salmonids keep the alewife population in check. Since its introduction from Eurasia around 1986, the ruffe has become the dominant species in terms of biomass in the St. Louis River estuary in western Lake Superior (Busiahn, pers. comm., 1995). Decreases in the number of native species is evident but cannot yet be directly attributed to the introduction of the ruffe. The rudd, another native of Eurasia, has been shown to hybridize with the North American native golden shiner (*Notemigonus crysoleucas*), which could lead to conservation implications for the native minnow (Burkhead and Williams, 1991). Another example, the brown trout, which was first introduced in Michigan in 1883 as a food fish, has had a negative impact on the native golden trout (*Oncorhynchus aguabonita*) in California (Courtenay and Williams, 1992). California is trying to eradicate the brown trout in hopes that the native golden trout will make a comeback.

In the early 1990s, it was the zebra mussel that triggered much of the interest concerning the impacts of nonindigenous aquatic species. In just a few years, the native unionid population of Lake St. Clair disappeared, coinciding with the arrival of the zebra mussel (Nalepa, 1994). The effects of zebra mussels on fish populations are not yet deemed a disaster, as was earlier suspected when the mussel first came on the North American scene (Marsden et al., 1991). However, there have been changes in fish species composition and a significant increase in water clarity followed by an increase in macrophytes attributed directly to the zebra mussel's filtering capabilities (Nichols, pers. comm., 1997). The rusty crayfish (*Orconectes rusticus*), a native of the lower Ohio River valley, has been found in many places north of its native range. This range expansion has been attributed to bait bucket introductions (Capelli, 1982). A consequence of this introduction has been the displacement of two native species of crayfishes and the reduction of macrophytes in several Wisconsin lakes (Olsen et al., 1991).

Nonindigenous aquatic plants for many years have had significant ecological and economic impacts in North America. Purple loosestrife, a perennial imported from Eurasia as an ornamental in the early 1800s, is so invasive into wetlands that it results in the suppression of native plant communities (Stuckey, 1980). Hydrilla, Eurasian water-milfoil, and water hyacinths are examples of exotic plants that have also dramatically altered the native habitat and become a nuisance for water-related activities such as boating and fishing.

Nonindigenous diseases pose a great threat to the native fauna and flora of North America. Several diseases have had a large impact on many species of fishes. Whirling disease is a parasitic infection of salmonids caused by the myxosporean protozoan, *Myxobolus cerebralis*. It was found only in Europe until the late 1950s, whereafter it was discovered in the eastern U.S. Within a decade the disease spread to the western U.S. and it is thought to be the cause of the devastating decline in salmonid populations there (Nehring and Walker, 1996). *Aeromonas salmonicida*, commonly known as furunculosis, is a bacterial pathogen thought at first to affect only fish species in the salmonid family. The disease was later found in cyprinids (carp, dace, tench, goldfish), catfishes, lampreys, sculpins, and pike. Furunculosis was first diagnosed in Europe in 1894 and was not found in North America until 1902 on brown trout that had been shipped from Europe. There are conflicting reports, however, stating that the disease in Europe may have originated from a shipment of Shasta rainbow trout from western North America. Furunculosis spread through Europe quickly, suggesting that there was no disease resistance, as one would expect if the disease was native to Europe. Rainbow trout seem to be naturally resistant (Austin and Austin, 1987), supporting the contention that furunculosis may be native to the U.S. Regardless of its origin, brown trout along with furunculosis have been shipped and distributed all across North America. The disease has been devastating to hatchery-reared fish and to wild fish to a lesser degree.

ECONOMIC

We have come to depend on our natural resources as commodities having great commercial value. Exotics have threatened these resources as well as contributed to them. Commercial fisheries in the Great Lakes suffered immensely from the impacts of the sea lamprey on the lake trout popu-lation. The zebra mussel, which clogs water intake pipes of a variety of water users, has cost million of dollars in clean up and control measures. A total of over $69 million in expenses has been incurred at 339 facilities for zebra mussel prevention and remediation (O'Neill, 1996). Approxi-mately $10 million is spent annually on control and research on the sea lamprey to reduce its predation, and if left unchecked, the lamprey could cost the Great Lakes region over $500 millions annually in lost fishing revenues and other indirect economic impacts (U.S. Congress, 1993). Aquatic weeds are a problem especially in the southeast. Nearly $100 million is spent annually on the control of nonindigenous aquatic weeds (U.S. Congress, 1993).

If there are positive impacts to the issue of nonindigenous species, they are from the recreational value of aquariums and sportfishing, economic value of the jobs generated by the pet trade industry, the fish farming industry, and sportfishing. The aquarium pet trade industry reaps the benefits of buying and selling nonindigenous aquatic organisms, which has been estimated at an annual worldwide economic value of nearly $600 million wholesale and $2 billion retail (Shotts and Gratzek, 1984). Because of the aquaculture industry, we are able to have a plentiful variety of a nutritional source of food in fish and shellfish. Sportfishing has also played an important economic and recreational role for millions of people in the U.S. Many cities and towns depend on the business sportfishing provides to help support the local economy.

CONTROL

"Nuisance" is a political term used in connection with nonindigenous species that have significant economic impacts. Sometimes it is necessary to control a nonindigenous species if it becomes a nuisance or has the potential to become a nuisance. Once a species becomes established, that is, reproduces successfully, eradication is almost impossible in large aquatic systems. The best one can hope for is to contain the species to its current range and keep the population to a small, manageable number. There are two strategies that can be employed to eliminate or control nonin-digenous aquatic species; biological and nonbiological. Biological control is accomplished through the introduction of natural predators, diseases, and parasites that will help keep the nuisance species in balance with its new environment. The logical place to find a species' natural enemy is in its native habitat. Careful research must be done before a new nonindigenous species is intentionally released to control the other. The "biocontrol" organism must target the unwanted nonindigenous species specifically. This method shows some promise for the control of purple loosestrife. Research was conducted and five species of beetles have been found to have host specificity and no negative effects on native plants (Malecki, 1995). Since 1992, four species of European beetles have been released in North America with encouraging results (Skinner, 1997). As for the zebra mussel, several species of fishes and waterfowl already present have found them as suitable prey, but these predators are not in numbers large enough to control the population. So far, scientists are having little success finding effective natural predators of zebra mussels in their indigenous habitat. When carefully studied and managed, biological control can be a preferable alternative to chemical and mechanical control of nonindigenous animals and plants.

Nonbiological controls fall into two categories, chemical and physical. The objective of chem-ical control is to target specific organisms (e.g., piscicides, lampricides, molluscides, herbicides). Chemicals that do not discriminate among organisms should only be used as a last resort after considerable scrutiny. Such chemicals can and are being used in small, contained systems such as that of water treatment and power plants where chlorine is used to kill zebra mussels. In open water systems, this practice is not recommended and probably illegal in most jurisdictions.

It is well known that chemicals can do a lot of harm and also be very cost-prohibitive; therefore, more "environmentally friendly" options are available in the form of physical control. A great deal of research has been conducted on how to physically control zebra mussels. Some of the methods currently being researched include thermal (steam, hot water), acoustic vibration, electrical current, filters and screens, coatings (toxic and nontoxic), toxic constructed piping, carbon dioxide injection, ultraviolet light, ozone, and anoxia/hypoxia. In cases of severe infestations, manual removal of millions of zebra mussels was required before control procedures could be implemented. The mechanical harvesting also of aquatic weeds is employed in many lakes as an alternative to chemical treatment. No one method is the "silver bullet." Control is usually site-specific, and several methods used in combination may work the best.

DISTRIBUTION AND STATUS

Nonindigenous freshwater organisms are ubiquitous in North America. In the U.S., nonindigenous aquatic organisms are found in every state, including Alaska and especially Hawaii (Table 1.8). However, there are areas where higher concentrations of introduced aquatic species exist. In the U.S., these areas are the southeast, southwest, and the Great Lakes region, mirroring the high human populations. Because the number of species increases closer to the equator, it follows that there would be more successful introductions from warmer climates into the southern portion of North America. With more people moving to the temperate climates, the number of pathways and the number of opportunities to move nonindigenous species also increases. In the U.S., the southern region extending from California to Florida and north through the Carolinas suffers from the greatest number of nonindigenous freshwater species. In addition, the Great Lakes, one of the largest supplies of fresh water in the world, is home for at least 139 nonindigenous species of fishes, invertebrates, plants, algae, and disease pathogens (Mills et al., 1993).

The number of nonindigenous aquatic species has increased greatly over the past 100 years. Since 1980 alone, at least 52 fish species of foreign origin have been introduced into North America. A majority of these recent introductions are from ballast water, aquarium releases, escapes from research and aquaculture facilities, contaminated stockings, and illegal biocontrol (U.S. Congress, 1993). Most of these are considered to be harmful economically and ecologically. Of the nearly 200 fish species of foreign origin that have been introduced in the last 150 years, over 70 species have established populations (Fuller et al., in prep.).

Whether it is for management or research purposes, it is essential to know the status of a species of concern. There is information in the GIS on the overall status for all nonindigenous aquatic species documented in the database and limited information on local populations. A population is considered established when evidence of natural reproduction is observed. The zebra mussel is one of many species that found its new surroundings very hospitable and became established quickly. Densities as high as $700,000/m^2$ were observed at a power plant in Michigan (Kovalak et al., 1993). When there is no evidence of reproduction of the introduced species at the time of the collection or observation, it is considered "collected." Specimens may or may not have been taken at that time, but a positive identification was made and whether it will become established is not known. Introduced species that persisted for a period of time or became established and then died out are considered "extirpated." The nile perch (*Lates niloticus*), a voracious predator, was stocked into several reservoirs in Texas but failed to reproduce, and all the fish eventually died. Knowing the history of the nile perch from the devastation it caused to native species when it was introduced to other lakes in Africa, it was fortunate that it did not become established in Texas. When individuals or populations do become established and are considered a nuisance, eradication may become necessary. Eradication is the direct result of deliberate human intervention. After escaping from an aquaculture facility, blue tilapia (*Oreochromis aureus*) became established in warmwater effluents of the Susquehanna River in the mid-1980s. To eradicate this population, condenser cooling water (the source of the warmwater effluent) from a power plant was shut off in December 1986, killing the population by cold-shocking them (Skinner, 1987). Aside from nonindigenous species found in open waters in North America, there are many species being held at research and aquaculture faciities. Until they escape or are released into open waters, their status is not considered while in captivity.

TABLE 1.8
Number of Nonindigenous Species Introduced into Waters of each U.S. State

	Amphibians	Bryozoans	Coelenter.	Crustace	Fishes	Mammals	molluscs	Plants	Reptiles
Alabama	0	0	0	2	53	1	2	17	1
Alaska	1	0	0	0	9	0	0	0	0
Arizona	7	0	0	4	83	0	4	4	10
Arkansas	0	0	0	1	47	0	2	7	0
California	5	1	2	13	157	0	5	16	12
Colorado	6	0	0	1	107	0	4	10	7
Connecticut	0	0	0	5	60	0	2	5	0
Delaware	0	0	0	0	28	1	2	14	0
Florida	13	0	0	3	123	1	9	60	16
Georgia	2	0	0	4	63	1	2	17	0
Hawaii	13	2	1	2	63	0	4	10	3
Idaho	2	0	0	2	60	1	9	4	0
Illinois	0	0	0	1	56	0	4	8	0
Indiana	0	1	0	2	42	1	4	4	0
Iowa	0	0	0	1	36	0	2	8	0
Kansas	0	0	0	1	49	0	1	2	0
Kentucky	0	0	1	1	56	0	3	11	0
Louisiana	2	0	0	2	28	1	2	25	0
Maine	1	0	0	4	34	0	1	3	0
Maryland	0	0	0	3	53	1	3	8	1
Massachussets	9	0	0	4	59	0	3	7	17
Michigan	0	1	1	1	45	1	7	5	1
Minnesota	0	0	0	2	50	0	3	4	0
Mississippi	0	0	0	2	23	1	2	10	0
Missouri	0	0	0	1	49	0	2	8	0
Montana	0	0	0	1	53	1	4	5	0
Nebraska	0	0	0	0	64	0	2	4	1
Nevada	1	0	0	3	94	0	2	2	2
New Hampshire	1	0	0	3	32	0	1	4	0
New Jersey	1	0	0	2	41	0	5	8	2
New Mexico	1	0	0	4	76	0	2	4	4
New York	0	0	1	5	72	1	13	10	0
North Carolina	0	0	0	2	82	1	2	12	0
North Dakota	0	0	0	0	33	0	0	3	0
Ohio	0	2	1	6	61	1	7	7	2
Oklahoma	0	0	0	2	52	0	4	4	0
Oregon	1	0	0	3	65	1	3	7	2
Pennsylvania	0	0	0	6	77	0	6	5	2
Rhode Island	1	0	1	3	12	0	1	4	0
South Carolina	0	0	1	3	38	1	1	9	1
South Dakota	0	0	0	0	45	0	0	2	0
Tennessee	0	0	0	4	52	0	3	12	0
Texas	4	0	0	2	107	1	7	21	3
Utah	2	0	0	2	79	0	2	4	1
Vermont	1	0	0	4	26	0	4	8	0
Virginia	0	0	3	2	91	1	3	8	0
Washington	2	0	1	1	48	1	1	7	0
West Virginia	0	0	2	2	72	0	2	6	0
Wisconsin	0	0	0	4	66	0	4	7	1
Wyoming	1	0	0	1	59	0	2	3	0

SUMMARY

The poor documentation of nonindigenous species is noted in a 1993 report published by the U.S. Congress' Office of Technology Assessment. The report found that available information was scattered and of variable quality, providing an incomplete picture of the status of nonindigenous aquatic species. Since the issuance of that report, it has been the intent of the U.S. Geological Survey to use the database at Gainesville, Florida to seek out and document new and historic information about nonindigenous aquatic introductions. By centralizing the information, we hope to paint a clearer and more complete picture of the history and current status of nonindigenous aquatic species. The information maintained in this database can then be used for research and in the decision-making and planning processes regarding nonindigenous aquatic species issues.

The information in the nonindigenous aquatic species database is by no means all-inclusive; introductions are occurring continuously. Therefore, new sources of information on aquatic introductions to add to the database are continually being sought. The goal is to document the introduction, distribution, and status of nonindigenous aquatic species accurately and make this information readily available.

Hundreds of species have been introduced in the waters of North America. The intentional introduction of aquatic organisms will most likely have some benefits associated with it, but at what cost to the ecosystem? Aquaculture for food, aquarium trade, sportfishing, and nurseries are the sources of most of these introductions and will continue to be the sources as long as people are able to reap the benefits, economically or aesthetically. Intentional introductions often have led to unintentional introductions, including aquaculture escapes and stock contamination, neither of which have proved to have introduced species of any benefit. There will always be considerable pressure from the public who demand better fishing opportunities, and aquaculture opportunities assuming that a new exotic will fill that demand. It brings to mind the saying "the grass is always greener on the other side of the fence." But usually it is not, as we soon discover.

With known and new unforeseeable pathways of introduction, continued introductions seem inevitable. Ballast water will probably be a primary source for future unintentional introductions unless compliance is met with existing policies and voluntary guidelines regarding ballast water exchange. The greatest potential impacts from introductions to native species and their habitats are the threats of exotic diseases and parasites, and the global loss of aquatic community diversity.

ACKNOWLEDGMENTS

I wish to acknowledge Ann Foster, Colette Jacono, and Pam Fuller for compiling the data on amphibians, reptiles, plants, and fishes, and Charles Boydston for reviewing the manuscript. Thanks go to Alan Dextrase for providing information on the legislation of nonindigenous species in Canada.

REFERENCES

Aquatic Nuisance Species Task Force, Report To Congress: Findings, Conclusions, and Recommendations of the Intentional Introduction Policy Review, U.S. Government Printing Office, Washington, D.C., 1994.

Austin, B., and D.A. Austin, *Bacterial Fish Pathogens: Disease in Farmed and Wild Fish*, Ellis Horwood, Chichester, England, 1987, p. 112.

Burkhead, N.M., and J.D. Williams, An Intergeneric Hybrid of a Native Minnow, the Golden Shiner, and an Exotic Minnow, the Rudd, *Trans. Am. Fish. Soc.*, 120, 781–795, 1991.

Busiahn, T., personal communication, 1995.

Capelli, G.M., Displacement of northern crayfish by *O. rusticus* (Girard), *Limnol. Oceanogr.*, 27, 741–745, 1982.

Carlton, J.T., Dispersal of Living Organisms into Aquatic Ecosystems as Mediated by Aquaculture and Fisheries Activities, in *Dispersal of Living Organisms into Aquatic Ecosystems*, Rosenfield, A. and Mann, R., Eds., Maryland Sea Grant Publication, College Park, MD, 1992.

Counts, C.L. III, *Corbicula fluminea* (Bivalvia: Sphaeriacea) in British Columbia, *The Nautilus*, 95, 12–13, 1981.

Counts, C.L. III, *Corbicula fluminea* (Bivalvia: Corbiculidae) in the State of Washington in 1937 and in Utah in 1975, *The Nautilus*, 99, 18–19, 1985.

Courtenay, W.R. Jr., and G.K. Meffe, Small fishes in strange places: a review of introduced Poeciliids, in *Ecology and Evolution of Livebearing Fishes (Poeciliidae)*, Meffe, G.K. and Nelson, F.F., Jr., Eds., Prentice Hall, Englewood Cliffs, NJ, 1989.

Courtenay, W.R. Jr., and Williams, J.D., Dispersal of Exotic Species from Aquaculture Sources, with Emphasis on Freshwater Fishes, in *Dispersal of Living Organisms into Aquatic Ecosystems*, Rosenfield, A. and Mann, R., Eds., Maryland Sea Grant Publication, College Park, MD, 1992.

Davis, W.B., The Mammals of Texas, Texas Parks and Wildlife Department Bulletin 41.

Exotic Species Programs., Ecologically Harmful Exotic Aquatic Plant and Wild Animal Species in Minnesota: Annual Report for 1993. Minnesota Department of Natural Resources, St. Paul, MN, 1993.

Fuller, P.L., L.G. Nico, and J.D. Williams, Nonindigenous Fishes Introduced into Inland Waters of the United States, Am. Fisheries Soc. Spec. Pub. 27, Bethesda, MD, 613 pp., 1999.

Havel, J., personal communication, 1994.

Janik, J., personal communication, 1997.

Johnson, S.K., Summary of Key Virus Diseases of Shrimp Aquaculture, prepared for the Texas Senate Natural Resources Interim Subcommittee, Texas A&M University, 19 April 1996.

Keevin, T., Inadvertent Transport of Live Zebra Mussels on Barges — Experiences in the St. Louis District, Spring 1992. U.S. Army Corps of Engineers, Zebra Mussel Research Technical Note ZMR-1-07, 1992

Kovalak, W.P, G.D. Longton, and R.D. Smithee, Infestation of power plant water systems by the zebra mussel (*Dreissena polymorpha Pallas*) in *Zebra Mussels Biology, Impacts, and Control*, Nalepa, T.F. and Schloesser, D.W., Eds., Lewis Publishers, Boca Raton, FL, 1993.

Krueger, C.C., and B. May, Ecological and genetic effects of salmonid introductions in North America, *Can. J. Fish. Aquat. Sci.,* 48 (suppl. 1), 66–77, 1991.

Ludwig Jr., H.R. and J.A. Leitch, Interbasin transfer of aquatic biota via anglers' bait buckets, *Fisheries,* 21, pp.14–18, 1996.

Malecki, R., Purple Loosestrife, in *Our Living Resources: A Report to the Nation on the Distribution, Abundance, and Health of U.S. Plants, Animals, and Ecosystems*, LaRoe, E.T., G.S. Farris, C.E. Puckett, P.D. Doran, and M.J. Mac, Eds., U.S. Department of the Interior, National Biological Service, Washington, D.C., 1995.

Marsden, J.E., R.E. Sparks, and K.D. Blodgett, Overview of the Zebra Mussel Invasion: Biology Impacts, and Projected Spread in *Proc. of the 1991 Governor's Conference on the Management of the Illinois River System*, Holly Korab, Ed., Third Biennial Conference, University of Illinois, Urbana, Ill, 1991, pp. 88–95.

Miller, R.R., J.D. Williams, and J.E. Williams, Extinction of North American Fishes during the last century, *Fisheries,* 14, 22–38, 1989.

Mills, E.L., J.H. Leach, J.T. Carlton, and C.L. Secor, Exotic species in the Great Lakes: a history of biotic crises and anthropogenic introductions, International Association for Great Lakes Research, *J. Great Lakes Res.,* 19, 1–54, 1993.

Mills, E.L., J.H. Leach, J.T. Carlton, and C.L. Secor, Exotic species and the integrity of the Great Lakes, *Bioscience,* 44, 666–676, 1994.

Nalepa, T.F., Decline of native unionid bivalves in Lake St. Clair after infestation by the zebra mussel, *Dreissena polymorpha, Can. J. Fish. Aquat. Sci.,* 51, 2227–2233, 1994.

Nehring, R.B. and P.G. Walker, Whirling disease in the wild: the new reality in the intermountain West, *Fisheries,* 21:6, 28–30, 1996.

Nichols, S.J., personal communication, 1997.

Olsen, T.M., D.M. Lodge, G.M. Capelli, and R.J. Houlihan. Mechanisms of impact of an introduced crayfish (*Orconectes rusticus*) on littoral congeners, snails, and macrophytes, *Can. J. Aquat. Sci.,* 48, 1853–1861, 1991.

O'Neil, T., *The Muskrat in the Louisiana Coastal Marshes*, Louisiana Department of Wildlife and Fisheries, New Orleans, 1949.

O'Neill, C.R., Jr., National Zebra Mussel Information Clearinghouse Infrastructure economic impact survey — 1995, *Dreissena!* 7, 2, 1996.

Shotts, E.B., Jr., and J.B. Gratzek, Bacteria, parasites, and viruses of aquarium fish and their shipping waters in *Distribution, Biology, and Management of Exotic Fishes*, Courtenay, W.R. and J.R. Stauffer, Eds., Johns Hopkins University Press, Baltimore, MD, 1984.

Simpson, J.C. and R.L. Wallace, *Fishes of Idaho*, University Press of Idaho, Moscow, ID, 1978, p. 165.

Skinner, L., Biological Control of Purple Loosestrife, in *Book of Abstracts: Second Northeast Conference on Nonindigenous Aquatic Species*, April 18-19, 1997, Burlington, Vermont. Connecticut Sea Grant Marine College Program, Groton, CT, 1997, pp. 7.

Skinner, W.F., Report on the Eradication of Tilapia from the Vicinity of Brunner Island Electric Station, Pennsylvania Power and Electric Co., Allentown, PA, 1987.

Slobodkin, L.B. and P.E. Bossert, The freshwater Cnidaria-or Coelenterates, in *Ecology and Classification of North American Freshwater Invertebrates*, Thorp, J.H. and A.P. Covich, Eds., Academic Press, San Diego, CA, 1991.

Stuckey, R.L., Distributional history of *Lythrum salicaria* (purple loosestrife) in North America, *Bartonia*, 47, 3–20, 1980.

Taylor, J.N., W.R. Courtenay, Jr., and J.A. McCann, Known impacts of exotic fishes in the continental U.S., in *Distribution, Biology and Management of Exotic Fishes*, Courtenay, W.R. Jr. and Stauffer, J.R. Jr., Eds., Johns Hopkins University Press, Baltimore, MD, 1984.

Trautman, M.B., *The Fishes of Ohio*, Ohio State University Press, Columbus, OH, 1981, p. 263.

U.S. Congress, Office of Technology Assessment, *Harmful Non-Indigenous Species in the U.S.*, OTA-F-565, U.S. Government Printing Office, Washington, DC, 1993.

Williams, J.D. and D.P. Jennings, Computerized data base for exotic fishes: the western U.S. *Calif. Fish and Game*, 77, 86–93, 1991.

Wood, T.S., Bryozoans, in *Ecology and Classification of North American Freshwater Invertebrates*, Thorp, J.H. and A.P. Covich, Eds., Academic Press, San Diego, CA, 1991.

2 Annotated Checklist of Introduced Invasive Fishes in Mexico, with Examples of Some Recent Introductions

Salvador Contreras-Balderas

INTRODUCTION

As of 1984, the so-called exotic (introduced) fish species in Mexico numbered 55. In 1997, such introductions numbered 90, with many known as secondary colonizers taking advantage of the changing environment. The last couple of decades have seen growing interest in nonindigenous species. Successively they have been called exotic, foreign, transcontinental, non-native, transplants, transfaunal, exogenous and more recently, invasive. An important reference work on most of the concepts behind these names is that supported by the American Fisheries Society Committee on Introduced Species (1987).

For the purposes of this paper, we will classify the introduced species by the main vector of original introductions in the following categories:

- aquaculture — cultivated for any reason: sports, prey, or forage for larger species
- commercial — food or sale
- ornamental — kept for aesthetics
- pest control — to keep down pest organisms
- bait releases — surplus bait from sport fishing
- species protection — introduced to prevent extinction in native areas
- accidental and invasive not purposefully introduced, which may have invaded from a different environment, often as a result of human impacts
- unknown — purpose, situation, or points of introduction are not known to the writer.

Worldwide, international introductions of inland aquatic species have been described by Welcomme (1988); he also summarized the different purposes of introductions, and the reasons why some species may become successful, useful, or pests. In Mexico, information on exotic fishes, originating both outside and within the country, was summarized by Contreras and Escalante (1984). Additional reports of introduced fish appear from time to time, frequently as "new records" (e.g., Gaspar, 1988; Díaz and Páramo, 1984) in numerous faunistic surveys, and unusually as "new species" (Hernández-Rolón, 1990). Some have appeared as reports on physiology, genetics, aquaculture, fisheries and others. It has been observed that changes in the environment, besides human impacts, generally favor the invasive characteristics of introduced species (Contreras, 1975b; Welcomme, 1988; Edwards and Contreras, 1991; Contreras and Lozano, 1994).

0-15667-0449-9/99/$0.00+$.50
© 1999 by CRC Press LLC

METHODS

We have compiled our data from numerous literature reports appearing after, or missed by, Contreras and Escalante (1984), and our own records from specimens housed at Universidad Autónoma de Nuevo León Fish Collection. For the Mexican states we used the following acronyms: AGS Aguascalientes, BCA Baja California, BCS Baja California Sur, CHI Chiapas, CHU Chihuahua, DGO Durango, GTO Guanajuato, HGO Hidalgo, JAL Jalisco, MIC Michoacán, MOR Morelos, QRO Queretaro, QUI Quintana Roo, SIN Sinaloa, SON Sonora, TAB Tabasco, TAM Tamaulipas, VER Veracruz, ZAC Zacatecas. Where no additional information is known after Contreras and Escalante (1984), we omit repeating data and refer the reader to that paper. An asterisk (*) indicates Mexican origin.

ANNOTATED CHECKLIST OF INVASIVE FRESHWATER FISH SPECIES IN MEXICO

FAMILY CLUPEIDAE. SARDINES.

1 *Dorosoma petenense* (Gunther). Threadfin shad.
Reports: Contreras and Escalante (1984). NLE Presa La Boca (S. Contreras, 1984; 1995). SON Río Yaqui estuary (Varela, 1989); TAM. Intruders expanding natural range up the Rio Bravo.
Cause: Forage and Accidental (mistakes, probably stocked at La Boca with fishes from Tancol Fish Hatchery, TAM).

2 *Dorosoma cepedianum* (Lesueur). Gizzard shad.
Reports: NLE Presa La Boca (Contreras, 1984; 1995).
Cause: Forage and Accidental (mistake, probably stocked with fishes introduced from Tancol Fish Hatchery, TAM).

FAMILY OSTEOGLOSSIDAE. BONYTONGUES.

3 *Arapaima gigas* (Cuvier). Arapaima.
Reports: Welcomme (1988). No known authenticated record.

FAMILY SALMONIDAE. SALMONS AND TROUTS.

4 *Onchorhynchus mykiss gairdneri* (Richardson). Rainbow trout.
Reports: Contreras and Escalante (1984). Welcomme (1988), who recorded *O. tschawitscha*, a species not yet introduced or recorded from México. MIC Río Duero (Cochran, et al., 1996); CHA Lagunas de Montebello (Lozano and Contreras, 1987); NLE San Cristóbal creek and trout fish farm (this report); OAX Temazcal (this report).
Cause: Sports and Commercial.

5 *Salvelinus fontinalis* (Mitchill). Brook trout.
Reports: Contreras and Escalante (1984). Welcomme (1988). No recent reports.
Cause: Sports and Commercial.

FAMILY CYPRINIDAE. CARPS.

6 *Algansea lacustris* Steindachner. Acumara.
Reports: Numerous localities outside its natural range (Lake Pátzcuaro).
Cause: Commercial.

7 *Barbus conchonius* (Hamilton-Buchanan). Rosy barb.
8 *Barbus titteya* (Deraniyagala). Cherry barb.
Reports: Contreras and Escalante (1984). Welcomme (1988). NLE (Contreras, 1978). Several other species were stocked at the time.
Cause: Ornamental.

9 *Carassius auratus* (Linnaeus). Goldfish.
Reports: Contreras and Escalante (1984). AGS Rio Verde (Contreras and Contreras, 1969); and JAL Santiago, Chapala; Rio Lerma basin at many localities: GTO, JAL, MEX, MIC, QRO (Soto, et al., 1991); COA Parras (Contreras and Maeda, 1985); HGO several localities (Soria, et al., 1996); JAL Chapala (Elizondo and Fernández, 1995); SON Rio Yaqui (Hendrickson, 1984); Río Magdalena (Hendrickson and Juárez, 1990); MEX Rio Lerma (Hendrickson, 1984); MEX Danxho, La Goleta and Macúa (Navarrete, et al., 1997), and Valle de Bravo (Gaspar and Fernández, 1997); MIC Lago de Cuitzeo (Chacón and Alvarado, 1995); OAX Presa Yosocuta, San Baltasar Chichicalpan (Contreras, et al., MS).
Cause: Commercial and Ornamental.

10 *Cyprinus carpio* Linnaeus. Common carp.
Reports: Contreras and Escalante (1984). Welcomme (1988). AGS Rio Verde (Contreras and Contreras, 1969); COA Parras (Contreras and Maeda, 1985); Presa Venustiano Carranza (Contreras, 1985); Rio Lerma basin at many localities: GTO, JAL, MEX, MIC, QRO (Soto, et al., 1991); GRO Presa El Caracol (Carranza and López, 1995); HGO several localities (Soria, et al., 1996); JAL Chapala (Elizondo and Fernández, 1995); MEX Santa María del Llano (Navarrete and Sánchez, 1989, Navarrete, et al., 1997); Valle de Bravo (Gaspar and Fernández, 1997), San Andres Timilpa (Sánchez and Navarrete, 1987; Navarrete, et al., 1997); MIC Lago de Pátzcuaro (Orbe and Acevedo, 1995); Lago de Cuitzeo (Chacón and Alvarado, 1995); Presa Infiernillo (Juárez, 1995); Río Duero (Cochran, et al. 1996); Laguna de Zacapu (Moncayo, 1996), Río Angulo (Medina, 1997); PUE Ponds near Puebla (López, et al., 1997); OAX Presa B. Juarez (Contreras, et al., MS); SON Rio Yaqui (Hendrickson, 1984); TAM Lower Rio Grande (Edwards and Contreras, 1991); BCS several localities (Villavicencio and Reynoso, MS).
Cause: Commercial and Ornamental.

11 *Ctenopharyngodon idella* (Valenciennes). Amur.
Reports: Contreras and Escalante (1984). Welcomme (1988). HGO several localities (Soria, et al., 1996); JAL Chapala (Elizondo y Fernández, 1995); MEX Santa María del Llano (Navarrete and Sánchez, 1989); NLE Presa La Boca (Contreras, 1984; 1995), CHA Rio Lacanjá (Rodiles, et al., 1997); PUE Ponds near Puebla (López, et al., 1997).
Cause: Commercial and Pest Control.

12 *Hypophthalmichthys molitrix* (Valenciennes). Silver carp.
Reports: Contreras and Escalante (1984). MEX Santa María del Llano (Navarrete y Sánchez, 1989); MIC Presa Infiernillo (Juárez, 1995).
Cause: Commercial.

13 *Aristichthys nobilis* (Richardson). Bighead carp.

14 *Megalobrema amblycephala* Yih.

15 *Mylopharyngodon piceus* Richardson. Black roach.
Reports: Welcomme (1988). MEX Santa María del Llano (except #14; Navarrete and Salgado, 1989). Numerous unspecified localities in Central Mexico, as components of Chinese-type polyculture.

Cause: Commercial.
16 *Gila bicolor mohavensis* (Snyder). Mohave tui chub.

17 *Gila orcutti* (Eigenmann and Eigenmann). Arroyo chub.
Reports: Contreras and Escalante (1984). Welcomme (1988). A 1995–96 survey of the region did not result in collecting the species (Ruiz and Contreras, 1997).
Cause: Protection.

18 *Gila modesta* (Garman). Saltillo chub.
Reports: *G. modesta* was stocked in two places at a fishless creek in Cañon del Chiflón, 40 km W Saltillo in 1992 (Contreras, MS; to be reported elsewhere).
Cause: Protection. Apparently introductions have failed.

19 *Campostoma anomalum* (Rafinesque). Stoneroller.
Reports: TAM Garza Valdéz, Río Soto la Marina (Contreras, 1984).
Cause: Accidental (mistake).

20 *Notemygonus crysoleucas* (Mitchill). Golden shiner.
Reports: Contreras and Escalante (1984). Welcomme (1988). TAM, NLE Presa Marte R. Gomez (Barajas and Contreras, 1975); TAM Lower Rio Grande (Edwards and Contreras, 1991).
Cause: Bait Release.

21 *Notropis chihuahua* Woolman. Chihuahua shiner.
Reports: COA Lower Rio Nazas (Obregon and Contreras, MS; to be reported elsewhere).
Cause: Accidental (mistake).

22 *Notropis* sp. Unnamed shiner.
Reports: CHU Presa at Ricardo Flores Magón in Río Carmen interior drainage (Espinoza, et al., 1988). The species is an undescribed endemic known from Rio Conchos and Rio Nazas basins.
Cause: Accidental (mistake).

23 *Cyprinella lutrensis* (Baird and Girard). Redshiner.
Reports: Contreras and Escalante (1984). SON Río Yaqui.
Cause: Bait Release and Accidental.

24 *Pimephales promelas* (Rafinesque). Fathead minnow.
Reports: Contreras and Escalante (1984). CHU Rio Yaqui (Hendrickson, 1984).
Cause: Forage, Bait Release and Accidental (secondary invaders).

25 *Pimephales vigilax perspicuus* (Baird and Girard) — Bullhead minnow.
Reports: Contreras and Escalante (1984). CHI/TEX Middle Rio Bravo (Contreras, 1984).
Cause: Forage, Bait Release and Accidental (secondary invaders).
Comment: The Western form, *P. v. vigilax*, is native to the Lower Rio Grande. The Eastern form, introduced in the Middle Rio Grande, is easily recognized by its beak-like mouth.

FAMILY CATOSTOMIDAE. CARPSUCKERS.

26 *Carpiodes carpio* (Rafinesque). River carpsucker.
Reports: CHU Rio Yaqui (Hendrickson, 1984).
Cause: Commercial.

27 *Carpiodes cyprinus* (Lesueur). Quillback.
Reports: Welcomme (1988). Unspecified site or reference (Espinoza, et al., 1993).
Cause: Commercial.

Family COBITIDAE. Loaches.

28 *Misgurnus anguillicaudatus* (Cantor). Oriental weatherfish.
Reports: Contreras and Escalante (1984). Welcomme (1988). MEX Laguna de Texcoco near Chapingo (Contreras and Escalante, 1984).
Cause: Commercial.

Familia CHARACIDAE. Tetras.

29 *Astyanax mexicanus* (Filippi). Mexican tetra.
Reports: CHU Ojo de Agua de La Hacienda (Contreras, 1984). The irrigation canals opened the way for entrance of this species in the spring, where there are two endemic fishes; the endemics are at risk because of their lack of defensive abilities, and the voracious nature of this exotic.
Cause: Accidental (secondary invaders).

Family SERRASALMIDAE. Pacus and Piranhas.

30 *Colossoma macropoma* Cuvier. Pacu.
Reports: MOR and TAB several localities (Espinoza, et al., 1993).
Cause: Commercial: experimental food production.

Familia ICTALURIDAE. Bullhead catfishes.

31 *Ictalurus furcatus* (Lesueur). Blue catfish.
Reports: SON Rio Yaqui, CHU Rio Casas Grandes (Hendrickson, 1984).
Cause: Commercial.

32 *Ictalurus punctatus* (Rafinesque). Channel catfish.
Reports: CHU Rio Yaqui (Hendrickson, 1984), Rio Casas Grandes; SIN Rio Elota (Beltrán, et al., 1997).
Cause: Commercial.

33 *Ameiurus melas* (Rafinesque). Black bullhead.
Reports: Contreras and Escalante (1984). CHU Rio Yaqui (Hendrickson, 1984), Ojo de la Casa, and Rio Casas Grandes (Espinoza, et al., 1988).
Cause: Commercial.

34 *Ameiurus natalis* (Lesueur). Yellow bullhead.
Reports: NLE Arroyo s/n near Montemorelos (this report).
Cause: Commercial.

35 *Pylodictis olivaris* (Rafinesque). Flathead catfish.
Reports: CHU Rio Yaqui (Hendrickson, 1984).
Cause: Commercial.

Family LORICARIIDAE. Suckermouth catfishes.

36 *Liposarcus multiradiatus* (Hancock). Placastro.
Reports: MIC Río Mezcala, Balsas basin (Guzmán and Barragán, 1997).

Cause: Ornamental and Accidental (escape).

Family FUNDULIDAE. Killifishes.

37 *Fundulus zebrinus* Jordan and Gilbert. Plains killifish.
Reports: Contreras and Escalante (1984). Welcomme (1988). CHU Rio Conchos (Leal and Contreras, 1987).
Cause: Forage, Bait Release and Accidental (secondary invaders).

Family POECILIIDAE. Livebearers.

38 *Gambusia affinis* (Baird and Girard). Mosquitofish.
Reports: Contreras and Escalante (1984). COA Parras (Contreras and Maeda, 1985); CHU Rio Casas Grandes and Ojo de la Casa (Espinoza, et al., 1988), Rio Yaqui (Hendrickson, 1984). NLE Presa La Boca (Contreras, 1984; 1995); BCN, BCS several localities (Ruiz and Contreras, 1985).
Cause: Pest Control (for mosquitoes) or Accidental (La Boca: Mistake).

39 *Gambusia hurtadoi* Hubbs and Springer. Guayacón de Dolores.
Reports: CHU Rio Conchos near Jiménez; this endemic from Ojo de La Hacienda escaped via the same irrigation canals that allowed the entrance of *Astyanax mexicanus* to the spring (Contreras, 1984).
Cause: Accidental (Escape).

40 *Gambusia regani* Hubbs. Guayacón veracruzano.
Reports: Contreras and Escalante (1984) as *G. panuco*. SLP Media Luna (Aguilera, et al., 1996).
Cause: Accidental.

41 *Poecilia latipunctata* (Meek). Moly azul.
Reports: SLP Numerous springs in the Valle del Rio Verde (La Media Luna) (Aguilera, et al., 1996), threatening several endemic fishes.
Cause: Accidental (?). Source not known, may have been stocked by aquarists.

42 *Poecilia mexicana* (Steindachner). Mexican molly.
Reports: SLP Numerous springs in the Valle del Rio Verde (La Media Luna) (Aguilera, et al., 1996), threatening several endemic fishes. MIC Río Duero (Cochran, et al., 1996).
Cause: Accidental (?). Source unknown, may be spontaneous invader from lower reaches of Rio Pánuco or stocked by aquarists.

43 *Poecilia: latipinna* X *mexicana*. Hybrid black molly.
Reports: BCA/SON Lower Rio Colorado (Ruiz and Contreras, 1985). The aquarium strains, black and their regression to wild type are common in this area.
Cause: Ornamental.

44 *Poecilia reticulata* (Peters). Guppy.
Reports: Contreras and Escalante (1984). Welcomme (1988). COA Parras (Contreras and Maeda, 1985); Rio Lerma basin at many localities: GTO, JAL, MEX, MIC, QRO (Soto, Barragán and López, 1991); AGS Río Verde, JAL Rio Santiago (Hendrickson, 1984); HGO several localities (Soria, et al., 1996); MIC Río Duero (Cochran, et al., 1997); BCS several localities (Ruiz and Contreras, 1985), several oasis (Villavicencio and Reynoso, MS); MOR Ojo de Agua de Tetelco, fish hatchery at Zacatepec (Contreras, et al., In press); MIC Ojo de Agua La Minzita; OAX ca.

Huajuápan de Leon (this report); PUE Springs around Tehuacán (this report); NLE Ojo de Agua de Apodaca, Ojo de Agua de Cieneguilla, Ojo de Agua de Mezquital (this report).
Cause: Ornamental.

45 *Heterandria bimaculata* (Haeckel).
Reports: GRO Presa El Caracol (Carranza and López, 1995); MOR Rio Balsas several localities (Gaspar, 1988), Tequesquitengo (Hernandez, 1990).
Cause: Unknown, probably aquarists or Accidental.

46 *Poeciliopsis gracilis* (Meek). Porthole livebearer.
Reports: GRO Presa El Caracol (Carranza and López, 1995); OAX Rio Balsas (Mejía, 1992).
Cause: Accidental (?).

47 *Xiphophorus couchianus* (Girard). Monterrey platy.
Reports: NLE Rio Chapultepec (this report). Endemic. Escaped through overflow from springs at La Pastora and Colonia Libertad, hybridizing with *X. variatus* (Contreras, 1978).
Cause: Accidental (escapes).

48 *Xiphophorus gordoni* (Miller and Minckley).
Cuatro Ciénegas Platy.
Reports: COA Rio Salado at Cariño de la Montaña (Contreras, 1984). Endemic.
Cause: Accidental (Escaped from the Valle de Cuatro Cienegas via irrigation channels).

49 *Xiphophorus helleri* Haeckel. Green swordtail.
Reports: MEX (Soto, et al., 1991); JAL Lago Chapala (Hendrickson, 1984). BCS Several oasis (Rodriguez, MS), ca. Cd. Constitución (this report); COA Parras (Contreras and Maeda, 1985); NLE Río La Silla, Rio Chapultepec (this report); MIC Río Angulo (Medina, 1997).
Cause: Ornamental.

50 *Xiphophorus maculatus* (Gunther). Southern platyfish.
Reports: JAL Rio Ameca (Hendrickson, 1984); BCS several localities (Ruiz and Contreras, 1985), Oasis ca. Cd. Constitucion (this report); other oasis (Villavicencio and Reynoso, MS).
Cause: Ornamental.

51 *Xiphophorus variatus* (Meek). Variable platyfish.
Reports: NLE Ojo de Agua La Libertad (Contreras, 1975b), Rio La Silla (this report).
Cause: Ornamental.

FAMILY **ATHERINIDAE**. SILVERSIDES.

52 *Chirostoma aculeatum* Barbour. Charal cuchillo.

53 *Chirostoma consocium* Jordan and Hubbs. Charal blanco.

54 *Chirostoma estor* Jordan. Pescado blanco (Pátzcuaro).

55 *Chirostoma grandocule* (Steindachner). Charal ojón.

56 *Chirostoma jordani* Woolman. Charal cambrai.

57 *Chirostoma labarcae* Meek. Charal de La Barca.

58 *Chirostoma sphyraena* Boulenger. Blanco Picudo (Chapala).
Reports: CHU (Leal and Contreras, 1987), DGO (Macías and Contreras, 1982), HGO several localities (Soria, et al., 1996); NLE Presa La Boca (Contreras, 1984). Undetermined species stocked under Programa de Pescado Blanco, supposed to be *Ch. estor*, at numerous sites.
Cause: Commercial.

59 *Menidia beryllina* (Cope). Tidewater silversides.
Report: CHU Rio Conchos basin (Leal and Contreras, 1987). NLE Presa La Boca (Contreras, 1984; 1995), probably stocked with fish from Tancol Fish Hatchery, TAM. TAM, NLE Lower Rio Bravo and Rio San Juan (Contreras and Lozano, 1994). According to C. Hubbs, Río Bravo/Rio Grande populations are more likely introduced. Edwards and Contreras (1991) regarded it as an invader from coastal areas, through TAM and up to COA Hidalgo, NLE Los Aldama and Laguna Monfort (this report).
Cause: Accidental (mistake).

60 *Membras martinica* (Valenciennes). Rough silversides.
Reports: NLE Presa La Boca (Contreras, 1984; 1995), probably stocked with fish from Tancol Fish Hatchery, TAM. TAM, NLE and COA invasive from lower Rio Bravo up to Presa Marte R. Gomez (Edwards and Contreras, 1991).
Cause: Accidental (mistake).

Family PERCICHTHYIDAE. Temperate basses.

61 *Roccus chrysops* (Rafinesque). White bass.
Reports: Contreras and Escalante (1984). NLE Rio Alamo (Ruiz and Contreras, 1985), TAM Presa Falcon, CHU Upper Rio Bravo near Ciudad Juarez (Contreras, 1984).
Cause: Sports, Accidental (secondary invader).

62 *Roccus saxatilis* (Walbaum). Striped bass.
Reports: Contreras and Escalante (1984). Rio Bravo (this report).
Cause: Sports, Accidental (secondary invader).

Family CENTRARCHIDAE. Sunfishes and black basses.

63 *Ambloplites rupestris* (Rafinesque). Rockbass.
Reports: Contreras and Escalante (1984). CHU Welcomme (1988). No localities or sources given (Espinoza, et al., 1993).
Cause: Sports and Forage.

64 *Lepomis auritus* (Linnaeus). Redbreast sunfish.
Reports: Contreras and Escalante (1984). Welcomme (1988). TAM Lower Rio Grande (Edwards and Contreras, 1991); MOR Tequesquitengo (Hernández, 1990).
Cause: Forage.

65 *Lepomis cyanellus* (Rafinesque). Green sunfish.
Reports: Contreras and Escalante (1984). MEX Rio Lerma (Soto, et al., 1991); MEX Valle de Bravo (Gaspar and Fernández, 1997); SON Rio Yaqui (Hendrickson, 1984); Río Magdalena (Hendrickson and Juárez, 1990).
Cause: Forage.

66 *Lepomis gulosus* (Cuvier). Warmouth.

Reports: Contreras and Escalante (1984). Welcomme (1988). AGS Rio Verde (Contreras and Contreras, 1969); COA Presa Venustiano Carranza (Contreras, 1985); TAM Lower Rio Grande (Edwards and Contreras, 1991).
Cause: Forage.

67 *Lepomis macrochirus* (Rafinesque). Bluegill sunfish.
Reports: Contreras and Escalante (1984). HGO several localities (Soria, et al., 1996); SON Rio Yaqui (Hendrickson, 1984); AGS Rio Verde; MEX Rio Lerma, JAL Laguna el Rosario (Hendrickson, 1984). Río Magdalena (Hendrickson and Juárez, 1990); MEX Valle de Bravo (Gaspar and Fernández, 1997).
Cause: Forage.

68 *Lepomis marginatus* (Holbrook). Dollar sunfish.
Reports: CHIH Lower Río Conchos (Leal and Contreras, 1987).
Cause: Forage.

69 *Lepomis megalotis* (Rafinesque). Longear sunfish.
Reports: Contreras and Escalante (1984). SON Rio Yaqui (Hendrickson, 1984). Native in the Rio Bravo basin.
Cause: Forage.

70 *Lepomis microlophus* (Gunther). Redear sunfish.
Reports: Contreras and Escalante (1984). Welcomme (1988). SON Rio Yaqui (Hendrickson, 1984); TAM Lower Rio Grande (Edwards and Contreras, 1991).
Cause: Forage. Regarded by some ichthyologists as native in the Rio Bravo basin. We consider it to be introduced, due to its absence from collections earlier than 1953-4, when Falcon Dam was closed and several fish introduced by Texas Fish and Game Commission (letter from M. Toole to F. Obregón-F., 1954).

71 *Lepomis punctatus* (Valenciennes). Spotted sunfish.
Reports: Contreras and Escalante (1984). Listed without localities or source (Espinoza, et al., 1993).
Cause: Forage.

72 *Micropterus dolomieui* (Lacépède). Smallmouth bass.
Reports: Contreras and Escalante (1984). Welcomme (1988). NLE Presa El Cuchillo (this report). Newspaper reports. Confirmed in a 1996 fishing tournament organized by Unión de Campistas de México, A.C.
Cause: Sports.

73 *Micropterus salmoides* (Lacépède). Largemouth bass.
Reports: Contreras and Escalante (1984). AGS Rio Verde (Contreras and Contreras, 1969); HGO several localities (Soria, et al., 1996); SIN Rio Fuerte (Hendrickson, 1984); MEX Valle de Bravo (Gaspar and Fernández, 1997); MIC Lago de Camécuaro (Guzmán, et al., 1979), Río Angulo (Medina, 1997); TAM Laguna de Champayán (Arteaga, 1985), Presa Vicente Guerrero (Blanco, 1990; Elizondo, et al., 1995); GTO, JAL, MEX, MIC Rio Lerma basin at several localities (Soto, et al., 1991); SIN Rio Elota (Beltrán, et al., 1997); SON Rio Yaqui (Hendrickson, 1984); Río Magdalena (Hendrickson and Juárez, 1990); VER Lago de Catemaco (Torres and Pérez, 1995).
Cause: Sports.

74 *Pomoxis annularis* Rafinesque. White crappie.

Reports: Contreras and Escalante (1984). Welcomme (1988). SON Rio Yaqui (Hendrickson, 1984); TAM Lower Rio Grande (Edwards and Contreras, 1991).
Cause: Sports.

75 *Pomoxis nigromaculatus* (Lesueur). Black crappie.
Reports: Contreras and Escalante (1984). Welcomme (1988). SON Rio Yaqui (Hendrickson, 1984).
Cause: Sports.

Family CICHLIDAE. Cichlids.

76 *Cichlasoma cyanoguttatum* (Baird and Girard). Rio Grande cichlid.
Reports: Contreras and Escalante (1984). MOR Lake Tequesquitengo (Hernández-Rolón 1983); later described as *Parapetenia cyanostigma* n sp. Hernández 1990.
Cause: probably Ornamental.

77 *Cichlasoma ellioti* (Meek).
Reports: VER Several localities NW from Papaloapam basin (Obregón, et al., 1994).
Cause: Commercial.

78 *Cichlasoma octofasciatum* (Regan). Jack Dempsey.
Reports: VER Several localities NW from Papaloapam basin (Obregón, et al., 1994).
Cause: Commercial.

79 *Cichlasoma managuense* (Gunther). Tiger guapote.
Reports: CAM Laguna de Términos (Olvera, et al., 1994); TAB unspecified locality (Arias and Durán, 1997).
Cause: Accidental (?); may be invasive from aquaculture in Guatemala.

80 *Cichlasoma motaguense* (Gunther).
Reports: TAB Rios Mezcalapa and González (Díaz and Páramo, 1984).
Cause: Accidental; may be invasive from aquaculture in Guatemala.

81 *Cichlasoma nigrofasciatum* (Gunther). Convict cichlid.
Reports: MIC, MOR, several localities in the Río Balsas basin (T. Contreras, In press).
Cause: Ornamental.

82 *Cichlasoma pearsei* (Hubbs)
Reports: Contreras and Escalante (1984). No recent records.
Cause: Commercial.

83 *Cichlasoma urophthalmus* (Gunther)
Reports: Contreras and Escalante (1984). VER Several localities in southern part (Obregon, et al., 1994); OAX Rio Usila (Rodiles, et al., 1997); Presa Cerro de Oro (Rodríguez, et al., 1995).
Cause: Commercial.

84 *Petenia splendida* (Gunther). Tenhuayaca.
Reports: Contreras and Escalante (1984). VER Rio Antigua (Obregón, et al., 1994); OAX Rio Usila (Rodiles, et al., 1997); Tuxtepec or Presa Cerro de Oro (Rodríguez, et al., 1995; Santiago, Jardón and Jaramillo, 1997).
Cause: Commercial.

85 *Oreochromis aureus* (Steindachner). Blue tilapia.
Reports: Contreras and Escalante (1984). Welcomme (1988). BCN several localities (Ruíz and Contreras, 1985); AGS Rio Verde, SIN Rio Fuerte, Rio Sinaloa, JAL Rio Santiago; MEX Rio Lerma, Laguna Zacoalco, JAL & MIC Rio Armeria (Hendrickson, 1984); CHA Río Lacanjá (Gaspar, 1996); COA Presa Venustiano Carranza (Contreras, 1985); JAL Sierra de Manatlán (Lyons and Navarro, 1990); MEX Valle de Bravo (Gaspar and Fernández, 1997); MIC Presa Zicuirán (Ramírez, et al., 1997), Río Angulo (Medina, 1997); MIC/GUE Presa Infiernillo (Jiménez, et al., 1997); MOR Tequesquitengo (Hernández, 1990); NLE Presa La Boca (Contreras, 1984); TAM Presa Vicente Guerrero (Elizondo, et al., 1995); JAL Chapala (Elizondo and Fernández, 1995); TAM Río Bravo en Presa Falcón (Contreras, 1995), Lower Rio Grande (Edwards and Contreras, 1991); TAM/NLE Río San Juan en Presa Marte R. Gómez (this report); SON Rio Yaqui (Hendrickson, 1984); Río Magdalena as Tilapia sp. (Hendrickson and Juárez, 1990), most probably representing O. aureus; VER Lago de Catemaco (Torres and Pérez, 1995); COA Los Mesquites (this report).
Cause: Commercial.

86 *Oreochromis mossambicus* (Peters). Mozambique mouthbrooder.
Reports: Contreras and Escalante (1984). Welcomme (1988). Numerous localities in central Mexico: CHA several localities (Lozano and Contreras, 1987), Presa La Angostura (Olmos, 1995); GTO, JAL, MEX, MIC, Rio Lerma basin at several localities (Soto, et al., 1991); HGO several localities (Soria, et al., 1996); MOR Tequesquitengo (Hernández, 1990).
Cause: Commercial.

87 *Oreochromis niloticus* (Linnaeus). Nile tilapia.
Reports: Welcomme (1988). Numerous localities in central and southern México. CHA Presa La Angostura (Olmos, 1995); HGO Tezontepec (Salvadores, 1980); OAX: Rio Usila (Rodiles, et al., 1997), and Tuxtepec (Santiago, et al., 1997); MOR Tequesquitengo (Hernández, 1990); MEX Valle de Bravo (Acereto, 1983), Laguna de Coatetelco (Gómez, et al., 1997); MIC Lago de Cuitzeo (Chacón and Alvarado, 1995), Río Angulo (Medina, 1997); OAX Presa Cerro de Oro (Rodríguez, et al., 1995); VER Laguna de Chila (Basurto, 1984); BCS stocked at La Purisima-San Isidro (Villavicencio and Reynoso, MS).
Cause: Commercial.

88 *Oreochromis hornorum* Trewavas. Wami tilapia.
Reports: Welcomme (1988). MOR Tequesquitengo (Hernández, 1990); OAX Presa Benito Juarez (this report).
Cause: Commercial.

89 *Tilapia rendalli* (Boulenger). Redbreast tilapia.
Reports: Welcomme (1988). CHA several localities (Lozano and Contreras, 1987); HGO several localities (Soria, et al., 1996); JAL Sierra de Manantlán (Lyons and Navarro, 1990); MIC Río Angulo (Medina, 1997); SIN Rivers Presidio, Piaxtla, and Baluarte (Beltrán, et al., 1997); OAX Presa Benito Juarez (this report).
Cause: Commercial.

90 *Tilapia zilli* (Gervais). Redbelly tilapia.
Reports: Contreras and Escalante (1984). Welcomme (1988). BCN several localities (Ruíz and Contreras, 1985).
Cause: Commercial.

DISCUSSION AND CONCLUSIONS

The invasive species in Mexico have increased from 55 in 1984 (Contreras and Escalante, 1984) to 90 in 1997, a 61.1% increase. The vectors of introductions are as follows: 67 (74.4%) from aquaculture; 9 (10.0%) for sports; 15 (16.7%) as forage species, 38 (42.2%) commercial (usually for food), 11 (12.2%) ornamental, 2 (2.2%) pest control, 5 (5.6%) bait releases; attempts for species protection account for 3 (3.3%); and finally, accidental represents 23 (25.6%). Only 9 (10.0%) appear to have multiple vectors of introduction.

Biogeographically, 47 (52.2%) are nearctic, 27 (30%) neotropical, 8 (8.9%) palearctic, 6 (6.7%) ethiopic, and 2 (2.2%) oriental. Geopolitically and using the standard concepts, 46 (51.1%) are national transplants and 44 (48.9%) are foreign. Ten of the Mexican species are shared with the U.S. and seem to have entered from this country. This changes the national count to 36 (40%) and the foreign count to 54 (60%). A breakdown of the foreign count shows 30 (33.3%) species being from U.S. (20 not shared), 8 (8.9%) Asian, 6 (6.8%) African, 3 (3.3%) each Central and South American, and 2 (2.2%) European. In a fine analysis of the distribution within Mexico, the Northern Mexico has most of the nearctic (U.S.) and palearctic (European) invasives, and the Southern regions have most of the other species recorded.

Using an approach different from that of Welcomme (1988), the types of invasions by fish species may be broken down as follows.

A. Accidental, stocking with very small sizes of fry from hatcheries, which may be invaded by native fish, has the risk of nonintentional introductions. This is the case of *Dorosoma cepedianum*, *D. petenense*, *Menidia beryllina*, and *Membras martinica* at Presa La Boca NLE, and *Notropis* sp. at Presa R.F. Magon CHU.
B. Expansive, highly efficient competitors, predators, or breeders, that easily displace other fishes. They usually expand from the site of original introduction into many localities. Examples are the tilapines and mosquito fish (*Gambusia affinis*), as well as *Pomoxis annularis*, *Lepomis gulosus*, in the lower Rio Bravo/Grande.
C. Escapes — in some cases endemic fishes have used irrigation channels to go beyond the original barriers, and are more or less successfully colonizing nearby areas. This is the case of *Gambusia hurtadoi* collected in the Rio Florido (Río Conchos) at Jiménez CHU. Having highly specific requirements, such species may not represent risks to the wide-ranging species or complex fish communities.
D. Invasive, where invasion takes place as a consequence of channels connecting springs and rivers. This is the case of *Astyanax mexicanus* coming in, and *Gambusia hurtadoi* escaping from, at Ojo de la Hacienda CHU (see above). Another cause of such invasion may be the change in the environment. For example, increased salinity in the lower Rio Bravo/Grande, resulted in ever increasing number of species being found there: from 8 in mid-19th century to 75 by the end of the 20th (Rodriguez, 1976; Edwards and Contreras, 1991). Some species penetrate 600 km inland (Contreras and Lozano, 1994), well above tidal zone. Most of them are members of the tropical rather than temperate communities (Castro and Sobrino, 1997). These fishes are not discussed in this paper, but they do deserve a mention as a special case of invasive species expanding, possibly as a result of water overuse or global warming.

See Table 2.1 for further information.

NATIONAL ENDEMICS AS INVADERS

There are a few cases known where endemic species considered to be at risk have become pests when introduced in other localities where they function as "exotics." The expansion of *Poecilia*

TABLE 2.1

Causes, Biogeographical Origin, and Country or Continental Sources of Invasive (Introduced) Fish Species in Mexico

Species	Common Name	Aquaculture				Pest Control	Bait Releases	Protection	Invaders, Accidental, and Escapes	Biogeographical Orgin	Country or Continent
		Sports	Forage	Commercial	Ornamental						
		Family Clupeidae. Sardines.									
1 *Dorosoma petenense	Threadfin shad	–	X	–	–	–	–	–	X	NA	MX+
2 *Dorosoma cepedianum	Gizzard shad	–	X	–	–	–	–	–	X	NA	MX+
		Family Osteoglossidae. Birstlemoths.									
3 Arapaima gigas.	Arapaima	–	–	X	–	–	–	–	–	NT	SA
		Family Salmonidae									
4 Onchorhynchus mykiss	Rainbow trout	X	–	X	–	–	–	–	–	NA	US
5 Salvelinus fontinalis	Brook trout	X	–	–	–	–	–	–	–	NA	US
		Family Cyprinidae. Carps									
6 *Algansea lacustris	Acúmara	–	–	X	–	–	–	–	–	NA	MX
7 Barbus conchonius	Rosy barb	–	–	–	X	–	–	–	–	OR	AS
8 Barbus titteya	Cherry barb	–	–	–	X	–	–	–	–	OR	AS
9 Carassius auratus	Goldfish	–	–	X	X	–	–	–	–	PA	EU
10 Cyprinus carpio	Common carp	–	–	X	X	–	–	–	–	PA	EU
11 Ctenopharyngodon idella	Amur	–	–	X	–	X	–	–	–	PA	AS
12 Hypophthalmichthys molitrix	Silver carp.	–	–	X	–	–	–	–	–	PA	AS
13 Aristichthys nobilis	Striped bighead	–	–	X	–	–	–	–	–	PA	AS
14 Megalobrema amblycephala	?	–	–	X	–	–	–	–	–	PA	AS
15 Mylopharyngodon piceus	Black roach	–	–	X	–	–	–	–	–	PA	AS
16 Gila bicolor mohavensis	Mohave tui chub	–	–	–	–	–	–	X	–	NA	US
17 Gila orcutti	Arroyo chub	–	–	–	–	–	–	X	–	NA	US
18 * Gila modesta	Saltillo chub	–	–	–	–	–	–	X	–	NA	MX
19 * Campostoma anomalum	Stoneroller	–	–	–	–	–	–	–	X	NA	MX

TABLE 2.1 (CONTINUED)
Causes, Biogeographical Origin, and Country or Continental Sources of Invasive (Introduced) Fish Species in Mexico

Species	Common Name	Sports	Forage	Commercial	Ornamental	Pest Control	Bait Releases	Protection	Invaders, Accidental, and Escapes	Biogeographical Orgin	Country or Continent
20 *Notemygonus crysoleucas*	Golden shiner	–	–	–	–	–	X	–	–	NA	US
21 * *Notropis chihuahua*	Chihuahua shiner	–	–	–	–	–	–	–	X	NA	MX
22 * *Notropis sp.*	Unnamed shiner	–	–	–	–	–	–	–	X	NA	MX
23 * *Cyprinella lutrensis*	Red shiner	–	–	–	–	–	X	–	X	NA	MX+
24 * *Pimephales promelas*	Fathead minnow	–	X	–	–	–	X	–	X	NA	MX+
25 * *Pimephales vigilax perspicuus*	Bullhead minnow	–	X	–	–	–	X	–	X	NA	US
Family Catostomidae. Carpsuckers.											
26 * *Carpiodes carpio*	River carpsucker	–	–	X	–	–	–	–	–	NA	MX
27 * *Carpiodes cyprinus*	Quillback	–	–	X	–	–	–	–	–	NA	US
Family Cobitidae. Loaches.											
28 *Misgurnus anguillicaudatus*	Oriental weatherfish	–	–	X	–	–	–	–	–	PA	AS
Family Characidae. Tetras.											
29 **Astyanax mexicanus*	Mexican tetra	–	–	–	–	–	–	–	X	NT	MX
Family Serrasalmidae. Pacus and Piranhas.											
30 *Colossoma macropoma*	Pacu	–	–	X	–	–	–	–	–	NT	SA
Family Ictaluridae. North American Catfishes.											
31 **Ictalurus furcatus*	Blue catfish	–	–	X	–	–	–	–	–	NA	MX+
32 **Ictalurus punctatus*	Channel catfish	–	–	X	–	–	–	–	–	NA	MX+
33 *Ameiurus melas*	Black bullhead	–	–	X	–	–	–	–	–	NA	US
34 *Ameiurus natalis*	Yellow bullhead	–	–	X	–	–	–	–	–	NA	US
35 **Pylodictis olivaris*	Flathead catfish	–	–	X	–	–	–	–	–	NA	MX+

No.	Species	Common name			Status	Origin
Family Loricariidae. Suckermouth Catfishes.						
36	*Liposarcus multiradiatus*	Placastro	–	X	NT	SA
Family Fundulidae. Killifishes.						
37	*Fundulus zebrinus*	Plains killifish	X	–	NA	US
Family Poeciliidae. Livebearers.						
38	*Gambusia affinis*	Mosquitofish	–	X	NT	MX
39	*Gambusia hurtadoi*	Guayacón de Dolores	–	X	NT	MX
40	*Gambusia regani*	Guayacón jarocho	–	X	NT	MX
41	*Poecilia latipunctata*	Moly azul	–	X	NT	MX
42	*Poecilia mexicana*	Mexican molly	–	X	NT	MX
43	*Poecilia hybrids*	Black molly	X	–	NT	MX+
44	*Poecilia reticulata*	Guppy	X	–	NT	SA
45	*Heterandria bimaculata*	?	–	X	NT	MX
46	*Poeciliopsis gracilis*	Pothole livebearer	–	X	NT	MX
47	*Xiphophorus couchianus*	Monterrey platy	–	X	NT	MX
48	*Xiphophorus gordoni*	Cuatro Ciénegas platy	–	X	NT	MX
49	*Xiphophorus helleri*	Green swordtail	X	–	NT	MX
50	*Xiphophorus maculatus*	Southern platy	X	–	NT	MX
51	*Xiphophorus variatus*	Variable platy	X	–	NT	MX
Family Atherinidae. Silversides.						
52	*Chirostoma aculeatum*	Charal cuchillo	X	–	NA	MX
53	*Chirostoma consocium*	Charal blanco	X	–	NA	MX
54	*Chirostoma estor*	Pescado blanco Pátzcuaro	X	–	NA	MX
55	*Chirostoma grandocule*	Charal ojón	X	–	NA	MX
56	*Chirostoma jordani*	Charal cambrai	X	–	NA	MX
57	*Chirostoma labarcae*	Charal de La Barca	X	–	NA	MX
58	*Chirostoma sphyraena*	Charal picudo	X	–	NA	MX
59	*Menidia beryllina*	Tidewater silversides	X	–	NA	MX
60	*Membras martinica*	Rough silversides	–	X	NA	MX

TABLE 2.1 (CONTINUED)
Causes, Biogeographical Origin, and Country or Continental Sources of Invasive (Introduced) Fish Species in Mexico

Species	Common Name	Aquaculture				Pest Control	Bait Releases	Protection	Invaders, Accidental, and Escapes	Biogeographical Orgin	Country or Continent
		Sports	Forage	Commercial	Ornamental						
Family Percichthyidae. Temperate Basses.											
61 *Roccus chrysops*	White bass	X	–	–	–	–	–	–	–	NA	US
62 *Roccus saxatilis*	Striped bass	X	–	–	–	–	–	–	–	NA	US
Family Centrarchidae. Sunfishes.											
63 *Ambloplites rupestris*	Rock bass	X	X	–	–	–	–	–	–	NA	US
64 *Lepomis auritus*	Redbreast sunfish	–	X	–	–	–	–	–	–	NA	US
65 *Lepomis cyanellus*	Green sunfish	–	X	–	–	–	–	–	–	NA	MX+
66 *Lepomis gulosus*	Warmouth	–	X	–	–	–	–	–	–	NA	US
67 *Lepomis macrochirus*	Bluegill sunfish	–	X	–	–	–	–	–	–	NA	MX+
68 *Lepomis marginatus*	Dollar sunfish	–	X	–	–	–	–	–	–	NA	US
69 *Lepomis megalotis*	Longear sunfish	–	X	–	–	–	–	–	–	NA	MX+
70 *Lepomis microlophus*	Redear sunfish	–	X	–	–	–	–	–	–	NA	US
71 *Lepomis punctatus*	Spotted sunfish	–	X	–	–	–	–	–	–	NA	US
72 *Micropterus dolomieui*	Smallmouth bass	X	–	–	–	–	–	–	–	NA	US
73 *Micropterus salmoides*	Largemouth bass	X	–	–	–	–	–	–	–	NA	MX+
74 *Pomoxis annularis*	White crappie	X	–	–	–	–	–	–	–	NA	US
75 *Pomoxis nigromaculatus*	Black crappie	X	–	–	–	–	–	–	–	NA	US
Family Cichlidae. Cichlids.											
76 *Cichlasoma cyanoguttatum*	Rio Grande cichlid	–	–	–	–	–	–	–	X	NT	MX
77 *Cichlasoma ellioti*	–	–	–	X	–	–	–	–	–	NT	MX
78 *Cichlasoma octofasciatum*	Jack Dempsey	–	–	X	–	–	–	–	–	NT	MX
79 *Cichlasoma urophthalmus*	–	–	X	–	–	–	–	–	–	NT	MX
80 *Cichlasoma managuense*	Tiger guapote	–	–	X	–	–	–	–	–	NT	CA
81 *Cichlasoma motaguense*	–	–	–	–	–	–	–	–	X	NT	CA

No.	Scientific name	Common name									Origin	Continent
82	*Cichlasoma nigrofasciatum*	Convict cichlid	–	–	–	X	–	–	–	–	NT	CA
83	**Cichlasoma pearsei*	–	–	–	X	–	–	–	–	–	NT	MX
84	**Petenia splendida*	Tenhuayaca	–	–	X	–	–	–	–	–	NT	MX
85	*Oreochromis aureus*	Blue tilapia	–	–	X	–	–	–	–	–	ET	AF
86	*Oreochromis mossambicus*	Mozambique mouthbrooder	–	–	X	–	–	–	–	–	ET	AF
87	*Oreochromis niloticus*	Nile tilapia	–	–	X	–	–	–	–	–	ET	AF
88	*Oreochromis hornorum*	Wami tilapia	–	–	X	–	–	–	–	–	ET	AF
89	*Tilapia rendalli*	Redbreast tilapia	–	–	X	–	–	–	–	–	ET	AF
90	*Tilapia zilli*	Redbelly tilapia	–	–	X	–	–	–	–	–	ET	AF
	TOTAL		9	15	38	11	2	5	3	23		

Symbols: A = Aquaculture (S = Sports, F = Forage, C = Commercial, O = Ornamental), P = Pest Control, B = Bait Releases, P = Protection, I = Invaders, accidental, and escapes; BG = Biogeographical origin: NA = Nearctic, NT = Neotropical, PA = PAlearctic, OR = ORiental, ET = Ethiopic, CO = COuntry or COntinent: MX = MeXico, US = United STates, AF = AFrica thru Middle East, SA = South America, CA = Central America. * = Mexican, + = Mexican obtained thru U.S. Sources.

latipunctata, an endemic at risk from the Rio Tamesí (= Guayalejo) transferred to La Media Luna
SLP, together with the introduction of *Poecilia mexicana* and *Gambusia regani*, is a case in point.
In the new environment it is closely associated with the dwindling populations of the Media Luna
endemics *Dionda mandibularis*, *Cualac tessellatus*, and *Ataeniobius toweri*. These species are also
considered at risk. The matter is further complicated by the decline of the local cichlids, associated
with the introduction of tilapias.

Little is known of the biological characteristics of such endemic species. Several of the observed
cases have multiple vectors of introduction, such as expanding beyond stocking sites, aquarist
actions, and accidents. A high number are species stocked intentionally by federal programs,
including national species (*Algansea*, *Chirostoma*, and *Cichlasoma*) or foreign species (*Onchorhynchus*, *Salvelinus*, *Cyprinus*, *Carassius*, *Tilapia*, *Oreochromis*, and others). A number are accidental transfers beyond their native ranges (*Campostoma*, *Notropis*, *Menidia*, and *Membras*). It is
interesting to note that the invasive species used by the national programs are supported in whole
or in part by international programs, especially FAO. This is the case with several carps and tilapias.
U.S. Fish and Wildlife, and some U.S. states' fish and game agencies (Arizona, Texas), have
supported the introduction of several sports fishes. For example, at the Rio Bravo/Rio Grande, the
U.S. (Texas Fish and Game) agencies have introduced different black basses, sunfishes, striped and
yellow basses, as well as other species. The Mexican side has retaliated with introduction of carps
and tilapias. Neither side has shown any regard for endemic fishes or consulting each other. This
is a false concept of sovereignty and independence, given that there is only a single basin, where
no fully independent actions are possible.

The differences go beyond the political and linguistic levels; they are also cultural and
economic. U.S. people look for recreation and sports, whereas Mexicans are in search of food
resources. The cases are subject to complex interactions of wild organisms between themselves,
and the socioeconomic aspects of humans.

ACKNOWLEDGMENTS

The people who have helped with field work to collect the fishes described in this chapter between
1958 and 1997 are too numerous to list. Financial support also came from a number of Mexican
and U.S. agencies. M. L. Lozano-Vilano allowed me access to Universidad Autónoma de Nuevo
León Fish Collection. Francisco Reynoso-Mendoza and Carlos Villavicencio-Garayzar (Universidad Autónoma de Baja California Sur), T. Contreras-Macbeath (Universidad Autónoma de Morelos), and A.F. Guzmán (Instituto Politécnico Nacional) allowed me to use their manuscripts and
in-press articles and sent copies immediately after publication. An anonymous reviewer did metic-
ulous work checking literature and indicating mistakes and omissions; his work is reflected heavily
in the revised version and contributed to its quality. Renata Claudi gave support, encouragement,
and allowed me time to rewrite the paper. To all of them, my sincere and deep appreciation.

REFERENCES

Acereto-S., N., Contribución al conocimiento de la pesquería de *Sarotherodon niloticus* en el embalse de Valle
    de Bravo, Edo. de México. Tesis Profesional, ENEP Iztacala, UNAM, México., 1983.
Aguilera-González, C., J. Montemayor-Leal, and S. Contreras-Balderas. Fishes of the Upper Río Verde and
    its disrupted pluvial basin, San Luis, Potosí, México. *Desert Fishes Council, Abstracts, Ann. Meet.* 1996.
Arias-Rodríguez, L., and A. Durán-Rodríguez, Estudio citogenético en el guapote tigre *Cichlasoma man-
    aguense* (Gunther, 1869) (Pisces: Cichlidae). *Abstracts, V Congreso Nacional de Ictiología, Mazatlán,
    México.* 1997.
Arredondo-Figueroa, J.L., Axalapascos de la Cuenca Oriental, Puebla. En: *Lagos y Presas de México.* De la
    Lanza-Espino, G. and J. L. García-Calderón, Eds. Centro de Ecología y Desarrollo. I:65–87. 1995.

Arteaga-M, J., Crecimiento, Reproducción, y hábitos alimenticios de lobina negra *Micropterus salmoides* (Lacépède) en la Laguna de Champayán, Altamira, Tamps. Tesis Profesional, Facultad de Ciencias Biológicas, Universidad Autónoma de Nuevo León, 1985.

Barajas-Martínez, L., and S. Contreras-Balderas, Morfología ecológica y niveles tróficos de peces en la Presa Marte R. Gómez, Noreste de México. *Memorias, IV Congreso Nacional de Zoología, México*, 1980.

Basurto, M., Estudio preliminar al conocimiento biológico y pesquero de la *Tilapia nilotica* (Linneo), en la Laguna de Chila, Veracruz, México. Tesis Profesional, Escuela de Ciencias Biológicas, Universidad del Noreste, México, 1984.

Beltrán-Álvarez, R., J. Sánchez-Palacios, G. Arroyo-Bustos, J.P. Ramírez, and H. Galavíz, La Ictiofauna dulceacuícola de las corrientes fluviales del Sur de Sinaloa, México. *Abstracts, V Congreso Nacional de Ictiología, Mazatlán, México*, 1997.

Blanco, H., Algunos parámetros biológicos y pesqueros de Tilapia (*Oreochromis aureus*: Steindachner, 1864) en la Presa Vicente Guerrero, Tamps., México. Tesis Profesional, Escuela de Ciencias Biológicas, Universidad del Noreste, México. 54 pp., 1990.

Carranza-Fraser, J., and M. López-Hernández, Presa El Caracol (Ingeniero Carlos Ramírez Ulloa). En: *Lagos y Presas de México*. de la Lanza-Espino, G., and J.L. García-Calderón, Eds. Centro de Ecología y Desarrollo, I:225–241. 1995.

Castro-Aguirre, J. L., and A. S. Sobrino-Figueroa, Análisis Ecológico-Geográfico de los Peces marinos que penetran hacia las aguas continentales de México. *Abstracts, V Congreso Nacional de Ictiología*, México, Mazatlán. 1997.

Chacón-Torres, A., and J. Alvarado-Díaz, El Lago de Cuitzeo. En: *Lagos y Presas de México*. de la Lanza-Espino, G., and J.L. García-Calderón, Eds., Centro de Ecología y Desarrollo, I:117–127. 1995.

Cochran, P.A., J. Lyons, and E. Merino-Nambo, Notes on the biology of the Mexican lampreys *Lampetra spadicea* and *L. gemminis* (Agnatha: Petromyzontidae). *Ichthyological Exploration of Freshwaters*, 7(2):173–180.

Contreras-Balderas, A.J., and S. Contreras-Balderas, Lista de Peces de Aguascalientes, México: Zoogeografía y Ecología. *Publ. Biol., FCB., UANL.* 3(1). 1989.

Contreras-Balderas, S., Cambios de Composición de especies en comunidades de peces en zonas semiáridas de México. *Publicaciones Biológicas, Universidad Autónoma de Nuevo León*, 2(7):181–194. 1975a.

Contreras-Balderas, S., *Impacto Ambiental de Obras Hidráulicas*. Informe, Plan Nacional Hidráulico, México. 1975b.

Contreras-Balderas, S., Lista zoogeográfica y ecológica de los peces de Coahuila, México. *Memorias del VIII Congreso Nacional de Zoología (México)*, 1:156–174. 1985.

Contreras-Balderas, S., New Records and Notes of Exotic Fishes in Northeastern Mexico. *Desert Fishes Council*, XVI:46–49, 1984.

Contreras-Balderas, S. Las Presas del Noreste de México. En: *Lagos y Presas de México*. de la Lanza-Espino, G., and J.L. García-Calderón, Eds. Centro de Ecología y Desarrollo, I:277–289. 1995.

Contreras-Balderas, S., and M.A. Escalante, Distribution and Known Impacts of Exotic Species in Mexico. Chapter 6, In: *Distribution, Biology, and Management of Exotic Fishes*. Courtenay Jr., W.R, and J.R. Stauffer, Eds. J. Hopkins University Press. 1984.

Contreras-Balderas, S., M.L. Lozano-Vilano, and M.E. García-Ramírez, Tercera Lista Anotada y Revisada de los Peces de Nuevo León, México. Cap. 6:71-78. En: *Listado Preliminar de la Fauna del Estado de Nuevo León, México*. S. Contreras-Balderas, F. González-Saldívar, D. Lazcano-Villarreal, y A. Contreras-A. Eds. Subcomisión de Fauna del Estado de Nuevo León, México. 1995.

Contreras-Balderas, S., and M.L. Lozano-Vilano, Water, Endangered Fishes, and Development Perspectives in Arid Lands of Mexico. *Conservation Biology*, 8(2):379–387. 1994.

Contreras-Balderas, S., M.L. Lozano-Vilano, and M.E. García-Ramírez, Peces del Centro y Sur de Oaxaca, México. *Abstracts, V Congreso Nacional de Ictiología*, México, Mazatlán, 1997.

Contreras-Balderas, S., and A. Maeda-Martínez, Estado Actual de la Ictiofauna Nativa de la cuenca de Parras, Coah., Méx. con Notas sobre algunos Invertebrados. *Memorias VIII Congreso Nacional de Zoología, México*, 1:59–69. 1985.

Contreras-MacBeath, T., H. Mejía-Mojica, and R. Carrillo-Wilson, Shifts in the composition of the fish fauna inhabiting waters of the State of Morelos, Mexico: a good example of bad management. *Journal of Aquatic Ecosystem Health* (Canada). In press.

Díaz-Pardo, E., and S. Páramo-Delgadillo, Dos nuevos registros para la ictiofauna dulceacuícola mexicana. *Anales, Escuela Nacional de Ciencias Biológicas, México,* 28:19–28. 1984.

Edwards, R. J., and S. Contreras-Balderas, Historical changes in the Ichthyofauna of the Lower Rio Grande (Rio Bravo del Norte), Texas and Mexico. *Southwestern Naturalist,* 36(2):201–212. 1991.

Elizondo-Garza, R., and J. Fernández-Méndez, Caracterización biológico-pesquera del Lago de Chapala, Jalisco-Michoacán, México, con un análisis de las capturas de charal, *Chirostoma chapalae* (Jordan y Snyder, 1900), en redes mangueadoras y atarrayas. *Publicaciones Biológicas, Facultad de Ciencias Biológicas, Universidad Autónoma de Nuevo Leon,* México, 8(1–2):62–96. 1995.

Elizondo-Garza, R., M. Martínez-Zavala, I. Roque-Villada and J. Fernandez-Mendez, Caracterización biológico-pesquera de la Presa "Vicente Guerrero" (Las Adjuntas), Tamaulipas, México, con un análisis de las capturas de tilapia, *Oreochromis aureus* (Steindachner, 1864) y lobina negra, *Micropterus salmoides* (Lacépède). 1802. *Publicaciónes Biológicas, Universidad Autónoma de Nuevo León,* 8(1–2):25–61. 1995.

Espinoza-Aguilar, A.D., M.L. Lozano-Vilano and S. Contreras-Balderas, Ictiofauna del Complejo Smalayuca, Chihuahua, México. *Desert Fishes Council, Abstracts, Ann. Meet.* 1988.

Espinoza-Pérez, H., T. Gaspar-Dillanes, and P. Fuentes-Mata, Los Peces Dulceacuícolas Mexicanos. En *Listados Faunísticos de México.* Instituto de Biología, Universidad Nacional Autónoma de México. III:i-iv, 1–99. 1993.

Gaspar-Dillanes, M.T., Nuevo registro de *Heterandria (Pseudoxiphophorus) bimaculatus* (Haeckel, 1848) en la Vertiente del Pacífico Mexicano. *Anales del Instituto de Biología, Universidad Nacional Autónoma de México, (Serie Zoología, 2)*:933–938. 1988.

Gaspar-Dillanes, T., Aportación al conocimiento de la ictiofauna de la Selva Lacandona, Chiapas. *Zoología Informa* (Instituto Politécnico Nacional, México), 33:41–54.1996.

Gaspar-Dillanes, M.T., and J.I. Fernández-Meléndez, Aspectos biológico-pesqueros de especies de importancia comercial en la Presa Miguel Alemán, Valle de Bravo, Méx. *Abstracts, V Congreso Nacional de Ictiología, Mazatlán, México,* 1997.

Gómez-Márquez, J.L., B. Peña-Mendoza, I.H. Salgado-Ugarte and M. Garduño-Paredes, Aspectos reproductivos de la tilapia *O. niloticus. Abstracts, V Congreso Nacional de Ictiología, Mazatlán, México,* 1997.

Guzmán, A.F., and J. Barragán-S, 1997. Presencia de bagre sudamericano (Osteichthyes: Loricariidae) en el Río Mezcala, Guerrero, México. *Vertebrata Mexicana,* 3:1–4, 1997.

Guzmán-Arroyo, M., J.L. Rojas-Galavíz, and F. Vera-Herrera, Crecimiento y aspectos poblacionales de la lobina negra *Micropterus salmoides* Lacépède en el Lago de Camécuaro, Michoacán) (Pisces: Centrarchidae). *Anales del Centro de Ciencias del Mar y Limnología, Universidad Nacional Autonóma de México,* 6(1):53–68, 1979.

Hendrickson, D.A., Distribution records of native and exotic fishes in Pacific drainages of Northern Mexico. *Journal, Arizona-Nevada Academy of Sciences, for 1983,* 18:33–38, 1984.

Hendrickson, D.A., and L. Juárez-Romero, Los peces de la cuenca del Río de la Concepción, Sonora, México, y el estatus del charalito sonorense, *Gila ditaenia,* una especie amenaza de extinción. *Southwestern Naturalist,* 35(2):177–187. 1990.

Hernández-Rolón, A., The cichlids of Lake Tequesquitengo. *Buntbarsche Bulletin,* 95:2–4, 1983.

Hernández-Rolón, A., Un noveau Cichlidé du système du Rio Balsas, Mexique (Pisces, Teleostei). *Revue Française du Cichlidophiles,* 1990:4–13, 1990.

Jiménez-Badillo, M.L., P. Trujillo-Jiménez and C. Ramírez-Camarena, Aspectos reproductivos de *Oreochromis aureus* (Pisces; Cichlidae) de la Presa Infiernillo, Michoacán-Guerrero, México. *Abstracts, V Congreso Nacional de Ictiología, Mazatlán, México,* 1997.

Juárez-Palacios, J.R., Presa Infiernillo (Adolfo López Mateos). En: *Lagos y Presas de México.* de la Lanza-Espino, G., and J.L. García-Calderón, Eds., I:211–223. 1995.

Leal-Sotelo, H. and S. Contreras-Balderas, Fish Fauna of the Río Conchos, Subbasin of Río Bravo, North Central México. *Desert Fishes Council, Abstracts, Ann. Meet.* 1987.

López-González, M.C., A. Guzmán-Sosa and J.S. Hernández-Aviles, Evaluación de un policultivo en estanques rústicos. *Abstracts, V Congreso Nacional de Ictiología, Mazatlán, México,* 1997.

Lozano-Vilano, M.L., and S. Contreras-Balderas, Lista Zoogeográfica y Ecológica de la Ictiofauna Continental de Chiapas, México. *Southwestern Naturalist,* 32(2):223–236, 1987.

Lyons, J., and S. Navarro-Pérez, Fishes of the Sierra de Manantlán, West-Central México. *Southwestern Naturalist,* 35(1):32–46. 1990.

Macías-Chávez, L.J., and S. Contreras-Balderas, Los Peces del Estado de Durango, México. *Desert Fishes Council, Abstracts, Ann. Meet.* 1982.

Medina-Nava, M., Ictiofauna de la subcuenca del Río Angulo, Cuenca del Lerma-Chapala, Michoacán. *Zoología Informa* (Instituto Politécnico Nacional, México), 35:25–52. 1997.

Moncayo-E., R., Estructura y función de la comunidad de peces de la Laguna de Zacapu Michoacán. Tesis de grado, Maestría en Ciencias. Centro Interdisciplinario de Ciencias del Mar, Instituto Politécnico Nacional. 1996.

Navarrete-Salgado, N.A., and R. Sánchez-Merino, El sistema de policultivo de peces en el medio rural mexicano. *Revista Latinoamericana de Acuicultura*, Lima, 39:45–68, 1989.

Navarrete-Salgado, N.A., G. Contreras-Rivero and G. Elías-Fernández, Ictiofauna de cuatro embalses del Estado de México y su relación con parámetros ambientales. *Abstracts, V Congreso Nacional de Ictiología, Mazatlán, México,* 1997.

Navarrete-Salgado, N.A., G. Elías-Fernández and G. Contreras-Rivero, Crecimiento y producción de carpa común y herbívora en un estanque rústico de San Andrés Timilpa, Edo. de México. *Abstracts, V Congreso Nacional de Ictiología, Mazatlán, México,* 1997.

Obregón-Barbosa, H., S. Contreras-Balderas, and M.L. Lozano-Vilano, The fishes of northern and central Veracruz, Mexico. *Hydrobiologia,* 286:79–95, 1994.

Olvera, M.A., I. Piña, I. Cu, and E. Chávez, *Proceedings, World Fisheries Congress,* 3:278–281, 1994.

Olmos-Tomasini. E., Presa La Angostura (Doctor Belisario Domínguez). En: *Lagos y Presas de México.* de la Lanza-Espino, G., and J.L. García-Calderón, Eds. *Centro de Ecología y Desarrollo,* I:262–267. 1995.

Orbe-Mendoza, A., and J. Acevedo-G., El Lago de Pátzcuaro. En: *Lagos y Presas de México.* de la Lanza-Espino, G., and J.L. García-Calderón, Eds., Centro de Ecología y Desarrollo, I:89–108. 1995.

Ramírez-Camarena, C., M.L. Jiménez-Badillo and C. Osuna-Paredes, Aspectos hidrológicos y pesqueros de la Presa Zicuirán, Mpio. La Huacana, Mich. *Abstracts, V Congreso Nacional de Ictiología, Mazatlán, México,* 1997.

Rodiles-Hernández, E. Díaz-Pardo, and A. Safa, *Estudio sobre la actividad pesquera en la cuenca del Río Usila, Oaxaca. situacion actual y perspectivas.* Programa de Aprovechamiento Integral de Recursos Naturales/Universidad Nacional Autónoma de México, Oaxaca. Informe, 1995.

Rodiles-Hernández, R., S. Domínguez-Cisneros and E. Velázquez-Velázquez, Diversidad ictiofaunística en el Rio Lacanjá, Selva Lacandona. *Abstracts, V Congreso Nacional de Ictiología, Mazatlán, México,* 1997.

Rodríguez-López, T., M. Morales-Román and A. Sánchez-Vázquez, Presa Cerro de Oro (Miguel de la Madrid Hurtado). En: *Lagos y Presas de México.* de la Lanza-Espino, G., and J.L. García-Cañderón, Eds. Centro de Ecología y Desarrollo, I:243–267. 1995.

Rodríguez-Olmos, Gloria, Cambios en la composición de especies de peces en comunidades del Bajo Río Bravo, México, Estados Unidos. Tesis profesional, Facultad de Ciencias Biológicas, Universidad Autónoma de Nuevo León. 1976.

Ruíz-Campos; G., and S. Contreras-Balderas, Peces del Río Alamo, Subcuenca del Río Bravo. I.Ictiofauna e Ictiogeografía. *Desert Fishes Council Proceed.* XVI:14–35. 1984.

Ruíz-Campos, G., and S. Contreras-Balderas, Ecological and Zoogeographical Check-List of the Continental Fishes of the Baja California Península, México. *Desert Fishes Council, Proceed.* XVII:105–117. 1985.

Ruiz-Campos, G., S. Contreras-Balderas, M.L. Lozano-Vilano and S. González-Guzmán, Estatus ecológico y distribucional de los peces continentales del Noroeste de Baja California, México: Distrito Sandieguense. *Abstracts, V Congreso Nacional de Ictiología, Mazatlán, México,* 1997.

Salvadores-B., M.L., Estudio de la Biología y Aspectos Poblacionales de la Tilapia (*Sarotherodon aureus,* Steindachner, 1964) (Pisces: Cichlidae) En la Presa Vicente Guerrero, Gro., México. Tesis Profesional, Facultad de Ciencias, Universidad Nacional Autónoma de México, 1980.

Sánchez-Merino, R., and N.A. Navarrete-Salgado, Rendimiento de carpa espejo (*Cyprinus carpio specularis*) en bordos del Estado de México. *Revista Latinoamericana de Acuicultura*, Lima, 33:35–44, 1987.

Sánchez-M., R., and Norma Navarrete-S., Rendimiento de la Carpa Espejo (*Cyprinus carpio specularis*) en bordos del Estado de México. *Revista Latinoamericana de Acuicultura,* 33:35–44, 1977.

Santiago, L.M.C., O.J. Jardón, S.G. Jaramillo, A.J.E. Reyes and V.A. Sánchez, Crecimiento y hábitos alimenticios de *Cichlasoma urophthalmus* (Gunther), *Oreochromis niloticus* (Linneo) y *Petenia splendida* (Gunther), Presa Miguel de la Madrid H. "Cerro de Oro" Tuxtepec, Oax. *Abstracts, V Congreso Nacional de Ictiología, Mazatlán, México,* 1997.

Soria-Barreto, M., L. Alcántara-Soria, and E. Soto-Galera, Ictiofauna del estado de Hidalgo. *Zoología Informa* (Instituto Politécnico Nacional, México), 33:55–78. 1996

Soto-G., H., J. Barragán, and E. López-L., Efectos del deterioro ambiental en la distribución de la ictiofauna lermense. *Universidad: Ciencia y Tecnología,* 1(4):61–68, 1991.

Torres-Orozco B., R.E., and A. Pérez-Rojas, El Lago de Catemaco. En: *Lagos y Presas de México.* de la Lanza-Espino, G,. and J.L.

García-Calderón. Centro de Ecología y Desarrollo, I:155–175. 1995.

Varela-Romero, A., *Dorosoma petenense* (Gunther), un nuevo registro para la cuenca del Río Yaqui, Sonora, México (Pisces; Clupeidae). *Ecologica* (Centro Ecológico de Sonora, México), 1(1):23–25. 1989.

Villavicencio-Garayzar, C.J., and F. Reynoso-Mendoza, Registro de *Tilapia* sp. en aguas interiores de Baja California Sur, Mexico, y su efecto en el sistema lacustre de La Purísima-San Isidro. MS.

Welcomme, R.L., International introductions of inland aquatic species. *FAO Fisheries Technical Paper* 294:1–318. 1988.

Section I

Intentional Introductions

3 Intentional Introductions: Are the Incalculable Risks Worth It?

Charles K. Minns and John M. Cooley

SECTION INTRODUCTION

As they grow and disperse, species unavoidably carry their parasites and diseases with them but not their predators, competitors, and prey. Unlike most species, *Homo sapiens* have often taken their preferred food, sport, and ornamental species with them as they colonized the Earth. We are familiar with stories of the introduction of rabbits and foxes into Australasia for food and sport, of Canadian beavers into the wilds of Patagonia to create a fur trade, and of Himalayan rhododendrons into Victorian English gardens. Many species have been similarly introduced to terrestrial and aquatic ecosystems.

An FAO survey documented intentional introduction of 291 freshwater species (mainly fish) into 148 countries (Welcomme 1992). Many of those introductions have been for aquaculture. Many of those species have escaped from their cages and ponds and invaded natural ecosystems. In fresh waters throughout the world, salmonid and carp species are the most common introductions. Salmonids have been introduced for sport and food and carp for ornament and food. The consequences of most aquatic introductions have been poorly documented or ignored (Welcomme 1992). Both intentional and accidental introductions have had their successes and failures. Success usually means that apparently benign ecosystem effects are accompanied by direct benefits. Failure brings disastrous ecosystem impacts regardless of the benefits. Overall introductions have produced a biological pollution around the world, degrading the fruits of evolution. As the levels of both intentional and accidental introductions escalate, nations are struggling to devise rules and regulations to minimize the risk of new negative effects and to reduce the impact of exotics already present.

Carlton and Geller (1993) described the expanding introduction of marine species via the ballast water of cargo ships as "ecological roulette." Magnuson (1976) characterized the use of exotics in lake fish management as "a game of chance." Meffe (1992) charged those using salmon hatcheries to compensate for overfishing and habitat loss on the Pacific coast with "techno-arrogance." Some question whether the evidence of negative effects is available (Crivelli 1995), although the most likely explanations are poor monitoring and inadequate assessment (by analogy with biological control agents reviewed in Simberloff & Stiling 1996). Thomas (1994) reported that aquatic vertebrate extinctions in North America were mainly caused by habitat loss (50%) and introductions (37%). In Britain, they have the "tens" rule (Williamson and Fitter 1996): 1 in 10 introduced species will appear in the wild, 1 in 10 of those species will become established, and 1 in 10 of those will become a pest. This rule implies that 1 in 1000 introduced species will cause negative impacts. The base figure of 10 seems far too high for freshwater species. Judging by North American experiences, especially in the Great Lakes basin, a rule of "twos" or "threes" seems more appropriate. The gambling imagery highlights the contrast of risks taken with accidental and intentional introductions. With accidental introductions we take risks through neglect and inattention. With intentional introductions we deliberately take the risks, often on the basis of inadequate information.

How we view introductions depends on our individual perceptions of the relationships between humans and nature. Some revere nature, respect the existence rights of other species, and see our responsibility to act as benign stewards. Others believe we own nature, have superior rights compared to all other species, and are free to exploit and alter nature as we see fit and necessary.

Various labels such as "deep ecology" and "wise use" are placed on the varying viewpoints, and all shades of opinion often invoke divine authority for their positions. We are faced with the dilemma of reconciling our diverse value systems with the ambiguities and inconsistencies of simultaneously controlling undesirable accidental invasions, promoting the use of imported species in aquaculture and agriculture, and conserving a dwindling store of natural biodiversity.

We cannot resolve those complex philosophical and ethical issues here, but we think there are clear patterns that point to the need for changes in our attitudes and behaviors toward nature:

- Humans, a technological species, have a long history of tinkering with species and ecosystems to direct more of nature's productivity to our end uses;
- Substantial evidence exists that we are overreaching the earth's sustainable capacity and creating conditions potentially detrimental to all forms of life, including humans;
- The frequency and the extent of negative impacts due to introductions are increasing.

There is ample evidence that we need to proceed more cautiously. FAO (1995) recently produced guidelines promoting the use of a precautionary approach to fish species introductions: "Because of the high probability that impacts of species introductions be of irreversible and unpredictable impacts, many species introductions are not precautionary. Therefore a strictly precautionary approach would not permit deliberate introductions and would take strong measures to prevent unintentional introductions."

Let us consider some familiar intentional aquatic successes and failures more closely. Crossman (1991) described the history of fish species introductions in North America. About 100 species have been introduced, most intentionally. Salmonid introductions are considered to be among the most successful throughout the world. Brown and rainbow trout have been introduced to lakes and rivers in many countries. West Coast Pacific salmon runs have been established in New Zealand.

As with the introduction of intensive agriculture to many lands, the effects have often been judged benign, with the benefits to humans far outweighing any negative effects on indigenous biota. The effects on endemics have largely been ignored, unmeasured. In the Great Lakes of North America, Pacific salmonids were first introduced as a stop-gap management action to control alewives, an invader, and provide fishing opportunities. This action came after native lake trout stocks succumbed to exploitation and sea lamprey, another accidental introduction. Now the introduced salmonids support sport fisheries worth billions of dollars, and some are questioning the continued efforts to restore lake trout. Even if stocking ceased now, runs of Pacific salmonids will persist, as these stocks have become naturalized. On the Pacific Coast, salmonid aquaculture using cages is increasing. Because of market-place preferences, Atlantic salmon are the more popular species for cage culture, and escapes are increasing. Similarly some culture, and escape, of Pacific salmonids occurred on the Atlantic coast. While stocking can artificially enhance fishing opportunities and aquaculture can mimic high-input agriculture to produce fish flesh, both often compromise the integrity of the receiving ecosystems and divert attention from the continuing degradation and abuse of nature. While many socioeconomic benefits derive from exotic salmonids, environmental disbenefits are treated as nonmonetary externalities and usually ignored.

The carps are another group that has been widely distributed around the world both for food and ornament. Elaborately colored specimens of Koi carp command astonishing prices. The Chinese intensively cultivate many species of carp in managed ponds. The pond culture is usually integrated with other intensive agricultural practice. Common carp were introduced into North America, particularly into the Great Lakes, in the late 19th century (Crossman 1991) to provide cheap food. They quickly spread and increased in abundance causing significant habitat damage and loss in near shore vegetated habitats. Their habits of uprooting plants and increasing turbidity have degraded many coastal wetlands in the lower Great Lakes.

There are other stories of success and failure. In this section on intentional introductions, several authors have provided a number of case studies both from North America and Mexico. What becomes clear from reading this section is that despite our long experience of the successes and failures associated with introductions, we still have much to learn about the management of introductions. Meanwhile, we think there are several complementary actions that can improve the situation:

- Consistent, Anticipatory Policy: All introductions, intentional and accidental, should be examined with the same magnifying glass. Both the reactive stance to present risks and contradictions in our dealings with intentionals versus accidentals have impeded the development of stronger policy. Policy needs to be coordinated at regional, national and international levels. Promotion of aquaculture must be balanced with conservation and protection of ecosystems and biodiversity.
- Quantitative Risk Assessment: Agencies need to develop and implement objective, quantitative risk-assessment frameworks. Existing subjective, nonquantitative frameworks are too vulnerable to the optimism and enthusiasm of those proposing introductions. This requires that governments invest in the development of the necessary scientific knowledge.
- Apply the Precautionary Principle: The precautionary approach is increasingly being advocated in many renewable resources sectors (disposal of chemical wastes, the management of fisheries and fish habitat, the conservation of rare and endangered species).The onus of proof must pass to those proposing actions affecting natural ecosystems. The proponents must show their actions will not cause negative effects before approval is given. If the knowledge base is insufficient to reduce the uncertainties to acceptable levels, the action should not be allowed to proceed. When the uncertainties and risks are high, say NO.
- Pass Effective Legislation: Laws and regulations are needed to enforce policies. We cannot rely on commercial self-regulation and the free market to ensure the natural common bio-wealth. Our socioeconomic system still treats most aspects of natural ecosystems as nonmonetary externalities, despite the efforts of ecological economists. Governments must act for the common, future good of the biosphere.

REFERENCES

Carlton, J.T. and J.B. Geller, Ecological roulette: the global transport of nonindigenous marine organisms, *Science*, 261:78–82, 1993.

Crivelli, A.J., Are fish introductions a threat to endemic freshwater fishes in the northern Mediterranean region? *Biological Conservation*, 72:311–319, 1995.

Crossman, E.J., Introduced freshwater fishes: a review of the North American perspective with emphasis on Canada, *Can. J. Fish. Aquat. Sci.*, 48(Suppl. 1):46–57, 1991.

FAO, Precautionary approach to fisheries. Part 1: Guidelines on the precautionary approach to capture fisheries and species introductions. FAO Rome, FAO Fisheries Technical Paper 350 (Part 1), 1995.

Magnuson, J.J., Managing with exotics — a game of chance, *Trans. Amer. Fish. Soc.*, 105(1):1–9,1976.

Meffe, G.K., Techno-arrogance and halfway technologies: salmon hatcheries on the Pacific coast of North America, *Conservation Biology*, 6(3):350–354, 1992.

Simberloff, D. and P. Stiling, How risky is biological control? *Ecology*, 77(7):1965–1974,1996.

Thomas, C.D., Extinction, colonization, and meta-populations: environmental tracking by rare species, *Conservation Biology*, 8(2):373–378, 1994.

Welcomme, R.L., A history of international introductions of inland aquatic species, *ICES Marine Sciences Symp.*, 194:3–14, 1992.

Williamson, M. and A. Fitter., The varying success of invaders, *Ecology*, 77(6):1661–1666, 1996.

4 Intentional Introductions of Nonindigenous Freshwater Organisms in North America

Alan J. Dextrase and Mark A. Coscarelli

INTRODUCTION

No country on Earth has seen more fish introductions than the United States. Almost 200 native species and at least 70 fishes from other continents have been introduced and become established in new ecosystems (Courtenay, 1993, 1995). Although numbers of introduced fish species in Canada and Mexico are smaller than in the U.S., they are still significant (Contreras and Escalante, 1984; Crossman, 1991). Hundreds of species of aquatic plants and invertebrates have also been introduced (Mills et al., 1993; Office of Technology Assessment, 1993). Although most aquatic plants and invertebrates have been introduced through unintentional sources, a large proportion of North America's fish introductions have been intentional. These have included both government-sanctioned introductions as well as unauthorized and usually illegal introductions conducted by private citizens. Introductions have been so commonplace, and in many instances have been occurring for such a long period of time, that the original distribution of many organisms is difficult, if not impossible, to discern.

Intentional introductions of freshwater organisms have a long history in North America. Early introductions were often the result of citizens trying to recreate the ambiance of their home countries by introducing familiar species of fish. The introduction of goldfish (*Carassius auratus*) in New York State in the 17th century followed by common carp (*Cyprinus carpio*) in 1831 are among the earliest recorded introductions (DeKay, 1842). Government-sponsored introductions date back to the 1870s in the U.S. and Mexico, and to the 1880s in Canada when numerous hatcheries were built to raise fish for stocking purposes (Contreras and Escalante, 1984; Crossman, 1991; Courtenay, 1995). The intent of many of these early stocking programs was to "improve" fisheries through the introduction of new species often considered superior to native species. As a result, introduced sport fishes are now the dominant fauna in many lakes and rivers in North America (Li and Moyle, 1993). The number of North American and global transfers has increased dramatically with improvements to transportation (Welcomme, 1988), and intentional introductions are an important and ongoing part of today's management programs.

This chapter provides an overview of the impacts of aquatic freshwater organisms that have been introduced intentionally to new ecosystems in North America and the legislative and policy approaches that have been put in place to control and guide these activities. We consider intentional introductions in the strictest sense, i.e., cases where humans have deliberately introduced an organism with the attempt to establish that species in the new environment. Although a portion of introductions resulting from aquarium releases, fish culture escapes, and bait-bucket dumping may meet this definition, they are not included in this chapter, as they are addressed in subsequent chapters of this book.

0-15667-0449-9/99/$0.00+$.50
© 1999 by CRC Press LLC

OVERVIEW OF INTENTIONAL INTRODUCTIONS IN NORTH AMERICA

Some freshwater fishes were undoubtedly introduced intentionally to North America very early during the European colonization of the continent, but large-scale government-sponsored introductions did not start until the late 1800s. During this period many species were widely introduced, quite often with little success, as fish were placed into unsuitable habitats. The current widespread North American distributions of common carp, rainbow trout (*Oncorhynchus mykiss*), and brown trout (*Salmo trutta*) are largely due to introductions that occurred in this era. It appears that 202 species of fish have been introduced into new ecosystems in North America through intentional introductions (Crossman and Cudmore, this volume). Not all of these introductions have resulted in established populations. Although most of these species are native to North America, 24 have originated from other continents. Many of the intentionally introduced species have also been introduced through accidental avenues such as bait-bucket and aquarium releases, and escapes from aquaculture facilities, but have developed at least one established population in a new water body from an intentional release. Not all of these species represent introductions that have been approved by government authorities. Moving fishes between water bodies seems to be a popular pastime for some citizens, and many species have been moved for various purposes as a result of these unauthorized introductions.

Unlike accidental introductions, intentional introductions are conducted with a purpose in mind and presumably with some form of forethought into the likely success (and impacts) of the introduced organisms. Fish species have been intentionally introduced for a variety of reasons. In the U.S. and Canada, most species have been introduced to establish sport fisheries or to create angling opportunities for additional species. The practice of supplementing existing stocks with hatchery-reared stocks is also a common practice to augment sport fisheries. The intentional introduction of invertebrates and fishes as forage items for sport fish (usually exotic species themselves) has also been a widespread practice. In Mexico most intentional introductions have been made to establish new protein sources for an impoverished society (Contreras and Escalante, 1984; Gonzalez and Garcia, this volume). Introductions to establish new food and commercial fisheries have also been made in Canada and the U.S., but to a much lesser extent. Additional species have been introduced in all three countries as a means of controlling other organisms (biological control). Some species may have been intentionally introduced to North America for reasons of sentiment or ornament (unauthorized), as is probably the case with many introductions of goldfish (Li and Moyle, 1993). Finally, several species have been intentionally introduced for conservation purposes, particularly in the American southwest. These are normally endangered fishes facing extinction due to threats in their native habitats. The intentional introduction of fish and other aquatic organisms has also resulted in the introduction of several nontarget organisms that were mistakenly included with the intentionally introduced target species. These different types of intentional introductions, and the impacts of some of the major species associated with them are discussed below.

IMPACTS OF INTENTIONAL INTRODUCTIONS

The world is rife with case histories illustrating the disastrous ecological and sometimes economic consequences associated with the establishment of introduced species. Over the last decade, considerable attention has been directed toward the impacts of introduced species in aquatic ecosystems (Courtenay and Stauffer, 1984; Stroud, 1986; Rosenfield and Mann, 1992; DeVoe, 1992; Schramm and Piper, 1995). There are few studies that clearly demonstrate cause and effect relationships between introductions and ecological impacts in aquatic systems (Taylor et al., 1984). Such relationships can only be demonstrated with an experimental approach that includes appropriate controls and replications. As a result, much of the evidence for impacts of introduced fishes has been anecdotal and circumstantial. Nevertheless, evidence for harmful ecosystem impacts associated with many introductions is compelling, and several recent studies have demonstrated negative effects as a direct result of fish introductions (Courtenay, 1993). Introduced species can negatively impact native species and fundamentally alter aquatic ecosystems through trophic alterations (predation, competition, food web alteration), genetic impacts (hybridization and indirect effects), the introduction of diseases and parasites, habitat alterations, and spatial alterations (aggressive effects and overcrowding) (Taylor et al., 1984). The isolation of freshwater lakes from one another (like islands in a sea of land) makes aquatic ecosystems particularly vulnerable to the impacts of introduced species (Magnuson, 1976).

Negative ecological impacts are often associated with unintentional or accidental introductions such as the sea lamprey (*Petromyzon marinus*) and zebra mussel (*Dreissena polymorpha*) in the Great Lakes. However, a recent review of introduced species in the U.S. revealed that intentional introductions are just as likely to have harmful impacts as are accidental introductions (Office of Technology Assessment, 1993). This suggests a history of poor species choices and complacency regarding their potential harm (Aquatic Nuisance Species Task Force, 1994). This may in part be due to the fact that several of the intentional introductions on which this analysis was based occurred earlier in the century (or before), when little thought was given to the potential impacts of the introduced organism on the ecosystem. However, negative impacts have been documented or are anticipated for many organisms that have recently been introduced intentionally (Courtenay and Robins, 1989).

There are two main reasons that intentionally introduced fishes often have unwanted and unanticipated ecological impacts. First, decisions related to proposed introductions are often politically driven by the demands of client groups and they are too often evaluated on what they can do for the parts of the ecosystems that humans value (Regier, 1968; Courtenay, 1993; Heidinger, 1993). Heidinger (1993) listed 12 reasons why fish are intentionally introduced. Most of these relate directly to consumptive uses by humans. Insufficient consideration is given to ecosystem integrity and resources that humans do not use, such as non-game fishes and invertebrates. This often leads to short-term, localized benefits related to the harvest of the introduced organism, but negative impacts over the long term as the introduced organism disperses and its impacts on various trophic levels become apparent. Moyle et al. (1986) described this phenomenon as the "Frankenstein effect." Intentionally introduced organisms can also spread far beyond the target area for their introduction, through accidental introductions and natural dispersal mechanisms (Lasenby et al., 1986; Courtenay and Robins, 1989). The second reason for unanticipated impacts is that our understanding of ecosystems and the ways they function are incomplete, making it difficult to predict the changes that will occur when a new species is introduced (Li and Moyle, 1993; Courtenay, 1995).

A common misconception is that exotic species from distant lands will cause problems for native biota, while the transfer of native species to nearby ecosystems is safe. While this dogma may be an adequate generalization regarding introduced species (Wagner, 1993) in that many well-known North American "pest species" have originated from other continents, there are also many examples of native North American species that have caused substantial ecosystem disruption when transferred within the continent. The introduction of the sea lamprey to the upper Great Lakes through the Welland Canal is a well-known example (Smith, 1972). The introduction of several species of native North American fish in the American west has also had substantial impacts on sometimes highly endemic fauna (Zuckerman and Behnke, 1986). Introduced species native to North America have contributed to the extinction of several North American fish taxa (Miller et al., 1989) and have been a causative factor in declines or are a continuing threat to many American fishes currently listed under the U.S. Endangered Species Act (Lassuy, 1995).

Ecosystem responses to species introductions are not dependent on the geographical origin of the introduced species. Similarly, it matters not whether the introduced species was transferred by accident or with intent. What does matter is the suitability of the new environment for the survival and reproduction of the new immigrant and how it interacts with the biotic and abiotic elements of its new ecosystem. Stable assemblages of species are more likely to result when the introduced species shares some evolutionary history with the recipient community (Ryder and Kerr, 1984; Li and Moyle, 1993). This is not necessarily true in all instances, as demonstrated by northern pike (*Esox lucius*) introductions into lakes containing muskellunge (*E. masquinongy*). Although there are numerous examples of naturally occurring and apparently stable sympatric populations of these two species, northern pike introductions are often detrimental to native muskellunge (Crossman, 1991). This may be due to reproductive adaptations in muskellunge populations that live sympatrically with northern pike, which may be absent in allopatric populations.

Many intentional introductions are rationalized on the assumption that the introduced species will occupy a "vacant niche" and thereby provide benefits to humans without affecting the ecosystem (Moyle et al., 1986; Li and Moyle, 1993). However, any introduced species that becomes established in a new ecosystem will have an impact on the recipient community (Ney, 1981; Li and Moyle, 1993; Courtenay, 1995). New species must exploit some resource to survive. There may be resources that are not fully exploited or "trophic opportunities" (see Leach, 1995), but not vacant niches. A niche as defined by Hutchinson (1957) is an attribute of an organism. Each organism has a fundamental niche that defines the boundaries of its potential life history. These boundaries are constrained by the abiotic environment as well as interactions with other species producing a realized niche for that organism in a particular environment. Because abiotic factors as well as species interactions will determine the realized niche of an introduced organism, its impacts will vary in different ecosystems. The introduction of a new species will similarly impose restraints (or expand) the realized niches of native species. Strongest interactions can be expected when the niches of an introduced species and a native species are similar.

INTRODUCTIONS FOR SPORTFISHING

Almost half of the nonindigneous freshwater fishes that have been intentionally introduced in North America were introduced to establish sport fisheries and to diversify angling opportunities (Crossman and Cudmore, this volume). Most of these introductions have taken place in the U.S. and Canada, but there are also examples in Mexico. These introductions typically involve keystone predators that are popular as sport fish. Many salmonids and centrarchids in particular now enjoy continent-wide distributions because of this practice.

Sport fishes are introduced to solve a variety of management "problems." In some cases, there is simply an absence of sport fish species in a particular water body. In other cases, there are species present, but there is a demand to introduce more desirable and familiar sport species. Human perturbations have made habitats unsuitable for native species in some instances and there is a call to introduce hardier nonnative sport fish species that can survive better in these altered habitats (sometimes for the short-term until habitats have been rehabilitated). This is commonly the case when new reservoirs are created, which are generally detrimental to the native lotic fish community (Courtenay, 1993). Several species are usually introduced into these systems to "fill the void" (Kohler et al., 1986), often producing a virtual predator "smorgasbord." Introductions of predatory fishes are often accompanied by the introduction of prey fish species in these systems.

Many introductions of sport fishes have been extremely successful in that numerous popular and economically beneficial sport fisheries have become established, providing recreation and income for millions of North Americans. Total economic output associated with sportfishing in the U.S. was estimated at $69 billion US in 1991 (Sport Fishing Institute, 1994). Moyle et al. (1986) conservatively estimated that 25%–50% of all freshwater sport fish harvested in the U.S. are from populations that were introduced. In Canada, expenditures related to sportfishing in 1990 totaled $8.1 billion CAN (Department of Fisheries and Oceans, 1994), but introduced species probably contribute to a smaller percentage of the total harvest than in the U.S.

The benefits that accrue to humans from sport fish introductions have come with considerable cost. Moyle et al. (1986) identified the reduced abundance of many native species, a lack of public understanding of native fishes, and management problems associated with unstable fish communities as negative consequences of sport fish introductions. The transfer of bait fish and its associated ecological impacts (Litvak and Mandrak, 1993 and in this volume) can also be attributed to many newly developed sport fisheries. The introduction of sport fish may also mask real management problems and therefore delay corrective actions.

In most cases, the positive benefits of sport fish introductions are immediately realized, while negative ecological impacts can go unnoticed or take many years to become apparent. Short-term goals to maximize fishing opportunities have led to numerous introductions of sport fishes. Decades of stocking fish have created expectations and demands (political pressure) from clients that sport fish will continually be available to provide recreational angling opportunities. Management for fishing has taken priority over management for fish in many instances due to this political pressure. The impacts of some of the more widespread fish species that have been introduced for sportfishing purposes are discussed below.

Salmonids

Salmonids have been the most widely cultured and stocked group of fishes in North America and are the target species of many popular sport fisheries. The brown trout is the only salmonid species not native to North America that has become established through these efforts. Although the brown trout was initially introduced to North America as a food fish, many subsequent introductions and its current widespread distribution relate to its use as a sport fish (Courtenay, 1993). Brown trout have established self-sustaining populations in many parts of Canada and the U.S. (Courtenay et al., 1984; Crossman, 1984) and are stocked on an ongoing basis in many areas where reproduction does not occur or is limited. The brown trout is widely regarded as one of the success stories resulting from exotic species introductions (Courtenay and Kohler, 1986). Brown trout are generally prized by anglers, and introductions have resulted in the development of numerous popular fisheries in streams, reservoirs, and the Great Lakes. The economic benefits of brown trout introductions have been significant in some instances (Zuckerman and Behnke, 1986). Despite their widespread success and acceptance as a sport fish in North America, negative impacts have been associated with brown trout introductions in several cases. Predation and competition with introduced brown trout has probably contributed to the displacement and disappearance of native brook trout (*Salvelinus fontinalis*) in many eastern North American streams (Krueger and May, 1991). Habitat changes associated with human settlement that have occurred in many of these stream environments may provide brown trout with a competitive advantage over brook trout (Krueger and May, 1991). Competition from brown trout has also been implicated in the declines of several salmonids native to western North America, including golden trout (*Oncorhynchus aguabonita*), cutthroat trout (*O. clarki*), gila trout (*O. gilae*), and Dolly Varden (*Salvelinus malma*) (Krueger and May, 1991; Courtenay, 1993). Predation by introduced brown trout is a major threat to the endangered Apache trout (*O. apache*) (Carmichael et al., 1995) and can also affect the abundance of non-game species (Krueger and May, 1991). The threat posed by brown trout to native salmonids has led to brown trout eradication programs in some instances, usually by the same agency that originally introduced them (Zuckerman and Behnke, 1986).

Introductions of native North American salmonids to new waters to enhance sport fishing have been extensive. The rainbow trout has enjoyed much success and popularity with anglers and is ubiquitous in suitable habitats in some parts of North America outside its native range due to widespread introductions dating back to the 19th century. Similar to brown trout, competition from introduced rainbow trout, may have been responsible for the displacement of native brook trout from streams in eastern North America (Larson and Moore, 1985; Krueger and May, 1991). Many stocks of cutthroat trout have been eliminated through interspecific introgression with introduced rainbow trout which often results in the production of hybrid swarms (Gerstung, 1988; Leary et al., 1995). Predation by introduced rainbow trout has a significant impact on the distribution of the threatened Little Colorado spinedace (*Lepidomeda vittata*) (Blinn et al., 1993). As rainbow trout were being distributed in eastern North America, brook trout introductions were taking place in the west. Introduced brook trout have replaced several native stocks of cutthroat trout in the American west (Gerstung, 1988; Varley and Gresswell, 1988), presumably through competition. Hybridization between introduced brook trout and native bull trout (*Salvelinus confluentus*) is apparently common (Leary et al., 1995). Since these hybrid offspring are normally sterile, hybridization does not represent a threat to the genetic integrity of bull trout, but it may pose a threat to the continued existence of bull trout in systems where it occurs frequently. Although competition from introduced salmonids seems to have been particularly detrimental to cutthroat trout, the mechanisms by which this occurs are not clear, and there are many examples of cutthroat trout populations that have persisted in the face of introductions. In some cases the replacement of cutthroat trout by nonnative salmonids may have been facilitated by habitat changes that favor the nonnative species, combined with angler harvest to which cutthroat trout are generally more vulnerable than are other salmonids (Griffith, 1988).

Perhaps the largest-scale introductions of nonnative salmonids have occurred in the St. Lawrence Great Lakes, where millions of salmonids are introduced each year to provide sport fishing opportunities. These introductions have been largely regarded as successful, as they have created a spectacular sport fishery and have added some stability to a degraded ecosystem (Radonski and Martin, 1986; Li and Moyle, 1993). Rainbow trout and brown trout were established from early stocking efforts in the late 1800s, but many early attempts to introduce Pacific salmons (*Oncorhynchus* spp.) failed or were not maintained (Emery, 1985). The impetus for large-scale stocking of exotic salmonids came in the 1960s after years of overharvest, habitat destruction, and introductions of exotic species (particularly the sea lamprey) left the lakes virtually devoid of top predators (Smith, 1972). At this time, the lakes were overpopulated with nonnative forage fishes (alewife (*Alosa pseudoharengus*) and rainbow smelt (*Osmerus mordax*)), which often littered beaches after large-scale die-offs. Clearly this ailing ecosystem provided little if any benefits to society.

Large-scale Pacific salmon stocking was initiated in Lake Michigan in 1966. Coho salmon (*Oncorhynchus kisutch*) was intentionally stocked in part to control the enormous numbers of alewives that were dying and fouling beaches, but also to provide a sport fishery and restore some stability to the Lake Michigan ecosystem (Emery, 1985). Chinook salmon (*O. tshawytscha*) were added to the stocking program in the following year. The coho salmon and chinook salmon survived very well in Lake Michigan, and an outstanding and very popular sport fishery quickly developed (Hansen et al., 1990). Although there was not unanimous support for these exotic species stocking programs around the lakes, it was not long before all management agencies on the Great Lakes followed Michigan's lead. The stocking of several nonnative salmonid species continues today to support these popular fisheries, as there is limited natural reproduction for most of these species throughout the basin (with the exception of Lake Superior and parts of Lake Michigan). Annual plantings of salmonids have been as high as 16 million fish in Lake Michigan and 8 million fish in Lake Ontario (Eck and Wells, 1987; Jones et al., 1993). Sharp declines in the alewife prey base that were probably climate-related (Eck and Wells, 1987) eventually led to the collapse of the Lake Michigan chinook salmon fishery, and have caused similar concerns for Lake Ontario (Jones et al., 1993). Stocking rates have been reduced accordingly. Lake trout (*Salvelinus namaycush*) stocking has occurred concurrently with these nonnative fish introductions in an ongoing effort to rehabilitate lake trout populations in the Great Lakes.

The benefits of the massive stocking program on the Great Lakes have been substantial. A tremendous sport fishery has risen from the ashes of the Great Lakes that provides substantial recreational opportunities and economic benefits. Annual expenditures related to sport fishing in the New York waters of Lake Ontario (most of which is targeted at nonnative salmonids) have been recently estimated to be about $300 million U.S. (Lange et al., 1995). Overpopulation of prey fish has also been successfully controlled (perhaps too well), and the presence of a revitalized fishery has probably increased public awareness and support for environmental protection and restoration on the Great Lakes. At the same time, the popular nonnative salmonid fisheries may be jeopardizing opportunities to rehabilitate native lake trout populations. Many anglers perceive attempts to restore lake trout as a threat to management programs for nonnative salmonids, which they prefer (Lange and Smith, 1995). This may lead to an erosion of public support for lake trout rehabilitation efforts.

An interesting paradox evident with some salmonids is that the same species can be a highly valued component of an ecosystem that is in decline in its native range, but become a "pest" by causing harmful ecological impacts when introduced elsewhere. A good example of this is the lake trout. Lake trout have strict habitat requirements, are long-lived, and have low-reproductive potential (Martin and Olver, 1980). Because of this, lake trout are particularly sensitive to habitat disruptions, exploitation, and introductions of introduced species, resulting in population declines and extirpations in several lakes throughout their native range. This has led to substantial rehabilitation efforts in some areas (particularly in the Great Lakes). At the same time, lake trout introduced into new water bodies have often had undesirable consequences on native fish species (Crossman, 1995). The much publicized recent introduction of lake trout into Yellowstone Lake poses a significant threat to the Yellowstone cutthroat trout (*O. clarki bouvieri*) in the core of its remaining undisturbed range (Kaeding et al., 1996). Brook trout are another example of a highly valued species that have declined throughout much of their native range, but are often "undesirable species" when introduced into the habitats of native salmonids in the west.

Centrarchids

At least 21 centrarchid species have been introduced to new water bodies in North America to establish sport fisheries (Crossman and Cudmore, this volume). The largemouth bass (*Micropterus salmoides*) has probably been the most widely introduced centrarchid, and introductions have created many popular sport fisheries in all three North American countries. Smallmouth bass (*M. dolomieu*), crappies (*Pomoxis* spp.) and bluegill (*Lepomis macrochirus*) have also been widely introduced.

Largemouth bass is the most popular and widely stocked warm-water species in the U.S. (Smith and Reeves, 1986). Early stockings were made principally into natural waters where largemouth bass were absent. The practice of introducing largemouth bass to farm ponds became popular in the middle of the 20th century, and today, largemouth bass introductions are widely used in newly created reservoirs and to enhance existing fisheries. Although introduced largemouth bass populations have generated some spectacular sport fisheries, negative ecological impacts have accompanied many introductions. Predation by largemouth bass has been cited as a factor in the extinction of several native North American fish taxa (Miller et al., 1989). Further, largemouth bass is cited as a contributing factor in the decline or a continuing threat to 21 native species listed under the U.S. Endangered Species Act (Lassuy, 1995). Largemouth bass is the most commonly cited introduced fish in such listings. Predation from introduced largemouth bass in Mexico has led to the displacement of native fishes in several locations with severe impacts on some endemic fish species (Contreras and Escalante, 1984; Gonzalez and Garcia, this volume).

Impacts of other centrarchids introduced as sport fish have also been reported. Introgression between introduced smallmouth bass and Guadalupe bass (*Micropterus treculi*) is considered a serious threat to Guadalupe bass within their highly restricted native range, as it has the potential to produce hybrid swarms (Morizot et al., 1991). Although declines in native walleye populations have been reported in some instances after smallmouth bass introductions, the interactions between these two species are poorly understood, and may be related to habitat characteristics or long-term climate trends (Johnson and Hale, 1977; Inskip and Magnuson, 1983). The bass tapeworm (*Proteocephalus ambloplites*) was introduced to Lake of the Woods (Minnesota–Ontario border) in the

1920s when smallmouth bass were introduced for sport fishing (Armstrong, 1985). Although this cestode now infects walleye and several other fish species in the lake and may be harmful to some individual fish, there is no evidence of population- or community-level impacts. Several sunfish (*Lepomis* spp.) have been widely introduced for sport fishing, but their impacts have not been well documented. Introduced bluegill may compete with and displace Sacramento perch (*Archoplites interruptus*) in California (Moyle, 1976). Bluegill introductions have also been associated with the replacement of native fishes in Mexico (Contreras and Escalante, 1984). Introduced green sunfish (*Lepomis cyanellus*), apparently stocked accidentally with bluegills, are a threat or have been implicated in the declines of several threatened and endangered species in the U.S. (Lassuy, 1995). Sunfish introductions have also probably led to the production of intergeneric hybrids. Unauthorized introductions of black crappie (*Pomoxis nigromaculatus*) in New York have resulted in the elimination of walleye in shallow, productive lakes through food web alterations and direct predation on young-of-the-year walleye (Schiavone, 1985). In deeper, less productive lakes black crappie did not have much of an impact on walleye populations.

Esocids

The esocids, particularly the muskellunge and northern pike, are popular sport fishes. These fish are highly piscivorous predators that grow to a large size, which makes them a popular species for the development of trophy fisheries, but also gives them the potential to drastically alter community structure. Intentional introductions of northern pike and muskellunge began in the 1880s in California (Webster et al., 1978). The ranges of both species were also extended in the East during this period. The extent of programs to introduce these species has not been as great as it has for salmonids and centrarchids; however, a 1983–1984 survey by Conover (1986) revealed that more than 42 million northern pike and more than 2 million muskellunge were stocked per year at that time.

The introduction of northern pike and muskellunge has created some popular sport fisheries, but has also had unwanted negative effects. Declines in native muskellunge populations have often been observed after the introduction of northern pike (Inskip, 1986). Possible mechanisms for these declines include predation of newly hatched muskellunge by northern pike fry, competition for food, and perhaps hybridization. The unauthorized introduction of northern pike into the northwestern U.S. has created some popular sport fisheries, but unplanned introductions and spread from introduced waters are a recurring problem (McMahon and Bennett, 1996). Few studies have documented the impacts of northern pike in the region, but there are concerns related to depletion of prey fish populations, and predation on native salmonids such as the westslope cutthroat trout (*Oncorhynchus clarki lewisi*) and bull trout. McMahon and Bennett (1996) concluded that the introduction of northern pike (and walleye (*Stizostedion vitreum*)) into the Pacific northwest had been a "boost" in that they created popular sport fisheries and enhanced local economies, but had been a "bane" in terms of undesirable ecosystem effects. Northern pike introductions in two Minnesota lakes resulted in declines in native walleye and yellow perch (*Perca flavescens*), presumably because of predation on yellow perch and competition with walleye (Colby et al., 1987). This occurred despite the fact that walleye and northern pike normally co-exist well (Johnson et al., 1977) and are key components of "harmonic percid communities" (Ryder and Kerr, 1978). Muskellunge have not been as widely introduced as northern pike, but several important sport fisheries have developed from these introductions. Negative impacts of muskellunge introductions on native sport fish have not been documented, although this is a concern of managers. Tiger muskellunge (*Esox lucius* X *E. masquinongy*) have been stocked in many U.S. waters to produce trophy fisheries and to control prey fishes. Most of these introductions have been reported as being successful and are dependent on hatchery production to sustain them (Conover, 1986).

Percids

Walleye have been widely introduced to establish new sport fisheries in the United States and Canada. Other percids including sauger (*S. canadense*), saugeye (*S. vitreum* X *S. canadense*), and yellow perch have been introduced on a much smaller scale.

The walleye is one of the most popular sport fish in North America, which has led to its introduction into numerous water bodies. More than 1 billion walleye were stocked in the U.S. and Canada in 1983–84 (Conover, 1986). Although much of this stocking effort is directed toward supplementing and rehabilitating existing stocks, the introduction of walleye to new water bodies remains an important objective for many North American management agencies (Fenton et al., 1996). Once again, many popular walleye sport fisheries have been developed, but there have been undesirable side effects of these introductions in several instances. In the northwestern U.S., a suite of authorized and unauthorized introductions of walleye has resulted in the establishment of walleye populations in numerous reservoirs in the Columbia and Missouri River basins (McMahon and Bennett, 1996). Their range continues to expand as they colonize and are introduced into additional water bodies. Concerns for prey depletion and predation on native and introduced salmonids have led to prohibitions on walleye introductions in many watersheds (Colby and Hunter, 1989; McMahon and Bennett, 1996). Despite these prohibitions and risk assessments to guide introductions, managers fear that precautionary efforts are being destroyed through unauthorized introductions by "bucket biologists." It is interesting to note that managers did not anticipate negative effects of walleye on salmonid fisheries, since reports of walleye preying on salmonids are uncommon (McMahon and Bennett, 1996). Different thermal regimes in northwestern reservoirs have allowed for more of a habitat overlap between these species. Walleye introductions into several Michigan and Wisconsin lakes have been implicated in declines of bass (*Micropterus* spp.) and prey fish (Kempinger and Carline, 1977; Norcross, 1986). Saugeyes (sauger X walleye) have been introduced into three states. In Ohio, the success and popularity of reservoir saugeye fisheries has recently resulted in the release of saugeyes in river systems in the Ohio River basin to expand saugeye fishing opportunities. Concerns have been expressed about the impacts of this introduction on native percid species in the basin (walleye and sauger) as hybrid reproduction has now been documented in the Ohio River (White and Schell, 1995). Reproduction of saugeyes has also been documented in Normandy Reservoir, Tennessee (Fiss et al., 1997). Yellow perch have been introduced to 14 states outside of their native range, where they are often considered undesirable because of their small size, potential impacts on native anadromous salmonids, and questionable value as a prey fish (Conover, 1986). Eradication programs have been considered for this species in California. A proposal to introduce zander (*Stizostedion lucioperca*) in North Dakota created considerable controversy among neighboring states and Canada due to concerns over potential ecological impacts and the introduction of new diseases (Office of Technology Assessment, 1993). Although zander were experimentally stocked into an isolated lake in the late 1980s, the program has since been halted and the fish extirpated.

Striped Bass

Striped bass (*Morone saxatilis*), native to the Atlantic coast of North America, were successfully introduced to the Pacific coast in the late 1800s and have been an important component of inland stocking programs in the U.S. in the latter half of the 20th century. Most of these introductions have been into newly impounded reservoirs to create fisheries for top predators as well as to control prey fish populations in some cases (Axon and Whitehurst, 1985; Harper and Naminga, 1986). Striped bass have become a popular sport fish in the southeastern and midwestern U.S., but current programs have not been as aggressive as in the 1970s due to difficulties in raising striped bass, variable survival, and thermal stresses in some reservoirs (Smith and Reeves, 1986). However, the economic impact associated with the development of successful striped bass fisheries can be substantial. Schorr et al. (1995) estimated that expenditures from striped bass anglers fishing Lake Texoma contributed more than $22 million US to the regional economy annually. In many cases striped bass introductions have been part of trial-and-error, multiple species introductions into reservoirs after attempts with other species have not been successful (Kohler et al., 1986). There are several examples of striped bass predation depleting populations of prey fish, competing with introduced salmonids, and preying upon introduced salmonids (normally rainbow trout) in "two story" reservoir fisheries (Axon and Whitehurst, 1985; Keith, 1986; Courtenay and Robins, 1989). The development and distribution of white bass (*M. chrysops*) X striped bass hybrids has become increasingly popular due to success of this fish where striped bass fisheries have failed (Smith and Reeves, 1986). The impacts of this hybrid are not well known, even though it has been stocked by at least 26 U.S. states. The introduction of this hybrid to an Illinois reservoir successfully "cropped" gizzard shad (*Dorosoma cepedianum*) populations, resulting in improved growth of white crappie (*Pomoxis annularis*) without detectable effects on the resident sport fish (Jahn et al., 1987). However, hybrid striped bass may replace native white bass (Keith, 1986), and there is some potential for this hybrid to introgress with its parental species (Avise and Van Den Avyle, 1984).

Supplemental Stocking

Although often not considered in discussions of exotics or nonindigenous species, the practice of introducing hatchery-reared fish into water bodies containing native populations of the same species has been widely used in North America. This practice has been used to supplement or enhance existing native stocks, usually in response to high levels of fishing effort or habitat problems. It has also been used to introduce new strains that supposedly have superior growth characteristics to native strains and will return more and bigger fish to anglers. There has been increasing concern that this practice is detrimental to native stocks. In some cases it may be one of the major problems hindering rehabilitation of native stocks (Ferguson, 1990; Hindar et al., 1991; Waples, 1991), although direct evidence supporting this concern is often absent or circumstantial (Campton, 1995).

Freshwater fish populations exhibit a high degree of reproductive isolation, and as a result many strains or stocks have developed that are locally adapted to their particular environments (MacLean and Evans, 1981). Introducing fish from a different genetic strain ignores this evolutionary concept. Even if the native strain is used as the source for supplementation, fish raised in hatchery environments often develop genetic characteristics that make them particularly well suited to survival in the hatchery as opposed to survival in the wild (Waples, 1991). Intraspecific introgression in either case can be detrimental to the locally adapted native population. Recent evidence suggests that the practice of supplemental stocking can have negative genetic effects on locally adapted stocks (Phillip, 1991; Waples, 1991), and extirpation of native stocks can occur through supplemental stocking even if the stocked fish do not reproduce or interbreed with the native stock (Evans and Willox, 1991). In many cases supplemental stocking is also ineffective in that hatchery fish do not survive well when strong native populations of the same species are present (Laarman, 1978).

These concerns have led to the development of genetic conservation programs by many government agencies that raise fish for supplemental stocking purposes (e.g., Cottrell et al., 1995). These programs aim to match recipient and donor genetic stocks where possible to minimize the potential genetic impacts of hatchery fish on locally adapted native strains. In few instances have these concerns actually led to the cessation of supplemental stocking programs.

Introductions of Prey Species

The goal of many North American fisheries management programs has been to maximize or optimize the production of sport fish (Noble, 1981). These fish are usually top predators whose production may be limited by the amount of food that is available. This is especially true in new impoundments where there may not be suitable food resource for introduced predatory fishes. To compensate for these (often perceived) inadequacies in the natural food supply, prey fishes (often called forage fishes) and other organisms have been introduced into numerous North American lakes, reservoirs, and rivers since the 1950s. The management of prey fish in these systems has developed haphazardly and has been described as "trial and error" and "hope for the best" approaches (Ney, 1981; Wydoski and Bennett, 1981). Such introductions of exotic species have the potential for enduring, high-profile mistakes (Ney, 1981).

Ney (1981) identified six characteristics of the ideal prey species: prolific, stable, trophically efficient, vulnerable to predation, nonmigrating, and innocuous. The first four characteristics will determine if the prey introduction achieves the objectives of increasing fish production, while the last two will determine if there are unwanted ecosystem effects. Unfortunately it is difficult to find species that have all of these characteristics.

Few widely introduced forage species have an entirely positive record (Ney and Orth, 1986) and some have clearly had negative ecosystem impacts (Lasenby et al., 1986). The introduction of prey species has the potential to produce unanticipated system-wide effects that can undermine even the best-intentioned efforts (DeVries and Stein, 1990). These unanticipated responses are usually due to the complex nature of recipient systems and a lack of understanding of these systems. Too often predator–prey relationships have been considered in a vacuum prior to an introduction taking place (Noble, 1986). A simple consideration of the predator species and the prey prior to the introduction (that the prey is a suitable size and that fish have eaten it in other systems) is not sufficient (Ney and Orth, 1986). A more holistic approach that considers competition and predation at all trophic levels, spatial refuges in the recipient water body, behavioral responses, indirect food web effects, and ontogenetic shifts in diet and habitat of prey and target species is required to gain some predictive insight into forage fish introductions (Ney and Orth, 1986; DeVries and Stein, 1990). The ability of managers to apply this approach is hampered by the complexity of systems, the instability of newly created systems (reservoirs), and by the lack of proper documentation of the impacts of previous prey species introductions (Wydoski and Bennett, 1981; DeVries and Stein, 1990). DeVries and Stein (1990) concluded that most biologists believe that prey fish introductions will increase the growth of predators based on the ongoing prey fish introductions across the U.S.

Fish species from at least 10 families have been intentionally introduced as prey species into North American fresh waters. All but one species, the wakasagi (*Hypomesus nipponensis*), are native to North America. The opossum shrimp (*Mysis relicta*) has been the most commonly introduced invertebrate prey species, although others such as crayfish have probably been introduced as well. The success and impacts of the most widespread introduced prey species are discussed below.

Clupeids

A review of published studies on North American prey fish introductions by DeVries and Stein (1990) revealed that introductions involving clupeids were 10 times more common than for any other group. Most of these introductions have involved threadfin shad (*Dorosoma petenense*) and gizzard shad, although alewife, blueback herring (*Alosa aestivalis*), and American shad (*A. sapidissima*) have also been introduced as prey fish.

The threadfin shad has been the most commonly introduced clupeid because of its small maximum size, which makes it available to predators at all life history stages (Noble, 1981; DeVries and Stein, 1990). Although gizzard shad have been introduced as a prey fish into many systems, they are often problematic. They can develop large populations of large individuals that are unavailable to predators and impact production of other elements of the fish community (Noble, 1981). The introduction of shad (*Dorosoma* spp.) as a prey species has produced inconsistent results. In many cases, the growth of predators has increased (largemouth bass and crappies *Pomoxis* spp.), but cases where neutral or negative responses have occurred are also common (Noble, 1981; DeVries and Stein, 1990). DeVries and Stein (1990) also noted that the impact of shad introductions on a presumed competitor, the bluegill, were just as likely to be positive or neutral as negative. In some lakes introductions of threadfin shad have resulted in reduced growth of young of the year sport fish (largemouth bass and crappies) through direct and indirect competition for zooplankton food resources (Wydoski and Bennett, 1981; Mosher, 1984; DeVries et al., 1991). These variable responses demonstrate the different impacts the same species can have when introduced into different ecosystems and also elude to the danger of widespread extrapolation and "copycat" introductions (Courtenay and Robins, 1989).

Alewives have been intentionally introduced as a prey fish in several New England lakes, where they have developed into stable prey bases (Kircheis and Stanley, 1981). Alewife introductions appeared to be beneficial to the growth of adult salmonids, but may suppress growth of younger fish (Kircheis and Stanley, 1981). In the Great Lakes, accidentally introduced alewives now form a major component of the prey base for salmonid fishes. However, introduced alewives have also been implicated in recruitment declines of coregonines in some parts of the Great Lakes. They may have negatively affected yellow-perch spawning through aggressive behavior, and have been cited as a factor in the suppression of recovery of lake trout and depressed walleye stocks (Smith, 1972; Wells, 1977; Schneider et al., 1991; Krueger et al., 1995).

Cyprinids

Although several species of cyprinids have been intentionally introduced as prey fish in the U.S. and Canada, there are few studies that have documented their success and impacts. Golden shiners (*Notemigonus crysoleucas*) have been introduced to many smaller water bodies as prey for large-mouth bass in the southern U.S., with mixed results. Although the growth of largemouth bass may be increased by these introductions, golden shiners may interfere with native prey and reduce bass production, and frequently overpopulate at sizes too big for consumption similar to gizzard shad (Noble, 1981). Fathead minnows (*Pimephales promelas*) have been used successfully to provide sustainable prey in small channel catfish (*Ictalurus punctatus*) ponds (Noble, 1981).

The introduction of the red shiner (*Cyprinella lutrensis*) into the Colorado River drainage as a prey fish in the 1950s provides an example of how unanticipated consequences can occur far from the site of introduction. The red shiner has subsequently spread upstream to the Moapa and Virgin rivers through bait bucket transfers where they have had adverse impacts on endemic endangered and threatened species (Deacon, 1988; Courtenay and Robins, 1989). The Asian tapeworm (*Bothriocephalus opsarichthydis*), which may have negative impacts on the endangered woundfin (*Plagopterus argentissimus*), was likely transported into the basin with red shiners. It is suspected that red shiners obtained this cestode from culture facilities in Arkansas holding grass carp (*Ctenopharyngodon idella*).

Salmonids

The introduction of salmonids to provide prey for predators has not been widely practiced, with the possible exception of kokanee (*Oncorhynchus nerka*) in the western U.S. and Canada. These introductions have generally been regarded as successful. These planktivorous salmonids are preyed upon extensively by larger salmonids with resultant growth increases, and negative impacts on predators have not been reported (Wydoski and Bennett, 1981). However, target predators have not fed on kokanee in some instances, resulting in the development of large kokanee populations. The impact of kokanee as a competitor with other planktivores including juvenile salmonids has not been evaluated, although competition between introduced kokanee and native cutthroat trout for zooplankton food resources has been suggested (Marnell, 1988). The introduction of lake herring (*Coregonus artedi*) into Lake Opeongo, Ontario was beneficial to the lake trout population and improved the lake trout sport fishing yield on the lake. Paradoxically, this has probably heightened the sensitivity of lake trout to exploitation, as the changes produced by the introduction have resulted in a later age to maturity for lake trout and an earlier age of capture (Matuszek et al., 1990).

Smelts

Rainbow smelt (*Osmerus mordax*) have been widely introduced as a prey fish into coldwater lakes. Introductions of rainbow smelt have increased growth rates of Atlantic salmon (*Salmo salar*) (Kircheis and Stanley, 1981), lake trout (Evans and Loftus, 1987), and walleye (Jones et al., 1994). However, introduced rainbow smelt populations have been associated with recruitment declines in several species, most notably lake whitefish (*Coregonus clupeaformis*), lake herring, and walleye (Loftus and Hulsman, 1986; Colby et al., 1987; Evans and Loftus, 1987). There are also indications that predators that switch from indigenous prey to rainbow smelt may accumulate higher levels of mercury (MacCrimmon et al., 1983; Mathers and Johansen, 1985; Evans and Loftus, 1987). Rainbow smelt have a eurythermal life history, are opportunistic feeders, and can occupy the roles of prey, predator, and competitor in aquatic systems (Evans and Loftus, 1987). Therefore, rainbow smelt have the potential to impact a broad range of species at different trophic levels. Rainbow smelt populations are also subject to extreme fluctuations in abundance, which may detract from their utility as a prey species (Kircheis and Stanley, 1981).

The dispersal of rainbow smelt within and outside of the Great Lakes basin provides a good illustration of how a single intentional introduction can lead to widespread dispersal of an organism. Rainbow smelt were originally introduced into the upper Great Lakes basin in 1912 when they were stocked as a prey fish in Crystal Lake, Michigan (Scott and Crossman, 1973). They dispersed from there downstream into Lake Michigan and rapidly colonized Lakes Michigan, Huron, and Superior. Rainbow smelt have since colonized at least 200 inland lakes in Ontario and Manitoba, including portions of the Winnipeg–Nelson River watershed giving them access to waters of Hudson and James Bay (Evans and Loftus, 1987; Franzin et al., 1994). This dispersal has been due to a combination of natural invasive movements and downstream dispersal as well as the inadvertent release of fertilized eggs during the processing of rainbow smelt captured during spring spawning runs in other water bodies (Evans and Loftus, 1987). The use of fresh rainbow smelt as bait in spring lake trout fisheries may have also contributed to their dispersal (Franzin et al., 1994). The tendency for rainbow smelt to spread and their potential to produce undesirable ecosystem impacts led Courtenay and Robins (1989) to question the wisdom of a recent proposal to introduce rainbow smelt in a Colorado River reservoir to serve as an alternate prey fish for introduced striped bass.

The wakasagi is a Japanese smelt that has been introduced into several California reservoirs as prey for trout. This species has the potential to hybridize with the endangered delta smelt (*Hypomesus transpacificus*) (Courtenay et al., 1986).

Other Fishes

Several sunfish (*Lepomis* spp.), and particularly bluegills, have been introduced as prey species, usually for largemouth bass in pond situations. Much literature has focused on maintaining an appropriate balance between bluegills and their predators in these situations (Flickinger and Bulow, 1993). Inland silversides (*Menydia beryllina*) and brook silversides (*Labidesthes sicculus*) have frequently been introduced as prey fishes with inconsistent results. They are largely consumed by littoral predators, but their interactions with native species are poorly understood (Noble, 1981; Boxrucker, 1986). Introduced inland silversides in Clear Lake, California represented one of a series of assaults that led to the extinction of the Clear Lake splittail (*Pogonichthys ciscoides*) (Miller et al., 1989).

Invertebrates

The opossum shrimp (*Mysis relicta*) has been intentionally introduced into more than 130 oligotrophic lakes in western North America, and into at least 70 lakes in Scandinavia (Lasenby et al., 1986). Recognition of the opossum shrimp as an important component of the diet in several fish species prompted these introductions to enhance fish production (primarily salmonids) in oligotrophic lakes where food was perceived as limiting. The species was first introduced in 1949 to Kootenay Lake, British Columbia (along with the deepwater amphipod *Pontoporeia hoyi*) to enhance the growth of rainbow trout (Lazenby et al., 1986). Although the growth of rainbow trout was not enhanced, the growth and size of planktivorous kokanee increased dramatically. This "success" led to the widespread introduction of the opossum shrimp in western North America. Many of these introductions produced undesirable results as mysids depleted zooplankton food resources, resulting in decreased size and abundance of kokanee and their virtual disappearance from some lakes (Martinez and Bergersen, 1989; Spencer et al., 1991). The introduction of the opossum shrimp to Flathead Lake, Montana led to widespread changes that cascaded through the food web (Spencer et al., 1991). Mysids were first found in the lake in 1981 and probably originated from intentional upstream introductions in the 1970s. Spawning runs of the large population of kokanee (established 50 years earlier) had become an important food source for several species of birds and mammals. The kokanee population quickly collapsed after the opossum shrimp introduction, which resulted in the displacement of migrating bald eagles (*Haliaeetus leucocephalus*) and grizzly bears (*Ursus arctos*), and may have contributed to increased mortality of bald eagles (Spencer et al., 1991; Li and Moyle, 1993).

There have been some positive effects of opossum shrimp introductions on other fish species. Lake trout, lake whitefish, and pygmy whitefish (*Prosopium coulteri*) have benefitted from some mysid introductions, perhaps because they have co-adapted and share some evolutionary history (Wydoski and Bennet, 1981; Lasenby et al., 1986) or perhaps because of their benthic feeding modes. Mysids are normally on the bottom of the hypolimnion during the day, but make nocturnal vertical migrations when they are not available to daytime planktivorous fishes such as kokanee and rainbow trout (Martinez and Bergersen, 1989). Unusual hydrologic conditions in part of Kootenay Lake made opossum shrimp available to kokanee during the daytime (Martin and Northcote, 1991). This demonstrates that abiotic factors can produce different results when species are introduced into apparently similar biotic communities. The downstream dispersal capabilities of opossum shrimp are impressive, especially for a species thought of as profundal. Lasenby et al. (1986) suggested that introduced mysids will probably eventually reach every downstream lake in a watershed. Introduced mysids have dispersed as far downstream as 240 km past many seemingly insurmountable barriers.

INTRODUCTIONS TO PROVIDE NEW FOOD SOURCES

Many aquatic organisms have been intentionally introduced to North America for the provision of a new food source and occasionally to develop commercial fisheries. Some fishes such as the brown trout (discussed earlier) were originally introduced as both a food and a sport fish. Introductions to establish new food sources have been much more common in Mexico than in Canada and the U.S., where many of these types of introductions have not occurred since the 1800s (Courtenay, 1993). In Mexico, importance has been placed on increasing the production and consumption of fish protein for what is a largely protein-deficient society (Contreras and Escalante, 1984; Contreras-Balderas, this volume). Unfortunately, this has been attempted through the widespread release of exotic fishes, which has often had unfortunate consequences for native species including endemic and endangered fishes (Contreras and Escalante, 1984; Courtenay, 1993). Such introductions are officially viewed by government agencies as justifiable and generally as positive in social and economic terms, despite the associated negative impact on native fauna. This motivation has led to the widespread release of exotic fishes in many protein-deficient nations for the development of subsistence and perhaps commercial fisheries (Courtenay and Williams, 1992).

Common Carp

The common carp, native to Eurasia, has been widely introduced around the world and was originally released in North America in 1831 by a private citizen (DeKay, 1842). Common carp were imported into government hatcheries in the U.S. and Canada in the 1870s and 1880s, from which they enjoyed a couple of decades of widespread introductions as a food and possibly a sport fish (Courtenay et al., 1984; Crossman, 1984). Although there are no records of the Canadian government intentionally releasing common carp, introductions by private citizens were assisted by government agencies (Crossman, 1984). In Canada and the U.S., government-sanctioned releases of common carp apparently ceased by the turn of the century, as the widespread introduction of this species was recognized as a mistake by the early 1900s (Courtenay and Williams, 1992). The range of the common carp has expanded ever since through secondary vectors such as natural dispersal, accidental introductions, and perhaps unauthorized releases. In Canada and the U.S. a limited amount of commercial and sport fishing has resulted from common carp introductions. In Mexico, common carp were first released in 1872–1873, and are apparently widespread (Contreras and Escalante, 1984; Balderas, this volume; Gonzalez and Garcia, this volume). The grass carp was also introduced into the open waters of Mexico in part as a food fish.

Contreras and Escalante (1984) reported that the introduction of common carp was associated with the disappearance of native fishes at several locations. The feeding activities of common carp result in the direct removal of aquatic macrophytes through consumption and uprooting (Taylor et al., 1984). The common carp's habit of digging through the substrate for food often results in increased turbidity that can indirectly decrease the abundance of aquatic macrophytes. These changes in habitat are often associated with declines in native fish populations and sometimes in waterfowl populations, but a direct cause and effect relationship to common carp has not been established (Taylor et al., 1984). However, reductions in common carp abundance through intensive control efforts have been associated with recovery of native fish populations suggesting that common carp are directly responsible for the observed impacts. Recent attempts to rehabilitate degraded wetlands in the Great Lakes have demonstrated that large populations of common carp may impede successful rehabilitation. Efforts to control common carp have been extensive and have included chemical methods, physical removal, physical exclusion, and biological control (Wiley and Wydoski, 1993; Krishka et al., 1996). Effective control is difficult to achieve in large open systems without intensive and ongoing efforts (Krishka et al., 1996).

Tilapia

Recent introductions of tilapia (primarily blue tilapia (*Oreochromis aureus*)) as a food fish by Mexican governments have been widespread. Courtenay (1993) suggested that many government officials are dedicated to spreading tilapia throughout Mexico's inland waters as a new source of protein. These efforts have not considered biological consequences, and despite good intentions, will probably prove detrimental to many native fishes in Mexico based on experiences in the U.S. and elsewhere in the world (Taylor et al., 1984; Courtenay, 1993). It is ironic that tilapia and carp are rarely included in the diet of Mexican people as a function of tradition (Chacon-Torres and Rosas-Monge, 1995). Although most intentional tilapia introductions in the southern U.S. have been made for biological control purposes, there are limited examples of unauthorized introductions for food purposes and for the development of commercial fisheries (Courtenay, 1993).

Other Fishes

In addition to the common carp and tilapia, 34 fish species have been intentionally introduced in Mexico, primarily to establish new food sources (Gonzalez and Garcia, this volume). The impacts of these introductions on native fauna and habitats have received little study. However, there are several examples of the replacement of native fishes by introduced species, usually in combination with other factors such as habitat stresses (Contreras and Escalante, 1984; Gonzalez and Garcia, this volume; Contreras-Balderas, this volume).

Bullfrogs

The bullfrog (*Rana catesbeiana*) has been widely introduced to the western and central U.S. and parts of northern Mexico (Bury and Whelan, 1984). Bullfrogs were first introduced to western North America in the late 1800s apparently to serve as a replacement for populations of the red-legged frog (*R. aurora*), which had declined from commercial exploitation (Jennings and Hayes, 1985). Their range has also expanded through unauthorized introductions (Moyle, 1973) and perhaps through escapes from ornamental ponds and other holding facilities. The decline or elimination of several species of frogs native to western North America has been attributed to predation and competition from introduced bullfrogs, often in association with habitat alterations (Moyle, 1973; Bury and Whelan, 1984; Hayes and Jennings, 1986; Orchard, this volume). Hayes and Jennings (1986) reviewed the declines of two of these species (yellow-legged frog (*R. boylii*) and red-legged frog) and suggested that predation from introduced sport fishes, particularly catfishes and basses, was more likely a causative factor. The native western frogs evolved with native fishes that are not particularly effective tadpole predators and therefore are vulnerable to fish predation. Some of the introduced fishes are effective tadpole predators, while bullfrog tadpoles may be unpalatable to these same predators. The review of this situation by Hayes and Jennings (1986) provides an excellent testament to the complex nature of ecosystem problems and to the difficulties in identifying causal relationships between introduced species and ecological impacts. Predation by introduced bullfrogs may have played a role in the extinction of Pahranagat spinedace (*Lepidomeda altivelis*) and the Ash Meadows poolfish (*Empitrichthys merriami*) (Miller et al., 1989).

INTRODUCTIONS OF BIOLOGICAL CONTROL ORGANISMS

Classical biological control is defined as the importation and release of an organism outside its natural range for the purpose of controlling a pest species. This is often viewed as an inexpensive and environmentally safe way to control undesirable species when compared to the traditional methods of mechanical control and the application of pesticides. Because biological control agents are self-replicating, there is no requirement for ongoing treatments. This makes biological control attractive from an economic standpoint. However, many of the indirect costs of biological control are usually not factored into accounting (Howarth, 1991). Although biological control is often considered to be environmentally safe and risk free, documentation of the results of biological control programs has been poor (Dahlsten, 1986), and few workers have studied the effects on nontarget species and other environmental aspects (Howarth, 1991). Unlike pesticides, biological control agents are usually enduring and have the ability to spread. It is interesting to note that biological control agents have been strongly implicated in the extinction of almost 100 animal species around the world, while there are no documented examples of extinctions from the use of pesticides (although many species and human health have been negatively affected) (Howarth, 1991). There have been some remarkable successes with the use of biological control agents. Equally, there are many cases where biological control agents have become "pests" themselves by enhancing the target species, affecting human health and attacking nontarget organisms (Howarth, 1991).

Success with the use of predaceous, parasitic, and phytophagous insects as biological control agents has led to the widespread use of fishes for this purpose, often with undesirable consequences

(Courtenay, 1993). Unlike many of the successful insect biological control agents that have very specific life history requirements and are normally monophagous, fishes tend to be more polyphagous and opportunistic, and experience ontogenetic shifts in their feeding behaviors (Courtenay, 1993). Consequently numerous ecosystem problems have been created from the use of fish as biological control agents. Howarth (1991) found that a large portion of vertebrates used as biological control agents had become disruptive and that freshwater habitats and islands seemed to be particularly vulnerable to the negative impacts of biological control. The relative low cost and "environmental safety" of biological control has led to the release of several biological control organisms by unregulated private interests who rarely document introductions or show environmental concerns (Howarth, 1991).

Several species of fish have been intentionally introduced as biological control agents in North American waters. Two of these, the grass carp and the mosquitofish (*Gambusia afffinis* and *G. holbrooki*), have been particularly widespread and have had mixed results. Other species have often been intentionally introduced to "thin out" overpopulated prey fish while providing sport fisheries at the same time. The use of biological control agents does not have a long history of use in aquatic ecosystems in North America, but there are some recent examples.

Mosquitofish

The mosquitofish, native to the southern U.S. and northern Mexico, was perhaps the first fish species used as a biological control agent. It has been widely introduced for mosquito control into at least 21 countries around the world (Welcomme, 1988). As livebearers with relatively broad physiological tolerances, these fish do not have any particular spawning requirements, other than perhaps appropriate water temperature and photoperiod, and consequently they have become established and spread from many locations where they have been introduced (Courtenay and Meffe, 1989). Although mosquitofish introductions have been labeled as successful, there has been much debate about their utility and impacts on aquatic systems (Welcomme, 1988; Li and Moyle, 1993). A review of mosquitofish introductions by Courtenay and Meffe (1989) found that the positive impacts of mosquitofish were outweighed by their negative impacts on other organisms due to their omnivorous feeding habits. Mosquitofish are particularly aggressive and prey upon the eggs, larvae, and juveniles of larger fish species, as well as on the adults of smaller fish species. Attempts at mosquito control using mosquitofish may actually enhance mosquito populations through the elimination of larvivorous invertebrates and fishes, which are often more effective at mosquito control than mosquitofish (Courtenay and Meffe, 1989; Li and Moyle, 1993). Introduced populations of mosquitofish have been associated with the extinction of two native North American poeciliids (*G. amistadensis* and *G. georgei*) (Miller et al., 1989). Negative impacts on water quality have also been associated with mosquitofish introductions. Courtenay and Meffe (1989) recommended that introductions of mosquitofish should cease and that biological control efforts should be based upon the use of native species.

Grass Carp

The grass carp, native to eastern Asia, has been introduced into at least 28 countries around the world, largely to control aquatic vegetation (Welcomme, 1988). The first introduction of grass carp into the U.S. occurred in the 1960s, and it has since been widely introduced through authorized and unauthorized stockings as well as escapes from culture facilities and ponds (Cassani, 1995). The first introductions of this species in Mexico also occurred in the 1960s; it has been introduced there both for vegetation control and as a food fish (Contreras and Escalante, 1984). In Canada, grass carp have not been intentionally released into open waters, but individual specimens have been collected from the Great Lakes (Crossman et al., 1987). Concern for the potential establishment of naturalized grass carp populations led to the development of techniques to induce triploidy.

Triploids are functionally sterile (Allen et al., 1986) and are widely used for vegetation control programs in most U.S. states. Eight U.S. states have no restrictions as to ploidy for grass carp introductions, while 12 states prohibit the use of grass carp (Cassani, 1995). Despite restrictions to triploid use, diploid fish have appeared in many systems. Grass carp have been subject to several illegal unauthorized introductions that have contributed to their present distribution (Guillory and Gasaway, 1978; Courtenay, 1993).

Grass carp spawn in rivers and require adequate current over a prolonged stretch (50–180 km, depending on temperature) of river to keep their semipelagic eggs afloat until hatching (Shireman and Smith, 1983; Opuszynski and Shireman, 1995). These exacting spawning requirements led the initial proponents of grass carp introductions in the U.S. to claim that it would not be possible for grass carp to reproduce in U.S. waters (Shelton and Shireman, 1984; Courtenay, 1993). Stanley et al. (1978) predicted that grass carp would be able to spawn in several large river systems in the U.S. and their prognostications have been realized. Natural reproduction of grass carp in North America was first reported in Mexico in 1974, and subsequent reproduction of grass carp has been reported in several U.S. rivers, including the Mississippi River and the Missouri River (Opuszynski and Shireman, 1995). Recent reports of reproduction in the Illinois River (Raibley et al., 1995) and the Trinity River (Howells, 1992) suggest that the range of naturalized grass carp is continuing to expand.

The use of grass carp is often preferred over traditional vegetation management tools such as mechanical harvesting and herbicides, because grass carp are less expensive and do not have the side effects associated with pesticides in particular. However, a review by Cassani (1995) revealed that the introduction of grass carp often results in the elimination of vegetation from target areas. This has resulted in various undesirable impacts including changes in water quality, productivity, community composition of fish and invertebrates, and reduction of waterfowl food sources. Grass carp are also assumed to be the source of the North American introduction of the Asian tapeworm *Bothriocephalus opsarichthydis*, which has infected several native species including endangered species (Hoffman and Schubert, 1984; Deacon, 1988). Although elimination of aquatic vegetation is normally not the goal of control programs, excessive stocking rates have often led to the eradication of vegetation (Cassani, 1995). Grass carp also preferentially feed on certain macrophyte species and may feed on native nontarget species as opposed to target species such as Eurasian watermilfoil (*Myriophyllum spicatum*) (McKnight and Hepp, 1995). Cassani (1995) recommended that several factors be considered before introducing grass carp for biological control and that control efforts should be directed toward macrophyte suppression rather than elimination to balance the potential long-term effects. Impacts of naturalized grass carp populations have not been documented (Cassani, 1995). Opuszynski and Shireman (1995) suggested that it is unlikely that naturalized grass carp will become abundant enough to have a "destructive" environmental effect due to the predation on juvenile grass carp by other fishes. It should be noted, however, that the impacts of introduced organisms may take many years to become apparent (Office of Technology Assessment, 1993) and that grass carp have the potential to be powerful modifiers of aquatic habitat.

Three additional species of Asian carps have been promoted as potential biological control agents. The bighead carp (*Hypothalmichthys nobilis*), used for zooplankton control, has escaped from aquaculture facilities, been illegally introduced, or been introduced as a stock contaminant with grass carp and has been found in open waters of all three countries. It is now established in the Missouri River (Courtenay et al., 1991). The silver carp (*H. molitrix*), promoted for control of phytoplankton, has also escaped into several natural waterways in the U.S., but is not yet established. The black carp (*Myllopharyngodon piceus*) is molluscivorous and has been promoted asa potential biological control agent for zebra mussels. Nico and Williams (1996) concluded that the black carp posed a high risk to native North American molluscs (many of which are threatened and endangered) and that it should not be introduced into open systems. All of these species, including the grass carp, are currently being imported live into Canadian cities for the growing live food fish market. This has raised significant concern over potential accidental or unauthorized release into aquatic ecosystems.

Tilapia

Several species of tilapia have been promoted and introduced for the biological control of aquatic vegetation as well as insects (mosquitoes and chironomid midges) in the southern United States (Courtenay et al., 1984). The most widespread of these species, blue tilapia, Mozambique tilapia (*Oreochromis mossambicus*), and redbelly tilapia (*Tilapia zilli*) have also been intentionally introduced legally and illegally for a variety of additional purposes including sport fishing, as a prey fish, and as a food fish. All three species are established in the southern U.S. from various introductions (Courtenay et al., 1984), and the blue tilapia and redbelly tilapia are established in Mexico (Contreras and Escalante, 1984). All species of tilapia are omnivorous and prolific spawners. As a consequence, many introduced populations have undergone explosive invasions and have impacted native species by reducing aquatic macrophytes, increasing turbidity, altering prey bases, preying on eggs, and displacement through aggression and overcrowding (Taylor et al., 1984). The ability for tilapia to successfully control aquatic plants and insects has been limited, while their ability to negatively impact native fauna has been substantial. The northward spread of most tilapias is limited by cold winter temperatures. Most species have lower lethal incipient temperatures of 6–12°C (Shafland and Pestrak, 1982). However, blue tilapia became abundant and were able to overwinter in the Susquehanna River, Pennsylvania by seeking winter refuge in the thermal effluent of a power generation facility (Stauffer et al., 1988). The Mozambique tilapia has also become established as far north as Idaho (Courtenay et al., 1991).

Other Fishes

The recognition of negative impacts from introduced fishes and invertebrates has often led to suggestions for the introduction of additional fish species for biological control purposes. Martinez and Bergerson (1989) suggested that the deepwater sculpin (*Myoxocephalus thompsoni*) could be used to control *Mysis relicta* in Lake Tahoe. Hepworth and Dufield (1987) suggested that another nonnative fish may have to be introduced into a Utah lake to control the crayfish *Orconectes virilis*, which was negatively impacting a fishery based upon introduced rainbow trout. As mentioned above, black carp have been touted as a potential biological control for zebra mussels. The omnivorous feeding habits of most fishes suggests that alternatives besides nonnative fish introductions should be considered. Attempts to control ruffe (*Gymnocephalus cernuus*) through enhancing native predators (walleye and northern pike) in western Lake Superior have not been successful (Ogle et al., 1996).

Predatory fishes have been routinely introduced to control overpopulations of prey fishes, which are often introduced species themselves. These types of introductions normally involve the use of a popular sport fish as the predator, thereby accomplishing two objectives. The introduction of nonindigenous Pacific salmons to the Great Lakes was largely successful in this regard. A survey of Canadian and U.S. resource management agencies by Wiley and Wydoski (1993) revealed that largemouth bass, walleye, northern pike, striped bass and hybrids, and muskellunge are most often used as "biological control agents" for the control of prey fishes and undesirable species. Gizzard shad, yellow perch, sunfish, common carp, and other cyprinids are the primary targets of these control programs. Although there are some examples of success with predator stocking (Harper and Naminga, 1986), there are also many examples with ineffective or negative results, and fish community responses have not been adequately studied in most instances (Noble, 1981; DeVries and Stein, 1990; Wiley and Wydoski, 1993). The management of systems based on unstable assemblages of exotic predators and prey is generally much more expensive and unpredictable than the management of self-sustaining, stable assemblages of native predators and prey (Christie et al., 1987).

Insects

As mentioned above, there have been some remarkable successes using insects as control agents for insect pests and weeds in agricultural systems (Dahlsten, 1986). However, there are also several examples where insects introduced for biological control purposes have had negative ecosystem consequences, including increasing the abundance of target pests and causing species extinctions (Howarth, 1991). Although most biological control efforts using insects have taken place in terrestrial ecosystems, several North American aquatic macrophytes have been the target or are currently being considered as the target for insect biological control programs. The control of alligatorweed (*Alternanthera philoxeroides*) in the southeastern U.S. using three different insect species (a leaf beetle, a stem boring moth, and a thrip) first released in the 1960s and 1970s is considered as one of the major biological control successes (Buckingham, 1996). Successful biological control of water hyacinth (*Eichornia crassipes*) has also been achieved in the southern U.S. (and many other countries) by the introduction of three south American insect species (two weevils and a pyralid moth) (DeLoach, 1991). The potential of insect biological control agents is also being investigated for hydrilla (*Hydrilla verticllata*), water lettuce (*Pistia stratiodes*), and Eurasian watermilfoil (DeLoach, 1991). Water lettuce has become a serious nuisance weed in many areas where alligatorweed and water hyacinth have been successfully controlled. A native North American weevil (*Euhrychiopsis lecontei*) has been identified as a potential biological control agent for Eurasian watermilfoil, but it may also impact on native species of watermilfoil (*Myriophyllum* spp.) (Creed and Sheldon, 1994). Much attention has been focussed on the biological control of purple loosestrife (*Lythrum salicaria*) in the U.S. and Canada because of the widespread impact of this showy plant on native species and wetland habitats (Thompson et al., 1987; Malecki et al., 1993). Three European weevils (*Hylobius transversovittatus* and *Nanophyes* spp.) and two leaf beetles (*Galerucella* spp.) have been imported to North America. The leaf beetles and *Hylobius* have been released at numerous sites across North America, and there are initial signs of success at some of the release sites (J. Corrigan, University of Guelph, Guelph, Ontario, Canada, personal communication).

It should be noted that while the introduction of any organism for biological control purposes will have some impact, and there is the potential for negative impacts with the use of insects, the import of insects for biological control is now subject to rigorous screening, experimental trials, and consultation before imports are approved, followed by a quarantine period (DeLoach, 1991). The intentional introduction of other organisms would benefit from the rigorous application of similar protocols.

INTRODUCTIONS FOR CONSERVATION PURPOSES

Many North American freshwater fishes have been introduced into new water bodies for conservation purposes, and at least 20 species have established new populations through these efforts (Crossman and Cudmore, this volume). Most of these species are small, non-sport species such as pupfish (*Cyprinodon* spp.), but new populations of Apache trout and Gila trout have also been established for this purpose (Courtenay, 1995). Introductions for conservation purposes are conducted to provide a refuge for species that are threatened with extirpation in their native habitats. Recent efforts of this type have usually resulted in introductions into refuges near native waters with the goal of reintroduction into native habitats when they become suitable (Courtenay, 1993), usually as part of a formal recovery plan. The conservation of some species at risk has also relied on using artificial propagation in hatcheries for supplementing existing populations or reintroducing species into formerly occupied habitats (Williams et al., 1988). Courtenay (1993) suggested that introductions for conservation purposes were one of the few introductions where concern is generally expressed against extending the historic ranges of fishes. The American Fisheries Society's guidelines (Williams et al., 1988) for introductions of fishes at risk recommend that introductions be restricted to native habitats whenever possible. These guidelines are intended to increase the success of these introductions, while protecting the genetic integrity of the species at risk and avoiding unwanted ecosystem impacts.

INTRODUCTIONS OF NONTARGET SPECIES

An astonishing array of fish and other freshwater aquatic organisms have been introduced as "contaminants" or "by-products" associated with the intentional introduction of other species. Although these introductions are not intentional in themselves (i.e., there is no deliberate attempt to establish the nontarget species), they are integrally linked to intentional introductions and so are considered here. In most cases these introductions have been the result of nontarget species being accidentally included in the transport container through carelessness when the target species was being introduced, but there are also examples where misidentifications have resulted in the wrong species being introduced. Some inconspicuous nontarget organisms may be unwittingly transported in water or packing materials, or be parasites or pathogens in or on the target organism. Carlton (1992) provides an excellent overview of examples of these nontarget introductions, for which there are many freshwater as well as marine examples.

Approximately 48 species of North American freshwater fish have in part been introduced to new water bodies as "contaminants" associated with the stocking of other species (Crossman and Cudmore, this volume). Some species such as the rainwater killifish (*Lucania parva*) and small sunfish (*Lepomis* spp.) have become particularly widespread through these "mistakes." The introduction of the exotic cladoceran *Daphnia lumholtzi* may have been the result of the intentional introduction of the Nile perch (*Lates niloticus*) from Africa into Texas reservoirs (Havel and Hebert, 1993). This cladoceran has spread to numerous reservoirs in the south-central U.S. where it has the potential to disrupt community structure (Havel et al., 1995). The potential of microscopic zebra mussel veligers to be introduced as nontarget organisms with fish stocking activities has led to the development of specific guidelines for hatchery and aquaculture operations (North Carolina Sea Grant, 1995). The release of nontarget organisms has the potential to create adverse ecosystem impacts, as there is no screening or evaluation of organisms prior to their introduction (similar to many accidental introduction pathways). Therefore, it is particularly important that efforts be made to minimize the risks of these types of introductions whenever target organisms are intentionally introduced.

LAWS, POLICY, AND REGULATION THAT CONTROL INTENTIONAL INTRODUCTIONS

Summaries of jurisdictional legislation to control and prevent intentional and unwanted introductions of fish species (Leach and Lewis, 1991; Stanley et al., 1991; Office of Technology Assessment, 1993) point to a maze of regulations and policies established across the U.S. and Canada. These regulations and policies have developed over several decades and now exist in a complex and fragmented manner. Prior to the 1960s, little concern was expressed regarding the possible detrimental effects of transfer of fishes (Stanley et al., 1991). Agencies with the responsibility to protect biotic resources themselves were introducing exotic species. In the past, agency introductions have been among the most harmful (Wingate, 1991).

Only about 25 years ago, the potential for damage was generally recognized and U.S. federal government and state agencies began developing policies and regulations for the control or prohibition of the importation of aquatic organisms (McCann, 1984). These measures, however, were difficult to enforce because of wide-open interstate commerce and a patchwork of state regulations in potential receiving states. In addition, user groups argued that tough preventive measures would stifle the pet trade and aquaculture. Furthermore, many professionals believed that exotic introductions were beneficial and ecological and genetic effects were vague and undocumented (Stanley et al., 1991).

Distinguishing between "good" and "bad" nonindigenous species is not easy. Some species produce both positive and negative consequences, depending on the location and the perceptions of the observers (Aquatic Nuisance Species Task Force, 1996). The number and impact of harmful nonindigenous species are chronically underestimated, especially for species that do not damage agriculture, industry, or human health. Harmful species cost millions to perhaps billions of dollars annually (Office of Technology Assessment, 1993).

As a result, the governments of Canada and the U.S. have begun to closely scrutinize intentional fish introductions while recognizing the need to improve our understanding and control of their effects. Unfortunately, Mexico does not yet recognize the problems posed by intentionally introduced species. However, there is considerable evidence from all three countries that some of our valuable natural resources are being lost because of competition or predation by introduced species. In some watersheds the entire indigenous fish fauna has disappeared over the last 50 years and has been replaced by species such as the Chinese carps and several tilapia species that are common around the world (Workshop on North American Nonindigenous Freshwater Fish, 1996). These losses in biodiversity are permanent.

UNITED STATES

A variety of state and federal authorities exist that address the use of nonindigenous species. In the U.S., federal and state governments presently divide responsibilities for introductions of fish, wildlife, and their diseases. Federal authority is concentrated on several existing laws that provide guidance in the use of nonindigenous species. For various reasons, these laws have not been effectively used with regard to nonindigenous species or have not been implemented (Office of Technology Assessment, 1993).

Federal authority can be found under the Lacey Act, National Environmental Policy Act, National Invasive Species Act, Executive Order 11987, federal funding authorities, (e.g., Federal Aid in Sport Fish Restoration Act, Sea Grant, Corps of Engineers grants), and the Federal Noxious Weed Act. The Nonindigenous Aquatic Nuisance Prevention and Control Act of 1990 and the reauthorized National Invasive Species Act (1996) address only unintentional sources of introductions, even though a review of intentional introductions policy in the U.S. was commissioned by this legislation (Aquatic Nuisance Species Task Force, 1994).

Federal Legislation

> The Lacey Act of 1900 (16 U.S.C. 701, 702: 31 Stat. 187, 32 Stat. 285) and 1981 amendments
> (Public Law 97-79, Nov 16, 1981: 95 Stat. 1073)

This Act provides the basis for existing federal regulations on species introductions (Clugston, 1986). The overall purpose of this Act as is "to provide for the control of illegally taken fish and wildlife." This Act makes it a federal offense to import, export, sell, acquire, or purchase fish, wildlife or plants taken, possessed, transported, or sold in violation of state, foreign, or other federal law.

The Lacey Act also prohibits importation of wild vertebrates and other "injurious" animals listed in the Act or declared by the Secretary of the Interior by regulation to be potentially harmful to human beings or the interests of agriculture, horticulture, forestry, or fishes and wildlife of the U.S. Few species have been listed under this provision. Only three aquatic species (zebra mussels, mitten crabs (*Eriocheir sinensis*), and walking catfish (*Clarias batrachus*)) and four viruses affecting salmonid fishes (VHSV, IHNV, IPNV, and OMV) are currently on the list. However, Lacey Act listing is a slow process, and many species listed under the act were only added after they had already become established. Despite the lack of legislative authority, the U.S. Fish and Wildlife Service (USFWS) is concerned about possible interaction of introduced fishes with native species and the resultant changes in the natural ecosystems. At the same time, the USFWS recognizes the needs and opportunities for use of nonindigenous fishes in selected aquatic ecosystems.

Felony criminal sanctions are provided for violations involving imports or exports, or violations of a commercial nature in which the value of the wildlife is in excess of $350. A misdemeanor violation was established, with a fine of up to $10,000 and imprisonment of up to 1 year, or both. Civil penalties up to $10,000 were provided. However, the Criminal Fines Improvement Act of 1987 increased the fines under the Lacey Act for misdemeanors to a maximum of $100,000 for individuals and $200,000 for organizations. Maximum fines for felonies were increased to $250,000 for individuals and $500,000 for organizations. The law covers all fish and wildlife and their parts or products, and plants protected by the Convention on International Trade in Endangered Species (CITES) and those protected by state law.

The effectiveness of the Lacey Act in influencing the movement of nonindigenous species is thus in large part a reflection of the strengths or weaknesses of the laws of other jurisdictions. The federal government's responsibilities are to support state or foreign regulations when such regulations are violated. Regulations that control the introduction of fish into the open waters of the U.S. are each state's responsibility.

National Environmental Policy Act of 1969 (42 U.S.C. 4321-4347; 83 Stat. 852)

This Act requires each federal agency to prepare an environmental impact statement on any major action they propose that may significantly affect the environment. The introduction of an exotic species could be determined to be such an action and require appropriate evaluations before the introduction takes place.

Endangered Species Act of 1973 (16 U.S.C. 1531-1543; 87 Stat. 884)

The Endangered Species Act prohibits the importation of animals that are endangered or threatened in their native lands. Entry of specific exotic fishes could be prevented by this legislation.

The Federal Noxious Weed Act (7 U.S.C. 2801 et. seq.; 88 Stat. 2148)

This Act delegates to the Secretary of Agriculture the authority to designate plants (including aquatic plants) as noxious weeds and prohibits the movement of such species in interstate and foreign commerce except by permit. Permits are obtained through the U.S. Department of Agriculture's Animal and Plant Health Inspection Service. Authority is also given to inspect, seize, and destroy products, or quarantine areas and to develop cooperative programs as needed to control, eradicate, or prevent the spread of noxious weeds.

Federal Policy

Executive Order 11987

Signed by President Carter on May 24, 1977, this order directs federal agencies, to the extent permitted by law, to prevent the introductions of exotic species into the waters that they own or control and to encourage states, local governments, and private citizens to prevent the introduction of exotic species into the nation's aquatic systems. The Order also restricts the use of federal funds, programs, or authorities for such introductions or for export of native species into ecosystems outside the U.S. This Order does not apply to importations into or from the U.S., if the Secretaries of Agriculture and Interior find that such introductions will not adversely affect natural ecosystems.

Section 3 of Executive Order 11987 directs the Secretary of Interior, in consultation with the Secretary of Agriculture and the heads of other appropriate federal agencies, to develop regulations to implement this Order. The Department of Interior has not issued formal regulations; however, draft regulations prepared by the USFWS in 1978 serve as internal guidelines to carry out federal responsibilities under President Carter's directive. These draft regulations place the responsibility on each executive agency to identify proposed federal introductions or federally related introductions, and to convey a written request for a biological opinion to the USFWS. It would be the responsibility of the requesting agency to conduct appropriate studies and provide the biological information needed for an adequate review of possible effects on a natural ecosystem. Within 90 days after receipt of such a request, the USFWS would issue a biological opinion that could include recommendations for modifications in the proposed introduction to ensure that no adverse effects are imposed on natural ecosystems. The USFWS could request additional studies or surveys, but would not be obligated to fund them.

State Laws and Regulation

The authority of the federal and state governments not only varies with the type of organism regulated, but also depends on the particular federal and state laws and agencies involved. States generally control the entry of species across state borders and release of these species within the state (Office of Technology Assessment, 1993). Standards of individual states vary considerably regarding which species and groups are regulated and how carefully they are regulated; many state efforts to regulate importation, possession, introduction, and release are inadequate (Wingate, 1991; Office of Technology Assessment, 1993). In some cases, the weaknesses of state programs stem from incomplete legal authority. The Lacey Act leaves decisions on almost all intentional introductions of fish and wildlife to the states; only the relatively few organisms on the list of injurious wildlife are prohibited (Clugston, 1984; Stanley et al., 1991). However, federal programs support many state-sponsored introductions, so the federal government has a strong interest in this area.

A survey of the 50 states by Wingate (1991) concludes that all states are concerned with introductions of new fish species by anyone outside their agencies. Introductions of exotic species into the waters of 39 responding states are controlled by statute or internal policy. Iowa is an example of a state that has no statutory authority, but it is able to prohibit the importation of various species deemed harmful to the environment by administrative policy. However, states lack the power to stop the importation and release of a potentially invasive species in a neighboring state. Since few federal laws compel states to cooperate with each other and states have differing priorities, conflicts can and do occur (Office of Technology Assessment, 1993).

Many states also restrict in-state movement of species such as walleye, muskellunge, largemouth and smallmouth bass, rainbow trout, and brook trout by watershed to reduce the possibility of contamination of the gene pool (Wingate, 1991). Florida and Alaska have very specific policies governing genetic integrity of fish. Twenty other states have a written or internal policy on in-state fish stocking.

CANADA

There are many national, provincial, regional, and international regulations, policies, and guidelines that apply to intentional introductions and transfers of aquatic organisms in Canada. The umbrella legislation governing introductions of fish and fish products to the provinces and territories of Canada is the Fisheries Act of Canada, which provides for the making of regulations concerning coastal and inland fisheries (Leach and Lewis, 1991). Most of the jurisdictions have enacted fishery regulations under the federal Fisheries Act that control movement of fish into their territories and between water bodies within their boundaries.

Federal Legislation

Fisheries Act of Canada

This legislation deals with the conservation and protection of fisheries resources. It also mandates the development of a national policy (currently in draft form) on introductions and transfers of aquatic organisms. Fisheries resources include organisms propagated in the aquaculture industry.

Fish Health Protection Regulations — The Fish Health Protection Regulations, promulgated under Section 43(b) of the Fisheries Act, require that fish imported to Canada or transferred between provinces must be accompanied by an import license. Import licenses are issued if sources of stocks have been inspected and shown to be free of selected infectious disease agents. The regulations apply only to salmonid species at present, but they are being amended to cover all other finfish, molluscs, and crustaceans. These regulations are administered by the Department of Fisheries and Oceans Sciences Directorate and by local fish health officers in each province.

Fishery (General) Regulations — The Fishery Regulations are a consolidation of common aspects of fisheries regulations in each province that come under the Fisheries Act. Part VIII of the Fisheries Regulations applies to the release of live fish into fish habitat and to the transfer of live fish to a fish rearing facility.

Provincial Fisheries Regulations — Provincial Fisheries Regulations promulgated under the federal Fisheries Act in each province (Leach and Lewis, 1991) are administered by the agency responsible for fisheries resource management in the province. The regulations generally require that organisms released into the waters of a province have no disease agent that may be harmful to fish and that the organism will have no adverse effect on the genetic characteristics or size of fish populations. The Maritime Provinces Fishery Regulations apply to fishing in the provinces of Nova Scotia, New Brunswick, and Prince Edward Island and adjacent tidal waters. Section 18 of the regulations prohibits the use of live fish for bait that were imported from another province. The Quebec Fisheries regulations specify which salmonids can be propagated or released in three zones. The zones and applicable control measures were established to minimize the impact of resource enhancement activities and aquaculture on wild resources.

Wild Animal and Plant Protection and Regulation of International and Interprovincial Trade Act (WAPPRIITA)

The WAPPRIITA applies to animals that are listed in appendices to CITES. The Act is administered by the Canadian Wildlife Service of the federal Department of the Environment. It permits provinces to make regulations that prohibit the import of endangered animals and plants as well as animals and plants that may be harmful to the environment. To date, only a small number of terrestrial organisms have been listed as environmentally harmful under this legislation.

Provincial Laws and Regulations

Laws and regulations promulgated by provincial governments by means other than the Fisheries Act can affect introductions and transfers of aquatic organisms. These include aquaculture acts, game and fish acts and associated regulations, environmental assessment acts and regulations, fish transportation regulations, and noxious weed acts.

Federal Policies

Wildlife Policy for Canada

The Wildlife Ministers Council of Canada has endorsed a Wildlife Policy for Canada (Anonymous, 1990). This policy recommends that introduction of a nonindigenous or genetically engineered species should be considered only if

- No indigenous species is suitable for the purpose of the introduction;
- Clear and well-defined benefits to human or natural communities are foreseen; and
- No known adverse environmental impacts is foreseen, and some means of controlling the introduced population exists (such as predators or climate).

Proposed national policy — Introductions and Transfers of Aquatic Organisms

The Department of Fisheries and Oceans (DFO) has developed a draft national policy to establish principles and standards for the intentional introduction and transfer of aquatic organisms. "Intentional" refers to the deliberate movement of live aquatic organisms into Canada or between provinces and territories. DFO's purpose in developing the National Policy is to minimize the impacts of introductions and transfers in recognition of the Department's responsibility to protect aquatic resources and, at the same time, permit environmentally sound fisheries resource enhancement and development of aquaculture.

Proposed national policy — Transgenic Aquatic Organisms: Policy Guidelines for Research with, or for Rearing in Natural Aquatic Ecosystems in Canada

The Department of Fisheries and Oceans has developed proposals for a national policy on research with transgenic aquatic organisms and rearing them in natural aquatic ecosystems. The policy addresses concerns of escapement of transgenic organisms from research facilities, and potential impacts if transgenic organisms are used for aquaculture or enhancement. The policy will provide the basis for promulgating a regulation on transgenics under the Fisheries Act.

It is anticipated that the policy will address the following items: inspection of research facilities using transgenic organisms, criteria for conducting research, risk assessment for the use of these organisms in or near natural ecosystems, a formal review and approval process, implementation of an incremental process for use of transgenics with provision for assessment at each step, use of only reproductively sterile organisms in other than the research facility, a requirement for annual reporting, and public hearings prior to release of transgenic organisms outside of a research facility.

Department of Fisheries and Oceans Policy on Importation of Aquatic Organisms Outside Canada

DFO has a policy requiring review, by DFO's Assistant Deputy Minister, Science of proposals to import aquatic organisms to Canada. The purpose of this policy is to ensure that decisions to import aquatic organisms are consistent across Canada, and that decisions are in compliance with international standards. Where there is a history of importation of certain species from sources in the U.S., and there has been minimal impact on local fisheries resources, habitat, or aquaculture, authorization prior to importation is not required; however, annual summaries of imported organisms must be submitted to the Assistant Deputy Minister, Science.

Canadian Biodiversity Strategy

This strategy urges the provinces to eliminate or reduce the risks associated with the intentional introduction of aquatic organisms.

Policies in Individual Provinces

DFO and/or provincial governments have established Transplant/Introductions Committees in some provinces. These committees assess proposals to introduce or transfer organisms to a province and provide advice to the approving authority on the acceptability of individual proposals. These committees do not have legislated or regulatory authority at present.

A range of policies on introductions and transfers of aquatic organisms have been or are being prepared in individual provinces. Leach and Lewis (1991) provide detailed information on provincial policies related to introductions and transfers. These policies provide for health protection, and preservation of genetic diversity and ecological integrity of wild fish populations. The policies serve as valuable building blocks with which to develop the national policy on introductions and transfers.

MEXICO

Mexico had at least 55 exotic fishes, 26 of which were foreign and 29 transplants, within the country in 1984 (Courtenay and Stauffer, 1984). In 1997, the count for introduced species was up to 90 (Balderas, this volume). The distribution of exotic fishes in Mexico is extensive for most common species. Most introductions were made without previous thought of negative impacts or methods of control. Many species are imported from the U.S., including fishes that share drainages in both nations.

Mexican environmental legislation states that biological conservation should include the protection of both evolutionary processes and genetic diversity. The only firm reference is Article 85, LGEEPA (Ley General del Equilibrio Ecologio y Proteccion Ambiental), which states that for (native) species protection, SEMARNAP (Secretario del Medio Ambiente, Recursos Naturales y Pesca) will promote that the Secretaria de Industria y Comercio establishes procedures for regulation or restriction, total or partial, to exportation or importation of specimens of flora and fauna, and even restrict movement from within or through Mexico for shipments to other countries (S. Contreras, Universidad Autónoma de Nuevo León, San Nicolás de los Garza, Nuevo León, Mexico, pers. com.). Nothing has been done legally to comply with this article.

The Mexican government is understandably concerned with the low-protein diet of many Mexicans, particularly country people. The government, therefore, measures success of introduced fishes in economic and social terms and not on real or potential adverse impacts to native biota or habitats (Contreras and Escalante, 1984). In addition, it must be noted that introduced fishes may not be the major cause of declines of native fishes in all localities. Areas of major development, including intense agriculture, and urban highway construction are responsible for losses of both native and introduced fish species.

BINATIONAL AGREEMENTS AND ORGANIZATIONS

Established within Canada and the U.S. are many organizations to coordinate efforts aimed at minimizing nonindigenous species impacts to fisheries resources. Examples include

- North American Commission of the North Atlantic Salmon Conservation Organization (NASCO), made up with representatives from Canada and the U.S., has guidance for the introduction and transfer of salmonids within the Atlantic Coast region to protect native Atlantic salmon populations;
- Great Lakes Fishery Commission (GLFC) is composed of representatives from Illinois, Indiana, Michigan, Minnesota, New York, Ohio, Pennsylvania, Wisconsin, Ontario, and the Canadian and U.S. federal governments. The commission oversees a binational program to control sea lamprey populations in the Great Lakes. In addition, the GLFC coordinates policies on fish disease control and a policy on introduction of exotic species, including a procedure for consensus on intentional introductions (Great Lakes Fishery Commission, 1980);

- Pacific Northwest Fish Health Protection Committee is composed of representatives from Alaska, California, Idaho, Montana, Oregon, Washington State, the U.S. federal government, and observers from British Columbia. The committee sets standards to protect the health of salmonid fisheries;
- Representatives from North Dakota, South Dakota, Minnesota, Montana, the U.S. federal government, provinces of Manitoba, Saskatchewan, Alberta, and the Canadian federal government have a compact to carry out activities for controlling introductions of harmful nonindigenous species.

INTERNATIONAL ORGANIZATIONS AND AGREEMENTS

On a global scale there are many organizations addressing the transport and introduction of aquatic organisms:

- International Council for Exploration of the Sea (ICES) has a Code of Practice for Introduction and Transfer of Marine Organisms;
- l'Office international des epizooties (OIE) developed an Animal Health Code for Finfish, Molluscs and Crustacea;
- General Agreement on Tariffs and Trade (GATT) retains standards for sanitary measures considered acceptable for international trade;
- Canada–United States Free Trade Agreement (FTA) and North American Free Trade Agreement (NAFTA) retains standards for sanitary measures considered acceptable for trade between Canada, United States, and Mexico;
- United Nations Conference on the Environment and Development (UNCED) Convention for the Preservation of Biodiversity. In 1992, the three countries participated and signed the Convention of Biological Diversity under the United Nations. The countries committed to "develop national strategies, plans or programs for the conservation and sustainable use of biological diversity."

RISK ANALYSIS

Restricting the introduction of harmful nonindigenous species requires an assessment of the risk posed by each new taxon or vector. Through decisions about the deliberate release of potentially beneficial nonindigenous species, we have opportunities to control and reduce the harm associated with many biological invasions. Explicit consideration of a species risk and of the values impinging on the decision, the inherent conflict between nonindigenous species benefits and the value of unaltered biota, are absent from much North American policy regarding species introductions (Ruesink et al., 1995).

Canada and the U.S. have recently prepared generic risk assessment protocols to guide intentional introductions of exotic species. Risk analysis is a standarized process for evaluating the risk of introducing nonindigenous organisms into a new environment, while determining the management steps necessary to ensure that the risk is minimized.

The risk assessment process determines types of factors that may be encountered with intentional introductions. The process utilizes qualitative and quantitative data in a systematic and consistent fashion. The objective is to produce scientifically defensible assessments on types of nonindigenous organisms and their specific pathways into the environment. The assessments are theoretical but carried out in a formal and systematic fashion (Aquatic Nuisance Species Task Force, 1996).

The assessment should estimate the overall risk. In addition, effective communications are an important part of the overall process. The process should be able to clearly explain uncertainties inherent to the analysis and to avoid design of a process that reflects a prearranged conclusion.

Quantitative risk assessments can be a valuable tool to understand and estimate overall risk. At the same time, assessments are subject to unknown factors and should therefore be tempered with best professional judgment.

Acceptable risk levels cannot be determined with risk assessments. The type and level of acceptable risk depends on how a person or agency interprets that risk and are characterized by value judgments. Determining how or when a specific organism will become established is not possible. Moreover, it is difficult to estimate the likelihood and potential of an organism to cause damage in its new environment, because it involves a combination of biological and ecological variables. Many scientists believe that the synergistic effects make it impossible to predict future ecological events.

While much of the information used in performing a risk assessment is scientifically valid, much of the process is subject to theory and interpretation. At the same time, an assessment provides an estimation based on the best information available. Management actions will rely on these estimations of risk to carry out actions aimed at restricting or prohibiting high risk pathways.

Obstacles to overcome when carrying out risk assessments, especially for government agencies, include historical precedent, legal parameters, operational procedures, and political influences. It is best to focus the assessment as much as possible on the biological and ecological factors of risk in an atmosphere free of regulatory and political influence.

Although the U.S. government recognizes the risks posed by nonindigenous species and for nearly a century has had legislation in place aimed at reducing harmful biological introductions, there is still a relentless increase in the number of nonindigenous species in North America. This influx might be curtailed by changing regulation to adequately reflect the risks involved and by shifting the burden of proof to those most likely to benefit from an introduction.

SUMMARY

Intentional introductions of freshwater aquatic organisms have a long history in North America. Early introductions were often the result of the actions of private citizens who wanted to introduce familiar fishes from abroad. The construction of hatcheries and government-sponsored stocking programs were widespread by the end of the 19th century. The introduction of sport fish to establish new fisheries and expand angling opportunities has been the principal reason for intentional introductions in Canada and the U.S. Introductions of prey fishes have often accompanied these introductions. In Mexico, most intentional introductions have been conducted with the intent of establishing new food sources for a largely protein deficient society. Aquatic organisms have also been introduced in all three countries for the purposes of biological control. In the U.S. several endemic fishes threatened with extinction in their native habitats have been intentionally introduced into new habitats for conservation purposes. Intentional introduction of aquatic organisms has also resulted in the accidental transfer of several species of nontarget organisms.

The ecological impacts of intentionally introduced organisms in North America have been extremely variable and have spanned the range of impacts that can be expected from introduced organisms (Taylor et al., 1984). Intentional introductions of sport fish have often resulted in the development of spectacular sport fisheries that provide immediate social and economic benefits to humans, but these have often come at a cost to native organisms and biodiversity. Intentionally introduced sport fish have contributed to the extinction of several North American fish species, and to the declines of many threatened and endangered fishes. Often, the negative impacts of these introduced species are not apparent until long after the initial introduction. Intentionally introduced organisms often disperse far beyond the target area for their introductions (through natural dispersal and secondary accidental introductions) giving them the potential to have widespread impacts. The introduction of prey species has had mixed results and has proceeded on a largely trial-and-error basis. The introduction of fish as new food sources has not necessarily benefitted the citizens of Mexico, but has had undesirable effects on several native species. The introduction of fish as biological control organisms has generally caused more problems than it solves because of the omnivorous feeding habits of the species involved. The use of sterile grass carp to control aquatic vegetation holds promise, but remains highly controversial. The use of nonnative phytophagous insects to control specific aquatic exotic plant species has been successful in some instances. Many species have provided benefits to humans while at the same time having unwanted ecosystem impacts. This has led to opposing views on the uses of such species in aquatic ecosystems.

In recent decades, increasing concern over the impacts of intentionally introduced aquatic organisms has led to the development of numerous laws and policies to control intentional introductions in the U.S. and Canada. In the U.S. most intentional introductions are governed by individual state laws and policies, although the federal government exercises control over federal waters and over programs that use federal funds. In Canada, intentional introductions of aquatic organisms are largely controlled by the provinces through regulations made under the federal Fisheries Act and provincial policies. Although intentional introductions have been subject to increased scrutiny in recent years in both countries, there is evidence that existing laws and policies and differing standards between adjacent jurisdictions have led to problems with respect to intentional introductions. In Mexico there is little if any legislation that regulates intentional introductions, and government-sponsored introductions of potentially harmful fishes such as tilapia and Chinese carps are ongoing. Concerns over the impacts of introduced species have been incorporated into numerous binational (Canada and the U.S.) and international agreements in recent years. These agreements generally lay out standards and protocols for the transfer of live organisms as a matter of policy.

Explicit consideration of the risks to native biota posed by species proposed for introduction versus their potential benefits is absent from much North American policy related to intentional introductions. The U.S. and Canada have recently developed generic risk assessment protocols to provide a standardized process for evaluating the risks associated with individual nonindigenous organisms and the pathways by which they are introduced. The use and refinement of such protocols for the review of proposed intentional introductions will help to reduce the undesirable ecosystem impacts that all too often accompany intentional introductions.

REFERENCES

Allen, S.K., Jr., N.T. Hagstrom, and R.G. Thiery, Cytological evaluation of the likelihood that triploid grass carp will reproduce, *Transactions of the American Fisheries Society*, 115, pp. 841–848, 1986.

Anonymous, A wildlife policy for Canada. Wildlife Ministers Council of Canada, Canadian Wildlife Service, Ottawa, Ontario, 1990.

Aquatic Nuisance Species Task Force, Findings, conclusions, and recommendations of the intentional introductions policy review. Aquatic Nuisance Species Task Force, Intentional Introductions Policy Review Committee, Report to U.S. Congress, Washington, DC, 1994.

Aquatic Nuisance Species Task Force, Generic nonindigenous aquatic organisms risk analysis review process. Aquatic Nuisance Species Task Force, Risk Assessment and Management Committee, Washington, DC, 1996.

Armstrong, K.B., The biology and histopathology of *Proteocephalus ambloplites* Leidy (cestoda: proteocephalidae) infecting walleye (*Stizostedion vitreum vitreum*) and yellow perch (*Perca flavescens*) in Lake of the Woods, Ontario, thesis presented to Lakehead University, Thunder Bay, Ontario, in partial fulfillment of the requirements for the degree of Master of Science, 1985.

Avise, J.C., and M.J. Van den Avyle, Genetic analysis of reproduction of hybrid white bass x striped bass in the Savannah River, *Transactions of the American Fisheries Society*, 113, pp. 563–570, 1984.

Axon, J.R., and D.K. Whitehurst, Striped bass management in lakes with emphasis on management problems, *Transactions of the American Fisheries Society*, 114, pp. 8–11, 1985.

Blinn, D.W., C. Runck, D.A. Clark, and J.N. Rinne, Effects of rainbow trout predation on Little Colorado spinedace, *Transactions of the American Fisheries Society*, 122, pp. 139–143, 1993.

Boxrucker, J., Evaluation of stocking inland silversides as supplemental forage for largemouth bass and white crappie, in *Reservoir Fisheries Management: Strategies for the 80's*, Hall, G.E., and M.J. Van Den Avyle, Eds., American Fisheries Society, Bethesda, MD, 1986.

Buckingham, G.R., Biological control of alligatorweed, *Alternanthera philoxeroides*, the world's first aquatic weed success story, *Castanea*, 61, pp. 232–233, 1996.

Bury, R.B., and J.A. Whelan, Ecology and management of the bullfrog, United States Department of the Interior, Fish and Wildlife Service, Resource Publication 155, 1984.

Campton, D.E., Genetic effects of hatchery fish on wild populations of Pacific salmon and steelhead: what do we really know?, in *American Fisheries Society Symposium 15: Proceedings of the International Symposium and Workshop on the Uses and Effects of Cultured Fishes in Aquatic Ecosystems,* American Fisheries Society, Bethesda, Maryland, pp. 337–353, 1995.

Carlton, J.T., Dispersal of living organisms into aquatic ecosystems as mediated by aquaculture and fisheries activities, in *Dispersal of Living Organisms into Aquatic Ecosystems*, Rosenfield, A., and R. Mann, Eds., Maryland Sea Grant College, College Park, MD, 1992.

Carmichael, G.J., J.N. Hanson, J.R. Novy, K.J. Meyer, and D.C. Morizot, Apache trout management: cultured fish, genetics, habitat improvements, and regulations, in *American Fisheries Society Symposium 15: Proceedings of the International Symposium and Workshop on the Uses and Effects of Cultured Fishes in Aquatic Ecosystems,* American Fisheries Society, Bethesda, MD, pp. 112–121, 1995.

Cassani, J.R., Problems and prospects for grass carp as a management tool, in *American Fisheries Society Symposium 15: Proceedings of the International Symposium and Workshop on the Uses and Effects of Cultured Fishes in Aquatic Ecosystems,* American Fisheries Society, Bethesda, MD, pp. 407–412, 1995.

Chacon-Torres, A., and C. Rosas-Monge, A restoration plan for pez blanco in Lake Patzcuaro, Mexico, in *American Fisheries Society Symposium 15: Proceedings of the International Symposium and Workshop on the Uses and Effects of Cultured Fishes in Aquatic Ecosystems,* American Fisheries Society, Bethesda, MD, pp. 122–126, 1995.

Christie, W.J., G.R. Spangler, K.H. Loftus, W.L. Hartman, P.J. Colby, M.A. Ross, and D.R. Talhelm, A perspective on Great Lakes fish community rehabilitation. *Canadian Journal of Fisheries and Aquatic Sciences.* 44 (Supplement 2), pp. 486–499, 1987.

Clugston, J.P., Strategies for reducing risks from introductions of aquatic organisms: the federal perspective, *Fisheries,* 11(2), pp. 26–29, 1986.

Colby, P.J., and C. Hunter, Environmental assessment of the introduction of walleye beyond their current range in Montana. Prepared by OEA Research for Montana Department of Fish, Wildlife and Parks, Helena, MT, 1989.

Colby, P.J., P.A. Ryan, D.H. Schupp, and S.L. Serns, Interactions in north-temperate lake fish communities, *Canadian Journal of Fisheries and Aquatic Sciences,* 44 (Supplement 2), pp. 104–128, 1987.

Conover, M.C., Stocking cool-water species to meet management needs, in *Fish Culture in Fisheries Management,* Stroud, R.H., Ed., American Fisheries Society, Bethesda, MD, 1986.

Contreras-B., S., and M.A. Escalante-C., Distribution and known impacts of exotic fishes in Mexico, in *Distribution, Biology, and Management of Exotic Fishes*, Courtenay, W.R., Jr., and J.R. Stauffer, Jr., Eds., Johns Hopkins University Press, Baltimore, MD, 1984.

Cottrell, K.D., S. Stuewe, and A. Brandenburg, Incorporating the stock concept and conservation genetics in an Illinois stocking program, in *American Fisheries Society Symposium 15: Proceedings of the International Symposium and Workshop on the Uses and Effects of Cultured Fishes in Aquatic Ecosystems.* American Fisheries Society, Bethesda, MD, pp. 244–248, 1995.

Courtenay, W.R., Jr., Biological pollution through fish introductions, in *Biological pollution: the control and impact of invasive exotic species: Proceedings of a Symposium held at the University Place Conference Center, Indiana University-Purdue University at Indianapolis on October 25 & 26, 1991*, McKnight, B.N., Ed., Indiana Academy of Science, Indianapolis, IN, pp. 35–61, 1993.

Courtenay, W.R., Jr., The case for caution with fish introductions, in *American Fisheries Society Symposium 15: Proceedings of the International Symposium and Workshop on the Uses and Effects of Cultured Fishes in Aquatic Ecosystems,* American Fisheries Society, Bethesda, MD, pp. 413–424, 1995.

Courtenay, W.R., Jr., D.A. Hensley, J.N. Taylor, and J.A. McCann, Distribution of exotic fishes in the continental United States, in *Distribution, Biology, and Management of Exotic Fishes*, Courtenay, W.R., Jr., and J.R. Stauffer, Jr., Eds., Johns Hopkins University Press, Baltimore, MD, 1984.

Courtenay, W.R., Jr., D.A. Hensley, J.N. Taylor, and J.A. McCann, Distribution of exotic fishes in North America, in *The Zoogeography of North American Freshwater Fishes*, Hocutt C.H., and E.O. Wiley, Eds., John Wiley and Sons, New York, 1986.

Courtenay, W.R., Jr., D.P. Jennings, and J.D. Williams, Exotic fishes, in *Common and Scientific Names of Fishes from the United States and Canada*, Robins, C.R., R.M. Bailey, C.E. Bond, J.R. Brooker, E.A. Lachner, R.N. Lea, and W.B. Scott, Eds., American Fisheries Society Special Publication 20, American Fisheries Society, Bethesda, MD, 1991.

Courtenay, W.R., Jr., and C.C. Kohler, Exotic fishes in North American fisheries management, in *Fish Culture in Fisheries Management,* Stroud, R.H., Ed., American Fisheries Society, Bethesda, MD, 1986.

Courtenay, W.R., Jr., and G.K. Meffe, Small fishes in strange places: a review of introduced poeciliids, in *Evolution and Ecology Of Livebearing Fishes (Poeciliidae)*, Meffe, G.K., and F.N. Snelson, Jr., Eds., Prentice-Hall, New York, 1989.

Courtenay, W.R., Jr., and C.R. Robins, Fish introductions: good management, mismanagement, or no management? *Reviews in Aquatic Sciences*, 1, pp. 159–172, 1989.

Courtenay, W.R., Jr., and J.R. Stauffer, Jr., Eds., *Distribution, Biology, and Management of Exotic Fishes*, Johns Hopkins University Press, Baltimore, MD, 1984.

Courtenay, W.R., Jr., and J.D. Williams, Dispersal of aquatic species from aquaculture sources, with emphasis on freshwater fishes, in *Dispersal of Living Organisms into Aquatic Ecosystems*, Rosenfield, A., and R. Mann, Eds., Maryland Sea Grant College, College Park, MD, 1992.

Creed, R.P., Jr., and S.P. Sheldon, Potential for a native weevil to serve as a biological control agent for Eurasian watermilfoil, U.S. Army Corps of Engineers Waterways Experiment Station Technical Report A-94-7, 1994.

Crossman, E.J., Introduction of exotic fishes into Canada, in *Distribution, Biology, and Management of Exotic Fishes*, Courtenay, W.R., Jr., and J.R. Stauffer, Jr., Eds., Johns Hopkins University Press, Baltimore, MD, 1984.

Crossman, E.J., Introduced freshwater fishes: a review of the North American perspective with emphasis on Canada, *Canadian Journal of Fisheries and Aquatic Sciences*, 48 (Supplement 1), pp. 46–57, 1991.

Crossman, E.J., Introduction of the lake trout (*Salvelinus namaycush*) in areas outside its native distribution: a review, *Journal of Great Lakes Research*, 21 (Supplement 1), pp. 17–29, 1995.

Crossman, E.J., P. Krause, and S.J. Nepszy, The first record of grass carp (*Ctenopharyngodon idella*) in Canadian waters, *The Canadian Field-Naturalist*, 101, pp. 584–586, 1987.

Dahlsten, D.L., Control of invaders, in *Ecology of Biological Invasions of North America and Hawaii (Ecological Studies; v. 58)*, Mooney, H.A., and J.A. Drake, Eds., Ecological Studies 58, Springer-Verlag, New York, 1986.

Deacon, J.E., The endangered woundfin and water management in the Virgin River, Utah, Arizona, Nevada, *Fisheries,* 13(1), pp. 18–24, 1988.

DeKay, J.E., *Zoology of New York IV: Fishes*, W. & A. White and J. Visscher, Albany, NY, 1842.

DeLoach, C.J., Past successes and current prospects for biological control of weeds in the United States and Canada, *Natural Areas Journal*, 11, pp. 129–142, 1991.

Department of Fisheries and Oceans, 1990 survey of recreational fishing in Canada. *Department of Fisheries and Oceans Economic and Commercial Analysis Report Number 148*, 1994.

DeVoe, M.R., Ed., Introductions and Transfers of Marine Species: Achieving a Balance Between Economic Development and Resource Protection — *Proceedings of the Conference and Workshop, October 30-November 2, 1991, Hilton Head Island, South Carolina*, South Carolina Sea Grant Consortium, Hilton Head, SC, 1992.

DeVries, D.R., and R.A. Stein, Manipulating shad to enhance sport fisheries in North America: an assessment, *North American Journal of Fisheries Management,* 10, pp. 209–223, 1990.

DeVries, D.R., R.A. Stein, J.G. Miner, and G.G. Mittelbach, Stocking threadfin shad: consequences for young-of-year fishes, *Transactions of the American Fisheries Society*, 120, pp. 368–381, 1991.

Eck, G.W., and L. Wells, Recent changes in Lake Michigan's fish community and their probable causes, with emphasis on the role of alewife (*Alosa pseudoharengus*), *Canadian Journal of Fisheries and Aquatic Sciences,* 44 (Suppl. 2), pp. 53–60. 1987.

Emery, L., Review of fish species introduced into the Great Lakes, 1819–1974, *Great Lakes Fishery Commission Technical Report Number 45*, 1985.

Evans, D.O., and D.H. Loftus, Colonization of inland lakes in the Great Lakes region by rainbow smelt, *Osmerus mordax*: their freshwater niche and effects on indigenous fishes, *Canadian Journal of Fisheries and Aquatic Sciences,* 44 (Suppl. 2), pp. 249–266, 1987.

Evans, D.O, and C.C. Willox, Loss of exploited, indigenous populations of lake trout, *Salvelinus namaycush*, by stocking of nonnative stocks, *Canadian Journal of Fisheries and Aquatic Sciences,* 48 (Suppl. 1), pp. 134–147, 1991.

Fenton, R., J.A. Mathias, and G.E.E. Moodie, Recent and future demand for walleye in North America, *Fisheries,* 21(1), pp. 6–12, 1996.

Ferguson, M.M., The genetic impact of introduced fishes on native species, *Canadian Journal Zoology,* 68, pp. 1053–1057, 1990.

Fiss, F.C., S.M. Sammons, P.W. Bettoli, and N. Billington, Reproduction among saugeyes (F_x hybrids) and walleyes in Normandy Reservoir, Tennessee, *North American Journal of Fisheries Management,* 17, pp. 215–219, 1997.

Flickinger, S.A., and F.J. Bulow, Small impoundments, in *Inland fisheries management in North America,* Kohler, C.C., and W.A. Hubert, Eds., American Fisheries Society, Bethesda, MD, 1993.

Franzin, W.G., B.A. Barton, R.A. Remnant, D.B. Wain, and S.J. Pagel, Range extension, present and potential distribution, and possible effects of rainbow smelt in Hudson Bay drainage waters of northwestern Ontario, Manitoba, and Minnesota, *North American Journal of Fisheries Management,* 14, pp. 65–76, 1994.

Gerstung, E.R., Status, life history, and management of the Lahontan cutthroat trout, in *American Fisheries Society Symposium 4: Status and Management of Interior Stocks of Cutthroat Trout,* American Fisheries Society, Bethesda, MD, pp. 93–106, 1988.

Great Lakes Fishery Commission. (1980), *A joint Strategic Plan for Management of Great Lakes Fisheries,* Great Lakes Fishery Commission, Ann Arbor, MI.

Griffith, J.S., Review of competition between cutthroat trout and other salmonids, in *American Fisheries Society Symposium 4: Status and Management of Interior Stocks of Cutthroat Trout,* American Fisheries Society, Bethesda, MD, pp. 134–140, 1988.

Guillory, V., and R.D. Gasaway, Zoogeography of the grass carp in the United States, *Transactions of the American Fisheries Society,* 107, pp. 105–112, 1978.

Hansen, M.J., P.T. Schultz, and B.A. Lasee, Changes in Wisconsin's Lake Michigan salmonid sport fishery, 1969-1985, *North American Journal of Fisheries Management,* 10, pp. 442–457, 1990.

Harper, J.L., and H.E. Namminga, Fish population trends in Texoma Reservoir following establishment of striped bass, in *Reservoir Fisheries Management: Strategies for the 80's,* Hall, G.E. and M.J. Van Den Avyle, Eds., American Fisheries Society, Bethesda, MD, 1986.

Havel, J.E., and P.D.N. Hebert, *Daphnia lumholtzi* in North America: another exotic zooplankter, *Limnology and Oceanography,* 38, pp.1823–1827, 1993.

Havel, J.E., W.R. Mabee, and J.R. Jones, Invasion of the exotic cladoceran *Daphnia lumholtzi* into North American reservoirs, *Canadian Journal of Fisheries and Aquatic Sciences,* 52, pp. 151–160, 1995.

Hayes, M.P., and M.R. Jennings, Decline of ranid frog species in western North America: are bullfrogs (*Rana catesbeiana*) responsible? *Journal of Herpetology,* 20 pp. 490–509, 1986.

Heidinger, R.C., Stocking for sport fisheries enhancement, in *Inland fisheries management in North America,* Kohler, C.C., and W.A. Hubert, Eds., American Fisheries Society, Bethesda, MD, 1993.

Hepworth, D.K., and D.J. Duffield, Interaction between exotic crayfish and stocked rainbow trout in Newcastle Reservoir, Utah, *North American Journal of Fisheries Management,* 7, pp. 554–561, 1987.

Hindar, K., N. Ryman, and F. Utter, Genetic effects of cultured fish on natural fish populations, *Canadian Journal of Fisheries and Aquatic Sciences,* 48, pp. 945–957, 1991.

Hoffman, G.L., and G. Schubert, Some parasites of exotic fishes, in *Distribution, Biology, and Management of Exotic Fishes,* Courtenay, W.R., Jr., and J.R. Stauffer, Jr., Eds., Johns Hopkins University Press, Baltimore, MD, 1984.

Howarth, F.G., Environmental impacts of classical biological control, *Annual Reviews in Entomology,* 36, pp. 485–509, 1991.

Howells, R.G., Confirmed grass carp spawning, *Fisheries,* 18(3), p. 36, 1992.

Hutchinson, G.E., Concluding remarks, *Cold Spring Harbor Symposia on Quantitative Biology,* 22, pp. 415–427, 1957.

Inskip, P.D., Negative associations between abundances of muskellunge and northern pike: evidence and possible explanations, in *Managing Muskies: A Treatise on the Biology and Propagation of Muskellunge in North America*, Hall, G.E., Ed., American Fisheries Society Special Publication 15, American Fisheries Society, Bethesda, MD, 1986.

Inskip, P.D., and J.J. Magnuson, Changes in fish populations over an 80-year period: Big Pine Lake, Wisconsin, *Transactions of the American Fisheries Society*, 112, pp. 378–389, 1983.

Jahn, L.A., D.R. Douglas, M.J. Terhaar, and G.W. Kruse, Effects of stocking hybrid striped bass in Spring Lake, Illinois, *North American Journal of Fisheries Management*, 7, pp. 522–530, 1987.

Jennings, M.R., and M.P. Hayes, Pre-1900 overharvest of California red-legged frogs (*Rana aurora draytonii*): the inducement for bullfrog (*Rana catesbeiana*) introduction, *Herpetologica*, 41, pp. 94–103, 1985.

Johnson, F.H., and J.G. Hale, Interrelations between walleye (*Stizostedion vitreum vitreum*) and smallmouth bass (*Micropterus dolomieui*) in four northeastern Minnesota lakes, 1948–69, *Journal of the Fisheries Research Board of Canada*, 34, pp. 1626–1632, 1977.

Johnson, M.G., J.H. Leach, C.K. Minns, and C.H. Olver, Limnological characteristics of Ontario lakes in relation to associations of walleye (Stizostedion vitreum vitreum), northern pike (Esox lucius), lake trout (Salvelinus namaycush), and smallmouth bass (Micropterus dolomieui), *J. Fisheries Research Board of Canada*, 34, pp. 1592–1601, 1977.

Jones, M.L., J.F. Koonce, and R. O'Gorman, Sustainability of hatchery-dependent salmonine fisheries in Lake Ontario: the conflict between predator demand and prey supply, *Transactions of the American Fisheries Society*, 122, pp. 1002–1018, 1993.

Jones, M.S., J.P. Goettl, Jr., and S.A. Flickinger, Changes in walleye food habits and growth following a rainbow smelt introduction, *North American Journal of Fisheries Management*, 14, pp. 409–414, 1994.

Kaeding, L.R., G.D. Boltz, and D.G. Carty, Lake trout discovered in Yellowstone Lake threaten native cutthroat trout, Fisheries, 21(3), pp. 16–20, 1996.

Keith, W.E., A review of introduction and maintenance stocking in reservoir fisheries management, in *Reservoir Fisheries Management: Strategies for the 80s*, Hall, G.E., and M.J. Van Den Avyle, Eds., American Fisheries Society, Bethesda, MD, 1986.

Kempinger, J.J., and R.F. Carline, Dynamics of the walleye (Stizostedion vitreum vitreum) population in Escanaba lake, Wisconsin, 1955–72, *J. Fisheries Research Board of Canada*, 34, pp. 1800–1811, 1977.

Kircheis, F.W., and J.G. Stanley, Theory and practice of forage-fish management in New England, *Transactions of the American Fisheries Society*, 110, pp. 729–737, 1981.

Kohler, C.C., J.J. Ney, and W.E. Kelso, Filling the void: development of a pelagic fishery and its consequences to littoral fishes in a Virginia mainstream reservoir, in *Reservoir Fisheries Management: Strategies for the 80s*, Hall, G.E., and M.J. Van Den Avyle, Eds., American Fisheries Society, Bethesda, MD, 1986.

Krishka, B.A., R.F. Cholmondeley, A.J. Dextrase, and P.J. Colby, Impacts of introductions and removals on Ontario percid communities. Percid Community Synthesis, Introductions and Removals Working Group, Ontario Ministry of Natural Resources, Peterborough, Ontario, 1996.

Krueger, C.C., and B. May, Ecological and genetic effects of salmonid introductions in North America, *Canadian Journal of Fisheries and Aquatic Sciences*, 48 (Supplement 1), pp. 66–77, 1991.

Krueger, C.C., D.L. Perkins, E.L. Mills, and J.E. Marsden, Predation by alewives on lake trout fry in Lake Ontario: role of an exotic species in preventing restoration of a native species, *Journal of Great Lakes Research*, 21 (Supplement 1), pp. 458–469,1995.

Laarman, P.W., Case histories of stocking walleyes in inland lakes, impoundments, and the Great Lakes — 100 years with walleyes, in *Select Coolwater Fishes*, American Fisheries Society Special Publication 11, American Fisheries Society, Bethesda, MD, pp. 254–260, 1978.

Lange, R.E., G.C. Le Tendre, T.H. Eckert, and C.P. Schneider, Enhancement of sportfishing in New York waters of Lake Ontario with hatchery-reared salmonines, in *American Fisheries Society Symposium 15: Proceedings of the International Symposium and Workshop on the Uses and Effects of Cultured Fishes in Aquatic Ecosystems*, American Fisheries Society, Bethesda, MD, pp. 7–11, 1995.

Lange, R.E., and P.A. Smith, Lake Ontario fishery management: the lake trout restoration issue, *Journal of Great Lakes Research*, 21 (Supplement 1), pp. 470–476, 1995.

Larson, G.L., and S.E. Moore, Encroachment of exotic rainbow trout into stream populations of native brook trout in the southern Appalachian Mountains, *Transactions of the American Fisheries Society*, 114, pp. 195–203, 1985.

Lasenby, D.C., T.G. Northcote, and M. Fürst, Theory, practice, and effects of *Mysis relicta* introductions into North American and Scandinavian Lakes, *Canadian Journal of Fisheries and Aquatic Sciences*, 43, pp. 1277–1284, 1986.

Lassuy, D., Introduced species as a factor in extinction and endangerment of native fish species, in *American Fisheries Society Symposium 15: Proceedings of the International Symposium and Workshop on the Uses and Effects of Cultured Fishes in Aquatic Ecosystems*, American Fisheries Society, Bethesda, MD, pp. 391–396, 1995.

Leach, J.H., Non-indigenous species in the Great Lakes: Were colonization and damage to ecosystem health predictable? *Journal of Aquatic Ecosystem Health*, 4, pp. 117–128, 1995.

Leach, J.H., and C.A. Lewis, Fish introductions in Canada: Provincial views and regulations, *Canadian Journal of Fisheries and Aquatic Sciences*, 48 (Supplement 1), pp. 156–161, 1991.

Leary, R.F., F.W. Allendorf, and G.K. Sage, Hybridization and introgression between introduced and native fish, in *American Fisheries Society Symposium 15: Proceedings of the International Symposium and Workshop on the Uses and Effects of Cultured Fishes in Aquatic Ecosystems*, American Fisheries Society, Bethesda, MD, pp. 91–101, 1995.

Li, H.W., and P.B. Moyle, Management of introduced fishes, in *Inland Fisheries Management in North America*, Kohler, C.C., and W.A. Hubert, Eds., American Fisheries Society, Bethesda, MD, 1993.

Litvak, M.K,, and N.E. Mandrak, Ecology of freshwater baitfish use in Canada and the United States, *Fisheries*, 18(12), pp. 6–13, 1993.

Loftus, D.H., and P.F. Hulsman, Predation on larval lake whitefish (*Coregonus clupeaformis*) and lake herring (*C. artedi*) by rainbow smelt (*Osmerus mordax*), *Canadian Journal of Fisheries and Aquatic Sciences*, 43, pp. 812–818, 1986.

MacCrimmon, H.R., C.D. Wren, and B.L. Gots, Mercury uptake by lake trout, *Salvelinus namaycush*, relative to age, growth and diet in Tadenac Lake with comparative data from other Precambrian shield lakes, *Canadian Journal of Fisheries and Aquatic Sciences*, 40, pp. 114–120, 1983.

MacLean, J.A., and D.O. Evans, The stock concept, discreteness of fish stocks, and fisheries management, *Canadian Journal of Fisheries and Aquatic Sciences*, 38, pp. 1889–1898, 1981.

Magnuson, J.J., Managing with exotics — a game of chance, *Transactions of the American Fisheries Society*, 105, pp. 1–9, 1976.

Malecki, R.A., B. Blossey, S.D. Hight, D. Schroeder, L.T. Kok, and J.R. Coulson, Biological control of purple loosestrife, *BioScience*, 43, pp. 680–686, 1993.

Marnell, L.F., Status of the westslope cutthroat trout in Glacier National Park, Montana, in *American Fisheries Society Symposium 4: Status and Management of Interior Stocks of Cutthroat Trout*, American Fisheries Society, Bethesda, MD, pp. 61–70, 1988.

Martin, A.D., and T.G. Northcote, Kootenay Lake: an inappropriate model for *Mysis relicta* introduction in north temperate lakes, in *American Fisheries Society Symposium 9: Mysids in Fisheries: Hard Lessons from Headlong Introduction*, American Fisheries Society, Bethesda, MD, pp. 23–29, 1991.

Martin, N.V., and C.H. Olver, The lake charr, (*Salvelinus namaycush*), in *Charrs, Salmonid Fishes of the Genus Salvelinus*, Balon, E.K., Ed., Dr. W. Junk BV Publishers, The Hague, Netherlands, 1980.

Martinez, P.J., and E.P. Bergersen, Proposed biological management of *Mysis relicta* in Colorado lakes and reservoirs, *North American Journal of Fisheries Management*, 9, pp. 1–11, 1989.

Mathers, R.A., and P.H. Johansen, The effects of feeding ecology on mercury accumulation in walleye, *Stizostedion vitreum*, and pike, *Esox lucius*, in Lake Simcoe, *Canadian Journal of Zoology*, 63, pp. 2006–2012, 1985.

Matuszek, J.E., B.J. Shuter, and J.M. Casselman, Changes in lake trout growth and abundance after the introduction of cisco into Lake Opeongo, Ontario, *Transactions of the American Fisheries Society*, 119, pp. 718–729, 1990.

McCann, J.A., Involvment of the American Fisheries Society with exotic species, 1969-1982, in *Distribution, Biology, and Management of Exotic Fishes*, Courtenay, W.R., Jr., and J.R. Stauffer, Jr., Eds., Johns Hopkins University Press, Baltimore, MD, 1984.

McKnight, S.K., and G.R. Hepp, Potential effect of grass carp herbivory on waterfowl foods, *Journal of Wildlife Management*, 59, pp. 720–727, 1995.

McMahon, T.E., and D.H. Bennett, Walleye and northern pike: boost or bane to northwest fisheries? *Fisheries*, 21(8), pp. 6–13, 1996.

Miller, R.R., J.D. Williams, and J.E. Williams, Extinctions of North American fishes during the past century, *Fisheries*, 14(6), pp. 22–38, 1989.

Mills, E.L., J.H. Leach, J.T. Carlton, and C.L. Secor, Exotic species in the Great Lakes: a history of biotic crises and anthropogenic introductions, *Journal of Great Lakes Research*, 19, pp. 1–54, 1993.

Morizot, D.C., S.W. Calhoun, L.L. Clepper, M.E. Schmidt, J.H. Williamson, and G.J. Carmichael, Multispecies hybridization among native and introduced centrarchid basses in central Texas, *Transactions of the American Fisheries Society*, 120, pp. 283–289, 1991.

Mosher, T.D., Responses of white crappie and black crappie to threadfin shad introductions in a lake containing gizzard shad, *North American Journal of Fisheries Management*, 4, pp. 365–370, 1984.

Moyle, P.B., Effects of introduced bullfrogs, *Rana catesbeiana*, on the native frogs of the San Joaquin Valley, California, *Copeia*, 1973, pp. 18–22, 1973.

Moyle, P.B., Fish introductions in California: history and impact on native fishes, *Biological Conservation*, 9, pp. 101–118, 1976.

Moyle, P.B., H.W. Li, and B.A. Barton, The Frankenstein effect: impact of introduced fishes on native fishes in North America, in *Fish Culture in Fisheries Management*, Stroud, R.H., Ed., American Fisheries Society, Bethesda, MD, 1986.

Ney, J.J., Evolution of forage-fish management in lakes and reservoirs, *Transactions of the American Fisheries Society*, 110, pp. 725–728, 1981.

Ney, J.J., and D.J. Orth, Coping with future shock: matching predator stocking programs to prey abundance, in *Fish Culture in Fisheries Management*, Stroud, R.H., Ed., American Fisheries Society, Bethesda, MD, 1986.

Nico, L.G., and J.D. Williams, Risk assessment on black carp (Pisces: cyprinidae), final draft. Report to the Risk Assessment and Management Committee of the Aquatic Nuisance Species Task Force, Washington, DC, 1996.

Noble, R.L., Management of forage fishes in impoundments of the southern United States, *Transactions of the American Fisheries Society*, 110, pp. 738–750, 1981.

Noble, R.L., Predator-prey interactions in reservoir communities, in *Reservoir Fisheries Management: Strategies for the 80's*, Hall, G.E. and M.J. Van Den Avyle, Eds., American Fisheries Society, Bethesda, MD, 1986.

Norcross, J.J., *The Walleye Fishery of Michigan's Lake Gogebic*. Michigan Department of Natural Resources Fisheries Technical Report 86-9, Ann Arbor, MI, 1986.

North Carolina Sea Grant, *Zebra Mussels And Aquaculture: What You Should Know*. North Carolina Sea Grant, Raleigh, NC, 1995.

Office of Technology Assessment, *Harmful Non-Indigenous Species in the United States*. Office of Technology Assessment-F-565, United States Congress, Office Technology Assessment, United States Government Printing Office, Washington, DC, 1993.

Ogle, D.H., J.H. Selgeby, J.F. Savino, R.M. Newman, and M.G. Henry, Predation on ruffe by native fishes of the St. Louis River estuary, Lake Superior, 1989–1991, *North American Journal of Fisheries Management*. 16, pp. 115–123, 1996.

Opuszynski, K., and J.V. Shireman, *Herbivorous Fishes: Culture and Use for Weed Management*, CRC Press, Boca Raton, FL, 1995.

Philipp, D.P., Genetic implications of introducing Florida largemouth bass, *Micropterus salmoides floridanus*. *Canadian Journal of Fisheries and Aquatic Sciences*, 48 (Suppl. 1), pp. 58–65, 1991.

Radonski, G.C., and R.G. Martin, Fish culture is a tool, not a panacea, in *Fish Culture in Fisheries Management*, Stroud, R.H., Ed., American Fisheries Society, Bethesda, MD, 1986.

Raibley, P.T., D. Bodgett, and R.E. Sparks, Evidence of grass carp (*Ctenopharyngodon idella*) reproduction in the Illinois and upper Mississippi Rivers, *Journal of Freshwater Ecology*, 10, pp. 64–74, 1995.

Regier, H.A., The potential misuse of exotic fish as introductions, in *A Symposium on Introductions of Exotic Species*, Loftus, K.H., Ed., Ontario Department of Lands and Forests, Toronto, Ontario, pp. 92–111, 1968.

Rosenfield, A., and R. Mann, Eds., *Dispersal of Living Organisms into Aquatic Ecosystems*, Maryland Sea Grant College, College Park, MD, 1992.

Ruesink, J.L., I.M. Parker, M.J. Groom, and P.M. Kareiva, Reducing the risks of nonindigenous species introductions, *Bioscience*, 45, pp. 465–75, 1995.

Ryder, R.A., and S.R. Kerr, The adult walleye in the percid community — a niche definition based on feeding behaviour and food specificity, in *Select Coolwater Fishes,* American Fisheries Society Special Publication 11, American Fisheries Society, Bethesda, MD, pp. 39–51, 1978.

Ryder, R.A., and S.R. Kerr, Reducing the risk of fish introductions — a rational approach to the management of integrated cold-water communities, in *Proceedings of the European Inland Fisheries Advisory Committee Symposium, EIFAC Technical Paper 42,* Food and Agriculture Organization of the United Nations, Rome, pp. 510–533, 1984.

Schiavone, A., Response of walleye populations to the introduction of the black crappie in the Indian River Lakes, *New York Fish and Game Journal,* 32(2), pp. 114–140, 1985.

Schneider, J.C., T.J. Lychwick, E.J. Trimberger, J.H. Peterson, R. O'Neal, and P.J. Schneeberger, Walleye rehabilitation in Lake Michigan, 1969–1989, in *Status of Walleye in the Great Lakes: Case Studies Prepared for the 1989 Workshop,* Colby, P.J., C.A. Lewis, and R.L. Eshenroder, Eds., Great Lakes Fishery Commission Special Publication 91-1, Great Lakes Fishery Commission, Ann Arbor, MI, pp. 23–62, 1991.

Schorr, M.S., J. Sah, D.F. Schreiner, M.R. Meador, and L.G. Hill, Regional economic impact of the Lake Texoma (Oklahoma-Texas) striped bass fishery, *Fisheries,* 20(5), pp. 14–18, 1995.

Schramm, H.L., and R.G. Piper, Eds., *American Fisheries Society Symposium 15: Proceedings of the International Symposium and Workshop on the Uses and Effects of Cultured Fishes in Aquatic Ecosystems,* American Fisheries Society, Bethesda, MD, 1995.

Scott, W.B., and E.J. Crossman, *Freshwater Fishes of Canada.* Fisheries Research Board of Canada Bulletin 184, Ottawa, Ontario, 1973.

Shafland, P.L., and J.M. Pestrak, Lower lethal temperature for fourteen nonnative fishes in Florida, *Environmental Biology of Fishes,* 7, pp. 149–156, 1982.

Shelton, W.L., and J.V. Shireman, Exotic fishes in warmwater aquaculture, in *Distribution, Biology, and Management of Exotic Fishes,* Courtenay, W.R., Jr., and J.R. Stauffer, Jr., Eds., Johns Hopkins University Press, Baltimore, MD 1984.

Shireman, J.V., and C.R. Smith, *Synopsis of Biological Data on the Grass Carp Ctenopharyngodon idella (Cuvier and Valenciennes 1844).* FAO Fisheries Synopsis 135, Food and Agriculture Organization of the United Nations, Rome, 1983.

Smith, S.H., Factors in ecological succession in oligotrophic fish communities of the Laurentian Great Lakes, *Journal of the Fisheries Research Board of Canada,* 29, pp. 717–730, 1972.

Smith, B.W., and W.C. Reeves, Stocking warm-water species to restore or enhance fisheries, in *Fish Culture in Fisheries Management,* Stroud, R.H., Ed., American Fisheries Society, Bethesda, MD, 1986.

Spencer, C.N., B.R. McClelland, and J.A. Stanford, Shrimp stocking, salmon collapse, and eagle displacement: cascading interactions in the food web of a large aquatic ecosystem, *Bioscience,* 41, pp. 14–21, 1991.

Sport Fishing Institute, *Economic Impact of Sportfishing in the United States.* Prepared by the Sport Fishing Institute, Washington, DC., 1994.

Stanley, J.G., W.W. Miley, II, and D.L. Sutton, Reproductive requirements and likelihood for naturalization of grass carp in the United States, *Transactions of the American Fisheries Society,* 107, pp. 119–128, 1978.

Stanley, J.G., R.A. Peoples, and J.A. McCann, U.S. Federal policies, legislation, and responsibilities related to importation of exotic fishes and other aquatic organisms, *Canadian Journal of Fisheries and Aquatic Sciences,* 48 (Suppl. 1), pp. 162–166, 1991.

Stauffer, J.R., Jr., S.E. Boltz, and J.M. Boltz, Cold shock susceptibility of blue tilapia from the Susquehanna River, Pennsylvania, *North American Journal of Fisheries Management,* 8, pp. 329–332, 1988.

Stroud, R.H., Ed., *Fish Culture in Fisheries Management,* American Fisheries Society, Bethesda, MD, 1986.

Taylor, J.N., W.R. Courtenay, Jr., and J.A. McCann, Known impacts of exotic fishes in the continental Unites States, in *Distribution, Biology, and Management of Exotic Fishes,* Courtenay, W.R., Jr., and J.R. Stauffer, Jr., Eds., Johns Hopkins University Press, Baltimore, MD, 1984.

Thompson, D.Q., R.L. Stuckey, and E.B. Thompson, Spread, impact, and control of purple loosestrife (*Lythrum salicaria*) in North American wetlands. *United States Department of the Interior, Fish and Wildlife Service, Fish and Wildlife Research 2,* 1987.

Varley, J.D., and R.E. Greswell, Ecology, status, and management of the Yellowstone cutthroat trout, in *American Fisheries Society Symposium 4: Status and Management of Interior Stocks of Cutthroat Trout,* American Fisheries Society, Bethesda, MD, pp. 13–24, 1988.

Wagner, H.W., Jr., Problems with biotic invasives: a biologist's viewpoint, in *Biological Pollution: The Control and Impact of Invasive Exotic Species: Proceedings of a Symposium held at the University Place Conference Center, Indiana University-Purdue University at Indianapolis on October 25 & 26, 1991.* McKnight, B.N., Ed., Indiana Academy of Science, Indianapolis, IN, pp. 1–8, 1993.

Waples, R.S., Genetic interactions between hatchery and wild salmonids: lessons from the Pacific northwest, *Canadian Journal of Fisheries and Aquatic Sciences*, 48 (Suppl. 1), pp. 124–133, 1991.

Webster, J., A. Trandahl, and J. Leonard, Historical perspective of propagation and management of coolwater fishes in the United States, in *Select Coolwater Fishes*, American Fisheries Society Special Publication 11, American Fisheries Society, Bethesda, MD, pp. 161–166, 1978.

Welcomme, R.L, *International Introductions of Inland Aquatic Species.* FAO Fisheries Technical Paper 294, Food and Agriculture Organization of the United Nations, Rome, 1988.

Wells, L., Changes in yellow perch (*Perca flavescens*) populations in Lake Michigan, 1954–75. *Journal of the Fisheries Research Board of Canada,* 34, pp. 1821–1829, 1977.

White, M.M., and S. Schell, An evaluation of the genetic integrity of Ohio River walleye and sauger stocks, in *American Fisheries Society Symposium 15: Proceedings of the International Symposium and Workshop on the Uses and Effects of Cultured Fishes in Aquatic Ecosystems*, American Fisheries Society, Bethesda, MD, pp. 52–60, 1995.

Wiley, R.W., and R.S. Wydoski, Management of undesirable fish species, in *Inland Fisheries Management in North America*, Kohler, C.C., and W.A. Hubert, Eds., American Fisheries Society, Bethesda, MD, 1993.

Williams, J.E., D.W. Sada, C.D. Williams, J.R. Bennett, J.E. Johnson, P.C. Marsh, D.E. McAllister, E.P. Pister, R.D. Radant, J.N. Rinne, M.D. Stone, L. Ulmer, and D.L. Withers, American Fisheries Society guidelines for introductions of threatened and endangered fishes, *Fisheries*, 13(5), pp. 5–11, 1988.

Wingate, P.J.. U.S. State's view and regulations on fish introductions. *Canadian Journal of Fisheries and Aquatic Sciences*, 48 (Suppl. 1), pp. 167–170, 1991.

Workshop on North American Nonindigenous Freshwater Fish, Statement of concern. Prepared by international panel of scientists from Mexico, Canada and the United States at the Gainesville Workshop, Commission for Environmental Cooperation, Montréal, Québec, 1996.

Wydoski, R.S., and D.H. Bennett, Forage species in lakes and reservoirs of the western United States, *Transactions of the American Fisheries Society*, 110, pp. 764–771, 1981.

Zuckerman, L.D., and R.J. Behnke, Introduced fishes in the San Luis Valley, in *Fish Culture in Fisheries Management,* Stroud, R.H., Ed., American Fisheries Society, Bethesda, MD, 1986.

5 Summary of Fishes Intentionally Introduced in North America

E.J. Crossman and B.C. Cudmore

A term often used for this category is "stocking," and for simplicity it can be defined as introductions of fishes that can be to said to have been "the will of the people," carried out by duly appointed government agencies, and in authorized locations. There are, however, known cases of introductions by authorized personnel of authorized species into unauthorized locations. The majority of cases of stocking are considered to have been of species and locations authorized at the time. Appropriate consideration today would suggest many of these were ecologically unwise. This would include nonindigenous fishes introduced in the past into government parks. It has been suggested that smallmouth bass, *Micropterus dolomieu*, was introduced into trout habitat in such a park in Ontario because the indigenous salmonid species were thought to be too difficult for tourists to catch in summer (see detailed account by Dextrase and Coscarelli, Chapter 4 of this volume).

In this treatment we consider that 214 forms (Table 5.1) have been introduced by this vector. The present status in regard to natural reproduction (established) of each form has been provided in the table. The reasons for these introductions, largely as recorded in the NAS database, are also provided (Table 5.1). It is not surprising that the largest number of introductions (91) were for sport, or recreation (Figure 5.1). Some of the earliest introductions into North America (e.g., brown trout, *Salmo trutta*) were for recreational purposes. Some of those for which the reason was listed as unknown may be in this category as well. The next highest number for which a reason was recorded is forage, and it would seem likely that these were mostly for food for recreational species, and possibly for commercial species. It is highly likely that, in many cases, the introduced forage was to support another introduced sportfish (e.g., rainbow smelt, *Osmerus mordax*, for Atlantic salmon, *Salmo salar*, in Michigan). Some of the 23 cases said to have been to create additional populations could have been recreational species, but they also included attempts to maintain populations of endangered species (e.g., pupfish, *Cyprinodon* spp.). It is interesting to note that only 24 forms were said to have been intentionally introduced for food, but this would include the common carp, *Cyprinus carpio*, one of the species considered now to have had a very negative impact on the aquatic ecosystems of North America. It is likely that some or all of those for which aquaculture (5) was listed as the reason were also for food or for biocontrol and would be additional to the fishes so listed. The biocontrol category includes the interesting attempt to introduce Alaska blackfish, *Dallia pectoralis*, for mosquito control in Ontario. The much smaller catch-all category — other (2 species) — includes redeye piranha, *Serrasalmus rhombeus*, introduced as a tourist attraction.

Looking at the recorded reasons for authorized introductions of fishes reveals that 71% of them could be said to be, in some way, anthropocentric. Only 28%, with possibly some contribution from those listed as unknown, could be said to benefit the ecosystem or biodiversity.

The term "accidental" (or "stock contaminant") is used in the NAS database; it includes 48 fishes (Table 5.2). That is a relatively high level of incidence compared with those introduced in ballast water. The reasons given in the database for these introductions indicate that over half of

TABLE 5.1
List of Fishes Introduced in Ecological North America (Excluding Mexico) Outside Their Native Range through Intentional Stocking Including Reason(s) for the Stocking

Scientific Name	Common Name	Reason(s)	Status[a]
Acipenseridae — sturgeons			
Acipenser transmontanus	white sturgeon	unknown	little to no reproduction
Amiidae — bowfins			
Amia calva	bowfin	unknown	established in most areas
Hiodontidae — mooneyes			
Hiodon alosoides	goldeye	forage	established in some areas
Megalopidae — tarpons			
Megalops atlanticus	tarpon	unknown	not established
Anguillidae — freshwater eels			
Anguilla anguilla	European eel	aquaculture	not established
Anguilla rostrata	American eel	aquaculture	established in most areas
Clupeidae — herrings			
Alosa aestivalis	blueback herring	forage	established
Alosa pseudoharengus	alewife	unknown	established in most areas
Alosa sapidissima	American shad	forage, food, sport, commercial	established in most areas
Dorosoma cepedianum	gizzard shad	forage	established in most areas
Dorosoma petenense	threadfin shad	forage	established in some areas
Chanidae — milkfishes			
Chanos chanos	milkfish	unknown	not established
Cyprinidae — carps and minnows			
Agosia chrysogaster	longfin dace	unknown	unknown
Campostoma ornatum	Mexican stoneroller	unknown	unknown
Carassius auratus	goldfish	unknown	established
Carassius carassius	crucian goldfish	unknown	possibly established in some areas
Couesius plumbeus	lake chub	forage	established in many areas
Ctenopharyngodon idella	grass carp	biocontrol, research	established in many areas
Ctenopharyngodon idella X *Hypophthalmichthys nobilis*		unknown	not established
Cyprinella spiloptera	spotfin shiner	forage	established in some areas
Cyprinella lutrensis	red shiner	forage	established in many areas
Cyprinus carpio	common carp	food	established in most areas
Gila atraria	Utah chub	unknown	established in many areas
Gila bicolor	Tui chub	create additional population of a rare species	established in some areas
Gila ditaenia	Sonora chub	unknown	unknown
Gila purpurea	Yaqui chub	create addtional population of a species at risk	established
Hypophthalmichthys molitrix	silver carp	biocontrol, food	established in some areas
Hypophthalmichthys molitrix X *H. nobilis*		food, biocontrol	not established

TABLE 5.1 (CONTINUED)
List of Fishes Introduced in Ecological North America (Excluding Mexico) Outside Their Native Range through Intentional Stocking Including Reason(s) for the Stocking

Scientific Name	Common Name	Reason(s)	Status[a]
Hypophthalmichthys nobilis	bighead carp	biocontrol	established in some areas
Lepidomeda albivallis	White River spinedace	unknown	not established
Lepidomeda mollispinis	Virgin spinedace	create additional population of a threatened species	established in some areas
Leuciscus idus	ide	unknown	established in some areas
Medo fulgida	spike dace	create additional population of a threatened species	not established
Moapa coriacea	Moapa dace	create additional population of an endangered species	not established
Nocomis bigguttatus	hornyhead chub	unknown	established
Notemigonus crysoleucas	golden shiner	forage	established in some areas
Notropis atherinoides	emerald shiner	forage	established
Notropis formosus	Yaqui shiner	unknown	unknown
Notropis hudsonius	spottail shiner	forage	established in many areas
Notropis stramineus	sand shiner	unknown	established
Orthodon microlepidotus	Sacramento blackfish	unknown	established in many areas
Phoxinus eos	northern redbelly dace	unknown	not established
Pimephales notatus	bluntnose minnow	forage	established
Pimephales promelas	fathead minnow	forage, biocontrol	established
Plagopterus argentissimus	woundfin	create additional population of an endangered species	not established
Platygobio gracilis	flathead chub	forage	not established
Ptychocheilus grandis	Sacramento squawfish	unknown	unknown
Rhinichthys osculus	speckled dace	unknown	established in many areas
Richardsonius balteatus	redside shiner	forage	established
Richardsonius egregius	Lahontan redside	unknown	established in some areas
Scardinius erythrophthalmus	rudd	forage	established in most areas
Tinca tinca	tench	food, sport	established in most areas

Catostomidae — suckers

Catostomus bernardi	Yaqui sucker	unknown	unknown
Catostomus commersoni	white sucker	forage	established
Catostomus fumeiventris	Owens sucker	transplanted as precautionary measure for a population at risk	established
Catostomus platyrhynchus	mountain sucker	unknown	established
Erimyzon sucetta	lake chubsucker	forage	established
Ictiobus bubalus	smallmouth buffalo	sport	established in some areas
Ictiobus cyprinellus	bigmouth buffalo	sport	established in some areas
Ictiobus niger	black buffalo	sport	established in some areas

Cobitidae — loaches

Misgurnus anguillicaudatus	oriental weatherfish	aquaculture	established

TABLE 5.1 (CONTINUED)
List of Fishes Introduced in Ecological North America (Excluding Mexico) Outside Their Native Range through Intentional Stocking Including Reason(s) for the Stocking

Scientific Name	Common Name	Reason(s)	Status[a]
Characidae — characins			
Astyanax mexicanus	banded tetra	unknown	unknown
Serrasalmus rhombeus	redeye piranha	tourist attraction	not established
Ictaluridae — bullhead catfishes			
Ameiurus catus	white catfish	sport, food	established in most areas
Ameiurus melas	black bullhead	sport, food	established in most areas
Ameiurus natalis	yellow bullhead	unknown	established in most areas
Ameiurus nebulosus	brown bullhead	food, sport	established in most areas
Ameiurus platycephalus	flat bullhead	unknown	established in most areas
Ictalurus furcatus	blue catfish	food, sport	established
Ictalurus pricei	Yaqui catfish	unknown	not established
Ictalurus punctatus	channel catfish	sport, food	established in most areas
Pylodictis olivaris	flathead catfish	unknown	established in most areas
Clariidae — labyrinth catfishes			
Clarias batrachus	walking catfish	unknown	established in some areas
Esocidae — pikes			
Esox americanus americanus	redfin pickerel	unknown	established in some areas
Esox americanus vermiculatus	grass pickerel	sport	established in some areas
Esox lucius	northern pike	sport	established in most areas
Esox lucius X *E. masquinongy*	tiger muskellunge	sport	maintained
Esox lucius X *E. reicherti*		sport	not established
Esox masquinongy	muskellunge	sport	established in some areas
Esox niger	chain pickerel	unknown	established in most areas
Esox reicherti	Amur pike	sport	established in some areas
Umbridae — mudminnows			
Dallia pectoralis	Alaska blackfish	biocontrol	established
Osmeridae — smelts			
Hypomesus nipponensis	wakasagi	forage	established in most areas
Osmerus mordax	rainbow smelt	forage	established
Salmonidae — trouts			
Coregonus artedi	lake herring	food, forage	established in most areas
Coregonus clupeaformis	lake whitefish	food	established in some areas
Hucho hucho	huchen	sport	not established
Oncorhynchus aguabonita	golden trout	sport	established in most areas
Oncorhynchus apache	Apache trout	create additional population of a species at risk	unknown
Oncorhynchus chrysogaster	Mexican golden trout	unknown	not established
Oncorhynchus clarki	cutthroat trout	sport	established in many areas
Oncorhynchus clarki bouvieri	Yellowstone cutthroat trout	sport	established
Oncorhynchus clarki lewisi X *O. mykiss*		sport	not established

TABLE 5.1 (CONTINUED)
List of Fishes Introduced in Ecological North America (Excluding Mexico) Outside Their Native Range through Intentional Stocking Including Reason(s) for the Stocking

Scientific Name	Common Name	Reason(s)	Status[a]
Oncorhynchus clarki X *O. mykiss*	cutbow trout	sport	maintained
Oncorhynchus gilae	Gila trout	create additional population of an endangered species	established
Oncorhynchus gorbuscha	pink salmon	biocontrol, sport, commercial	established
Oncorhynchus keta	chum salmon	sport, biocontrol, commercial	not established
Oncorhynchus kisutch	coho salmon	sport, biocontrol, commercial, system rehabilitation	not established
Oncorhynchus masou	cherry salmon	experimental	unknown
Oncorhynchus mykiss	rainbow trout	sport, system rehabilitation	established in some areas
Oncorhynchus mykiss X *Hucho hucho*		sport	not established
Oncorhynchus nerka	kokanee salmon	sport, forage	established in some areas
Oncorhynchus tshawytscha	chinook salmon	sport, biocontrol, system rehabilitation	not established
Prosopium gemmifer	Bonneville cisco	unknown	not established
Salmo letnica	Ohrid trout	unknown	not established
Salmo salar	Atlantic salmon	sport, species rehabilitation	not established
Salmo salar X *S. trutta*	sambrown	sport	maintained
Salmo trutta	brown trout	sport	some natural reproduction
Salmo trutta stigmosa	brown trout	sport	not established
Salmo trutta X *Salvelinus fontinalis*	tiger trout	sport	maintained
Salvelinus alpinus	Arctic char	sport	established in some areas
Salvelinus alpinus marstoni	Quebec red trout	sport	established
Salvelinus alpinus X *S. fontinalis*		sport	not established
Salvelinus fontinalis	brook trout	sport	established in most areas
Salvelinus fontinalis timagamiensis	Aurora trout	sport	established
Salvelinus fontinalis X *S. malma* (or *S. confluentis*)		sport	not established
Salvelinus fontinalis X *S. namaycush*	splake	sport, system rehabilitation	maintained
Salvelinus fontinalis X *S.namaycush* X *S. namaycush*	splake backcross	sport, commercial	not established
Salvelinus malma	Dolly Varden	sport	established in some areas
Salvelinus namaycush	lake trout	sport, commercial, species rehabilitation	established
Thymallus arcticus	Arctic grayling	sport	established in most areas
Percopsidae — trout-perches			
Percopsis omiscomaycus	trout-perch	forage	established in some areas
Amblyopsidae — cavefishes			
Chologaster agassizi	spring cavefish	to established population near research facilities	not established

TABLE 5.1 (CONTINUED)
List of Fishes Introduced in Ecological North America (Excluding Mexico) Outside Their Native Range through Intentional Stocking Including Reason(s) for the Stocking

Scientific Name	Common Name	Reason(s)	Status[a]
Gadidae — cods			
Lota lota	burbot	sport	established in some areas
Atherinidae — silversides			
Chirostoma jordani	charal	food	not established
Labidesthes sicculus	brook silverside	forage	established in some areas
Membras martinica	rough silverside	forage	possibly established
Menidia beryllina	inland silverside	forage, biocontrol	established
Aplocheilidae — rivulines			
Cynolebias bellotti	Argentine pearlfish	biocontrol	not established
Cynolebias nigrippinis	blackfin pearlfish	biocontrol	not established
Cynolebias whitei	Rio pearlfish	biocontrol	not established
Rivulus harti	giant rivulus	unknown	established in some areas
Goodeidae — splitfins			
Crenichthys nevadae	Railroad Valley springfish	create additional population of a rare species	established
Empetrichthys latos	Pahrump killifish	create additional population of an endangered species	established in some areas
Poeciliidae — livebearers			
Gambusia affinis	western mosquitofish	biocontrol	established in some areas
Gambusia gaigei	Big Bend gambusia	create additional population of a species at risk	unknown
Gambusia holbrooki	eastern mosquitofish	biocontrol	established
Gambusia nobilis	Pecos gambusia	create additional population of an endangered species	established in some areas
Poecilia reticulata	guppy	unknown	established in most areas
Poeciliopsis occidentalis	Gila topminnow	create additional population of an endangered species	established in some areas
Cyprinodontidae — pupfishes			
Cyprinodon bovinus	Leon Springs pupfish	create additional population of a species at risk	unknown
Cyprinodon diabolis	Devils Hole pupfish	create additional population of an endangered species	not established
Cyprinodon elegans	Comanche Springs pupfish	create additional population of a species at risk	unknown
Cyprinodon macularius	desert pupfish	create additional population of an endangered species	established
Cyprinodon nevadensis	Amargosa pupfish	create additional population of an endangered species	established in some areas
Cyprinodon radiosus	Owens pupfish	create additional population of an endangered species	established

TABLE 5.1 (CONTINUED)
List of Fishes Introduced in Ecological North America (Excluding Mexico) Outside Their Native Range through Intentional Stocking Including Reason(s) for the Stocking

Scientific Name	Common Name	Reason(s)	Status[a]
Gasterosteidae — sticklebacks			
Culaea inconstans	brook stickleback	unknown	established in some areas
Gasterosteus aculeatus	threespine stickleback	unknown	established in most areas
Pungitius pungitius	ninespine stickleback	forage	established in some areas
Cottidae — sculpins			
Cottus bairdi	mottled sculpin	unknown	established in some areas
Centropomidae — snooks			
Lates angustifrons	Tanganyika lates	sport	not established
Lates mariae	bigeye lates	sport	not established
Lates niloticus	Nile perch	sport	not established
Moronidae — temperate basses			
Morone americana	white perch	sport	established in most areas
Morone americana X *M. saxatilis*	Virginia bass	sport	maintained
Morone chrysops	white bass	sport	established in most areas
Morone chrysops X *M. mississippiensis*		sport	maintained
Morone chrysops X *M. saxatilis*	whiper	sport	maintained
Morone mississippiensis	yellow bass	sport	established in some areas, maintained
Morone saxatilis	striped bass	sport	established, maintained
Centrarchidae — sunfishes			
Ambloplites cavifrons	Roanoke bass	sport	established in some areas
Ambloplites rupestris	rock bass	sport	established
Archoplites interruptus	Sacramento perch	sport	established
Centrarchus macropterus	flier	unknown	not established
Chaenobryttus gulosus	warmouth	sport, food	established in most areas
Lepomis auritus	redbreast sunfish	sport, forage, food	established
Lepomis cyanellus	green sunfish	unknown	established in most areas
Lepomis cyanellus X *L. macrochirus*		sport	established in some areas
Lepomis gibbosus	pumpkinseed	sport, forage	established in most areas
Lepomis macrochirus	bluegill	sport, forage, food	established in most areas
Lepomis megalotis	longear sunfish	forage, food	established
Lepomis microlophus	redear sunfish	sport	established in most areas
Lepomis miniatus	redspotted sunfish	sport	established
Lepomis punctatus	spotted sunfish	food, forage	unknown
Micropterus coosae	redeye bass	sport	established
Micropterus dolomieu	smallmouth bass	sport, food	established in most areas
Micropterus dolomieu X *M. treculi*		sport	not established

TABLE 5.1 (CONTINUED)
List of Fishes Introduced in Ecological North America (Excluding Mexico) Outside Their Native Range through Intentional Stocking Including Reason(s) for the Stocking

Scientific Name	Common Name	Reason(s)	Status[a]
Micropterus dolomieu X *M. punctulatus*		sport	not established
Micropterus punctulatus	spotted bass	sport	established in most areas
Micropterus salmoides	largemouth bass	sport, food	established in most areas
Micropterus sp. (undescribed)	shoal bass	sport	established
Micropterus treculi	Guadalupe bass	sport	established
Pomoxis annularis	white crappie	sport, food	established in most areas
Pomoxis annularis X *P. nigromaculatus*		sport	not established
Pomoxis nigromaculatus	black crappie	sport	established in some areas
Percidae — perches			
Etheostoma nuchale	watercress darter	create additional population of an endangered species	established in some areas
Perca flavescens	yellow perch	sport, forage	established in most areas
Percina tanasi	snail darter	create additional population of an endangered species	established
Stizostedion canadense	sauger	sport	established in most areas
Stizostedion canadense X *S. vitreum*	saugeye	sport	maintained
Stizostedion lucioperca	zander	research suitability for stocking as a sport fish	not established
Stizostedion vitreum	walleye	sport, commercial	established in most areas
Sciaenidae — drums			
Aplodinotus grunniens	freshwater drum	sport	established
Cynoscion nebulosus	spotted sea trout	sport	not established
Cynoscion nebulosus X *C. xanthulus*		sport	not established
Leiostomus xanthurus	spot	sport	not established
Micropogonias undulatus	Atlantic croaker	sport	not established
Pogonias cromis	black drum	unknown	not established
Pogonias cromis X *Sciaenops ocellatus*		sport	not established
Sciaenops ocellatus	red drum	sport	not established
Cichlidae — cichlids			
Astronotus ocellatus	oscar	sport	established in some areas
Cichla ocellaris	peacock cichlid	sport, biocontrol	briefly established in some areas
Cichla temensis	speckled pavon	sport	not established
Cichlasoma nigrofasciatum	convict cichlid	unkown	established in some areas
Oreochromis aureus	blue tilapia	sport, forage, food, biocontrol	established in some areas
Oreochromis mossambica	Mozambique tilapia	sport, food, commercial, biocontrol	established in many areas
Oreochromis niloticus	Nile tilapia	aquaculture	established in some areas
Saratherodon melanotheron	blackchin tilapia	commercial	established
Tilapia urolepis	Wami tilapia	biocontrol	established

TABLE 5.1 (CONTINUED)
List of Fishes Introduced in Ecological North America (Excluding Mexico) Outside Their Native Range through Intentional Stocking Including Reason(s) for the Stocking

Scientific Name	Common Name	Reason(s)	Status[a]
Tilapia zilli	redbelly tilapia	biocontrol, forage, food, aquaculture	established in some areas
Labridae — wrasses			
Tautoga onitis	tautog	unknown	not established
Gobiidae — gobies			
Gillichthys mirabilis	longjaw mudsucker	bait	established
Bothidae — lefteye flounders			
Paralichthys albiguttata	gulf flounder	sport	not established
Paralichthys lethostigma	southern flounder	sport	not established

[a] Status refers to the reproductive nature of "populations" of that fish in areas where it has been introduced, regardless of vector(s) used. Many statements on status were derived from information in the NAS database. Some notations may be open to criticism (e.g., hybrids).

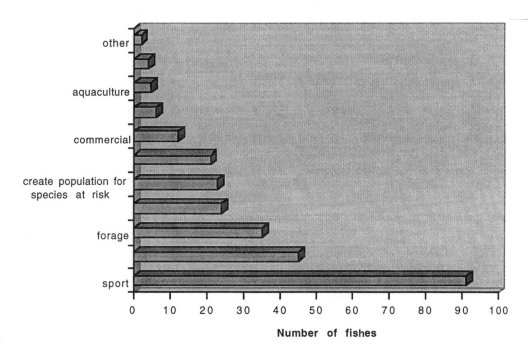

FIGURE 5.1 Reasons for intentional stocking of fishes in ecological North America (excluding Mexico) outside their native range.

them resulted from contamination of stocks of some other species being liberated in particular locations where the contaminating species was not indigenous, or entered such locations as a result of transportation problems. The reason for the largest number of the remainder was unknown. An

TABLE 5.2

Fishes Introduced in Ecological North America (Excluding Mexico) Outside their Native Range through Accidental Introduction Including Cause(s) and their Reproductive Status in the Nonindigenous Area(s)

Scientific Name	Common Name	Cause(s)	Status[a]
Anguillidae — freshwater eels			
Anguilla rostrata	American eel	accident during transport	established in most areas
Clupeidae — herrings			
Alosa sapidissima	American shad	lost during transport	established in most areas
Cyprinidae — carps and minnows			
Carassius auratus	goldfish	stock contaminant	established
Cyprinella spiloptera	spotfin shiner	stock contaminant	established in some areas
Gila orcutti	arroyo chub	stock contaminant	established
Hypophthalmichthys molitrix	silver carp	stock contaminant	established in some areas
Hypophthalmichthys nobilis	bighead carp	stock contaminant	established in some areas
Notropis atherinoides	emerald shiner	unknown	established
Notropis hudsonius	spottail shiner	stock contaminant	established in many areas
Notropis rubricroceus	saffron shiner	stock contaminant	established
Notropis stramineus	sand shiner	unknown	established
Pimephales notatus	bluntnose minnow	stock contaminant	established
Pimephales vigilax	bullhead minnow	stock contaminant	established in some areas
Richardsonius balteatus	redside shiner	unknown	established
Semotilus atromaculatus	creek chub	unknown	established in most areas
Catostomidae — suckers			
Carpiodes carpio	river carpsucker	stock contaminant	established
Catostomus catostomus	longnose sucker	stock contaminant	established
Catostomus commersoni	white sucker	stock contaminant	established
Ictaluridae — bullhead catfishes			
Ameiurus brunneus	snail bullhead	stock contaminant	established
Noturus flavus	stonecat	stock contaminant	not established
Noturus gyrinus	tadpole madtom	stock contaminant	established in most areas
Pylodictus olivaris	flathead catfish	stock contaminant	established in most areas
Esocidae — pikes			
Esox americanus vermiculatus	grass pickerel	misidentification, stock contaminant	established in some areas
Umbridae — mudminnows			
Dallia pectoralis	Alaska blackfish	unknown	established
Salmonidae — trouts			
Oncorhynchus clarki	cutthroat trout	unknown	established in some areas
Oncorhynchus mykiss	rainbow trout	unknown	some natural reproduction
Salmo salar	Atlantic salmon	unknown	not established
Salvelinus alpinus	Arctic char	unknown	established in some areas
Atherinidae — silversides			
Labidesthes sicculus	brook silverside	unknown	established in some areas
Fundulidae — topminnows and killifishes			
Fundulus diaphanus	banded killifish	stock contaminant	established in some areas

TABLE 5.2 (CONTINUED)
Fishes Introduced in Ecological North America (Excluding Mexico) Outside their Native Range through Accidental Introduction Including Cause(s) and their Reproductive Status in the Nonindigenous Area(s)

Scientific Name	Common Name	Cause(s)	Status[a]
Fundulus sciadicus	plains topminnow	unknown	not established
Fundulus zebrinus	plains killifish	stock contaminant	established in some areas
Lucania parva	rainwater killifish	stock contaminant	established in most areas
Gasterosteidae — sticklebacks			
Culaea inconstans	brook stickleback	stock contaminant	established in some areas
Gasterosteus aculeatus	threespine stickleback	unknown	established in most areas
Cottidae — sculpins			
Cottus bairdi	mottled sculpin	unknown	established in some areas
Centrarchidae — sunfishes			
Centrarchus macropterus	flier	stock contaminant	not established
Lepomis cyanellus	green sunfish	stock contaminant	established in most areas
Lepomis cyanellus X *L. macrochirus*		result from stocking one or both parent species outside their native range(s)	established in some areas
Lepomis gibbosus X *L. humilis*		result from stocking one or both parent species outside their native range(s)	established
Lepomis humilis	orangespotted sunfish	stock contaminant	established
Lepomis macrochirus	bluefill	unknown	established in most areas
Lepomis megalotis	longear sunfish	stock contaminant	established
Percidae — perches			
Etheostoma blennioides	greenside darter	unknown	established
Percina caprodes	logperch	stock contaminant	established
Percina macrolepida	bigscale logperch	stock contaminant	established
Gobiidae — gobies			
Acanthogobius flavimanus	yellowfin goby	eggs on imported oysters	established
Gobiosoma bosc	naked goby	stock contaminant	not established

[a] Status refers to the reproductive nature of "populations" of that fish in areas where it has been introduced, regardless of vector(s) used. Many statements on status were derived from information in the NAS database. Some notations may be open to criticism (e.g., hybrids).

example was eggs of the yellowfin goby, *Acanthogobius flavimanus*, which arrived in a shipment of imported oysters.

The recent appearance of the greenside darter, *Etheostoma blennioides*, in the lower Grand River, a tributary to the eastern end of Lake Erie, was suggested to have been a stock contaminant. That species was not known to occur in that section of the river but was abundant in the Thames River system tributary to Lake St. Clair. There had been an authorized transfer of walleyes, *Stizostedion vitreum*, from the Thames River to the Grand River. However, the darter was probably not a contaminant in that transfer, as the nets used to harvest the walleyes would not have captured the darter, and the darter is known upstream in the Grand River. In contrast, the Thames River walleye themselves might be considered an introduced form in this action, since the genetic makeup of that population differs significantly from the existing population in eastern Lake Erie.

The following are major sources for the information on fishes included here and in subsequent chapters by Crossman and Cudmore in this publication;

The Nonindigenous Aquatic Species Database (NAS), US Department of the Interior, US Geological Survey, Biological Resources Division, Florida–Caribbean Science Center, Bailey and Smith (1981), Crossman (1984,1991), Courtenay (1979, 1991, 1992, 1995), Courtenay and Robbins (1973), Courtenay and Taylor (1984), Courtenay and Williams (1992), Courtenay et al. (1984,1986, 1991), Emery (1985), Hocutt and Wiley (1986), Lachner et al. (1970), Mills et al.(1991, 1993a,b, 1994), records of the Royal Ontario Museum (ROM), and individuals working in the field who collect, or come in contact with unusual species. The Nonindigenous Aquatic Species database is a very large list (13,500 records of 454 "species", September 1995) for the United States and is, necessarily, a major source of information on fishes when the area to be treated is North America. As the words North America in the tile of this book are intended to apply to an ecological area and not a political one, we have included Alaska but not Hawaii or other US territories off the continent. Also since there is a separate treatment of Mexico (Salvadore Contreras-Balderas) in this publication, we have not included introductions into that country in our lists.

REFERENCES

Bailey, R.M., and G.R. Smith, Origin and Geography of the Fish Fauna of the Laurentian Great Lakes, *Canadian Journal of Fisheries and Aquatic Sciences*, 38, pp. 1539–1561, 1981.

Courtenay, W.R., Jr., The Introduction of Exotic Organisms, in *Wildlife in America*, Brokaw, H.P., Ed., US Government Printing Office, Washington, D.C., 1979.

Courtenay, W.R., Jr., Fish Conservation and the Enigma of Introduced Species, in *Introduced and Translocated Fishes and their Ecological Effects*, Pollard, D.A., Ed., Australian Bureau of Natural Resources, Canberra, 1990.

Courtenay, W.R., Jr., Biological Pollution Through Fish Introductions, in *Biological Pollution: The Control and Impact of Invasive Exotic Species*, Indiana Academy of Sciences, IN, 1991.

Courtenay, W.R., Jr., The Case for Caution with Fish Introduction, in Uses and Effects of Cultured Fishes in Aquatic Ecosystems, Schramm, H.C., Jr. and R.G.Piper, Eds., American Fisheries Society Symposium 15, Bethseda, pp. 413–425, 1995.

Courtenay, W.R., Jr., and C.R. Robins, Exotic Aquatic Organisms in Florida with Emphasis on Fishes: A Review and Recommendations, *Transactions of the American Fisheries Society*, 102, pp. 1–12, 1973.

Courtenay, W.R., Jr., and J.N. Taylor, The exotic ichthyofauna of the continental United States with preliminary observations on the intra-national transplants. EIFAC Technical Paper 42: 466–487, 1984.

Courtenay, W.R., Jr., and J.D. Williams, Dispersal of Exotic Species from Aquaculture Sources, with Emphasis on Freshwater Fishes, in *Dispersal of Living Organisms into Aquatic Ecosystems*, Rosenfield A. and R. Mann, Eds., 1992.

Courtenay, W.R., Jr., D.A. Hensley, J.N. Taylor, and J.A. McCann, Distribution of Exotic Fishes in the Continental United States, in *Distribution, Biology and Management of Exotic Fishes*, Courtenay, W.R., Jr., and J.R. Stauffer, Jr., Eds., Johns Hopkins University Press, Baltimore, 1984.

Courtenay, W.R., Jr., D.A. Hensley, J.N. Taylor, and J.A. McCann, Distribution of Exotic Fishes in North America, in *The Zoogeography of North American Freshwater Fishes*, Hocutt, C.H., and E.O. Wiley, Eds., John Wiley and Sons, New York, 1986.

Courtenay, W.R., Jr., D.P. Jennings, and J.D. Williams, Exotic Fishes, in *American Fisheries Society, Special Publication No. 20*, Robins,C.R., R.M Bailey, C.E. Bond, J.R. Brooker, E.A. Lachner, R.N. Lea, and W.B. Scott, Eds., American Fisheries Society, Bethesda, 1991.

Crossman, E.J., Introduction of Exotic Fishes into Canada, in *Distribution Biology and Management of Exotic Fishes*, Courtenay, W.R., Jr., and J.R. Stauffer, Jr., Eds., Johns Hopkins University Press, Baltimore, 1984.

Crossman, E.J., Introduced Freshwater Fishes: A Review of the North American Perspective with Emphasis on Canada, *Canadian Journal of Fisheries and Aquatic Sciences*, 48, Supplement 1, pp. 46–57, 1991.

Department of the Interior, United States Geological Survey, Florida–Caribbean Science Center(1997).*Nonin-digenous Aquatic Species Database*, http://www.nfrcg.gov/nas/nas.html).

Emery, L., Review of Fish Species Introduced into the Great Lakes, 1819–1974. *Great Lakes Fishery Commission*, Technical Report 45, Great Lakes Fishery Commission, Ann Arbor, 31 pp., 1985.

Hocutt, C.H., and E.O. Wiley, Eds., *The Zoogeography of North American Freshwater Fishes*, John Wiley and Sons, New York, 866 pp., 1986.

Lachner, E.A., C.R. Robins, and W.R. Courtenay, Jr., *Exotic Fishes and Other Aquatic Organisms Introduced into North America*, Smithsonian Contributions to Zoology, No. 59, 29 pp, 1970.

Mills, E.L., J.H. Leach, J.T. Carlton, and C.L. Secor, *Exotic Species in the Great Lakes: A History of Biotic Crises and Anthropogenic Introductions*. Great Lakes Fishery Commission, Research Completion Report, 117 pp., 1991.

Mills, E.L., J.H. Leach, J.T. Carlton, and C.L. Secor, What Next? The Predation and Management of Exotic Species in the Great Lakes. Great Lakes Fishery Commission, Report of the 1991 Workshop, 22 pp., 1993a.

Mills, E.L., J.H. Leach, C.L. Secor, and J.T. Carlton, Exotic Species in the Great Lakes: A History of Biotic Crises and Anthropogenic Introductions, *Journal of Great Lakes Research*, 19, pp. 1–54, 1993b.

Mills, E.L., J.H. Leach, J.T. Carlton, and C.L. Secor, Exotic Species and the Integrity of the Great Lakes, *Bioscience*, 44, 666–676, 1994.

6 Impact of Introduced Fish for Aquaculture in Mexican Freshwater Systems

Luis Zambrano and Constantino Macías-García

INTRODUCTION

Intentional introductions of fish into Mexican freshwater systems have been rather common through the years and have increased in the last few decades. However, knowledge of these activities is scant and dispersed because most of the introductions are the result of decisions of local people or mid-rank government officers whose activities normally are unrecorded.

Some attempts have been made to document the distribution of introduced fish and their effects on native species (i.e., Arredondo-Figueroa, 1983; Contreras and Escalante, 1984). However, the task of accurately describing fish populations, let alone the impact of nonnatives on them, has proved a difficult challenge. Within the country, there are only 69 sizable natural lakes, and 12 reservoirs with capacity higher than 4×10^9 m^3. However, these systems, plus 14,000 small lakes and reservoirs, are distributed in a complex topography under a patchwork of soils and climatic conditions. This has produced a poor biomass fish fauna, but one that is largely madeup of endemic species. The number of native species described is approximately 500 (Diaz-Pardo and Guerra-Magaña, 1996). The scarcity of studies on Mexican fish distribution and biology makes it difficult to draw general conclusions on the effects of exotic species on aquatic communities.

Current estimates suggest that native Mexican fish species are disappearing at an alarming rate. In a National Freshwater Fish List (Espinoza et al., 1993), more than 31% of the species in México are considered at risk, in danger, or already extinct. Two factors are generally believed to have contributed to the decline of Mexican native fish: the rapid deterioration of the habitat through pollution and desiccation, and invasion of nonnative species. Because the first factor does not show signs of abating (freshwater use for human consumption is increasing), and the second is largely unrecognized and poorly studied, the future of native species looks grim. There have been some attempts to solve pollution and desiccation in freshwater systems, but species introductions in rivers and lakes are not seen as a problem, and therefore there have been very few opportunities to study them or even to understand them.

In this chapter we first review the reasons for introducing nonnative fish in México, and then we provide evidence on some of the known effects that such introductions have on freshwater ecosystems, focusing on the effects of introductions of fish for human consumption.

NONNATIVE SPECIES

We define as nonnative an organism that is introduced into a system where it did not evolve (i.e., the introduced species never inhabited the system). This new organism can be from the same country and even from the same biogeographic area. We will refer to the introduction of species from the same country to a different ecosystem as translocation.

Because several sectors of society deal with freshwater ecosystems, often with different or even conflicting aims, introductions of nonnative fish can arise from a variety of reasons, although we can distinguish three main forms of introduction:

1. Regional translocations. This type of introduction is not usually documented, because most are conducted by local people. They usually involve nearby lakes or rivers and are the result of requirements of small groups or even individuals. These translocations are usually spontaneous and improvised and usually remain unrecorded. As several of these translocations took place before even basic knowledge of fish faunas had been gathered, it is impossible, without the use of genetic techniques, to determine whether a particular species is a native component of a certain community or not. This situation baffles any attempt to put precise figures to such translocations in México, and there may be a high proportion of cryptogenic species (of unknown origin; Carlton, 1996) in the country.

2. Pet trade. Introductions of ornamental fish occur in three ways:

    1. Aquarium owners and pondkeepers sometimes release their fish once they are too large or too numerous to be maintained. Because few fish are released in any one instance, and since they may be placed in the wrong habitat, most fish freed in this way will die. However, because large numbers of people do practice such releases, effective habitat colonization by some species has resulted.

    2. Purposeful introductions of "food fish" (species used as food for large pet cichlids and other pets such as turtles and garter snakes) are illegally practiced, and may account for the spread of a few poeciliid species in the Mexican High Plateau (including México City, where they compete with endangered native species). By their very nature, these translocations go unregistered.

    3. Ornamental fish are also produced outdoors in a few areas within México, notably in the states of Morelos and Hidalgo; fry and juveniles often escape fish farms and invade the surrounding watershed. This type of culture normally originates when other forms of profit from freshwater resources disappear, or when new freshwater habitats are created (i.e., ponds and irrigation channels). In Hidalgo (Central México, Figure 6.1), an ornamental goldfish (*Carassius auratus*) farm was established when commercial exploitation of worms (*Tubifex sp.*), which need poor-quality water to grow, ceased following the construction of a water treatment plant upstream. Even though the ornamental fish production in México is rather limited, it has already had some impact on native fish faunas. Some aquarium species have reputedly caused decreases in native fish populations. In the Amacuzac basin (State of Morelos), the ornamental *Cichlasoma nigrofaciatum* escaped from farms into lakes and rivers, where it is rapidly increasing in numbers (Diaz-Pardo, 1996), while the endemic (and endangered) *Cichlasoma istlanum* is disappearing (Espinoza et al., 1993; Carrillo-Willson, 1996).

3. Purposeful introductions and translocation for human consumption. Food shortages in México have not been solved. In rural zones, food with animal protein origin is scarce or too expansive for most people, and aquaculture has been seen as one of the solutions to increase animal-protein intake in rural communities. Benefits of aquaculture normally translate into fish production. However, the proponents of such introductions do not contemplate critical ecological and social factors that may decrease these benefits and create other problems. Introductions have not been preceded by studies on local use, and on productivity of native freshwater fauna. This often leads to drastic changes in local use and perception of freshwater resources. This activity is practiced in all the states of the Federation; it is supported by different breeding centers and research facilities throughout the country, and all the fish involved are nonnative, being mainly varieties of exotic fish, carp and tilapia (Table 6.1). Aquaculture National Programs are implemented in central offices focusing on maximizing the production of a few species, all of them nonnative. These programs never have enough information about the biology of native species and the ecology of the system. Under National Programs Policy, most of

FIGURE 6.1 Mexican regions of freshwater fish distribution. (1) North México, (2) Gulf of México, (3) Central México, (4) Chapala-Lerma Basin, (5) Balsas Basin, (6) South México, (7) Yucatan Peninsula. For more information, see Table 6.1.

TABLE 6.1
Number of Lakes and Reservoirs in the Country, Number of Native Fish Species in Each Region, and Production of Exotic Species

	Natural Lakes	Big Reservoirs	Native Species	Endemic Species	Endangered Species	Extinct Species	Production (tons)		
							Tilapia	Carp	Bass
1. North México	10	7	157	41	57	8	17187.7	9884.3	745.7
2. Gulf of México	10	0	90	23	20	0	18838.7	2801.1	187.6
3. Central México	1	0	28	15	7	3	2013.1	7046.3	9.6
4. Chapala-Lerma Basin	8	0	73	39	9	0	26360.0	5302.1	210.6
5. Balsas Basin	5	3	45	14	18	0	4753.9	513.3	126.6
6. South México	5	2	84	19	6	0	7504.6	103.7	3.1
7. Yucatan Peninsula	3	0	29	12	7	0	750.9	128.1	0.0
Total	42	12	506	163	124	11	77408.9	25779.0	1283.1

Note: Fish production = per-year average (1989–1995; SEMARNAP, 1995).

the specific activities dealing with freshwater environments are the result of individual actions answering local nutritional or economic needs.

Culture of translocated fish such as bass (*Micropterus salmoides*) is promoted with less enthusiasm than for other nonnative fish. However, this culture has high production yields (Table 6.1). The distribution of bass has changed from a strip along the northern portion of the Mexican–American border states of Sonora, Chihuahua, Coahuila, Nuevo León, and Tamaulipas, to a continuous area that goes as far south as the states of Tabasco and Oaxaca (Arredondo-Figueroa, 1983). Other translocated species for aquaculture are rainbow trout (*Salmo gairdneri*), translocated from northwest to central México in the last decade of the 19th century; the cyprinid *Algansea lacustris*, originally from Pátzcuaro Lake and introduced to sites throughout the country; three species of catfish (*Ictalurus furcatus*, *I. punctatus*, and *Pylodictis olivaris*), originally from the northeast and translocated to the northwest of México in 1933; and finally, different species of the whitefish (genus *Chirostoma*), originally from particular sites of the Chapala-Lerma basin, and later translocated to sites within and outside the basin (Arredondo-Figueroa, 1983).

However, outside of some reproduction centers of rainbow trout, very little research has been conducted on the economic potential of native fisheries, for instance, the silverside *Chirostoma estor* (whitefish). All these translocations have been made without any research on large-scale farming. The few opportunistic attempts to translocate whitefish have been carried out, but potential ecological effects (and likelihood that the species will become established) have not been investigated in advance. Small-scale farming of native fish has also been tried in the states of Tabasco (South México, Figure 6.1), Tamaulipas (Gulf of México, Figure 6.1) and Durango (North México, Figure 6.1), but these have been isolated efforts of little consequence at the national level. This leaves the nationwide system of production in big and structured breeding centers devoted only to multiply carp and tilapia, and the decisions on where and when to introduce these fish to the local officers.

A lack of coordination and communication between different sectors involved in aquaculture, such as scientists, producers, and government, provokes a vacuum of information about the activities that surround the management of freshwater ecosystems. Even for the most-used species, carp, it is difficult to know with any degree of precision where and when it has been officially introduced.

Of the three forms of introduction, that for the purpose of human consumption, creates the highest probability for an alien species to become a pest. This is because the best aquaculture species grow fast and normally can resist changes in the trophic web or in the environment better than native species. Also, if we follow the "tens" rule (Williamson and Fitter, 1996), the way an alien species reaches a system goes through three steps: (1) to escape, (2) to get established, and (3) to become a pest. Probabilities for each step are 1 in 10 (0.1). However, in the case of farm fish, probabilities for steps (1) and (2) become 1, because the practice of repopulating (consistent introduction of juveniles in the system) helps the alien species to establish themselves in the system. Even if fisheries in many national systems could control the nonnative fish populations, their yearly restocking maintains populations at high densities.

THE RISKS OF AQUACULTURE

Nonnative fish introduction for human consumption began with the common carp *(Cyprinus carpio)* late in the 19th century (Mujica, 1987), when it was mainly introduced to the lakes near México City. During the 1930s, a program for rural aquaculture and continental fisheries was developed, and it was at that time the bass was translocated from Northern México into the central High Plateau. A greater push for aquaculture occurred between the late 1970s and early 1980s, with the nationwide food program (SAM; sistema alimentario mexicano). This program spawned the creation of centers to produce fast-growing, easy-to-breed species such as common carp, herbivore carp (*C. idella*), and varieties of tilapia (*Oreochromis* sp.), which were then introduced to lakes and reservoirs throughout the country, requiring only an assessment of likelihood of the species becoming established.

As can be expected, the impact of aquaculture on the environment is different in different parts of the country (Table 6.1, Figure 6.1). The effect of nonnatives on biodiversity is higher in areas with a large number of endemic species. North México, the biggest area by far, has the highest number of native species, and it also has the highest number of species extinct or endangered. However, the Chapala-Lerma Basin that has the largest proportion of endemic species, and it is also the area with the largest production of fish introduced for human consumption. The increase in both volume and extent of aquaculture in recent decades is posing a serious risk of habitat deterioration and loss of biodiversity in the country. The risk is particularly grave in Central México (Lerma, Balsas, and Tula basins) where the proportion of endemics is the highest and where aquaculture has been increasing fastest.

Freshwater fisheries are valuable economic activities, as they produce more than 100 tons of fish per year, mostly tilapia and several varieties of carp that account for 86% of the total freshwater catch (SEMARNAP, 1995). These varieties are cultured in several places and their distribution runs throughout the country. Currently 36 introduced species are recognized in México, most of them (28) for aquaculture or fisheries (Table 6.2). The rest of the introduced species are either for use in the aquarium trade, or the motives for their introductions/translocations are unknown.

Nonnative fish are already present, and sustain fisheries, in virtually every natural lake in México. In addition, large numbers of dams have been built and populated with nonnative fish in recent years. These are either large hydroelectric ponds or irrigation reservoirs through the country, or small (1 – 10 ha) agricultural ponds built to supplement irrigation to small parcels during the dry season. Ponds in the last category have been built at an increasing speed in the last 20 years (Hernández et al. 1995): in Tlaxcala (Central México) alone, which is the smallest State in México and lacks natural lakes, the registered number of ponds is 935.

In spite of the different nature of reservoirs and lakes, nonnative fish introductions for fisheries are conducted in the same indiscriminate way, and at the same scale, in both types of environments. Partial ecological justification for these introductions is given by advancing the belief that carp and tilapia have negligible ecological impact in Mexican water bodies because, owing to their origin and feeding habits, they occupy hitherto "empty" niches. Economic considerations, which carry more weight in political decisions than environmental thinking, lend strong support for wholesale introductions of carp and tilapia. This is because both species are highly adaptable to a wide variety of environments, they grow rapidly, are cheap to produce, and the technology required to produce, collect, and process both species at an industrial scale is readily available, having been already developed in other countries. Clearly these features are attractive to politicians having to satisfy immediate nutritional needs of a large rural population; quick results are obtained without having to invest in developing technologies to commercially exploit native species.

The short-term economic success of the introduction of nonnative fish has discouraged research on the potential use of native species. Very few native species have been studied with the view to producing them at a commercial level. Only production of the whitefish has been investigated in some detail. However, because research has not been linked with actual large-scale production, programs to produce and to translocate whitefish in several parts of the country have not been carried out yet. Thus it remains as large a threat to the environment as carp and tilapia.

ECOLOGICAL IMPACT OF NONNATIVE FISH ON ECOSYSTEMS: DIFFERENT EFFECTS AND CASE STUDIES

Introductions do not operate in a vacuum; it is possible that their effect on the native ecosystems is obscured by effects of other practices that degrade the environment such as pollution and over-fishing. Thus putative consequences of the introduction of nonnative species should be investigated, paying due attention to potentially confounding variables. The following are examples of cases in which native faunas have been affected by introductions of nonnative fish in at least one of three possible ways: directly (through competition, predation, etc.) indirectly, and in the long term (usually involving changes in the environment brought about through over-fishing, pollution, etc.).

DIRECT EFFECTS

It is sometimes possible to correlate an increase in the populations of nonnatives with a decrease in the population(s) of native species; although seldom quantified, the causes of such a correlation probably lie in direct competition for food, predation, and/or the introduction of exogenous para-sites. There are many documented examples in different countries of Central America of piscivorous fish impact on native species (Fernando, 1991). There are few documented examples from Northern México at Casas Grandes, Peña del Aguila y Lerdo (Contreras and Escalante, 1984). Native cichlids

TABLE 6.2
Nonnative Species Introduced to México

Specie	Origin	Year of Introduction	Reason
Cyprinidae			
Aristichthys nobilis	China	1975	aquaculture
Barbus conchonius	India	1967	pet trade
Barbus titteya	Sri Lanka	1967	pet trade
Carassius auratus	Eurasiatic region	Before 1904	aquaculture
Ctenopharyngodon idella	Asia	1965	aquaculture
Cyprinus carpio	Eurasiatic region	1855	aquaculture
Gila bicolor mohavensis	U.S.	1955	ecological protection
Gila orcutti	U.S.	1955	ecological protection
Hypophtalmichthys molitrix	Asia	1965	aquaculture
Megalobrema amblycephala	China	1979	aquaculture
Mylopharyngodon piceus	Asia	1979	aquaculture
Notemigonus crysoleucas	U.S.	1973	angling
Catastomidae			
Carpoides cyprinus	U.S.	unknown	unknown
Carpoides carpio	U.S.	unknown	unknown
Cobitidae			
Misgurnus anguillicaudatus	East of Asia	1960s	aquaculture
Characidae			
Collosoma macropoma	Brasil	1980s	aquaculture
Ictaluridae			
Ictalurus melas	Canada and U.S.	unknown	aquaculture
Salmonidae			
Onchorhynchus clarki	Translocation		aquaculture
Onchorhynchus virginalis	Translocation		aquaculture
Onchorhynchus mykiss	Translocation		aquaculture
Salvelinus fontinalis	North America	Before 1957	aquaculture
Cyprinodontidae			
Fundulus zebrinus	U.S.	unknown	unknown
Poecilidae			
Poecilidae reticulata	South America	Before 1961	pet trade
Percichthyidae			
Morone chrysops	U.S.	unknown	aquaculture and angling
Morone saxatilis	U.S.	unknown	aquaculture and angling
Centrachidae			
Ambloplites rupestris	U.S.	unknown	aquaculture
Leopomis auritus	U.S.	unknown	aquaculture
Lepomis punctatus	U.S.	unknown	aquaculture
Micropterus dolomieui	Canada and U.S.	unknown	aquaculture
Pomoxis annularis	Canada and U.S.	Probably in the 1950s	aquaculture
Cichlidae			
Oreochromis aeureus	Northeast of Africa	1964	aquaculture

TABLE 6.2 (CONTINUED)
Nonnative Species Introduced to México

Specie	Origin	Year of Introduction	Reason
Oreochromis mossambicus	Africa	1964	aquaculture
Oreochromis niloticus	Africa	1978	aquaculture
Oreochromis urolepis hornorum	Africa	1981	aquaculture
Tilapia rendalli	Africa	1964	aquaculture
Tilapia zillii	Africa	unknown	aquaculture

Note: Compiled after Espinoza et al., 1993.

face the added threat of hybridization with nonnative species, which accelerates the displacement of native species (Contreras and Escalante, 1984). Sometimes distinctions between direct and indirect effects of nonnatives are unclear. Changes in the equilibrium or in the resilience of an aquatic community are suspected in a few cases. A native species (*Cichlasoma labridens*) was profitably exploited even 10 years after the introduction of bass in Atezca lagoon. Then tilapia was introduced, and in 4 years, total yearly catch dropped 70% (Diaz-Pardo, 1996). Although full details are lacking, it seems that the community withstood one introduction, but could not resist a second one, because its structure had been modified by the first introduction to a degree that made it more vulnerable to future changes.

INDIRECT EFFECTS: THE CASE OF CARP (*C. CARPIO*)

In spite of abundant examples in the literature on the undesirable effects carp has on aquatic ecosystems, and even though such effects are known for oligotrophic temperate environments, carp introduction effects are not quite clear in eutrophied systems. This is because the impact of carp introductions is not a direct cause–effect relation. Few examples show that carp adversely affects a particular species within the community, basically because effects are mediated through the feeding habits of carp. Being a benthivorous fish, it seldom damages native fish populations through predation. Instead, carp modifies the cycles of nutrients in such a way that native members of the community are affected through rather indirect paths. Because indirect effects are difficult to quantify, most studies on the impact of carp introductions follow the black-box approach: the initial and final conditions are known, but what lies in between is largely matter of speculation. Generally, producers do not associate system deterioration with carp introduction. Related to carp introductions there are many changes in the system provoked by the fish itself or by the human activities that surround its introduction. Producers often promote eutrophication in the system in order to increase carp production because cyprinid yield is directly related to lake productivity (Persson, 1990).

Because of a greater sensitivity to external influences, small lakes achieve eutrophication more rapidly than large ones. This factor makes small lakes useful experimental models to investigate the effect of common carp on aquatic ecosystems, with the added bonus that the variables involved are more efficiently and economically controlled.

The municipality of Acambay, in the State of México (border between Central México and Lerma-Basin, Figure 6.1), contains at least 50 small irrigation ponds, some of them rather old. Acambay is located at 2250 masl, and, although technically in a temperate area, experiences basically two major seasons (dry and rainy), as typical of the tropics. The ponds are mainly used to supplement irrigation of maize fields; thus, they are partially or totally drained at the end of the dry season, in March and April, and are filled by the rain in July and August. As in most of the country, in these ponds the culture of carp has been actively promoted (Hernández et al., 1995) on the assumption that no native freshwater resources were available. Largely ignored or neglected was the fact that the local people (many of whom are of Otomi ethnic origin) had a long tradition of incorporating crayfish (*Cambarellus montezumae*), Amarillo fish (*Girardinichthys multiradiatus*), axolotls (*Ambystoma* sp.), and the large tadpoles of the local frogs (*Rana berlandieri*) into their diet. Fortunately, not all land owners have introduced carp into their ponds. Where carp has been introduced, the stocking density was largely a matter of personal choice. This has resulted in similar ponds having different carp densities. This makes for an ideal study area to investigate the short-term consequences of carp introduction in relatively simple ecosystems. Climate and soil variables are the same for all the ponds; they differ only in their exposure to carp.

We used this fortuitous experimental setting to measure some variables deemed to be influenced by carp density in several ponds. We measured water turbidity, suspended solids, nutrients concentration, productivity of suspended algae, area covered by macrophytes, and abundance of zooplankton, fish, and benthic invertebrates. Measurements were taken in nine ponds with densities of carp varying between 0 and 4.5 in capture per unit effort (CPUE). The field work was conducted between February and March of 1995.

A preliminary analysis showed that the ponds can be divided into two large groups according to water turbidity (Zambrano et al., in press). We compared these two groups in terms of all the other measured variables using nonparametric tests (Table 6.3), and found that the ponds may be going through a dynamic process characterized by two points of equilibrium (Scheffer et al., 1993). Thus ponds can be categorized as

1. Transparent ponds, with high abundance of macrophyte cover, and benthic invertebrates
2. Turbid ponds, with few macrophytes.

Relative carp abundance (CPUE) was always >1 in turbid ponds, and always <1 in transparent ponds.

TABLE 6.3
Wilcoxon Matched-Pair Comparison of Different Variables in Acambay Ponds, Estado de México

Parameter	Ponds		Wilcoxon-U-test
	Turbid	Clear	p
Ammonium (mg/l)	0.63	0.05	0.17
Total phosophorus (mg/l)	0.11	0.27	0.10
Total nitrogen (mg/l)	1164.00	198.00	0.07
Suspended solids (mg/l)	1200.00	250.00	0.10
Macrophytes (% coverage)	9.03	38.56	0.03
Secchi depth (m)	0.19	0.72	0.01
Chlorophyll a (μg/l)	3.50	8.80	0.11
Zooplankton (org./l)	13.34	21.06	0.35
Invertebrates (org./m^2)	129.80	844.00	0.05
Carp no. (CPUE)	2.75	0.07	0.05

The points of equilibrium in the dynamics of ponds can be affected by changes in the fish community (Bronmark and Weisner, 1992); for instance, large numbers of benthivorous fish can result in a decrease in the abundance of macrophytes (Ten Winkel and Meulemans, 1984). Plant abundance, together with water turbidity, may determine whether a community remains in a particular point of equilibrium or not. Carp, being benthivorous, resuspend sediments and increase turbidity (Brewkelaar et al., 1994). Once suspended solids increase, suspended algae also increase, because macrophytes can no longer out-compete them. Higher plants are also weakened because carp browse in and around their roots; dying plants further promote algae growth as they decay by liberating nutrients into the water column. Benthic communities are affected in two ways; they are preyed on by the carp (Tatrai et al., 1994), and they lose their breeding refuges once the macrophytes are gone (Moss, 1990).

Even though the negative effects of carp on aquatic ecosystems are not limited to biotic components of the environment, the greatest danger posed by *C. carpio* still lies in the fact that its impacts are difficult to quantify. Without precise descriptions of the cascade of deleterious effects of carp on freshwater ecosystems, the task of convincing the authorities, producers, and consumers that carp should not be introduced in fragile environments looks daunting.

MEDIUM- AND LARGE-SCALE EFFECTS

It is possible that most of the damage done by nonnative species would be through indirect effects. Some records may show that the effects of introduction of nonnative species have not been spectacular, except in few cases in tropical America (Fernando, 1991). Changes in the systems after introduction are slow and almost unperceptible for a few months. But adding up of these effects over a long time results in changes to the freshwater community, sometimes greater than the few recorded spectacular short-term effects.

In larger ecosystems, the effects of introducing nonnative species may take a long time to manifest themselves, but by the same token such effects will be more difficult to reverse. Lake Pátzcuaro is found within a sub-basin of the Lerma-Chapala river basin, in the Neovolcanic Transverse axis, at 2035 masl (Chacon-Torres, 1993). The lake has no outlets; its basin is of volcanic origin, and is surrounded by pine–oak forests, grasslands, and crop fields (Orbe and Acevedo, 1995). Temperature in the area has an annual average of 16°C, with a maximum of 37°C and a minimum of –5°C. The population in the vicinity is of Purepecha ethnic origin, being mainly devoted to farming and fishing. In 1980 when the local population was about 80,000 (Rosas et al., 1984), there were 1000 fishermen who captured between 300 and 500 tons of fish per year (Toledo and Barrera-Bassols, 1984). The number has already increased to 1500 fishermen (Chacon-Torres and Rosas-Monge, 1995). Most fish present in the lake are exploited; this includes 11 native species (mostly of the Goodeidae and Atherinidae families, notably the whitefish) and 5 nonnative species, particularly carp, tilapia, and bass (Chacon-Torres, 1993).

Although whitefish has traditionally been the staple fish food for the locals, in 1933 the government promoted the translocation of a potential predator, bass, into Pátzcuaro from the north of the country (Toledo-Díaz, 1995). Effects of bass on native fish faunas were documented early in the 1940s (De Buen, 1941), but were never deemed to be strong enough to eliminate native-fish fisheries. Later (early 1970s), carp and tilapia were also massively introduced in Pátzcuaro. Even then, the price of whitefish in 1980 was nine times that of carp and three times that of tilapia (Toledo and Barrera-Bassols, 1984). However, catches of whitefish began to decline in the 1980s, forcing fishermen to concentrate on carp and tilapia for food and for sale (Figure 6.2). In the last 10 years, fishery shifts from whitefish to carp and tilapia meant that the price of whitefish dropped to a seventh of its value 10 years before, and catch has been reduced by 80% (Chacon-Torres and Rosas-Monge, 1995). In this decade there has been a general collapse of fisheries in Pátzcuaro, involving both native and nonnative species (Figure 6.2). The reasons for such a generalized decline are not clear, but the fact that several species seem to be equally affected suggest some explanations.

It is commonly assumed that over-fishing of all species is the reason for the decline in total catch. Consequently, laws have been passed to regulate fishing season and minimum net-mesh for each species. However, the decline in catch was so massive and rapid that over-fishing seems to provide an incomplete explanation, and other changes in Pátzcuaro's ecosystem deserve closer inspection. Eutrophication and sediment deposition are likely candidates; in the last 40 years the average depth decreased 2 m, and Secchi depth changed from 2 m to 0.5 m (Orbe and Acevedo, 1995). Pátzcuaro is a monomictic lake, but it has a patchy distribution in community and of water quality (Rosas et al., 1993); some patches are highly polluted, with low biological diversity, while

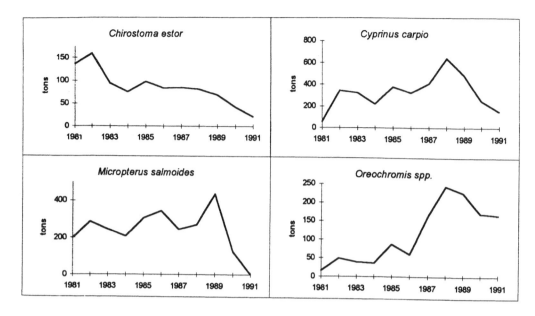

FIGURE 6.2 Production in metric tons of native (*Chirsotoma estor*) and three introduced fish in Pátzcuaro Lake from 1981 to 1991 (compiled after Orbe and Acevedo, 1995).

other patches have healthy populations of macrophytes and a diversity of benthic organisms (Rosas et al., 1984). Patchiness of this sort is bound to influence the demography of the different fish species through cascade effects at several levels of the ecosystem.

As usual in ecology, a complete explanation probably involves several processes acting simultaneously. We hypothesize that in Pátzcuaro, the rapid population growth of nonnative species must have modified some elements in the trophic web; these we presume meant a reduction in the availability of organisms consumed by the native species. The latter (and indeed all fish species but bass) were further subjected to a new predator (bass), to increased fishing pressures, and to an increment in the discharges of pollutants from agriculture and domestic waste. As the populations of its prey decreased, and because fishing pressure did not abate, bass populations then also collapsed. Fewer numbers of bass did not translate into an increase in prey-species numbers because these remained under pressure from fishing, competition, and pollution, a fate somewhat escaped by tilapia because of its relatively low ecological requirements.

We thus believe that the collapse of Pátzcuaro's ecosystem results from three intermingled factors: (1) habitat degradation due to eutrophication and pollution, (2) over-fishing, and (3) over-stressing the ecosystem through the introduction of aggressive nonnative fish; these damaged the trophic web and precipitated the crash of the ecosystem.

Unchecked introductions of nonnative fish may have helped satisfy the fishermen´s needs in the short term, during which both fish production (mainly of nonnative species) and fishermen population grew. In the long term, however, such growth provoked severe ecological and social problems. Because the inhabitants of Pátzcuaro depend on the lake for livelihood, and given that habitat restoration is difficult and costly, the indiscriminate introduction of nonnative fish appears to be a rather unsound policy.

CONCLUSION

Introduction of nonnative species without any information about the biology of the exotic organism and the ecology of the receiving system may accrue immediate benefits, but the drawbacks are a

Pandora's box. Socially, in communities near big lakes and reservoirs, it creates new resources and new expectancies for growth, both of which may suddenly crash, leaving numerous families deprived of their means to make a living. Ecologically, introductions without any knowledge threaten the country's biological diversity, and the health of all of its lakes, because no distinction is made of natural lakes and manmade reservoirs.

A keystone of the introduction of alien species in Mexican systems is the complete lack of knowledge about the ecology of the system and the biology of the species involved. Examples such as the repeated introduction of goldfish in the lakes of Zempoala National Park, or the signs indicating a yearly carp stocking as an ecological restoration in protected areas of Xochimilco (both in Central México), show the poor knowledge of limnology in the zone.

In addition, there is no knowledge or control on the dispersal capacities of farm fish to freshwater habitats. This in turn leads to accidents like that at the nature reserve Lagunas de Montebello in Chiapas where a carp (<5 year old) suddenly appeared or the dispersal of tilapia from a farm into all the water bodies of Tabasco following the hurricane in 1994. The very status of protected areas implies preservation of native fauna and flora. They are threatened in their role as sanctuaries by species introduced (or allowed to spread into them) for no reason.

Ecological thinking and respect for native cultures are newcomers in the arena of political planning, and may only slowly replace the entrenched short-term productivity views that led to wholesale introductions of nonnative fish throughout the country. While shortsightedness favors indiscriminate introduction of nonnative species to solve immediate problems, research on the potential (and often traditional) use of native species, which may not pay in the short term, is likely to generate sounder production practices in the long run. Such potential is clearly in jeopardy, because native species are being displaced by nonnatives throughout the country.

ACKNOWLEDGMENTS

We thank Dr. Martin Perrow for aiding in the analysis of carp indirect effects, and Beatriz Contreras for helping with the maps of fish distribution. Also, we thank Dr. Edmundo Diaz-Pardo, Monica Peña, and Francisco Nieto, who assisted by compiling information and commenting on the manuscript.

REFERENCES

Arredondo-Figueroa, J., Especies animales acuáticas de importancia nutricional introducidas en México, *Biotica*, 8(2), pp. 175–199, 1983.

Bronmark, C., and S. E. B. Weisner, Indirect effects of fish community structure on submerged vegetation in shallow, eutrophic lakes: an alternative mechanism, *Hydrobiologia*, 243/244, pp. 293–301, 1992.

Brewkelaar, A., E. Lammens, J. Breteler, and I. Tatrai, Effects of benthivorous bream (*Abramis brama*) and carp (*Cyprinus carpio*) on sediment resuspension and concentration of nutrients and chlorophyll *a*. *Freshwater Biology*, 32, pp. 112–121, 1994.

Carlton, J.T., Biological invasion and cryptogenic species, *Ecology*, 77(6), pp. 1653–1655, 1996.

Carrillo-Willson, C.R., "Variación en la esctructura de los ensambles icticos a lo largo del Rio Amacuzac, Morelos," Thesis presented to the School of Biology, Universidad Autónoma de Morelos, in requirements for the degree of BS, 1996.

Chacon-Torres, A., Lake Pátzcuaro, México: Watershed and water quality deterioration in a tropical high-altitude Latin American Lake, *Lake and Reserv. Manage*, 8(1), pp.34–47, 1993.

Chacon-Torres, A., and C. Rosas-Monge, A restoration plan for pez blanco in Lake Pátzcuaro, México, *American Fishery Society Symposium*, 15, pp. 122–126, 1995.

Contreras, B.S., and C.M.A. Escalante. Distribution and known impacts of exotic fishes in México. In *Distribution, Biology, and Management of Exotic Fishes, 1985*. Courtenay, W.R. and J.R. Stauffer, Eds., The Johns Hopkins University Press, Baltimore, MD, 1984.

De Buen, F., El *Micropterus* (Huro) *salmoides* y los resultados de su aclimatación en el Lago de Pátzcuaro, *Rev. Soc. Mex Hist. Nat.* 2, pp. 69–78. 1941.

Díaz-Pardo, E., personal communication, 1996.

Díaz-Pardo, E., and C. Guerra-Magaña, Cincuenta años de historia de la colección de peces dulceacuícolas mexicanos de la Escuela nacional de Ciencias Biológicas, IPN. *Zoologia Informa*, 33. 1996.

Espinoza P.H., M.T. Gaspar, and P. Fuentes, *Listados Faunísticos de México. III Los Peces Dulceacuícolas Mexicanos*. Instituto de Biología UNAM. México, 1993.

Fernando C.H., Impacts of fish introduction in Tropical Asia and America, *Can. J. Fish Aquat. Sci.*, 48, (suppl. 1), pp. 24–32. 1991.

Hernandez-Avilés, J.S., M.C. Galñindo de Santiago, and J. Lorea Pérez, Bordos o Microembalses, in *Lagos y Presas de México*. 1st ed., Lanza, E., and C.J. Garcia, Ed., Centro de Ecología y Desarrollo, México, 1995.

Moss, B., Engineering and biological approaches to the restoration from eutrophication of shallow lakes in which aquatic plant communities are important components, *Hydrobiologia*, 200/201, pp. 367–377, 1990.

Mujica Cruz, E., Los cuerpos de agua continetales adecuados para el cultivo de carpa, *Rev. Mex. de Acuac.*, 9, pp. 7–10, 1987.

Orbe, M.A., and J.G. Acevedo., El lago de Pátzcuaro, in *Lagos y Presas de México*, 1st ed., Lanza, E., and C.J. Garcia, Ed., Centro de Ecología y Desarrollo, México, 1995.

Persson, L., Interespecific interactions, in *Cyprinid Fishes, Systematic, Biology and Exploitation,* Winfield, I.J., and Nelson, J.S., Eds., Chapman and Hall, London, 1990.

Rosas, I., M. Mazari, J. Saavedra, and A.P. Baez., Benthic organisms as indicators of water quality in Lake Pátzcuaro, México, *Water, Air and Soil Pollution*, 25, pp. 401–414, 1984.

Rosas, I., A. Velazco, R. Belmont, A. Baez, and A. Martinez, The algal community as an indicator of the trophic status of Lake Pátzcuaro, México, *Environmental Pollution*, 80, pp. 255–262, 1993.

Scheffer, M., S.H. Hosper, M.-L. Meijer, B. Moss, and E. Jeppesen, Alternative equilibria in shallow lakes, *TREE*, 8, pp. 275–279, 1993.

SEMARNAP. Dir. Gral. Informática y Registro Pesquero, *Anuario Estadistico de Pesca 1995*, México D.F., 1995.

Tatrai, I., E.H. Lammens, A.W. Brewkelaar, and J.K. Beteler, The impact of mature cyprinid fish on the composition and biomass of benthic macroinvertebrates, *Arch. Hydrobiol.*, 131, pp. 309–320, 1994.

Ten Winkel, E.H., and J.T. Meulemans, Effects of fish upon submerged vegetation, *Hydrobiol. Bull.*, 18, pp. 157–158, 1984.

Toledo-Diaz, M.P., Consumo de aterínidos (*Chirostoma* spp.) por la lobina negra (*Micropterus salmoides*) en el lago de Pátzcuaro, Mich., México en 1986. *Ciencia Pesquera*. 11, pp. 71–74, 1995.

Toledo, M.V., and N. Barrera-Bassols, *Ecologia y Desarrollo Rural en Pátzcuaro*, Instituto de Biología, UNAM. México, 1984.

Williamson, M. and A. Fitter, The varying success of invaders, *Ecology*, 77(6), pp. 1661–1666, 1996.

Zambrano, L., M.R. Perrow, C. Macías-Garcia, and V.M. Aguirre Hidalgo, Impact of introduced carp (*Cyprinus carpio*) in subtropical shallow ponds, J. Aquatic Ecostystem Stress and Recovery (in press).

Section II

Aquarium and Water Garden Trade

7 Aquariums and Water Gardens as Vectors of Introduction

Walter R. Courtenay, Jr.

SECTION INTRODUCTION

Aquaria have become common fixtures in many homes and businesses, particularly during the latter half of this century. Fascination with trying to keep a goldfish or two alive in a glass bowl on a counter or table, once popular, transformed into maintaining other, more diverse nonindigenous aquatic species in larger, sometimes expensive, aquaria with special lighting and advanced filtration systems. One can purchase an increasingly wide diversity of fishes, plants, or mollusks from many different parts of this planet at local pet stores.

Explosive growth of the aquarium trade began shortly after World War II with the advent of jet cargo aircraft, plastic bags, and styrofoam shipping boxes that facilitated movement of nonindigenous aquatic organisms over many hundreds to thousands of miles in less than a day's time. Recognition that many imported species could be cultured in suitable climates led to development of culture facilities, particularly in subtropical regions of North America. Expansion of the trade resulted in one of the fastest growing hobbies in history.

While most hobbies are essentially without pollution potential, the aquarium hobby is not one of those. Importation of nonindigenous aquatic organisms became a large industry, at some ports-of-entry comprising (with shipping water) a major entering cargo by weight. Because there were mostly no restrictions on what could be imported, almost any species was legal by default. The sudden appearance of foreign fishes, sometimes followed by locally established populations, in waters near Miami International Airport, for example, suggested that import brokers released fishes not purchased or retrieved by secondary suppliers or retailers. Steadily increasing numbers of nonreproducing and establishing species of fishes and other aquatic species of foreign origin in waters near culture facilities clearly indicated leakage from those facilities, later documented through observation of nonfiltered effluents and active pumping of ponds in which fish species had become mixed. Overly successful culture of aquatic plants, such as *Hydrilla*, often led to dumping excess production into nearby waters. Thus, the trade itself became a biological polluter, introducing many species that became established as reproducing populations in hospitable habitats.

Responsibility also extends to some hobbyists who, in their zeal to enjoy their pet fish or aquarium snails, introduced them purposefully into warm-water environments such as thermal spring outflows, in which several species became established. These releases probably included intentional ones, just to test success of releases or to later return and harvest progeny for sale. Other hobbyists, doubtless the majority, simply tired of their aquarium inhabitants or the fact that one or more species were too prolific or grew too large for the container, and released them into open waters, believing that freedom was better than confinement. Several of these releases became established, but most remain recorded as incidental captures during unfortunately infrequent fish surveys or catches reported by anglers, but in alarmingly increasing numbers.

Of major concern is that the aquarium trade and many hobbyists fail to recognize or do not care that introductions of nonindigenous species can result in multiple negative effects on native species, their food webs, and habitats, as detailed in many chapters in this volume. Once a potentially damaging species is released and has become established, it typically cannot be effectively controlled or eradicated unless it is localized, often with great cost financially and to native species; most damage becomes apparent well after the invading species has expanded its range when control or eradication are impossible. So far, we seem to have failed in conveying this message to the public, regulators, and legislators.

8 Summary of North American Fish Introductions through the Aquarium/Horticulture Trade

E.J. Crossman and B.C. Cudmore

This vector is meant to refer to private aquariums normally in households, not public aquariums to which one pays an admission price. It may be the most novel vector, but it certainly contributes, in an insidious way, to the problem of introduced species. The earliest record of a fish species introduced into the U.S. is usually given as that for the goldfish, *Carrassius auratus*, in the 1660s (DeKay 1842). It may seem surprising that this vector can be ascribed to 97 species of introduced fishes (Table 8.1). It may not be so surprising if one considers that at least 2000 species, largely aquarium fishes, have been imported into the U.S., or will be in the near future (McCann 1984). In addition, in the late 1970s and 1980s, 200 million individual aquarium fishes were imported into the U.S., and they constituted only 20% of the aquarium trade at the time (Courtenay 1990). The percentage listed above for this vector is equal to that attributed here to aquaculture and considerably higher than that attributable to ballast water, which receives so much more notoriety and concern.

Individuals spend considerable time, effort, and money keeping aquarium fishes alive, and aquarium environments in good condition. Those individuals are reluctant to kill such specimens when fish get too large or they wish to change an aquarium setup. Sadly, the alternative taken too often is to liberate the fishes and plants in nearby public waters. The problem is often exacerbated by the notoriety given the discovery of such unusual species in public waters. The event is recorded in the newspapers, the specimen is often reinstalled in an aquarium, and the process is started over. The species of fishes listed in this vector are usually readily attributable to it on the basis of their being well-known and popular aquarium species or representing popular families, etc. The individuals found alive in public waters are most often cichlids, characins, livebearers, and exotic cyprinids and catfishes. Almost universally it is single individuals of a species.

An interesting example of results from this vector is the introduction of aquarium species in an unusual habitat in regard to altitude and latitude. Aquarists released four species into hotsprings in Banff National Park, Alberta. Two of them, sailfin molly, *Mollienesia latipinna*, and African jewelfish, *Hemichromis bimaculatus*, have become established and exist there today. An additional two other species, convict cichlid, *Cichlosoma nigrofasciatum*, and freshwater angelfish, *Pterophyllum scalare*, released into the same situation have since disappeared (Nelson and Paetz 1992).

A potential problem is the growing popularity amongst aquarists of "native" species and the trading of those species via a newsletter with wide circulation. The concern is for the rarer species (e.g., darters) with very limited distributions, small populations, and often endangered or vulnerable status. Such trading could be dangerous exploitation at one end and introduction at the other. As with fossil locations there is resistance now to the appearance in readily available scientific journals of precise localities for these species. Such information prepared in Canada

TABLE 8.1
Fishes Introduced in Ecological North America (Excluding Mexico) Outside their Native Range through Aquarium Releases or Escapes and their Reproductive Status in the Nonindigenous Area(s)

Scientific Name	Common Name	Status[a]
Acipenseridae — sturgeons		
Acipenser or *Scaphirhynchus* sp.	eastern species of sturgeon	not established
Lepisosteidae — gars		
Atractosteus spatula	alligator gar	not established
Osteoglossidae — bonytongues		
Osteoglossum bicirrhosum	arawana	not established
Cyprinidae — carps and minnows		
Brachydanio rerio	zebra danio	established in some areas
Carassius auratus	goldfish	established
Cyprinella lutrensis	red shiner	established in many areas
Danio malabaricus	Malabar danio	not established
Leuciscus ideus	ide	established in some areas
Morulius chrysophekadion	black sharkminnow	not established
Puntius nigrofasciatus	black ruby barb	not established
Puntius tetrazona	tiger barb	not established
Rhodeus sericeus	bitterling	established in some areas
Tinca tinca	tench	established in some areas
Cobitidae — loaches		
Misgurnus anguillicaudatus	oriental weatherfish	established
Characidae — characins		
Colossoma and/or *Piaractus* sp.	unidentified pacus	not established
Colossoma cf. *bidens*		not established
Colossoma macropomum	black pacu	not established
Gymnocorymbus ternetzi	black tetra	not established
Hyphessobrycon cf. *serpae*	serpae tetra	not established
Leporinus fasciatus	banded leporinus	not established
Metynnis maculatus	spotted metynnis	not established
Myleus cf. *rubripinnis*	redhook myleus	not established
Piaractus brachypomus	redbellied pacu	not established
Piaractus mesopotamicus	small-scaled pacu	not established
Pygocentrus and/or *Serrasalmus* sp.	unidentified piranhas	not established
Pygocentrus nattereri	red piranha	not established
Serrasalmus rhombeus	redeye piranha	not established
Clariidae — labyrinth catfishes		
Clarias batrachus	walking catfish	established in some areas
Doradidae — thorny catfishes		
Platydoras costatus	Raphael catfish	not established
Pseudodoras niger	ripsaw catfish	not established
Pterodoras granulosus	granulated catfish	not established
Pterodoras sp.	thorny catfish	not established
Trachycorystes fisheri	driftwood catfish	not established

TABLE 8.1 (CONTINUED)
Fishes Introduced in Ecological North America (Excluding Mexico) Outside their Native Range through Aquarium Releases or Escapes and their Reproductive Status in the Nonindigenous Area(s)

Scientific Name	Common Name	Status[a]
Pimelodidae — longwhiskered catfishes		
Perrunichthys perruno	leopard catfish	not established
Phractocephalus hemiliopterus	redtail catfish	not established
Callichthyidae — plated catfishes		
Callichthys callichthys	cascarudo	not established
Corydoras sp.	corydoras	not established
Loricariidae — suckermouth catfishes		
Hypostomus sp.	suckermouth catfish	established in some areas
Liposarcus multiradiatus	sailfin catfish	established
Liposarcus pardalis		not established
Otocinclus sp.		not established
Panaque nigrolineatus	royal panaque	not established
Melanotaeniidae — rainbowfishes		
Melanotaenia nigrans	black-banded rainbowfish	not established
Adrianichthyidae — ricefishes		
Oryzias latipes	Japenese medaka	not established
Fundulidae — topminnows and killifishes		
Fundulus diaphanus	banded killifish	established in some areas
Goodeidae — splitfins		
Ameca splendens	butterfly splitfin	not established
Poeciliidae — livebearers		
Belonesox belizanus	pike killifish	established in some areas
Gambusia affinis	western mosquitofish	established in some areas
Gambusia holbrooki	eastern mosquitofish	established
Poecilia latipinna	sailfin molly	established
Poecilia latipunctata	broadspotted molly	not established
Poecilia mexicana	shortfin molly	established
Poecilia petenensis	Peten molly	possibly established
Poecilia reticulata	guppy	established in most areas
Poecilia sphenops	liberty molly	established in most areas
Poeciliopsis gracilis	porthole livebearer	possibly established in some areas
Xiphophorus helleri	green swordtail	established in some areas
Xiphophorus maculatus	southern platyfish	established in most areas
Xiphophorus variatus	variable platyfish	established
Poecilia latipinna X *P. velifera*	sailfin molly X black molly	not established
Xiphophorus maculatus X *X. variatus*	southern platyfish X variable platyfish	not established
Xiphophorus helleri X *X. maculatus*	green swordtail X red swordtail	not established
Xiphophorus helleri X *X. variatus*	green swordtail X variable platyfish	not established
Synbranchidae — swamp eels		
Monopterus albus	swamp eel	established

TABLE 8.1 (CONTINUED)
Fishes Introduced in Ecological North America (Excluding Mexico) Outside their Native Range through Aquarium Releases or Escapes and their Reproductive Status in the Nonindigenous Area(s)

Scientific Name	Common Name	Status[a]
Centrarchidae — sunfishes		
Enneacanthus gloriosus	bluespotted sunfish	established in most areas
Cichlidae — cichlids		
Astronotus ocellatus	oscar	established in some areas
Cichlasoma beani	Sinaloan cichlid	not established
Cichlasoma citrinellum	midas cichlid	established in some areas
Cichlasoma cyanoguttatum	Rio Grande cichlid	established in some areas
Cichlasoma managuense	jaguar guapote	established in some areas
Cichlasoma meeki	firemouth cichlid	established in some areas
Cichlasoma nigrofasciatum	convict cichlid	established in some areas
Cichlasoma octofasciatum	Jack Dempsey	established in some areas
Cichlasoma salvini	yellowbelly cichlid	established
Cichlasoma synspilum	redhead cichlid	not established
Cichlasoma trimaculatum	threespot cichlid	not established
Cichlasoma urophthalmus	Mayan cichlid	established
Geophagus surinamensis	redstriped eartheater	possibly established in some areas
Hemichromis letournauxi	African jewelfish	established in most areas
Heros serverus	banded cichlid	not established
Labeotropheus sp.	mbuna	not established
Melanochromis auratus	golden mbuna	not established
Melanochromis johanni	blue gray mbuna	not established
Oreochromis aureus	blue tilapia	established in some areas
Oreochromis mossambica	Mozambique tilapia	established in many areas
Pseudotropheus sp.	African lake cichlid	not established
Pterophyllum scalare	freshwater angelfish	not established
Sarotherodon melanotheron	blackchin angelfish	established
Telmatochromis cf. bifrenatus	Lake Tanganyika dwarf cichlid	not established
Tilapia mariae	spotted tilapia	established in some areas
Helostomatidae — kissing gourami		
Helostoma temmincki	kissing gourami	not established
Belontiidae — gouramies		
Betta splendens	Siamese fightingfish	some reproduction
Colisa fasciata	banded gourami	not established
Macropodus opercularis	paradisefish	not established
Osphronemidae — giant gouramies		
Osphronemus goramy	giant gourami	not established
Channidae — snakeheads		
Channa micropeltes	giant snakehead	not established

TABLE 8.1 (CONTINUED)
Fishes Introduced in Ecological North America (Excluding Mexico) Outside their Native Range through Aquarium Releases or Escapes and their Reproductive Status in the Nonindigenous Area(s)

Scientific Name	Common Name	Status[a]
	Notopteridae — featherbacks	
Notopterus chitala	clown knifefish	not established

[a] Status refers to the reproductive nature of "populations" of that fish in areas where it has been introduced, regardless of vector(s) used. Many statements on status were derived from information in the NAS database. Some notations may be open to criticism (e.g., hybrids).

for status reports on rare and endangered species does not appear in the published version but is available to qualified individuals.

Another potential problem is the recent increase in the cyclical interest in water gardens. This development will stimulate the importation and distribution of nonindigenous fishes, amphibians, and aquatic plants. Near centers of high population, large businesses have appeared that provide construction materials, plants, invertebrates, and fishes for these gardens. In the North, the likelihood of dumping is even stronger as a result of winter freezing, with the only alternative of taking the species indoors. A shipment of goldfish into Ontario for the aquarium and water-garden trade probably included the first individual of the rudd, *Scardinius erythrophthalmus*, to reach Canada.

In the past the water temperature in the winter was thought in the North, to be the solution for this type of introduction, especially of piranhas (Characidae). We must be more careful today, as the outfall of cooling water from electric generating stations and other industrial situations may provide a haven overwinter for such species, and/or, for some, an opportunity to acclimate gradually to ambient winter temperatures.

The Aquatic Conservation Network (ACN), founded in 1991, is a charitable, nonprofit corporation dedicated to conserving aquatic life. It maintains "an Affiliate Club Program targeted at amateur aquarist societies to encourage a sense of stewardship in fish-keeping, — [and] a focus on the conservation of aquatic biodiversity" (Anon. 1996). We have suggested to the ACN that it adopt a program of educating amateur aquarists in regard to the potential harmful impact on aquatic biodiversity of releasing nonindigenous species, or aquarium fishes of any kind, into public waters. Another organization of aquarists, the North American Native Fishes Association (NANFA) already has such a program (pers. com. Robert Rice)

REFERENCES

Anon. Aquatic Conservation Network: Aquarists Dedicated to the Preservation of Aquatic Life, *Fisheries*, 12, p. 42, 1996.

Courtenay, W.R., Jr., Fish Conservation and the Enigma of Introduced Species, in *Introduced and Translocated Fishes and their Ecological Effects*, Pollard, D.A., Ed., Australian Bureau of Natural Resources, Canberra, 1990.

DeKay, J.E., *Zoology of New York — IV: Fishes*. W. and A. Whiteland and J. Vischer, Albany, NY, 1842.

McCann, J.A., Involvement of the American Fisheries Society with Exotic Species, in *Distribution, Biology, and Management of Exotic Fishes*, Courtenay, W.R., Jr., and J.R. Stauffer, Jr., Eds., Johns Hopkins University Press, Baltimore, MD, 1984.

Nelson, J.S., and M.J. Paetz, *The Fishes of Alberta*. (2nd Ed.) University of Alberta Press, Edmonton, Alberta, 1992.

9 Mollusc Introductions Through Aquarium Trade

Gerald L. Mackie

INTRODUCTION

Of the 22 species of molluscs that have been introduced to North America, eight appear to have been introduced via the aquarium trade. All but one of these are gastropods, the other being a fingernail clam. None have been demonstrated to have any kind of ecological impact; some have major quarantine significance, others have potential ecological benefits.

Each species introduced through the aquarium trade was examined for the following 10 attributes, when known, to establish what common characteristics, if any, these invaders share:

1. species characteristics
2. mode of life and habitat
3. most common dispersal mechanisms
4. morphological, behavioral, and physiological adaptations to their habitats
5. predators
6. parasites
7. reproductive potential and life cycle
8. food and feeding habits
9. ecological and/or socioeconomic impact potential
10. a summary of control measures used if the species is a biofouler.

Thorough description of the above characteristics is given by the author in Chapter 15, in the Ballast Water section of this book.

DETAILED REVIEW OF MOLLUSCA INTRODUCED BY THE AQUARIUM TRADE

Class: Gastropoda

Subclass: Prosobranchia (see features in Chapter 15 of this volume)

Family: Pilidae (Ampulariidae)
Mostly amphibious snails with large globose or subglobose shells. Mantle cavity divided into two compartments, the left containing a gill for extracting dissolved oxygen from water and the right serving as a lung for air-breathing.

Marisa cornuarietis (giant ram's-horn)

Originally from northern South America and southern Central America, the species was introduced into Florida, probably in the mid-1950s (Hunt 1958).

0-15667-0449-9/99/$0.00+$.50
© 1999 by CRC Press LLC

Species characteristics: Shell discoidal and relatively large, adults commonly 30–35 mm in diameter, occasionally up to 40 mm. Periostracum olive color with spiral reddish or brown bands. The most diagnostic features are the spire sunken below the body whorl and a very wide umbilicus (Burch 1989).

Mode of life and habitat: Typically in shallow areas of lakes and in ponds and slow streams with considerable macrophyte biomass.

Dispersal mechanisms: The species probably utilizes dispersal agents common to most other gastropods, including waterfowl, aquatic mammals and insects, at least for juveniles and younger stages.

Adaptations: Tolerance of gastropods to temperature extremes is an intra- and interspecific variable (Aldridge 1983). Some ampullariids can cool themselves 5–10°C below ambient by evaporation when heat-stressed, but this mechanism is available only to snails adapted to a terrestrial life (Aldridge 1983). Aquatic snails must either tolerate or acclimatize to changing temperatures. Of the exotic gastropods, only the pulmonates and ampullariids have the abilities to avoid anoxia by rising to the surface and respiring atmospheric oxygen. However, even they, like all other exotic species, are obligated to utilize dissolved oxygen in the water when submerged and are essentially hypoxia-intolerant. Ampullariids have a part of their mantle adapted into a gas-filled lung and can avoid anoxic conditions, but when submerged their ctenidial gill requires oxygenated water in order to survive.

Predators: The same organisms that disperse the species also probably prey on it.

Parasites: As with most freshwater snails, the species is probably an intermediate host for some digean trematodes.

Reproductive potential and life cycle: The species has ampullariid reproductive features. The sexes are separate and the female lays shelled eggs. The eggs hatch into young snails which, if water is absent, bury themselves in mud and respire using the right mantle compartment to breathe atmospheric oxygen or use dissolved oxygen extracted by a gill in the left mantle compartment.

Food and feeding habits: Cedeno-Leon and Thomas (1982) examined the feeding niches of both *Marisa cornuarietis* and *Biomphalaria glabrata* and found that juvenile *M. cornuarietis* fed more nonselectively than did *B. glabrata* on a variety of macrophytes, while adults fed on both macrophytes and eggs of *B. glabrata* under laboratory-controlled conditions. However, eggs of *B. glabrata* formed an insignificant part of the diet of *M. cornuarietis* of all ages, compared with plants. In nature, most gastropods, including *M. cornuarietis*, utilize fresh, senescent, or decaying plant materials as its main food source (Runham and Hunter 1970; Chatfield 1976; Cedeno-Leon and Thomas 1982). The species of plants ingested depend upon their texture, whether they are devoid of secondary plant factors that serve as repellents, feeding inhibitors, or toxicants and whether they contain attractants, arrestants, phagostimulants, feeding incitants, and nutrients (Linstedt 1971; Thomas 1982). Most gastropods are associated with aquatic plants because the plants detoxify environments by removing ammonia and carbon dioxide produced by the snails, they provide snails with shelter from predators, currents, ultraviolet radiation, and other inimical forces, the snails use plant surfaces for oviposition, a source of epiphytes, and for gaining access to the air–water interface, and the plants provide the snails with an increased supply of oxygen during the day, and resources such as calcium, amino acids and carboxylic acids in microhabitats with moribund plant tissue (Cdeno-Leon and Thomas 1982; Thomas 1982; Thomas et al. 1983). *Marisa cornuarietis* is also an efficient antagonist and predator of *Bulinus truncatus*, the transmitter of urinary bilharziasis (Demain and Kamel 1973; Demain and Lufty 1965, 1966).

Impact potential: Its ecological impact is unknown, but it is thought to be an efficient competitor for food resources (Cedeno-Leon and Thomas 1982). The species has potential to be of beneficial socioeconomic value because it acts as a predator on the eggs and juveniles of snails like *Biomphalaria glabrata* (see description below) that are intermediate hosts of trematodes causing schistosomiasis (Cedeno-Leno and Thomas 1982).

Control: None, since there is no evidence to suggest that the species requires immediate control.

Family: Thiaridae
The family shares features and can be confused with pleurocerids, but thiarids are parthenogenetic and their mantle edge has fleshy protuberances or papillae; snails of the family Pleuroceridae are dioecious and the mantle edge is smooth.

Melanoides tuberculata (red-rim Melania)

Native to much of Africa and the eastern Mediterranean countries, throughout India, Southeast Asia, Malaysia and southern China, northern Japan, northern Australia, and the New Hebrides, the species was introduced in the early 1960s into Texas (Murray 1964; Murray and Wopschell 1965), Arizona and Oregon, and Florida (Dundee 1974). Three other populations of the species have also been recorded from New Orleans, Louisiana (Dundee and Paine 1977). *Melanoides tuberculata* has a tropical distribution and apparently will not succeed in temperate climates. It has been recently introduced into the French West Indian islands of Guadeloupe and Martinique where it is now being used to help control *Biomphalaria glabrata*, the intermediate host of *Schistosoma mansoni,* the agent of intestinal schistosomiasis (Pointier et al. 1992). Brown (1979) has examined the biogeographical aspects of the species in Africa.

Considerable research has been carried out on this species because it is somewhat genetically unique, it is an intermediate host of the human liver fluke, *Opisthorchis sinensis*, and it has a high potential as a biological control agent for *B. glabrata*. The species is both parthenogenetic and bisexual and has the highest interpopulation Nei's distance (0.701) within any species of snail reported to date.

Species characteristics: Shell with rounded whorls sculptured with spiral threads and grooves and transverse lines which commonly develop into low costae. The intersections of spiral and transverse elements typically produce a reticulate or nodular pattern (Burch 1989). Most specimens within a population range from 20 – 30 mm, with a mean of 25 mm in shell height and 12 whorls; some specimens may reach 45 mm with 14.5 whorls (Dundee and Paine 1977)

Mode of life and habitat: The ecology of the species in North American waters has been described by Dundee and Paine (1977). Briefly, for Louisiana populations, it is most common in freshwater springs with low flow rates (depositional flow), a pH of 7.0 – 7.5, an alkalinity of 300 – 560 mg $CaCO_3/l$, water temperatures of 18 – 25°C, and soft mud or gravelly mud bottoms. The species is most commonly found on macrophytes, especially *Typha*, *Lysimachia* and grasses. This agrees with descriptions for tropical Indo-Pacific Islands, except the snail occurs in waters with pH as low as 6 (Starmuhlner 1979)

Dispersal mechanisms: Birds and mammals are probably the main dispersal vectors within a continent.

Adaptations: The species is adapted to tropical climates and has the ability to raise its upper lethal threshold temperature. Khot (1977) was able to raise the upper tolerance limit in warm acclimated *M. tuberculata* from 36.4°C to 39.0°C and in cold acclimated snails from 36.4°C to 33.0°C.

Predators: Waterfowl, aquatic mammals, and fish are probably the primary predators. In China, where raw fish is a delicacy, most of the liver fluke infections come from eating raw fish.

Parasites: The species is intermediate host of the human liver fluke (*Opisthorchis sinensis*) (Dundee and Paine 1977). Human infections are acquired through eating poorly cooked or raw fish in which metacercariae are found (Cheng 1973). The adult digean is found in biliary ducts and causes thickening of duct walls, leading to severe cirrhosis and ultimate death if not treated.

Reproductive potential and life cycle: The species typically reproduces by parthenogenesis, but apparently many populations are composed of both bisexual and parthenogenetic individuals (Jacobs 1957). *M. tuberculata* bears clusters of live young. Life history traits can vary greatly among morphs of parthenogenetic snails, indicating its ability to adapt to different demographic strategies (Pointier et al. 1992). The species tends to have *k*-strategy life history traits (McArthur and Wilson 1967) with long life spans (aver. 2–3.5 yr, max. 4–5 yr), relatively low fecundities (net reproduction rates of 8.6 – 35 young/day/female during the birth period, June to November) and an intrinsic rate of natural increase (0.20–0.30), low juvenile mortality (10–24% in first 2 years) but high adult mortality (60–100% in last 2 years), and slow growth rates (4–5 month to reach 10 mm, 14–17 month to reach 15 mm, >30 month to reach 20 mm, the average maximum size) (Leveque 1971; Berry and Kadri 1974; Dudgeon 1986; Pointier et al. 1992, 1993). Growth and fecundity rates may be slightly enhanced in warmer waters.

Food and feeding habits: Like all gastropods, the species is a grazer on attached algae.

Impact potential: Its greatest impact will probably be in relation to its role as intermediate host of the human liver fluke (*Opisthorchis sinensis*) (Dundee and Paine 1977).

Control: Since the species is a tropical species, freezing will probably be an effective control method, although it has not been tested. Molluscicides have been used in the past to control the species and eliminate or reduce its role as an intermediate host of the human liver fluke.

Tarebia (Thiara) granifera (quilted Melania)

Native to Malaysia and the Philippines, this species was introduced to Florida in the early 1950s (Abbott 1952) and Texas in the early 1960s (Murray 1964).

Species characteristics: Shell with flattened whorls, especially on the spire, sculpturing of spiral rows of beads and nodules which are generally aligned in transverse rows (Burch 1989).

Mode of life and habitat: The species has an ecology similar to *M. tuberculata* (Murray and Wopschall 1965).

Dispersal mechanisms: Dispersal agents are probably waterfowl and aquatic mammals. The shells are so large that currents probably play little role in even dispersing the species within a body of water.

Adaptations: Poorly known.

Predators: Probably fish, waterfowl, and aquatic mammals.

Parasites: The species is an intermediate host of the Oriental lung fluke, *Paragonimus westermani* (Abbott 1952), which causes paragonimiasis in humans. The fluke develops through its redial and cercarial stages in the snail, releasing the cercaria into the water. The cercariae enter the skin of humans wading or bathing in the water and develop into the adult worm in the lung or other tissues of the human, causing the disease known as paragonimiasis. The lung fluke is also a common parasite of mink in the U.S. Human infestation occurs through eating raw crabs or crayfish that act as second intermediate hosts of the flukes.

Reproductive potential and life cycle: Reproduction parthenogenetic, without males. Females have a brood pouch in the neck region, but completely separate from the uterus and enclosing part of the haemocoele (Burch 1989).

Food and feeding habits: Probably feeds on epiphytic algae and on leaf and stem tissues of plants large enough to support their weight.

Impact potential: The species has potential impact similar to *M. tuberculata*, but it is the interme-diate host of the Oriental lung fluke, *Paragonimus westermani* (Abbott 1952).

Control: Although the species is a potential intermediate host of the Oriental lung fluke, the fluke has not been reported in any North American populations and appears to require no immediate control. In fact, the species appears to have some beneficial value in that it has been used to biologically control *Biomphalaria glabrata* in marsh and stream complexes in Saint Lucia, West Indies (Prentice 1983).

Subclass: Pulmonata

Snails in this subclass lack an operculum, their mantle cavity serves as a lung for air-breathing and most are hermaphroditic.

Family: Physidae

Snails with a spiral shell but coiling is sinistral (coils to the left). Most physids have moderately small shells (<2 cm high) that are thin and semi-transparent.

Physella acuta (European Physa)

Native to Europe, Mediterranean regions, and Africa, the species has been introduced into Australia, Hawaii, and perhaps parts of continental U.S. (Burch 1989). Its current distribution is unknown, as is most of its biology and ecology.

Species characteristics: Adult shell about 1–1.5 cm high; spire tall and narrow with deep sutures, with 4–4.5 whorls; columella well developed; aperture elongate.

Mode of life and habitat: There is no information on its ecological requirements in North American habitats. Most physids prefer standing water with plenty of vegetation. They can live in enriched waters and, being pulmonates, they can avoid anoxia by rising to the surface to breathe atmospheric oxygen.

Dispersal mechanisms: Probably similar to most gastropods; waterfowl, aquatic mammals and large insects are common vectors.

Adaptations: There is no information on the species' morphological, behavioral, or physiological tolerances and requirements.

Predators: Fish, waterfowl, and aquatic mammals are common predators of all gastropods.

Parasites: None are known for *P. acuta*, however physids are well-known second intermediate host for a variety of digean trematodes. The cercariae develop in the snail intestine and are evacuated with the faeces. The faeces are then eaten or taken in with water consumed by the final host. At least four families of digean trematodes use *Physa* species for the second intermediate host, including the strigeid, *Cotylurus flabelliformis* (ducks are the final host), the echinostomatid, *Echinistoma revolutum* (ducks are the final host), the gorgoderid, *Gorgodera amplicava* (frogs are the final host), and the halipegid, *Halipegus eccentricus* (frogs are the final host). None of the physids appear to harbor digeans with quarantine significance.

Reproductive potential and life cycle: Not described for *P. acuta*. All physids are oviparous, most laying eggs in clusters on aquatic plants and submerged objects. Most have short life spans, typically less than a year.

Food and feeding habits: Probably epiphytic algae and leaf and stem tissues of submerged and floating macrophytes.

Impact potential: There is no information on the species ecological or socioeconomic impacts.

Control: None, since there is no evidence to suggest that the species requires immediate control.

Stenophysa marmorata (marbled aplexa)

Introduced into Texas and native to Brazil, Guatemala, Uruguay, Venezuela, and the West Indies (Burch 1989), but its biology, ecology, and ecological and socioeconomic significance are poorly understood.

Species characteristics: Stenophysa species are distinguished from *Physa* and *Physella* species by the mantle edges being serrated and extending beyond the edge of the shell apertural lip and partly overlapping the shell; in *Physella* the mantle edge is smooth or has digitations, only on the parietal side of the mantle and the mantle does not extend beyond the shell apertural lip; in *Physa* the mantle also has digitations but they occur on both sides of the mantle. *Stenophysa marmorata* is distinguished from *S. maugeriae* from its relatively small size (<16 mm high) and a horn to light tan color, the shell being translucent and rarely variegated (Burch 1989). See description of *S. maugeriae* for comparison.

Mode of life and habitat: It appears to be prefer warm water in ponds and slow-moving rivers.

Dispersal mechanisms: Unknown, but it probably relies on waterfowl, fish, aquatic mammals, and perhaps large insects as dispersal agents.

Adaptations: None known.

Predators: Probably waterfowl, fish, and aquatic mammals.

Parasites: None known.

Reproductive potential and life cycle: Not described, but all physids are oviparous, most laying eggs in clusters on aquatic plants and submerged objects. Life spans is probably less than a year, as with most physids.

Food and feeding habits: Probably epiphytic algae and leaf and stem tissues of submerged and floating macrophytes, as with most physids.

Impact potential: None reported to date, but probably very little.

Control: None, since there is no evidence to suggest that the species requires immediate control.

Stenophysa maugeriae (tawny aplexa)

The species is native to Mexico and is recorded only from Texas (Burch 1989). Its biology and ecological and socioeconomic significance in North America is also unknown.

Species characteristics: Shell relatively large, up to 30 mm or more high, tan to chestnut brown in color, opaque, commonly variegated (Burch 1989).

Mode of life and habitat: Like *S. marmorata* it appears to prefer warm water in ponds or slow-moving rivers.

Dispersal mechanisms: Unknown but it probably relies on waterfowl, fish, aquatic mammals, and perhaps large insects as dispersal agents.

Adaptations: None known.

Predators: Probably waterfowl, fish and aquatic mammals.

Parasites: None known.

Reproductive potential and life cycle: See *S. marmorata*.

Food and feeding habits: See *S. marmorata*.

Impact potential: None reported to date, but probably very little.

Control: None, since there is no evidence to suggest that the species requires immediate control.

Family: Planorbidae

Shells minute (<1 mm) to relatively large (>30 mm) and coiled in one plane (discoidal in shape). All animals are sinistral (but some shells may appear to be dextral, i.e., appear coiled to the right because aperture is tipped to left side making the lower side the spire side and the upper side the umbilical side) and have excretory, respiratory, and reproductive openings on the left side.

Biomphalaria glabrata (bloodfluke Planorb)

Introduced to Florida, the species is of quarantine significance because it is the intermediate host of *Schistosoma mansoni*. It is native to West Indies, Venezuela, Surinam, French Guiana, and Brazil (Burch 1989). Much of the research has been oriented toward methods to control its spread, such as developing attractants and arrestants (Thomas 1982; Thomas et al. 1975; Thomas et al.. 1983).

Species characteristics: Adult shell larger than 8 mm and up to more than 30 mm, shell thin, fragile, body whorl relatively depressed, adults larger than 15 mm with five or more whorls.

Mode of life and habitat: It appears to prefer warm, slow-moving waters or ponds. It is not common in streams with flows exceeding 0.3 m/s and prefers depositional substrates. The species is closely associated with aquatic macrophytes and is most abundant in water basins with a variety of vegetation. They feed on plants of nearly all habits, including submerged species with roots (*Nasturtium, Elodea, Potamogeton, Myriophyllum*), submerged species without roots (*Ceratophyllum*), and floating species with floating roots (*Eicchornia, Salvinia, Pistia, Lemna*), although they prefer *Nasturtium* and *Lemna* (Cedeno-Leon and Thomas 1982).

Dispersal mechanisms: Probably waterfowl on their feathers and feet, large beetles and hemipterans, amphibians, fish and aquatic mammals.

Adaptations: The species has an ability to secrete lamellae at the shell aperture and emigrate from water to enter a period of dormancy that may last for at least 512 days (Dannemann and Pieri 1993). Even nonlamellate snails can survive out of water for at least 256 days. Apparently the lamellae increase the shell thickness and decrease the area of the shell aperture, which reduces water loss and enhances survival under extreme conditions (Dannemann and Pieri 1993).

Predators: Mostly fish, waterfowl, and aquatic mammals, but *Marisa cornuarietis* has been shown to prey on eggs and young of the species.

Parasites: The deadly blood fluke, *Schistosoma mansoni*, uses *Biomphalaria glabrata* for its intermediate host. The fluke develops through a series of larval stages called sporocysts, rediae, and cercariae in the snail, which releases the cercarial stage that enters the skin of humans wading or swimming in the water. The cercariae find their way into intestinal veins of humans where they develop into adult worms and block the flow of blood in the veins, the disease being known as schistosomiasis.

Reproductive potential and life cycle: The species tends to have *r*-strategy life history traits, with rapid growth rates (18 mm in 7 month), high reproductive rates (e.g., an intrinsic rate of increase of 0.84–0.88 and a net reproduction rate of 101–153 eggs/day), a short life span (less than a year), a short mean generation time (121–124 days) and a high adult mortality rate, especially after 160 days (about half its life span), when 60% of the population dies (Pointier et al. 1991). The species is a facultatively self-fertile hermaphrodite and when isolated will reproduce readily by self-fertilization, but normally, in wild populations, cross-fertilization is the rule (Paraense 1976; Vernon et al. 1995). There is a plethora of literature, too voluminous to consider here, on genetics and reproductive behavior of *B. glabrata*, as indicated by Vernon and Taylor (1996), and references therein, who showed that mating occurs equally often between snails in male and female roles.

Food and feeding habits: Feeds on epiphytic algae and leaf and stem tissues of submersed and floating aquatic plants (Cedeno-Leon and Thomas 1982).

Impact potential: As with all species of *Biomphalaria*, there is a very high risk of infection with deadly schistosomes for any humans wading, swimming, or bathing in waters with the snail.

Control: Numerous studies have been done to try to control the population sizes and spread of *B. glabrata*. The most promising is a biological control using *Marisa cornuarietis*, for three reasons. First, it acts as a predator on the eggs and juveniles of *B. glabrata* and other snail hosts of schistosomiasis. However, Cedeno-Leon and Thomas (1982) showed that even though *Marisa* feeds on both eggs and juveniles of *B. glabrata*, they preferred plant material over snail diets. Second, *Marisa* may release allelopathic factors (Cedeno-Leon 1975). Third, *Marisa* is known to be an efficient competitor for food resources (Cedeno-Leon and Thomas 1982). Thomas (1982, and literature cited therein) also suggests that for *B. glabrata* two short-chain carboxylic acids, propionic and *n*-butyric acids, may act as pheromones applied in a controlled release formulation to attract conspecific snails and remove *B. glabrata* selectively from the environment. Pheromones are chemicals released by an individual that may act on conspecifics causing a behavioral response (releaser effect) or a physiological response (primer effect). Alternatively, aquatic plants may also be useful control agents. Some plant (and animal) species release allomones (chemicals released from one species to act as a repellent, feeding inhibitor, or toxicant to a different species, such as some snails). Others release kairomones (chemicals released from one species to act as an attractant or an arrestant to a different species). Bousefield (1978, 1979) found that certain macromolecules, such as proteins and polypeptides, in some aquatic plants can serve as potent attractants to *B. glabrata*. However, low molecular weight compounds such as amino and carboxylic acids have faster diffusion rates and are probably more important as allomones and kairomones (Wilson 1970; Uhazy et al. 1978). Thomas and Assefa (1979) and Thomas et al. (1980) suggest that *B. glabrata* can detect conspecifics by responding to gradients in the concentration of H^+, HCO_3^-, NH_4^+, Ca^{2+}, Mg^{2+}, amino acids, or carboxylic acids and changes in the Mg^{2+}/Ca^{2+} ratio. Ca/Mg ratios, in particular, are important in the physiology and cationic responses of organs and haemolymph of *B. glabrata* (Nduku and Harrison 1980a, b). Calcium alone is a key factor in determining the abundance and distribution of snails (Thomas et al. 1974). *Biomphalaria glabrata* can tolerate levels as low as 0.025 mM (1 mg) Ca/l.

More recently, *Melanoides tuberculata* has been used successfully as a competitor to control *Biomphalaria* in watercress beds in Martinique (Pointier et al. 1989, 1992) and ponds in Guadeloupe (Pointier 1989). Colonization by intentional introductions of *M. tuberculata* is usually successful in a variety of habitats, but the impact on the snail host may vary according to site (Pointier et al. 1993).

The variation in success of snail competitors to control *Biomphalaria* may depend in part on the affect of the control agent on growth and cercarial production in *Biomphalaria*. Mone (1991) found that the presence of nontarget species, including *Marisa cornuarietis* and *Melanoides tuberculata* may actually enhance growth and cercarial production in *Biomphalaria glabrata*. The growth in size of shell in *B. glabrata* was correlated with and increase in interior volume and space for *Schistosoma*. Mone (1993) proposed two mechanisms for this result: First, excretion by the nontarget species of additional nutritive substances that are ingested by *Biomphalaria*. This includes, indirectly, exogenous plant factors that act as phagostimulants or feeding stimulants, causing nontarget species to ingest more food. Second, secretion of growth-promoting factors by the nontarget species may enhance growth of *Biomphalaria*. However, these results are in contrast to those discussed above.

CLASS: BIVALVIA

Subclass: Heterodonta

Family: Sphaeriidae
Mostly tiny (<5 mm long) clams (genus *Pisidium*), but a few as large as 2.5 cm long (genus *Sphaerium*). Shell lacks a nacre and has a complex-crossed lamellar structure instead (Mackie 1978a). Hinge plate with cardinal and lateral teeth. Hermaphroditic, ovoviviparous with larvae maturing in brood pouches (Mackie 1978b).

Pisidium punctiferum (striate pea clam)

This rare species has been recorded only from Florida and Texas (Burch 1975b). Native to Mexico, Central America, and the Caribbean Islands, it appears to have advanced northward from Mexico into Texas and then into Florida. The date of its introduction is unknown. It appears to be a tropical species, but its ecological and socioeconomic impacts are unknown.

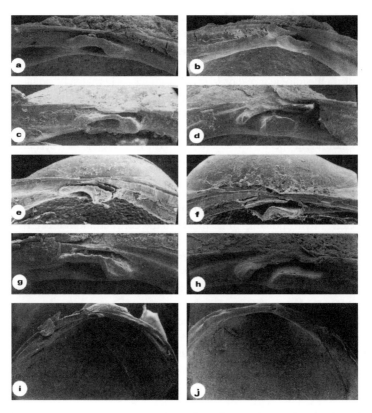

FIGURE 9.1   Hinge features of *Pisidium punctiferum* from Costa Rica (a–d and g–j) and Brazil (e-f). a, c, e, g show C3; b, d, f, h show C2 (lower tooth) and C4; i and j show right and left hinge plates, respectively.

Species characteristics: Adult shell moderately small (<3 mm long), rather high in outline. Walls rather thin, moderately inflated. Striae prominent, fine, irregularly spaced. Periostracum yellowish, surface dull to silky. Beaks low, broad, not very prominent. Anterior end rather short, with a rounded slope, forming a rounded angle with ventral margin. Posterior end rounded, joining the ventral margin imperceptibly. Ventral margin long and gently curved.

Mode of life and habitat: Burch (1975) suggests that standing or slow-moving water, aquatic vegetation, and muddy substrates are typical habitat features for the species. Sphaeriids generally inhabit shallow waters, although a few (e.g., *Pisidium conventus*, *Sphaerium nitidum*) are found, or are even restricted to, cold hypolimnia of oligotrophic lakes.

Dispersal mechanisms: Mackie (1979) reviewed the dispersal mechanisms for several species of fingernail clams, but there is no literature that describes dispersal mechanisms for *P. punctiferum*. Other sphaeriids are dispersed by waterfowl, both externally on feathers and feet and internally through the digestive tract, large insects, amphibians, fish, and aquatic mammals.

Adaptations: Of all the exotic species of molluscs, only a few appear to have adaptations for surviving prolonged periods of desiccation or exposure. Most species of *Pomacea* can burrow into the mud and survive several months without water. Other species of introduced gastropods cannot tolerate even short periods of desiccation. Although some native sphaeriids (e.g., *Sphaerium occidentale*, *Musculium securis*) have physiological and life history adaptations for life in ephemeral habitats (McKee and Mackie 1980, 1981, 1983), the exotic species of sphaeriids have been reported only from permanent bodies of water, in fact mostly large lakes or rivers.

Predators: Fish, waterfowl, and aquatic mammals, perhaps some amphibians such as mud puppies.

Parasites: None reported to date for the species. Mackie (1976) reviews trematode parasitism in the family, but none are of quarantine significance. All species of sphaeriids are second intermediate hosts, the final hosts being fish (trout, salmon, bullheads, bass) and amphibians (frogs and salamanders).

Reproductive potential and life cycle: Unknown for the species. All sphaeriids are hermaphrodites and ovoviviparous, producing young that are miniatures of the adult (Mackie 1978b). The larvae are brooded in marsupial sacs produced on filaments of the inner gill. Larvae develop through four stages (embryos, fetal larvae, prodissoconch larvae, and extra-marsupial larvae) before being born. Most sphaeriids have small broods (~1–10 per adult), some producing up to three broods per year. Most have a short life span, some as short as 4–8 months, others 12–24 months.

Food and feeding habits: All bivalves are filter feeders, with some evidence of deposit feeding. Food consists mostly of bacteria, detritus, and algae, especially diatoms.

Impact potential: Probably none. Most sphaeriids have a low productivity and burrow in the sediments. Their only competitors are unionid clams, which tend to inhabit shallow water, but they too have a low productivity.

Control: None, since there is no evidence to suggest that the species requires immediate control.

SUMMARY OF IMPACT POTENTIAL OF SPECIES INTRODUCED VIA THE AQUARIUM TRADE

SPECIES WITH LITTLE OR NO ECOLOGICAL IMPACT

Most species introduced via the aquarium trade will have little or no obvious impacts on the ecology of the ecosystem. Although most freshwater molluscs are intermediate hosts of numerous species of digean trematodes whose definitive hosts are fish or waterfowl, there are no studies that have assessed the added impact of trematodes in molluscs introduced via the aquarium trade on definitive hosts. It is not anticipated that the exotic species of molluscs will add significantly to the infestation of definitive hosts that are already infected by trematodes of native species of molluscs. As discussed in Chapter 15, all exotic species will have to compete with their native counterparts for food and space, but very few species introduced via the aquarium trade are known to have replaced native species; rather they seem to have added to the diversity of molluscan fauna. Included in this group are

Physella acuta
Stenophysa marmorata? (little known about its ecology*)*
Stenophysa maugeriae? (little known about its ecology)
Pisidium punctiferum

In summary, four of the eight species of molluscs introduced by the aquarium trade appear to be of little or no potential ecological risk.

SPECIES OF POTENTIAL SOCIOECONOMIC BENEFIT

Species that are known to have some beneficial uses (given in parentheses) include

> *Marisa cornuarietis* (as predator of eggs and young and competitor of adults of *Biomphalaria glabrata*)
> *Melanoides tuberculata* (as competitor of juvenile and adult *Biomphalaria glabrata*)
> *Tarebia granifera* (as competitor of juvenile and adult *Biomphalaria glabrata*)

In summary, three of the eight species of molluscs introduced by the aquarium trade have been demonstrated to have some beneficial use.

SPECIES OF POTENTIAL ECOLOGICAL AND/OR SOCIOECONOMIC HARM

There are several types of potentially harmful socioeconomic impacts. These include (i) Vectors of Human Parasites; (ii) Agricultural Impacts; (iii) Industrial and/or Utility Impacts; (iv) Navigational Impacts; and (v) Recreational Impacts. The ecological impacts can be divided into (i) Biological Impacts and (ii) Limnological Impacts.

Socioeconomic Impacts

(i) Vectors of Human Parasites:

Species known to act as intermediate hosts of parasites harmful to man include (parasites given in parentheses):

> *Melanoides tuberculata* (human liver fluke, *Opisthorchis sinensis*)
> *Tarebia granifera* (Oriental lung fluke, *Paragonimus westermani*)
> *Biomphalaria glabrata* (human blood fluke, *Schistosoma mansoni*)

These mollusc species, in particular, should be actively searched for and examined for parasites. To date, only *B. glabrata* has been demonstrated to carry parasites of human quarantine significance in North America and should be eliminated before it spreads. Ironically, two of the species (*Melanoides tuberculata*, *Thiara granifera*) have excellent potential to control the abundance and spread of *B. glabrata*.

SUMMARY

Table 9.1 summarizes the country of origin, date of entry, first documented location in North America, and their reported impact(s), when known, for molluscs introduced via the aquarium trade.

TABLE 9.1
Summary of Species Introduced Via the Aquarium Trade and Their Origin, Date of Entry, First Documented Location in North America and Reported Impact(s), When Known

Species	Origin	Date	First Location	Impact(s
Marisa cornuarietis	South America	late 1980s	Hawaii	Competition for food resources; predator of eggs and young of *Biomphalaria glabrata* a benefit
Melanoides tuberculata	Africa	early 1960s	Texas	Intermediate host of human liver fluke, *Opisthorchis sinensis*
Tarebia granifera	Malaysia/Philippines	early 1950s	Florida	Intermediate host of Oriental lung fluke, *Paragonimus westermani*
Physella acuta	Europe?	1980s?	Hawaii	None reported to date
Stenophysa marmorata	South America/ Mexico	1970s?	Texas	None reported to date
Stenophysa maugeriae	Mexico	1970s?	Texas	None reported to date
Biomphalaria glabrata	South America	1970s?	Florida	Intermediate host of *Schistosoma mansoni*, cause of schistosomiasis
Pisidium punctiferum	Mexico	1970s?	Florida/ Texas	None reported to date

REFERENCES

Abbott, R.T. 1952. A study of an intermediate snail host (*Thiara granifera*) of the Oriental lung fluke (*Paragonimus*), *Proc. U.S. Natl. Mus.*, 102: 71–116 pls. 8,9.

Aldridge, D.W. 1983. Physiological ecology of freshwater prosobranchs. In W.D. Russel-Hunter (Ed.), *The Mollusca*, Vol. 6, *Ecology*, Academic Press, New York. pp. 330–359.

Berry, A.J. and A.B.H. Kadri. 1974. Reproduction in the Malaysian freshwater cerithiacean gastropod *Melanoides tuberculata*, *J. Zool.*, London 172: 369–381.

Bousefield, J.D. 1978. Rheotaxis and chemoreception in the freshwater snail *Biomphalaria glabrata* (Say) — estimation of the molecular weight of active factors, *Biol. Bull.*, 154: 361–373.

Bousefield, J.D. 1979. Plant extracts and possibly triggered rheotaxis in *Biomphalaria glabrata* (Say) snail intermediate host of *Schistosoma mansoni* (Sambon)., *J. Appl. Ecol.*, 16: 681–690.

Brown, D.S. 1979. Biogeographical aspects of African freshwater gastropods, *Malacologia*, 18: 79–102.

Burch, J.B. 1975. *Freshwater Sphaericean Clams (Mollusca: Pelecypoda) of North America*. EPA Indentification Manual no. 35, Malacological Publications, Hamburg, MI.

Burch, J.B. 1989. *North American Freshwater Snails*, Malacological Publ., Hamburg, MI.

Cedeno-Leon, A. and J.D. Thomas. 1982. Competition between *Biomphalaria glabrata* (Say) and *Marisa cornuarietis* (L.): Feeding niches, *J. Appl. Ecol.*, 19: 707–721.

Chatfield, J.E. 1976. Studies on food and feeding in some European land molluscs, *J. Conch.*, 29: 5–20.

Cheng, T.C. 1973. *General Parasitology*. Academic Press, New York.

Dannemann, R.D.A. and O.S. Pieri. 1993. Prolonged survival out of water of polymorphic *Biomphalaria glabrata* (Say) from a seasonally drying habitat of North-East Brazil, *J. Moll. Stud.*, 59: 263–265.

Demain, E.S. and E.G. Kamel. 1973. Effects of *Marisa cornuarietis* on *Bulinus truncatus* populations under semi-field conditions in Egypt, *Malacologia*, 14: 439.

Demain, E.S. and R.G. Lufty. 1965a. Predatory activity of *Marisa cornuarietis* against *Bulinus truncatus*, the transmitter of urinary schistosomiasis, *Ann. Trop. Med. Parasit.*, 59: 331–337.

Demain, E.S. and R.G. Lufty. 1966. Factors affecting the predation of *Marisa cornuarietis* on *Bulinus truncatus*, *Biomphalaria alexandria* and Lymnaea caillaudi, *Oikos*, 17: 212–230.

Dudgeon, D. 1986. The life cycle, population dynamics and productivity of *Melanoides tuberculata* (Muller, 1974) (Gastropoda: Prosobranchia: Thiaridae) in Hong Kong, *J. Zool.*, London, 208: 37–53.

Dundee, D.S. 1974. Catalog of introduced molluscs of eastern North America (north of Mexico), *Sterkiana*, 55: 1–37.

Dundee, D.S. and A. Paine. 1977. Ecology of the snail, *Melanoides tuberculata* (Muller), intermediate host of the human liver fluke (*Opisthorchis sinensis*) in New Orleans, Louisiana, *Nautilus*, 91: 17–20.

Hunt, B.P. 1958. Introduction of *Marisa* into Florida, *Nautilus*, 72: 53–55.

Jacobs, J. 1957. Cytological studies of Melaniidae (Mollusca) with special reference to parthenogenesis and polyploidy. I. Oogenesis of the parthenogenetic species of Melanoides (Prosobranchia – Gastropoda), *Trans. Roy. Soc. Edinburgh*, 63: 341–352.

Khot, R.P. 1977. Studies on some physiological aspects and control of the snail, *Melanoides tuberculatus*. Ph.D thesis, Marathwada University, Aurangabad, India. (Not seen, cited by Lomte 1979).

Leveque, C. 1971. Equation de Von Bertalanffy et croissance des mollusques benthiques du lac Tchad, *Cahiers de l'O.R.S.T.O.M., serie Hydrobiolgie*, 5: 263–283. Not seen, cited by Poitier et al. 1992).

Lindstedt, K.J. 1971. Chemical control of feeding behaviour, *Comp. Biochem. Physiol.*, 39A: 553–581.

Lomte, V.S. 1979. Thermoregulation in the freshwater lamellibranch *Parreysia corrugata*, *Malacologia*, 18: 257–263).

Mackie , G.L. 1976. Trematode parasitism in the Sphaeriidae clams, and the effects on three Ottawa River species, *Nautilus*, 90: 36–41.

Mackie, G.L. 1978a. Shell structure in freshwater Sphaeriacea (Bivalvia: Heterodonta), *Canad. J. Zool.*, 56: 1–6.

Mackie, G.L. 1978b. Are sphaeriid clams ovoviviparous or viviparous? *Nautilus*, 92: 135–146.

Mackie, G.L. 1979. Growth dynamics in natural populations of Sphaeriidae clams (*Sphaerium, Musculium, Pisidium*), *Can. J. Zool.*, 57:441–456.

McKee, P.M. and G.L. Mackie. 1980. Desiccation resistance in *Sphaerium occidentale* and *Musculium securis* (Bivalvia: Sphaeriidae) from a temporary pond, *Can. J. Zool.*, 58: 1693–1696.

McKee, P.M. and G.L. Mackie. 1981. Life history adaptations of the fingernail clams *Sphaerium occidentale* and *Musculium securis* (Bivalvia: Sphaeriidae) to ephemeral habitats, *Can. J. Zool.*, 59: 2219–2229.

McKee, P. M. and G. L. Mackie. 1981. Respiratory adaptations of the fingernail clams *Sphaerium occidentale* and *Musculium securis* (Bivalvia: Sphaeriidae) to ephemeral habitats, *Can. J. Fish. Aq. Sci.*, 49: 783–791.

Mone, H. 1993. Influence of non-target molluscs on the growth of *Biomphalaria glabrata* infected with *Schistosoma mansoni*: Correlation between growth and cercarial production, *J. Moll. Stud.*, 57: 1–10.

Murray, H.A. 1964. *Tarebia granifera* and *Melanoides tuberculata* in Texas, *Ann. Rep. Am. Malacol. Union*, 1964. 31: 15–16.

Murray, H.A. and L.J. Wopschell. 1965. Ecology of *Melanoides tuberculata* (Muller) and *Tarebia granifera* (Lamarck) in south Texas, *Ann. Rep. Am. Malacol. Union*, 1965 32: 25–26.

Nduku, W.K. and A.D. Harrison. 1980a. Cationic responses of organs and haemolymph of *Biomphalaria pfeifferi* (Krauss), *Biomphalaria glabrata* (Say) and *Helisoma trivolvis* (Say) (Gastropoda: Planorbidae) to cationic alterations of the medium, *Hydrobiologia*, 68: 119–138.

Nduku, W.K. and A.D. Harrison. 1980b. Water relations and osmotic pressure in *Biomphalaria pfeifferi* (Krauss), *Biomphalaria glabrata* (Say) and *Helisoma trivolvis* (Say) (Gastropoda: Planorbidae) in response to cationic alterations of the medium, *Hydrobiologia*, 68: 139–144.

Paraense, W.L. 1976. the sites of cross- and self-fertilisation in planorbid snails, *Rev. Brazil. Biol.*, 36: 535–539.

Pointier, J.P. 1989. Comparison between two biological control trials of *Biomphalaria glabrata* in a pond in Guadeloupe, French West Indies, *J. Med. Appl. Malacol.*, 1: 83–95.

Pointier, J.P., B. Delay, J.L. Toffart, M. Lefevre, and R. Romero-Alvarez. 1992. Life history traits of three morphs of *Melanoides tuberculata* (Gastropoda: Thiaridae), an invading snail in the French West Indies, *J. Moll. Stud.*, 58: 415–423.

Pointier, J.P., A. Guyard and A. Mosser. 1989. Biological control of *Biomphalaria glabrata,* and *B. straminea* by the competitor snail *Thiara tuberculata* in a transmission site of schistosomiasis in Martinique, French West Indies, *Ann. Trop. Med. Parasit.*, 83: 263–269.

Pointier, J.P., A. Theron and G. Borel. 1993. Ecology of the introduced snail *Melanoides tuberculata* (Gastropoda: Thiaridae) in relation to Biomphalaria glabrata in the marshy forest zone of Guadeloupe, French West Indies, *J. Moll. Stud.*, 59: 421–428.

Pointier, J.P., J.L. Toffart, and M. Lefevre. 1991. Life tables of freshwater snails of the genus *Biomphalaria* (*B. glabrata, B. alexandrina, B. straminea*) and one of its competitors *Melanoides tuberculata* under laboratory conditions, *Malacologia*, 33: 43–54.

Prentice, M.A. 1983. Displacement of *Biomphalaria glabrata* by the snail *Thiara granifera* in field habitats in St. Lucia, West Indies, *Ann. Trop. Med. Parasitol.*, 77: 51–59.

Runham, N.W. and P.J. Hunter. 1970. *Terrestrial Slugs*, Hutchinson, London.

Starmuhlner, F. 1979. Distribution of freshwater molluscs in mountain streams of tropical Indo-Pacific islands (Madagascar, Ceylon, New Caledonia), *Malacologia*, 18: 245–255.

Thomas, J.D. 1973. Schistosomiasis and the control of the molluscan host of human schistosomes with particular reference to possible self-regulatory mechanisms, *Adv. Parasitol.*, 11: 307–394.

Thomas, J.D. 1982. Chemical ecology of the snail hosts of schistosomiasis: snail-snail and snail-plant interactions, *Malacologia*, 22: 81–91.

Thomas, J.D. and B. Assefa. 1979. behavioural responses of amino acids by juvenile *Biomphalaria glabrata*, a snail host of Schistosoma mansoni, *Comp. Biochem. Physiol.*, 466: 17–27.

Thomas, J.D., B. Assefa, C. Cowley and J. Ofosu-Barko. 1980. Behavioural responses to amino acids and related compounds, including propionic acid by adult *Biomphalaria glabrata* (Say) a snail host of *Schistosoma mansoni, Comp. Biochem. Physiol.*, 66C: 17–27.

Thomas, J.D., M. Benjamin, A. Lough and R.H. Aram. 1974. The effects of calcium in the external environment on the growth and natality rates of *Biomphalaria glabrata* (Say), *J. Ann. Ecol.*, 43: 839–860.

Thomas, J. D. G. J. Goldsworthy, and R. H. Aram. 1975. Studies on the chemical ecology of snails: The effect of chemical conditioning by the adult snails on the growth of juvenile snails, *J. An. Ecol.*, 44: 1–27.

Thomas, J.D., B. Grealy, and C.F. Fennell. 1983. The effects of varying the quantity and quality of various plants on feeding and growth of *Biomphalaria glabrata* (Gastropoda), *Oikos,* 41: 77–90.

Uhazy, S.L., R.T. Tanaka, and A.J. MacInnis. 1978. *Schistosoma mansoni*: Identification of chemicals that attract or trap its snail vector, *Biomphalaria glabrata, Science*, 201: 924–926.

Vernon, J.G., C.S. Jones, and L.R. Noble. 1995. Random amplified polymorphic DNA markers (RAPDs) reveal cross-fertilization in *Biomphalaria glabrata* (Pulmonata: Basommatophora), *J. Moll. Stud.*, 61: 455–465.

Vernon, J.G. and J.K. Taylor. 1996. Patterns of sexual roles adopted by the schistosome-vector snail *Biomphalaria glabrata* (Planorbidae), *J. Moll. Stud.*, 62: 235–241.

Wilson, E.O. 1970. Chemical communication within animal species. In: Sondheimer, E. and J.B. Simone, *Chemical Ecology*, pp. 133–155. Academic Press, N. Y.

10 Description, Biology, and Ecological Impact of the Screw Snail, *Thiara Tuberculata* (Müller, 1774) (Gastropoda: Thiaridae) in Mexico

*Alberto Contreras-Arquieta and
Salvador Contreras-Balderas*

INTRODUCTION

Freshwater snail studies are scarce in Mexico. The interest in knowledge of snails was high only at the end of the 19th century and early in the 20th century (Burch and Cruz-Reyes, 1987). Thus it is difficult to determine how many species of snails or clams exist in Mexico. Since 1960 there have been only sporadic papers, giving a limited vision of the Mexican freshwater snail fauna (Contreras-Arquieta, 1991).

Most non-marine molluscs have a well-defined range consisting of a single, usually continuous region. However, a small number of species have either been introduced or have naturally invaded many regions (Smith, 1989). Introductions of freshwater molluscs in Mexico probably started in the 1960s, when samples of the screw snail *Thiara tuberculata* (= *Melanoides tuberculata*) (Müller 1774; Morrison, 1954) and the *Corbicula manilensis* and *C. flaminea* asiatic clams appeared (Philippi, 1844; Hills and Mayden, 1985; Torres-Orozco and Revueltas-Valle, 1996). The screw snail is considered a travelling snail with high ecological significance in North, Central, and South America (Smith, 1989).

The aquarium trade and fans of aquaria may be the main agents of the introduction and distribution of molluscs and many other freshwater plant and animal species in Mexico via careless washing of tanks, and disposal of water and detritus into public sewage systems. The snail may survive a long time under those conditions.

The screw snail was described from the Coast of Coromandel, India, by Müller in 1774, and was first recorded in Mexico by Abbott (1973). Reddell (1981) reported a large population of screw snail as *M. maculata* from Cueva de Cantil Blanco, Veracruz, and also mentioned three species of unidentified thiarids from caves in Mexico. Jiménez-G., Guajardo-M. and Torres-G. (1980) reported *Oncomelania* sp., from Parras de La Fuente, Coahuila, Mexico, based on young screw snails; later Contreras-Balderas and Maeda-Martínez (1985) reported the same species as *Tarebia* sp. from the same locality. Contreras-Arquieta et al. (1995a) reidentified them as *Thiara*, and reported their distribution in Mexico in 12 states and 65 localities, 63 of which were new records. Contreras-Arquieta (1998) reported new records for Valle de Cuatro Ciénegas and other localities in the state of Coahuila, Mexico.

The impact of the screw snail in Mexico is little known, but Contreras-Arquieta et al. (1995b) presented data on the probable impacts in some areas of northeast Mexico, showing a 25% to 100% loss of native molluscs. To measure or predict the impact of the screw snail on the native molluscan fauna requires knowledge of its biology, geographical distribution, and environmental conditions under which it thrives.

DISTRIBUTION OF THIARIDAE

Worldwide: Many African countries — some endemic subgenera occur in Lakes Tanganyika and Malawi (Banarescu, 1990); Eastern Mediterranean; throughout India, Southeastern Asia, Malaysia and South China, North to Isla Ryukyu and Japan, South and East through Pacific islands, Australia and New Hebrides; southeastern Europe (Burch, 1960, 1982; Morrison, 1954; Taylor, 1981); introduced in U.S. in Texas (Murray, 1964), Oregon (Murray, 1971), Florida (Dundee, 1974), Louisiana (Dundee and Paine, 1977), Arizona (Taylor, 1981), and Nevada (Williams and Sada, 1985); and in Mexico (Abbott, 1973; Pointier and McCullough, 1989); several Caribbean Islands, Central America, and northern South America (Pointier and McCullough, 1989).

Mexico: The screw snail is known in 16 states: Chiapas, Chihuahua, Coahuila, Colima, Durango, Jalisco, Morelos, Nayarit, Nuevo León, Puebla, Quintana Roo, San Luis Potosí, Sinaloa, Tabasco, Tamaulipas and Veracruz (Figures 10.1 and 10.2) (Contreras-Arquieta et al., 1995a; unpublished data); possibly also in the state of Sonora, where artisan works including screw snail shells have been seen.

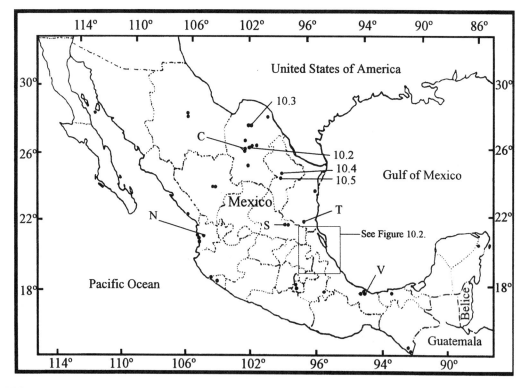

FIGURE 10.1 Distribution of the introduced freshwater screwsnail *Thiara tuberculata* in Mexico. Numbers indicate localities in Tables 10.2–10.5, and letters indicate vulnerable areas.

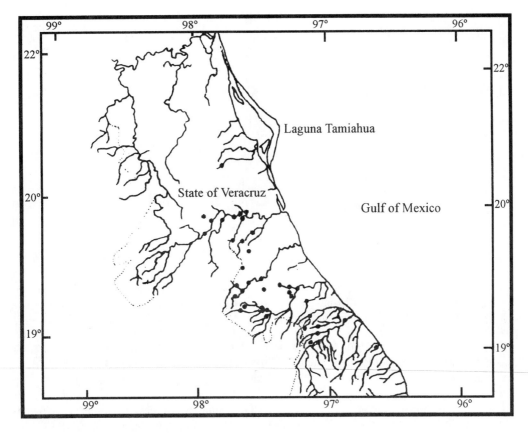

FIGURE 10.2   Distribution of the introduced freshwater snail *Thiara tuberculata* in north and central Veracruz state, Mexico.

DESCRIPTION

For recognition of the species, the shell is elongate conical, imperforate, with reticular sculpture and a paucispiral operculum; 11 to 13 whorls color maroon with irregular blotches; aperture oval, with a brown band generally in the columellar lip. For detailed descriptions of the shell character- istics in general, see Burch (1960, 1982), Gomez et al. (1986), Morrison (1954), and for Mexican populations see Contreras-Arquieta et al. (1995a).

Conchological studies in West Indian populations of *Thiara tuberculata* (Figure 10.3) showed the presence of four distinct morphs (Pointier, 1989). In Mexico, two morphs were found, Falaise and Gosier; some Mexican populations are of the decollate type (Contreras-Arquieta et al., 1995a).

The radula has 69 to 121 rows of 7 transversal teeth, with general formula (10-12), (10-11), (4-6)-1-1, R, 1-1-(4-6), (10-11), (10-12) (Contreras-Arquieta et al., 1995a); similar characters have been reported in populations from the Dominican Republic (Gómez et al., 1986). There are 12 digitiform papillae, rarely 11; the 3 or 4 closer to the collumelar lip more prominent (Contreras-Arquieta et al., 1995a).

The species is parthenogenetic and ovoviviparous, the brood pouch with 62 to 88 embryos (75% fully developed and 25% in different stages of development); frequency of birth is not known; unshelled embryos measure 0.15 to 0.2 mm, while shelled snails near birth are 2 to 4 mm, having 3 to 4 spires (Contreras-Arquieta et al., 1995a). The smallest specimen having embryos was 12.8 mm under laboratory conditions; it takes young snails 25 weeks or around 6 months to become fully mature and start producing young (Kruatrache et al., 1990).

(a)   (b)   (c)

FIGURE 10.3 Shells from three populations of *Thiara tuberculata* in Mexico, (a) Falaise morph from Coahuila (UANL-134:41.4), (b) Gosier morph from Nuevo León (UANL-413:33.5), and (c) Decollate type from Nayarit (UANL-499:15.3). Size in millimeters.

The intrinsic growth rate (r) of the screw snail is low, 0.18 to 0.25 in natural form and 0.13 to 0.23 under laboratory conditions, reaching 30.2 mm in 85.1 days (Pointier et al., 1991) or 1.2 mm/month (Pointier and McCullough, 1989). Its development is slow, and its life is long, 4 to 5 years (Pointier and McCullough, 1989; Pointier et al., 1989, 1991). Reproduction is attained at a size of 23.7 mm shell length, and brood size is 62 to 88 young (Contreras-Arquieta et al., 1995a). Mortality is relatively low, with 50% of individuals surviving 5 years of culture (Pointier et al., 1991).

THIARA AS VECTOR OF DISEASES

The screw snail has been introduced in many countries throughout the world as biological control of snail vectors due to its capacity for predation or competition, to reduce local populations, especially species of *Biomphalaria* — the intermediate host of bilharziosis — as in several countries in Central and South America (Gómez et al., 1986; Malek, 1985). In Mexico, little is known about its ability to transmit or harbor diseases or if its introduction may have carried any infective agents.

The screw snail is the intermediate host to the larval phases of the trematodes *Diorchitrema formosanum* and *Clonorchis sinensis* in China (Abbott, 1973; Malek, 1976) and *Opisthorchis sinensis* in Louisiana, U.S. (Dundee and Paine, 1977). In Mexico, Rodríguez-Vargas and Salgado-Maldonado (1991) reported *Melanoides tuberculata* as intermediate host of the trematode *Centrocestus formosanus*, the second intermediate host is the fish *Xiphophorus helleri*, at Manantial Las Estacas, Morelos, all three being introduced species.

ECOLOGY

In the state of Veracruz, 95% of the rivers sampled were invaded by this snail. Water temperatures ranged from 17–23°C, conductivity was variable (200–900 μmhos). In the state of Durango, the screw snail is found in waters at 33°C and pH 7, at densities of 1500 to 2500 snails/m^2 (Contreras-Arquieta et al., 1995a). Similar conditions were reported by Dundee and Paine (1977) for New Orleans, Louisiana, where pH was 7.0–7.5, temperatures 18–25°C, and population densities were up to 2700/m^2. Densities have been reported at 13,338 in Martinique, but the highest are the reports for Texas, where they reached 51,650 individuals/m^2 (Murray and Wopschall, 1965), and up to 37,500 individuals/m^2 in Florida (Roessler et al., 1977; Pointier et al. 1989).

Pointier and McCullough (1989) listed favorable and unfavorable conditions that influence population densities of Thiarid snails:

Favorable Factors	Unfavorable Factors
Permanent waters	Temporary waters
Shallow waters	Deep waters
Emergent plants	Floating plants
Well-oxygenated "niches"	Anoxic "niches"

These factors mostly agree with our observations. However, some localities have particular habitats. For example, in the Río Salado in Coahuila, the snails are found buried as deep as 10 cm in mud, while in Durango, at Ojo de Agua La Concha, a thermal spring poor in oxygen, the snails are abundant down to a 4-m depth. In Veracruz they have been found in creeks with high organic pollution, contrasting with the conditions given by Pointier and McCoullugh (1989). The snails stay buried in slightly muddy sand or gravel during daylight, while at night and early morning they climb to the air/water interface. Apparently they have little tolerance to some polluted waters, as in Guadalupe at Nuevo León, Mexico, where they were found only in a headspring flowing to the Río La Silla, and were absent from the river polluted with many domestic and industrial effluents.

The screw snail is very hardy. It survives extensive aquarium cleaning, even washing gravel with soap three to four times, rinsing with chlorinated water, draining the gravel and leaving it without water. Remaining humidity keeps the snails alive in the interstitial spaces. After the aquarium is filled with chlorinated water, rinsed, and filled with clean water, the young snails resume activities.

ECOLOGICAL IMPACTS

In Table 10.1 we present data on number of collecting localities for the screw snail, solitary or coexisting with native species of gastropods (Contreras-Arquieta et al., 1995b). Environmental conditions of air and water, especially temperature, salinity, hardness, and dissolved oxygen where *T. tuberculata* is found in Mexico are diverse. More frequently, recent records show *T. tuberculata* as the lone species, with no native species found.

The localities and vulnerable ecosystems referred to appear in Figure 10.1. Ecosystems were considered vulnerable due to being centers of high endemism, and having several species endangered or threatened.

In the Upper Río Salado, Coahuila, this snail is expanding its distribution and abundance. In Celemania and Estación Hermanas, only *T. tuberculata* is found (Contreras-Arquieta, 1998). In 1965 (Taylor, 1966) there were five native species of molluscs at El Cariño de la Montaña; in 1986 the 100% of the native snail fauna had been lost due to invasion of the exotic species. In 1987 only a single live specimen of *Helisoma anceps* was found. Densities of up to 2500 individuals/m^2 are associated with dwindling native snail fauna (Table 10.2), only their empty shells remaining to record their former presence.

In the Río Sabinas at Múzquiz, Coahuila, the probable place of introduction was near the bridge along the Boquillas road. Only *T. tuberculata* is found; the density is not high at this location, but it is increasing downstream, possibly competing with *Helisoma anceps* (Table 10.3).

From 1986 to 1988, the Río La Silla, Guadalupe, Nuevo León (Table 10.4), had abundant populations of four species of gastropods. Their population densities diminished nearly 100% in 1990. In 1991 the habitat was shared by native species and the exotic, but *Biomphalaria havanensis*

TABLE 10.1
Number of Localities Where *Thiara tuberculata* Was Found Alone or in Association With Other Snails, 1995 and 1996

	1995		1996	
	Loc.	**%**	**Loc.**	**%**
Only *T. tuberculata*	15	23.4	22	28.6
T. tuberculata and *Physa* sp.	22	34.4	23	29.9
T. tuberculata and *Pomacea* sp.	2	3.1	2	2.6
T. tuberculata, Physa sp. and 1 other	6	9.4	6	7.8
T. tuberculata and 2 or more species[a]	14	21.9	18	23.4
No data[b]	5	7.8	6	7.8

[a] Usually one is *Physa* sp.
[b] Localities not known to have contained other snail species.

Source: Contreras-Arquieta, A., G. Guajardo-Martínez, and S. Contreras-Balderas, Publicaciones Biológicas – F.C.B./U.A.N.L., México, 8(1 y 2):17–24, 1995; also, unpublished data.

TABLE 10.2
Biodiversity Losses in the Freshwater Snail Fauna at El Cariño de la Montaña, Sacramento, Coahuila

	1966	1986	1987[a]	1996
Native				
Physa virgata	X			
Cochliopina riograndensis	X			
Helisoma anceps	X		1	
Gundlachia excentrica	X			
Pisidium compressum	X			
Exotic				
Thiara tuberculata		31	178	204[b]
Species Ratio (native/introduced)	5/0	0/1	1/1	0/1
Species Loss (%)	0	100	80	100

[a] Empty shells of *N. minckleyi, D. coahuilae* and *P. manantialis* were also collected.
[b] Population density estimated at 2,500 specimens/m^2.

Source: Modified from Contreras-Arquieta et al., 1995b.

has not been collected since. In 1996, only populations of *Thiara* and *Physa*, the genus most tolerant to the presence of *Thiara* remained (Table 10.1). *Physa* has a life span of little less than 1 year, lays several masses of eggs, 30 to 50 at a time, and has a birth rate of 95%.

In the reservoir Presa Rodrígo Gómez "La Boca", Santiago, Nuevo León (Table 10.5), clams and snails were abundant prior to 1985. There were nine native species, of which seven were

TABLE 10.3
Biodiversity Losses in the Freshwater Snail Fauna at Río Sabinas, Múzquiz, Coahuila, Mexico, 1994

Locality	1	2	3
Native (Live)			
Helisoma anceps	0	23	81
Gundlachia excentrica	?	0[a]	0
Exotic			
Thiara tuberculata	41	80	132
Species Ratio (native/introduction)	2?/1	2/1	1/1
Species Loss (%)	100	50[a]	50

Note: 1 – Near bridge on Boquilla road; 2 – 1.5 km up river; 3 – 3 km down river.

[a] Only one recently dead shell found.

TABLE 10.4
Biodiversity Losses in the Freshwater Snail Fauna in the Río La Silla, Guadalupe, Nuevo León, México

	1986–1988	1990	1991	1996
Native				
Biomphalaria havanensis	X			
Planorbella trivolvis	X	?	X[a]	
Gundlachia radiata	X	?	X[a]	
Physa virgata	X	?	X[a]	X[a]
Exotics				
Thiara tuberculata	—	X	X	X
Species Ratio (N/I)	4/0	?/1	3/1	1/1
Species Loss (%)	0	100?	25[b]	75%

[a] Not found in the same microhabitat as *T. tuberculata*.
[b] Although loss is 25% of species, the remaining species, 75%, are strongly impacted negatively by the presence of *Thiara*.

common to abundant. In 1985, only one native live specimen was taken in the southern beach of the reservoir; the density of *Thiara* increased to 150 specimens/m^2 at that location and began to move to the northern beach. The populations are kept at a low level in the southern beach, and are fairly constant, if higher in the northern parts. The snail is now distributed almost throughout the reservoir, reaching densities up to 300 individuals/m^2. In 1996, the species present were similar to 1992, but became less abundant; *Cochliopina riograndensis*, *Pseudosuccinea columella*, and *Sphaerium* sp. were not found, while the once dominant species, *Biomphalaria* and *Planorbella*, are now scarce.

TABLE 10.5
Biodiversity Losses in the Freshwater Snail Fauna at La Presa
Rodrigo Gómez "La Boca", Santiago, Nuevo León, México

	>1984	1985	1987–1991	1992	1996
Native					
Pyrgophorus spinosus	a	a	c	r	r
Cochliopina riograndensis	c	r	—	—	—
Physa mexicana	a	a	c	r	e
Pseudosuccinea columella	r	e	—	—	—
Biomphalaria havanensis	a	c	e	e	e
Planorbella trivolvis	a	a	c	e	e
Gundlachia radiata	a	c	e	r	—
Sphaerium sp.	c	e	—	—	—
Anodonta imbecilis	a	a	c	e	r
Exotics					
Thiara tuberculata	—	r	c	a	a
Corbicula manilensis	c	c	e	r	e
Species Ratio (N/I)	9/1	9/2	6/2	6/2	5/2
Species Loss (%)	0	0	40	40	50

Note: a — abundant, c — common, e — scarce, and r — rare.

DISCUSSION

Introduction and spread of screw snails in Mexico is correlated with the decline of native snail populations. The screw snails have a rapid rate of establishment and are widely distributed in the country, as shown in Table 10.1. The number of localities where *Thiara* is the only remaining species has risen from 15 to 22, increasing 5% in 1 year. The impact is not always visible, due to lack of information on species communities and their distribution in Mexico; however, such impact may be deduced for some localities (Tables 10.2 – 10.5).

The life span of screw snails is 4 to 5 years, whereas for native species it is 1 or 2 years (depending on the group). The screw snails give birth to 62 to 88 embryos per brood, with a mortality of less than 50%. They either outcompete natives for space and/or food, or their population increases faster, or they have a wider tolerance range to environmental conditions, including pollution from diferent sources. Besides, being parthenogenetical, a single invader may start a local population very rapidly.

Finding the screw snail in vulnerable localities where there are endemic snails may increase threats to the endemics or be the cause of their endangered status (Pantanal spring, Nayarit in 1990; Río Sabinas at El Limón, tributary of Río Pánuco in Tamaulipas, and in Catemaco Lake, Veracruz, both in 1992, the Valley of Cuatro Ciénegas, Coahuila in 1994, Media Luna spring, San Luis Potosí and neighboring areas in 1996). For some of these sites there have been no collections after the first appearance of screw snails, hence there is no recent information about the impact on local snail communities.

The more vulnerable ecosystems are Cuatro Ciénegas valley with 12 endemics out of 21 species (Taylor, 1966; Minckley, 1969; Hershler, 1985); and the Río Pánuco System, where almost 20 species of the Pleuroceridae family live (Pilsbry and Hinkley, 1909; Pilsbry, 1956), all now considered endemic to the Panuco system. The Pleuroceridae family is the sister group of the Thiaridae family, with very similar ecological requirements, perhaps competing for food and space, with impact difficult to predict.

REFERENCES

Abbott, R.T., Spread of *Melanoides tuberculata, The Nautilus,* 87(1):29. 1973.

Banarescu, P., *Zoogeography of Fresh Water.* General Distribution and Dispersal of freshwater Animals. Vol 1. pp. 303–360. AULA – Verlag Wiesbaden, 1990.

Burch, J.B., *Some Snails and Slugs Of Quarantine Significance of the United State.* Agricultural Research Service, United States Departament of Agricultural. 1960.

Burch, J.B., *Fresh Water Snails (Mollusca : Gastropoda) of North America.* U. S. Enviromental Protection Agency, Cincinnati, OH. EPA 600/3-82-026. 1982.

Burch, J.B., and A. Cruz-Reyes, Clave genérica para la identificación de Gastrópodos de Agua Dulce en Mexico. *Instituto de Biología, U.N.A.M,* Mexico, 1987.

Contreras-Arquieta, A., *Caracoles Dulceacuícolas (Mollusca: Gastropoda) de la Subcuenca San Juan, tributario del Río Bravo, Noreste de Mexico.* Tesis de Licenciatura, Inédita. Facultad de Ciencias Biológicas, U.A.N.L., 1991.

Contreras-Arquieta, A., New records of the snails *Thiara tuberculata* (Müller, 1774) (Gastropoda: Thiaridae) exogen in the Cuatro Cienegas Basin, and its distribution in the State of Coahuila, Mexico. *The Southwestern Naturalist,* 43 (2): 283–286, 1998.

Contreras-Arquieta, A., G. Guajardo-Martinez, and S. Contreras-Balderas, Redescripción del Caracol Exógeno *Thiara (Melanoides) tuberculata* (Müller, 1774) (Gastropoda: Thiaridae) y su distribución en Mexico. *Publicaciones Biológicas — F.C.B./U.A.N.L., Mexico,* 8(1 y 2):1–16, 1995a.

Contreras-Arquieta, A., G. Guajardo-Martinez, and S. Contreras-Balderas, *Thiara (Melanoides) tuberculata* (Müller, 1774) (Gastropoda: Thiaridae), su probable impacto ecológico en Mexico. *Publicaciones Biológicas — F.C.B./U.A.N.L., Mexico,* 8(1 y 2):17–24, 1995b.

Contreras-Balderas, S., and A.M. Maeda-Martinez, Estado actual de la Ictiofauna nativa de la Cuenca de Parras, Coahuila, Mexico, con notas sobre algunos invertebrados. *Memorias, Octavo Congreso Nacional de Zoología, Escuela Normal Superior del Estado, Saltillo, Coahuila.* I:59–67, 1985.

Dundee, D.S., Catalog of introduced molluscs of eastern North America (North of Mexico), *Sterkiana,* 55:1–37, 1974.

Dundee, D.S. and A. Paine, Ecology of the snail, *Melanoides tuberculata* (Müller) intermediate host of the human liver fluke (*Opisthorchis sinensis*) in New Orleans, Louisiana, *The Nautilus,* 91(1):17–20, 1977.

Gomez, J.D., M. Vargas and E.A. Malek, *Moluscos de Agua Dulce de República Dominicana.* Publicación de la Universidad de Santo Domingo, Santo Domingo, República Dominicana. Colección Ambiente y Sociedad No. 2, Vol. 535:1–135, 1986.

Hershler, R., Systematic revision of the Hydrobiidae (Gastropoda: Rissoacea) of the Cuatro Ciénegas Basin, Coahuila, Mexico. *Malacologia,* 26(1/2):31–123, 1985.

Hills, D.M. and R.L. Mayden, Spread of the Asiatic clam *Corbicula* (Bivalvia: Corbiculacea) into the New World Tropics, *Southwestern Naturalist,* 30(3):454–456, 1985.

Jimenez-G., F., G. Guajardo-Martinez, and B.A. Torres-G, Hallazgo de *Fasciola hepatica* en el Municipio de Parras, Coahuila, Méx. *IV Congreso Nacional de Zoología, del 7 al 12 de Diciembre de 1980. Ensenada, Baja California, Mexico.* Resúmenes, p. 3., 1980.

Kruatrachue, M., E.S. Upatham, S. Vichasri, and V. Baidikul, Culture method for the Thiarid snails *Brotia costula, Tarebia granifera* and *Melanoides tuberculata* (Prosobranchia: Mesogastropoda), *Journal of Medical and Applied Malacology,* 2:93–99, 1990.

Malek, E.A., Medical important mollusks, Chapter 66, pp. 669–685; Section X Medical Important Animals. In *Tropical Medicine,* Hunter, G.W., J.C. Swartsewalder, and D.F. Clyde (Eds.), W. B. Saunders, 1976.

Malek, E.A., Snails host of Schistosomiasis and other snail – trasmited diseases in tropical America: A manual. *Pan American Health Organization,* Scientific Publication, (478):1–325, 1985.

Minckley, W.L., Enviroments of the Bolson of Cuatro Cienegas, Coahuila, Mexico; with special reference of the Aquatic Biota. *The University of Texas at El Paso Siences series,* (2):1–65, 1969.

Morrison, J.P.E., The Relationships of Old and New World Melanians. *Proceedings United States National Museum,* 103 (3325):357–394, 1954.

Murray, H.D., *Tarebia granifera* and *Melanoides tuberculata* in Texas. *Annual Report of American Malacological Union,* 31:15–16. 1964.

Murray, H.D., The introduction and spread of thiarids in the United States. *The Biologist,* 53(3):133–135, 1971.

Murray, H.D. and L.J. Wopschell, Ecology of *Melanoides tuberculata* (Müller) and *Tarebia granifera* (Lamarck) in South Texas. *American Malacological Union Incorporated,* Abstr. 25–26, 1965.

Müller, O.F., Vermium terrestrium et fluviatilium, seu animalium infusoriorum, helminthicorum, et testaceorum,
non marinorum. succinta historia. *Heineck et Faber, Havinae et Lipsiae,* 2:1–214, 1774.

Phillipi, R.A., Descriptiones testaceorum quorundam novorum, maxime chinensium. *Zeitschrift für Malako-
zoologie,* 1:161–167, 1844.

Pilsbry, H.A., Inland Mollusca of Northern Mexico. III. Polygyridae and Potadominae. *Proceedings of the
Academy Natural Sciences of Philadelphia,* 108:19–40; plates 2–4, 1956.

Pilsbry, H.A. and A.A. Hinkley, Melaniidae of the Panuco River System, Mexico. *Proceedings of the Academy
Natural Sciences of Philadelphia,* 60:519–531, plate 13–14, 1909.

Pointier, J.P., Conchological studies of *Thiara (Melanoides) tuberculata* (Mollusca: Gastropoda: Thiaridae) in
the French West Indies. *Walkerana,* 3 (10):203–209, 1989.

Pointier, J.P., A. Guyard, and A. Mosser, Biological control of *Biomphalaria grabarta* and *B. straminea* by
the competitor snail *Thiara tuberculata* in a transmission site of schistosoniasis in Martinique, French
West Indies. *Annals of Tropical Medicine and Parasitology,* 83(3):263–269, 1989.

Pointier, J. P. and F. McCullough,, Biological control of the snail hosts of *Schistosoma mansoni* in the
Carribbean area using *Thiara* spp. *Acta Tropica,* 46:147–155, 1989.

Pointier, J.P., J.L. Toffart, and M. Lefèvre, Life tables of freshwater snails of the genus *Biomphalaria (B.
glabrata, B. alexandrina, B. straminea)* and of one of its competitors *Melanoides tuberculata* under
laboratory conditions. *Malacologia,* 33(1–2):43–54, 1991.

Reddell, J.R., A Review of the Cavernicole Fauna of Mexico, Guatemala, and Belize. *Texas Memorial Museum
Bulletin* (The University of Texas) 27:1–327, 1981.

Rodriguez-Vargas, M.I. and G. Salgado-Maldonado, *Centrocestus formosanus* (Trematoda) parásito de *Melanoides tuberculata* (Prosobranchia) y *Xiphophorus helleri* (Poeciliidae) en el Manantial de las "Estacas"
Morelos. *XI Congreso Nacional de Zoología. del 28 al 31 de Octubre de 1991. Mérida, Yucatán. Resumen*
111, 1991.

Roessler, M.A., G.L. Beardsley, and D.C. Tabb, New Records of the introduced snail, *Melanoides tuberculata*
(Mollusca: Thiaridae) in South Florida. *Florida Science,* 40:87–94, 1977.

Smith, B.J., Travelling Snails. *Journal of Medical and Applied Malacology,* 1:195–204, 1989.

Taylor, D.W.A., Remarkable snails fauna from Coahuila, Mexico. *The Veliger,* 9(2):152–228, Plate 8 to 19;
25 text figures, 1966.

Taylor, D.W., Freshwater mollusks of California: A distributional checklist. *California Fish and Game,*
67(3):140–163, 1981.

Torres-Orozco, R. and E. Revueltas-Valle, New southern most record of the Asiatic Clam *Corbicula* flaminea
(Bivalvia: Corbiculacea) in Mexico. *Southwestern Naturalist,* 41(1):60–98, 1996.

Williams, J.E. and D.W. Sada, Status of two endangered fishes, *Cyprinodon nevadensis mionectes* and *Rhinichthys osculus nevadensis,* from two springs in Ash Meadows, Nevada. *Southwestern Naturalist,*
30(4):475–484, 1985.

Section III

Bait

11 Baitfish Trade as a Vector of Aquatic Introductions

M.K. Litvak and N.E. Mandrak

INTRODUCTION

The ecological and economic impacts of the freshwater baitfish industry have been severely underestimated in North America (Litvak and Mandrak 1993). The baitfish industry is the cause of ecological impacts in two locations: harvest areas (wild fishery) and areas of use (sportfishing). To distinguish between these areas of impact, we will term the ecosystem from which baitfish are harvested the *donor ecosystem*, and the ecosystem in which baitfish are used the *recipient ecosystem* (Litvak and Mandrak 1993).

This book examines introductions of nonindigenous aquatic species in North America. Various chapters provide graphic examples of the potential impacts of introductions of exotic species through anthropogenic vectors. Other chapters illustrate that nonindigenous introductions are not limited to intercontinental transfers of exotic species. Equally, we should *not* limit our definition of nonindigenous to the species level. For example, if baitfish caught in a donor ecosystem are released into a geographically isolated recipient ecosystem, even though the species may occur in the recipient ecosystem, the transfer of this potentially different genetic strain of baitfish should also be defined as a nonindigenous introduction. The justification for this reasoning is that the introduction of this fish could lead to genetic pollution (contamination of the native gene pool), as well as transfer of disease and parasites. Therefore, if we define nonindigenous in an evolutionary and ecological context, introduction of a nonindigenous organism would have to satisfy the following axiom: movement of an organism from one geographically and/or temporally isolated area (donor ecosystem) to another (recipient ecosystem). Degree of geographic and temporal isolation of the populations, both in terms of donor and recipient ecosystems, will determine level of risk of ecological impact. If we accept the definition of a nonindigenous organism at the population level rather than species level, we immediately recognize the tremendous potential for movement of nonindigenous organisms between populations and the potential for alteration of donor and recipient ecosystems.

The use of baitfish for sportfishing is widespread in Canada and the U.S. Previously, we examined the scope and potential impacts of baitfish use in Canada and U.S. in terms of ecology and economics (Litvak and Mandrak 1993). While determining the economic value of the baitfish industry is difficult (Nielson 1982), we estimated its value in 1991 to be in excess of US $1 billion annually for both U.S. and Canada. More recently, Meronek et al. (1995) found that the baitfish industry in six north central American states contributed U.S. $250 million to their economies. In all likelihood, we continue to underestimate the economic impact of the baitfish industry, particularly in light of difficulties in determining the actual number of fish bought and sold (Nielson 1982). Additionally, the economic value we calculated did not include spin-offs to the sportfishing industry that are important to many states and provinces, often representing a large portion of their tourist dollar.

The objectives of this chapter are: (1) to provide evidence of baitfish introduction; (2) to present the results of a survey of baitfish dealers in Ontario; (3) to determine the probability of baitfish introduction; (4) to identify potential impacts of the baitfish industry on donor and recipient ecosystems; (5) to summarize changes in provincial and state regulations governing the use of baitfish between 1956, 1988–91 and 1996–97; and, (6) to suggest strategies for mitigation of the incidental transfer of organisms through the baitfish industry in Canada and the U.S.

EVIDENCE FOR INTRODUCTIONS ASSOCIATED WITH THE BAITFISH INDUSTRY

Although circumstantial, there is compelling evidence that indicates the extent to which fish species have been introduced into new regions through the use of baitfish. Wild commercial baitfish harvesting techniques, angler behavior with regard to bait-buckets, and current range expansions all indicate the tremendous potential for bait-bucket transfers. Baitfish species caught by individual anglers are often transported from donor to recipient ecosystems. In all likelihood the baitfish species, or genes, are not native to the recipient ecosystem. Baitfish species obtained through retail sale are almost always transported to recipient ecosystems in which the species, or genes, are not native (Litvak, personal observation). Baitfish species might be accidentally released by spillage or escape from the hook, but are more commonly deliberately released by well-intentioned anglers. Many anglers feel that it would be cruel to kill rather than release unused baitfishes, or that released baitfishes are beneficial (e.g., forage, increased abundance) to the recipient ecosystems (Litvak and Mandrak 1993).

The harvesting process may also lead to fish species being introduced into waterbodies in which the species, or genes, are not native. Commercial harvesters have been known to "seed" ponds and small lakes with nonnative baitfish species to create a population that could be regularly, and often exclusively, harvested. Additionally, when multiple sites are harvested, individuals may be caught at one harvesting site but not removed from the net, and then inadvertently released at the next harvesting site when the net is reused. It is also common practice for commercial harvesters to remove unwanted fish species from their catch while traveling between sites and release the unwanted species at the next harvest site.

Harvesters are most likely responsible for the recent occurrence of a disjunct population of green sunfish (*Lepomis cyanellus*) in a northern tributary of Lake Ontario (Mandrak, unpubl. data). The nearest known native population is found in a western tributary of Lake Ontario, 100 km away. Both populations are found at locations frequented by the same harvesters. This type of range extension, while frequently observed, is difficult to document. This is because the actual transfer process has been rarely, if ever, observed, and the lag between introduction and establishment is usually lengthy. It is also difficult to determine the number of species introduced into nonnative waterbodies within their range, because it is difficult to determine if new species records are the result of increased sampling effort or introduction. Introduction as the result of baitfish use is usually inferred by the suitability and use of the introduced species as bait, and the presence of angling for piscivorous species in the recipient ecosystem.

In the U.S., 109 species have been documented, or hypothesized, to have been introduced through bait-bucket release into waterbodies in which they are not native (Table 11.1). These species are primarily in the family Cyprinidae (77 species). Most of these introductions have occurred east of the Rocky Mountains, where most piscivorous sportfishes (e.g., members of the Esocidae, Centrarchidae, Percidae) are native (Lee et al. 1980).

Similar statistics are not available for Canada, but 12 species have been hypothesized to have been introduced through bait-buckets into Ontario waterbodies in which the species are not native (Litvak and Mandrak 1993). Four introduced species and eight native species of baitfish exhibit unexpected disjunct distributions (Figure 11.1).

TABLE 11.1
Fish Species Hypothesized to Have Been Introduced by Bait-Bucket Transfer in North America by Province and State

Species	Location
Clupeidae	
Alosa pseudoharengus	NY[1], ON[1]
Cyprinidae	
Agosia chrysogaster	AZ, NM
Campostoma anomalum	CT, NC, NM, NY, VA , ON[2]
Carassius auratus	AL, AR, AZ, CA, CO, CT, DE, FL, HI, IA, ID, IL, IN, KS, KY, LA, MA, MD, MI, MN, MO, MT, NE, NV, NH, NJ, NM, NY, NC, OH, OK, ON[2], PA, RI, SD, TN, TX, UT, VA, WA, WV, WI, WY
Clinostomus elongatus	ON[2], OH[3]
Couesius plumbeus	WY, MT[4]
Cyprinella analostana	NY
Cyprinella galactura	KY, NC
Cyprinella lutrensis	AL, CA, CO, DE[5], GA, NC[5], NM, NV, SC[5], UT, VA[4]
Cyprinella whipplei	OK
Cyprinus carpio	ON[2]
Exoglossum laurae	WV
Exoglossum maxillingua	VA, WV
Gila atraria	ID, MT, NV
Gila bicolor	CA, ID, NV
Gila copei	UT, WY
Gila orcutti	CA
Hybognathus hankinsoni	BC[6], NY, TN
Hybognathus placitus	NM
Hybognathus regius	ME
Lepidomeda albivallis	CA
Leuciscus idus	AR
Luxilis albeolus	NC
Luxilis cardinalis	OK
Luxilis cerasinus	VA
Luxilis chrysocephalus	ON[2], WV
Luxilis coccogenis	NC
Luxilis cornutus	WV
Luxilis pilsbryi	OK[7]
Luxilis zonistius	GA
Lythrurus ardens	NC, VA
Lythrurus umbratilis	OH[3]
Macrhybopsis storeriana	AL[7]
Nocomis asper	OK
Nocomis biguttatus	KY, NY, ON[2]
Nocomis leptocephalus	NC, VA
Nocomis micropogon	GA, NC, SC, ON[2,8]
Nocomis raneyi	NC, VA
Notemigonus crysoleucas	AR, AZ, CA, CO, FL, IA, KS[9], KY, MN, MT, NC, NE, NM, NV, NY, OH, OR, SD, TN, TX, UT,VA, WV, WY
Notropis amblops	FL
Notropis amoenus	NC, NY

TABLE 11.1 (CONTINUED)
Fish Species Hypothesized to Have Been Introduced by Bait-Bucket
Transfer in North America by Province and State

Species	Location
Notropis ariommus	WV
Notropis atherinoides	MA, ME, NY, WV
Notropis baileyi	AL, GA, FL
Notropis bairdi	KS, OK
Notropis bifrenatus	NC
Notropis brazosensis	OK[10]
Notropis buccatus	VA
Notropis buccula	TX
Notropis buchanani	PA
Notropis chiliticus	NC, VA
Notropis girardi	NM, OK
Notropis harperi	FL
Notropis hudsonius	ME
Notropis hypsilepis	GA
Notropis leuciodus	NC, SC, VA
Notropis lutipinnis	NC, GA
Notropis oxyrhynchus	TX
Notropis ozarcanus	OK
Notropis potteri	OK[11]
Notropis procne	NC, VA, WV
Notropis rubricroceus	NC, VA
Notropis shumardi	LA, TX
Notropis telescopus	NC, VA, WV
Notropis texanus	TN
Notropis volucellus	MA, PA, WV
Notropis whipplei	OK[7]
Notropis xaenocephalus	GA
Phenacobius mirabilis	MI, NM, OH[3]
Phoxinus eos	OK[10]
Phoxinus oreas	NC, VA
Pimephales notatus	MA, SD, WV
Pimephales promelas	AL, AR, AZ, CA, CO, CT, DE, FL, GA, ID, KS[9], KY, LA, MA, MN, MO, MS, MT, NE, NC, NH, NM, NV, OH, OK, OR, PA, SC[5], TN, TX, UT, VA, WA, WV, WY
Pimephales vigilax	KS (N. Mandrak, unpubl. data), NE (N. Mandrak, unpubl. data), NM, OH[3], TX, UT
Rhinichthys cataractae	UT
Richardsonius balteatus	AB[12], AZ, MT, UT, WY
Scardinius erythopthalmus	AL, AR, CT, IL, KS, MA, ME, MO, MS, NE, NJ, NY, OK, ON[2], PA, SD, TX, VA, WI, WV
Semotilus atromaculatus	CO, UT
Semotilus corporalis	NY, VA
Catostomidae	
Catostomus catostomus	CO, WY
Catostomus commersoni	CO

TABLE 11.1 (CONTINUED)
Fish Species Hypothesized to Have Been Introduced by Bait-Bucket Transfer in North America by Province and State

Species	Location
Catostomus platyrhynchus	UT
Catostomus plebeius	NM
Hypentelium nigricans	MA, ON[2]
Thoburnia rhothoeca	VA

Characidae

Astyanax mexicanus	CA, LA, NM, OH, OK, TX

Ictaluridae

Noturus flavus	WV
Noturus insignis	MA, MD, MI, NH, NY, NC, PA, TN, VA, WV, ON[2,13], QU[14]

Osmeridae

Osmerus mordax	ON[2], MA[15], MN[15]

Cyprinodontidae

Cyprinodon variegatus	TX
Fundulus catenatus	KY, OK
Fundulus diaphanus	OR, WA
Fundulus grandis	NM, TX
Fundulus heteroclitus	NH, PA
Fundulus sciadus	CO
Fundulus stellifer	GA
Fundulus waccamensis	NC
Fundulus zebrinus	AR, CO, MO, NV, NM, SD, TX, UT, WY

Gasterosteidae

Culaea inconstans	AL, CA, CO, CT, KY, ME, NM, OK[10]
Gasterosteus aculeatus	AL[112], CA, ON[1,16], MI[16], WI[16]

Centrarchidae

Enneacanthus gloriosus	NY[1]

Percidae

Ammocrypta bifascia	FL
Etheostoma edwini	FL
Percina maculata	ON (N. Mandrak, pers. obs.)

Cichlidae

Cichlasoma nigrofasciatum	HI
Oreochromis aureus	AL, AZ, CA, FL
Oreochromis mossambicus	HI
Sarotherodon melantheron	HI
Tilapia rendalli	HI

Gobiidae

Acanthogobius flavimanus	CA
Gillichtys mirabilis	CA

TABLE 11.1 (CONTINUED)
Fish Species Hypothesized to Have Been Introduced by Bait-Bucket
Transfer in North America by Province and State

Species	Location

Note: Unless denoted by superscript, state data are from the website of the Biological Resources Division, USGS (1996). Scientific names and taxonomic order according to Robins et al. (1991).

[1] Mills et al. 1993

[2] Litvak and Mandrak 1993

[3] Trautman 1981

[4] Brown 1971

[5] Rodhe et al. 1994

[6] Carl et al. 1959

[7] Miller and Robison 1973

[8] Goodchild and Tilt 1976

[9] Cross and Collins 1995

[10] Heard 1956

[11] Hall 1956

[12] Nelson and Paetz 1992

[13] Rubec and Coad 1974

[14] McAllister and Coad 1974

[15] Franzin et al. 1994

[16] Stedman and Bowen 1985

SURVEY OF BAITFISH DEALERS AND USERS IN ONTARIO

We conducted a survey of retail baitfish dealers in Toronto in 1988 (Litvak and Mandrak 1993). We were given permission to examine the contents of the dealers' baitfish tanks. Tank contents indicated that they received fishes harvested from many different drainages in Ontario. Of the 28 fish species identified in the tanks, 6 species were illegal baitfish (Table 11.2). These results were similar to other studies. LoVullo and Stauffer (1993) found that 43% of species present in baitfish dealer tanks in Pennsylvania were illegal. Ludwig and Leitch (1996) also found a significant number of their samples (28.5%) contained a non-baitfish species in Minnesota and North Dakota.

We also surveyed a number of anglers at retail baitfish stores (Litvak and Mandrak 1993). Almost half the baitfish users released their unused baitfish into the water when finished fishing (Figure 11.2). This is particularly significant, since many of the baitfish were used outside of their native distributions. As well, almost all of these anglers thought they were doing something good for the ecosystem by releasing their unused live baitfish. This misconception is not limited to anglers. One angler told us that when a park warden saw him attempting to kill his leftover baitfish by dumping them on the beach, he was told to release his leftover baitfish into the water. This was despite the fact that release of live baitfish is prohibited in Ontario.

CALCULATION OF PROBABILITY OF RELEASE OF FISH FROM BAIT-BUCKETS

We feel there is a very high probability that many of the range expansions observed in Canada and the U.S. are a result of bait-bucket transfers. Ludwig and Leitch (1996) defined the probability of a bait-bucket transfer as a product of three independent event probabilities: (1) probability of transportation across a basin boundary (0.1; in their case Mississippi to Hudson Bay), (2) probability

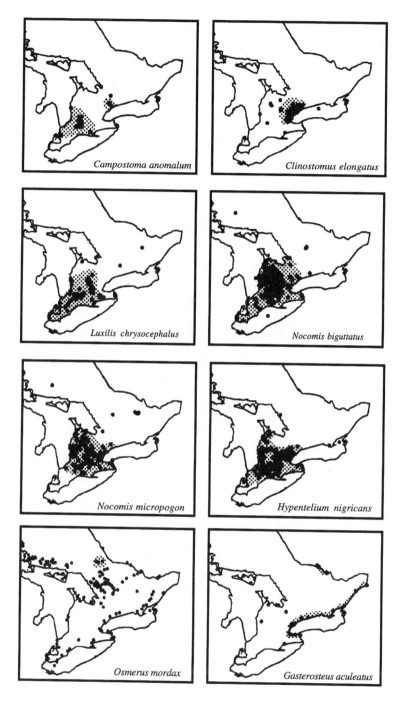

FIGURE 11.1a Distribution of eight native species of baitfish exhibiting disjunct distributions indicating possible bait-bucket introductions. Stippling represents native distribution. (After Litvak and Mandrak 1993).

that a bait-bucket contains a non-baitfish species (0.285), and (3) probability of release after use (they used our value of 0.42). In their study, this yielded a single event probability of a non-baitfish being released after use of 0.01 [(0.1) × (0.285) × (0.42)]. Considering there were over 10 million angler days in their study region, the cumulative probability (calculated with the binomial probability formula) of 10,000 non-baitfish releases approached 1. If they had not limited their impacts

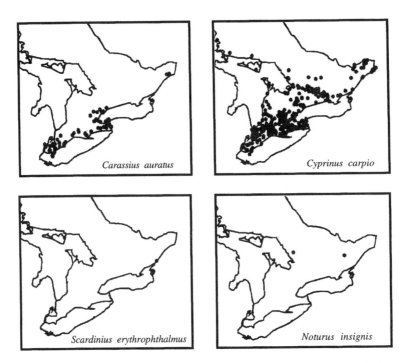

FIGURE 11.1b Distribution of four introduced species that are, in whole or in part, likely a result of bait-bucket transfers. (After Litvak and Mandrak 1993).

of introduction to non-baitfish species, but still limited the impact to interbasin transfer, their probability of a single event would have increased to 0.04 [(0.1) × (1) × (0.42)]; in other words, every 4 out of 100 angler events would lead to a release of a nonindigenous species or genes.

We can take their approach with data from our previous study to examine the probability of introduction of nonindigenous species or genes. In our study, almost all of the baitfish were transported to recipient ecosystems that were isolated from donor ecosystems (Figure 11.3); therefore, the probability of transportation is 1. The probability that a bait bucket contained a non-baitfish species is not relevant. Therefore, probability of a release event occurring is the product of the probabilities of transport and release. Based on our survey, the probability of release after use would still be 0.42 [(1) × (0.42)], or 42 out of 100 angler events.

It is probable that all commercially harvested baitfish released live are placed into recipient systems in which they are not indigenous. The reason is found in the harvest strategy employed by most commercial harvesters. Baitfish are at highest abundances in areas where there are no piscivorous predators. Therefore, most commercial baitfish harvesters do not collect baitfish in recipient ecosystems.

To truly assess the probability of baitfish being released live into a recipient ecosystem, we need to know not only the incidence of a release event, but the number of baitfish released by each angler, and the probability of survival after release. Clearly, if the probability of survival after a release is greater than zero, the potential for introductions of nonindigenous baitfish is very high.

ECOLOGICAL IMPACTS OF THE FRESHWATER BAITFISH INDUSTRY

As the result of the difficulty in capturing and identifying introduced baitfish in recipient ecosystems, the ecological impacts of the baitfish industry are not well documented. However, we can infer potential impacts from the known biology of baitfish species and known impacts of baitfish species introduced by other means (e.g., stocking of forage-fish).

TABLE 11.2
Composition and Frequency of Species Found in Holding Tanks at four Toronto Baitfish Dealerships

Scientific Name	Common Name	Number of Destinations
Luxilis cornutus	common shiner	—
Catostomus commersoni	white sucker	—
Semotilus atromaculatus	creek chub	—
Pimephales notatus	bluntnose minnow	—
Pimephales promelas	fathead minnow	1
Phoxinus eos	northern redbelly dace	2
Rhinichthys atratulus	blacknose dace	4
Ambloplites rupestris*	rock bass	—
Cyprinus carpio*	carp	8
Margariscus margarita	pearl dace	1
Nocomis biguttatus	hornyhead chub	18
Umbra limi	central mudminnow	—
Phoxinus neogaeus	finescale dace	7
Hypentelium nigricans	northern hog sucker	18
Culaea inconstans	brook stickleback	2
Lepomis gibbosus*	pumpkinseed	—
Micropterus dolemieu*	smallmouth bass	—
Campostroma anomalum	central stoneroller	19
Notropis atherinoides	emerald shiner	10
Couesius plumbeus	lake chub	17
Notemigonus crysoleucas	golden shiner	1
Rhinichthys cataractae	longnose dace	5
Moxostoma anisurum	silver redhorse	17
Ameiurus nebulosus*	brown bullhead	—
Cottus cognatus	slimy sculpin	4
Micropterus salmoides*	largemouth bass	—
Etheostoma nigrum	johnny darter	6
Percinas maculata	blackside darter	20

Note: Species are ordered in frequency of occurrence. Number of destinations refers to the destinations in Figure 11.2 that are outside the range of the species. Scientific names marked with (*) denote species that are not legal baitfish in Ontario.

EFFECTS ON DONOR ECOSYSTEMS

Fisheries scientists have long been concerned about the effects of baitfish harvesting on donor systems. In the early 1900s, fisheries managers were primarily concerned about depleting the baitfish supply (Evermann 1901). In the 1930s, it was recognized that harvesting baitfish not only affected the supply, but also affected sportfish abundance as the result of forage-fish depletion (Radcliffe 1931; Markus 1939). Declines in baitfish supply led to the development of baitfish culture operations (Markus 1939; Dobie et al. 1956). Litvak and Mandrak (1993) classified the ecological impacts on the donor ecosystem into three categories: (1) population alteration; (2) trophic alteration; and, (3) habitat alteration. We follow this scheme in our current discussion.

Population Alteration

Baitfish harvesting may directly alter the abundance of targeted baitfish species and nontargeted species. In addition to the intentional capture of targeted baitfish species, nontargeted species are

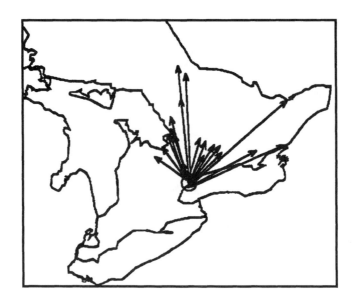

FIGURE 11.2 Destinations of customers surveyed buying baitfish from four Toronto baitfish retailers. Open circle represents Toronto and arrowheads are angler destinations. (After Litvak and Mandrak 1993).

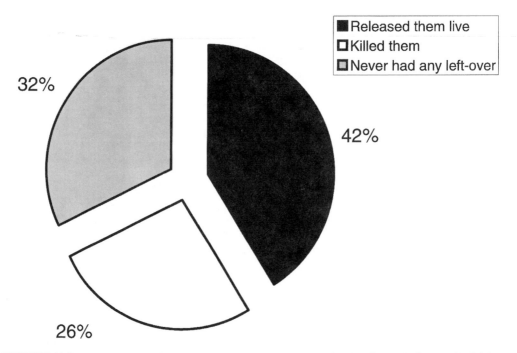

FIGURE 11.3 Response to angler survey (n = 34) questions regarding the disposal of unused baitfish.

often incidentally captured. By-catch is typically returned to the donor ecosystem, but if injured as the result of capture, individuals in the by-catch will have an increased likelihood of mortality. The high frequency of occurrence of nontargeted species found in tanks of retail bait dealers (Litvak and Mandrak 1993; LoVullo and Stauffer 1993; Ludwig and Leitch 1996) suggests that not all nontargeted fish are returned to donor ecosystems.

Many legal baitfish species are found on provincial, state, or federal endangered species lists. In Canada, 15 species of legal baitfish are listed as vulnerable or rare by the Committee on the

Status of Endangered Wildlife in Canada (Campbell 1995). In the U.S., over 200 species of baitfish are listed as of special concern, rare, or endangered in at least one state (Johnson 1987).

Opinions of the effect of harvesting on the abundance of targeted baitfish species vary in the literature. A study examining the effects of baitfish harvesting on the donor ecosystem in a West Virginia stream concluded that resulting decreases in forage-fish abundance and density were short-lived (Brandt and Shreck 1975). Conversely, Portt (1985) concluded that the removal of a substantial portion of the biomass of baitfish species could have short- and long-term effects on forage-fish abundance.

Trophic Alteration

Baitfish harvesting may indirectly alter the abundance of nontargeted fish species (e.g., sport-fishes) and other taxa (e.g., aquatic invertebrates, zooplankton, and waterfowl) by altering trophic interactions. Removal of forage-fish from lakes has been documented to change zooplankton species composition, decrease primary productivity, increase zooplankton size and abundance, and decrease the abundance and growth rates of other fish species (e.g., Carpenter et al. 1987; Litvak and Hansell 1990).

Habitat Alteration

Damage to physical and biological habitat in the donor ecosystem may occur as a result of the harvesting process. Physical damage may include substrate disturbance, removal of snag habitats (e.g., dead tree branches), and destruction of spawning habitat (e.g., Thomsen and Hasler 1944).

EFFECTS ON RECIPIENT ECOSYSTEMS

Ecologists have long been concerned about the effects of introduced species on native ecosystems (Elton 1958), and the effects of a wide range of introduced species have been well documented (e.g., Mooney and Drake 1986; Drake et al. 1989; Baltz 1990). Ecological impacts of introduced species on recipient ecosystems have been classified into five categories (after Kohler and Courtenay 1986): (1) habitat alteration; (2) trophic alteration; (3) spatial alteration; (4) gene pool deterioration; and, (5) introduction of disease.

Habitat Alteration

The feeding and spawning behavior of introduced baitfishes might impact habitat in recipient ecosystems. For example, destruction of aquatic vegetation by carps (e.g., *Ctenopharyngodon idella, Cyprinus carpio;* Terrell and Terrell 1975; Crevelli 1983) and increased turbidity caused by goldfish (*Carassius auratus*) is well documented (e.g., Richardson and Whoriskey 1992).

Trophic Alteration

In general, it has been shown that fish assemblages containing introduced species tend to be less stable as a result of predation and competition (e.g., Meng et al. 1994). Predation by introduced baitfishes on organisms at various trophic levels might alter trophic structure of the recipient ecosystem. Predation might alter the abundance of phytoplankton and zooplankton and other taxa (e.g., Carpenter et al. 1987; Litvak and Hansell 1990). Predation on eggs, fry, or adults might alter the abundance of native aquatic species. For example, California newt abundance declined in Southern California after the introduction of the mosquitofish (*Gambusia affinis*). The mosquitofish were found to prey heavily on newt larvae (Gamradt and Kats 1996). Rainbow smelt (*Osmerus mordax*) have been documented to feed on eggs of native sportfishes (Evans and Loftus 1987). It has been shown that red shiner (*Cyprinella lutrensis*) and fathead minnow (*Pimephales promelas*) exhibit interspecific aggression competition for food with the endangered Colorado squawfish (*Ptychocheilus lucius*) under experimental conditions (Karp and Tyus 1990). Karp and Tyus (1990) hypothesized the introduction of these two minnows will adversely affect the growth and survival of juvenile Colorado squawfish. Competition for food resources between redside shiner (*Richardsonius balteatus*), introduced into a British Columbia lake for forage, and rainbow trout (*Oncorhynchus mykiss*) led to decreased growth rates in the trout (Larkin and Smith 1954). Redside shiners were also documented to prey on trout fry.

Spatial Alteration

Competition between introduced baitfishes and native organisms might result in spatial displacement of native species in the recipient ecosystem. Magnan and Fitzgerald (1984) concluded that brook trout (*Salvelinus fontinalis*) were displaced by creek chub (*Semotilus atromaculatus*) as the result of competition for food resources. Baitfish species such as fathead minnow (*Pimephales promelas*) may displace native species from spawning habitat as the result of territorial aggression (Carlson 1967).

Gene Pool Deterioration

Intraspecific or interspecific hybridization between baitfish and fishes native to the recipient ecosystem might result in decreased reproductive fitness. The gene pool of native species evolved in response to the environment of the recipient ecosystem, whereas the introduced baitfishes may not be adapted to the environment of the recipient ecosystem. Intraspecific hybridization between a native species and an introduced individual of the same species might lead to fertile offspring with decreased fitness (e.g., Ferguson 1990; Philipp 1991). Interspecific hybridization between a native species and an introduced species would probably lead to fertile offspring with decreased fitness or infertile offspring (e.g., Burkhead and Williams 1991).

Introduction of Disease and Parasites

Disease may also be introduced into the recipient ecosystem through transport of water and baitfish from the donor ecosystem. Summerfelt and Warner (1970) documented the transfer of a protozoan parasite (*Plistophora ovariae*) from cultured golden shiners (*Notemigonus crysoleucas*) to native golden shiners in the wild. Ostland et al. (1987) demonstrated the transfer of a bacterial parasite (*Aeromonas salmonicida*) causing furunculosis from baitfish species to salmonines in laboratory experiments. Exposure to a ciliate protozoan, *Ichthyophthirius multifiliis*, caused the death of an estimated 18 million fishes native to Lake Titicaca, South America in December 1981 (Wurtsbaugh and Tapia 1988). It is hypothesized that this parasite, not native to Lake Titicaca, came from introduced sportfishes. This incidence exemplifies the potential severity of the impact of disease introduced into recipient ecosystems through nonindigenous introductions.

STRATEGIES FOR MITIGATION OF IMPACTS

The Quebec Department of Fish and Game conducted a survey of Canadian and American baitfish restrictions in 1956 (Prevost 1957). Data were collected from seven provinces and 37 states. We collected data on baitfish restrictions from the 1988–1991 fishing regulations for 45 American states and 10 Canadian provinces. More recently, we collected data on baitfish restrictions from 1996–1997 fishing regulations for 7 provinces and 43 states. We compared the regulations based on a series of eight questions (Table 11.3) that deal with the extent and strength of regulation. There is no doubt that there is an increase in restrictions over the past 31 years. However, we still find that many jurisdictions do not address the transport, import, export, or release of baitfish in their regulations. This suggests that governing bodies still do not recognize the potential ecological impacts associated with the baitfish industry.

TABLE 11.3

Summary and Comparison of 1956, 1988–1991 and 1996–1997 Baitfish Regulations for American States and Canadian Provinces And Territories

	1956 (n = 44)			1988–1991 (n = 57)			1996–1997 (n = 50)		
Questions	Y	N	NA	Y	N	NA	Y	N	NA
Are there baitfish regulations?	38	5	1	54	3	0	50	0	0
Is the use of baitfish legal?	32	6	0	45	9	0	42	8	0
Are species of legal baitfish limited?	8	1	23	32	2	11	31	3	9
Is the use of baitfish prohibited in certain areas (e.g., trout waters)?	16	0	16	35	0	10	30	4	9
Is the release of unused, live baitfish prohibited?	—			5	8	32	23	10	10
Is the transport of baitfish prohibited within the jurisdiction?	—			3	6(+6*)	30	14	9	19
Is the export of baitfish prohibited?	—			9	1(+4*)	31	7	16(+2*)	17
Is the import of baitfish prohibited?	—			6	6*	33	11	13(+3*)	14

Note: Y- yes; N - no; NA -not adressesed; *-limited by governing agency

When formulating regulations for natural resources, governments must consider the potential socioeconomic, ecological, and regulatory costs and benefits. This is not a simple decision path. In terms of regulating the use of baitfish, decisions are more difficult because of the dearth of information on the ecology of baitfish (Litvak and Mandrak 1993). We have established that the baitfish industry has a large economic impact. But can we effectively regulate this industry so that its ecological impact is limited?

Based on first-hand experience, obtaining information on activities of harvesters and dealers is difficult. While most are cooperative, the shear number of harvesters and dealers makes it difficult to determine the extent of harvest, method of harvest, and number of baitfish sold (Nielson 1982). Exacerbating this problem is the availability of technology (oxygen, aerators, and live wells) that allows for the long-distance transport of baitfish.

So how do we regulate the baitfish industry? Regulation in almost all instances is limitation of an activity. To help formulate a strategy for mitigation we have broken down the baitfish industry into a number of activities: wild harvest, aquaculture, export, import, retail sales, and use. Each of these activities can affect the donor and/or recipient ecosystems.

WILD HARVEST

If wild harvest is to be permitted, there are a number of ways in which regulations can be developed to limit ecological impacts on the donor ecosystem.

1. **Enforce existing regulations.** Many states and provinces already have restrictions imposed on the collection of baitfish.
2. **Limit species to be used.** Exclude those species of special concern. This could include the banning the collection of rare and endangered species. A ban could also be placed on species known to be opportunistic invaders, as those species that are likely to successfully colonize recipient ecosystems.
3. **Reduce the impact and efficiency of harvesting by limiting gear type, harvest method, and number of harvests made annually at a site.** A method often employed in the baitfish trade is to use two seine nets and two groups of seiners. Each group pulls their nets from opposite ends of a stream reach and seines toward a central area. This

method is efficient at catching most fishes in the harvested area. Limitation to one seine of a maximum length would prevent this strategy. The prohibition of bag seines would also reduce the efficiency of harvesting, and reduce physical damage to captured fishes and habitat. Also, non-baitfish must be released, baitfish must be graded, and nets cleaned of fish at the end of harvest at each site. This will prevent transfer of fishes between harvest sites. Alternatively, harvesting could be limited to passive gear (e.g., minnow traps, trapnets). For example, restriction to a short-set (e.g., 24 h) passive harvest gear would reduce harvest efficiency, reduce physical damage to captured fishes, and reduce the probability of fish transfer between harvest sites. The limitation of gear type, harvest strategy, and number of harvests would not only limit the number of fish harvested but also reduce the physical damage to the habitat. However, policing this type of restriction will be difficult.

4. **Prohibit harvest in ecologically sensitive regions, including areas containing species of special concern.** Restrict harvest to areas that have a lower probability of being damaged. Restriction of harvest to areas where enforcement of regulations will ensure that harvesting is conducted in a way which minimizes impacts.

5. **Eliminate commercial harvest of baitfish** and allow anglers to catch and use baitfish in waterbodies in which they are used.

6. **Prohibit all collection of baitfish.** If baitfish were to be used under this scheme, they would have to be provided through aquaculture production.

AQUACULTURE

The production of baitfish from aquaculture is a growing industry. The limitation of baitfish use to fishes produced through aquaculture would completely eliminate impacts to donor ecosystems. Many baitfish have already been successfully cultured and are produced on a large scale. For example, the annual production of golden shiners in Arkansas alone is in excess of U.S. $21 million wholesale (H.K. Dupree pers., comm.). Baitfish aquaculture could be used to lower the probability of deleterious effects from an introduction a number of ways:

1. **Restrict production to use of cultured fish.** The use of cultured fish must involve dependence on domesticated broodstock for eggs and sperm. No wild-caught fish should be used to provide gametes.

2. **Restrict production to sterile fish** through production of triploids or other genetic manipulation.

3. **Permit sale of baitfish** by only certified disease-free operations.

EXPORT AND IMPORT

The export and import of baitfish provides the vector for many nonindigenous introductions. Current range expansions indicate the importance of export and import of baitfish between drainages, states, and provinces as well as between Canada and the U.S.

1. **Enforce existing regulations.** Many states and provinces already have restrictions imposed on the import and export of baitfish.

2. **Prohibit import** of baitfish across country, state, and provincial boundaries.

3. **Prohibit export** of wild harvested baitfish to minimize the impact of harvest within a jurisdiction. Many southern states depend on harvest from other northern states because their baitfish stocks have been depleted. Although baitfish import can be illegal, the export may not be as regulated.

4. **Restrict use to drainage systems** in which the baitfish are caught.
5. **Restrict use** to waterbody in which bait are caught.
6. **Restrict use to preserved baitfish** in order to limit the number of potential introductions.

Retail Sale and Use

1. **Enforce existing regulations.** Many states and provinces already have restrictions imposed on sale and use of baitfish.
2. **Prohibit transport** of baitfish by anglers.
3. **Prohibit use of oxygen or aeration devices in retail packaging.** This would also have the effect of limiting transport.
4. **Prohibit sale and use of live baitfish.** Limitation of sale to preserved baitfish would eliminate the negative impacts on the recipient ecosystem.
5. **Prohibit sale of all baitfish.**
6. **Educate dealers and anglers.** The end-users in the baitfish industry are bait dealers and anglers. These two activities provide an excellent opportunity for regulation. Regular inspection and certification of dealers' tanks will limit the number of illegal baitfish. The dealers can also be a point source for the distribution of information on the proper use of baitfish. If licenses for retail sale of baitfish require the distribution of a fact sheet on the use of baitfish to anglers, the potential for release will be dramatically reduced. For example, a simple sticker could be applied to all baitfish bags indicating that it is illegal to release the baitfish after use. In our survey of anglers at the bait dealerships we found that anglers were very keen to protect their sportfishery. They just did not know the regulations.

EDUCATION

If a baitfish industry is to exist, the key to minimizing negative ecological impacts on both donor and recipient ecosystems is development of a very strong education program. Managers, harvesters, retailers, and anglers must all be informed on the restrictions on the collection, sale, and use of baitfish. The key to success in managing this industry is for the shareholders to participate in the process. If they recognize that the only way the baitfish industry will be permitted is through careful use and monitoring, they will help limit the negative ecological effects.

CONCLUSIONS

As mentioned previously, the baitfish industry in North America is economically important. Unfortunately, it is also recognized that the negative impact of the baitfish industry on the flora and fauna of North America has been great. The potential for ecological damage is further realized with the acknowledgment that introductions are not limited to the species level but also exist on the population level. Resource managers must ask themselves whether the economic value of this industry is worth more than the realized and future negative ecological impacts. Each jurisdiction must weigh the costs and benefits of the continuation of this industry. In this chapter we provided a number of activities that could be regulated in order to eliminate or lessen its impact. If the use of baitfish is to be permitted, the most important activity that can be undertaken, regardless of management strategy, is education of shareholders. Explanation for justification of the management strategy chosen is important. This will allow effective execution of the chosen strategy. We found that most anglers and dealers that we talked with are concerned with the sportfishing resource. Many would be receptive to the strategies outlined in the previous section providing they are brought into the decision making process.

REFERENCES

Baltz, D.M., Introduced fishes in marine systems and inland seas, *Biological Conservation*, 56:151–177, 1990.

Biological Resources Division, United States Geological Survey. *Nonindigeneous Aquatic Species (NAS)*. www.nfrcg.gov/nas. Accessed November 12, 1996.

Brandt, T.M. and C.B. Shreck, Effects of harvesting aquatic bait species from a small West Virginia stream, *Trans. Am. Fish. Soc.*, 104:446–453, 1975.

Brown, C.J., *Fishes of Montana*, Big Sky Books. Montana State University, Bozeman, MT, 1971.

Burkhead, N.M. and J.D. Williams., An intergeneric hybrid of a native minnow, the golden shiner, and an exotic minnow, the rudd, *Trans. Am. Fish. Soc.*, 120:781–795, 1991.

Campbell, R.R., Rare and endangered fishes and marine mammals of Canada: COSEWIC Fish and Marine Mammal Subcommittee status reports IX, *Can. Field-Nat.*, 109:395–401, 1995.

Carl, G.C., W.A. Clemens, and C.C. Lindsey, *The Fresh-Water Fishes of British Columbia*. British Columbia Provincial Museum, Department of Education, Handbook No. 5. 1959.

Carlson, D.R., Fathead minnow, *Pimephales promelas* Rafinesque, in the Des Moines River, Boone County, Iowa, and the Skunk River drainage, Hammond and Story counties, Iowa, *Iowa State J. Sci.*, 41:363–374, 1967.

Carpenter, S.R., J.F. Kitchell, J.R. Hodgson, P.A. Cochran, J.J. Elser, M.M. Elser, D.M. Lodge, D. Kretcher, X. He, and C.N. von Ende, Regulation of lake primary productivity by food web structure, *Ecology*, 68:1863–1876, 1987.

Crevelli, A.J., The destruction of aquatic vegetation by carp, *Hydrobiologia*, 106:37–41, 1983.

Cross, F.B., and J.T. Collins, *Fishes in Kansas*, University of Kansas Museum of Natural History Publication Series 14:1–315, 1995.

Dobie, J., O.L. Meehan, S.F. Snieszko, and G.N. Washburn, *Raising Baitfishes*, U.S. Dept. of Interior. Fish and Wildlife Service. Circular 35, 1956.

Drake, J.A. et al. (Eds.), *Biological Invasions: A Global Perspective*, John Wiley and Sons, Chichester, MA, 1989.

Elton. C.S., *The Ecology of Invasions*, John Wiley and Sons, New York, 1959.

Evans, D.O. and D.H. Loftus, Colonization of inland lakes in the Great Lakes region by rainbow smelt, *Osmerus mordax*: their freshwater niche and effects on indigenous fishes, *Can. J. Fish. Aquat. Sci.* (suppl. 2):249–266, 1987.

Evermann, B.W., Bait minnows, *Sixth Annual Report of New York Forest, Fish and Game Commission*, 1900:307–356, 1901.

Ferguson, M.M., The genetic impact of introduced fishes on native species, *Can. J. Zool.*, 68:1053–1057, 1990.

Franzin, W.G., B.A. Barton, R.A. Remnant, D.B. Wain, and S.J. Pagel, Range extension, present and potential distribution, and possible effects of rainbow smelt in Hudson Bay drainage waters of northwestern Ontario, Manitoba, and Minnesota, *N. Amer. J. Fish. Manag.*, 14:65–76, 1994.

Gamradt, S.C. and L.B. Kats, Effect of introduced crayfish and mosquitofish on California newts, *Cons. Biol.*, 10:1155–1162, 1996.

Goodchild, G.A. and J.C. Tilt, A range extension of *Nocomis micropogon*, the river chub, into eastern Ontario, *Can. Field-Nat.*, 90:491-492, 1976.

Hall, G.E. Additions to the fish fauna of Oklahoma with a summary of introduced species, *Southwestern Naturalist*, 1:16–26, 1956.

Heard, W.R., Live bait imports: *Chrosomus eos* and *Eucalia inconstans* as potential additions to Oklahoma's fish fauna, *Proc. Okla. Acad. Sci.*, 1956, 47–48.

Johnson, J.E., *Protected Fishes of the United States and Canada*, American Fisheries Society, Bethesda, MD. 42 pp, 1987.

Karp, C.A. and H.M. Tyus, Behavioral interactions between young Colorado squawfish and six fish species, *Copeia*, 1990:25–34, 1990.

Kohler, C.C. and W.R. Courtenay, Jr., American Fisheries Society position on introductions of aquatic species, *Fisheries*, 11:39–42, 1986.

Larkin, P.A. and S.B. Smith, Some effects of introduction of the redside shiner on the Kamloops trout in Paul Lake, British Columbia, *Trans. Am. Fish. Soc.*, 83:161–175, 1954.

Lee, D.S., C.R. Gilbert, C.H. Hocutt, R.E. Jenkins, D.E. McAllister, and J.R. Stauffer Jr., *Atlas of North American Freshwater Fishes*, North Carolina Biological Survey Publication 1980–12, North Carolina State Museum of Natural History, 1980.

Litvak, M.K. and R.I.C. Hansell., Investigation of food habit and niche relationships in a cyprinid community, *Can. J. Zool.*, 68:1873–1879, 1990.

Litvak, M.K. and N.E. Mandrak, Ecology of freshwater baitfish use in Canada and the United States, *Fisheries*, 18:6–13, 1993.

LoVullo, T.J. and J.R. Stauffer, The retail bait-fish industry in Pennsylvania — source of introduced species, *J. Penn. Acad. Sci.*, 67:13–15, 1993.

Ludwig, H.R. Jr. and J.A. Leitch, Interbasin transfer of aquatic biota via angler's bait buckets, *Fisheries*, 21(7):14–18, 1996.

Magnan, P. and G.J. Fitzgerald, Mechanisms responsible for the niche shift of brook charr; *Salvelinus fontinalis* Mitchell, when living sympatrically with creek chub, *Semotilus atromaculatus* Mitchell, *Can. J. Zool.*, 63:1548–1555, 1984.

Markus, H.C., *Propagation of Bait and Forage Fish*, U.S. Bureau of Fisheries, Fisheries Circular 28. 1939.

McAllister, D.E. and B.W. Coad, *Fishes of Canada's National Capital Region*, Natl. Mus. Nat. Sci. Misc. Spec. Pub. 24, 1974.

Meng, L., P.B. Moyle, and B. Herbold, Changes in abundance and distribution of native and introduced fishes of Suisun Marsh, *Trans. Amer. Fish. Soc.*, 123:498–507, 1994.

Meronek, T.G., F.A. Copee, and D.W. Coble, A summary of bait regulations in the north central United States, *Fisheries*, 20(11):16–23, 1995.

Miller, R.J. and H.W. Robison, *The fishes of Oklahoma*, Oklahoma State University Press, Stillwater, OK, 1973.

Mills, E.L., J.H. Leach, J.T. Carlton, and C.L. Secor, Exotic species in the Great Lakes: a history of biotic crises and anthropogenic introductions, *J. Great Lakes Res.*, 19:1–54, 1993.

Mooney, H.A. and J.A Drake (Eds.), *Ecology of Biological Invasion of North America and Hawaii*, Ecological Studies 58, Springer-Verlag, New York, 1986.

Nielson, L.A., The bait-fish industry in Ohio and West Virginia, with special reference to the Ohio River sport fishery, *N. Am. J. Fish. Manage.*, 2:232–238, 1982.

Nelson, J.S. and M.J. Paetz, *The Fishes of Alberta*, The University of Alberta Press, Edmonton, 1992.

Ostland, V.E., B.D. Hicks, and J.G. Daly, Furunculosis in baitfish and its transmission to salmonids, *Diseases of Aquatic Organisms*, 2:163–166, 1987.

Philipp, D.P., Genetic implications of introducing Florida largemouth bass, *Micropterus salmoides floridanus*, *Can. J. Fish. Aquat. Sci.*, (suppl. 1):58–65, 1991.

Portt, C. B., The effect of depth and harvest on baitfish in southern Ontario streams, *Ont. Fish. Tech. Rept. Ser.*, 15, 1985.

Prevost, G., Livebait minnow restrictions, Quebec Biological Bureau, University of Montreal, Montreal, PQ (unpublished ms), 1957.

Radcliffe, L., Propagation of minnows, *Trans. Am. Fish. Soc.*, 61:131–138, 1931.

Richardson, M.J. and F.G. Whoriskey, Factors influencing the production of turbidity by goldfish (*Carrasius auratus*), *Can. J. Zool.*, 70(8):1585–1589, 1992.

Robins, C.R., R.M. Bailey, C.E. Bond, J.R. Brooker, E.A. Lachner, R.N. Lea, and W.B. Scott, *Common and Scientific Names of Fishes from the United States and Canada*, 5th ed. American Fisheries Society Special Publication 20. 1991.

Rohde, F.C., Arndt, R.G., Lindquist, D.G., and J. F. Parnell, *Freshwater Fishes of the Carolinas, Virginia, and Delaware*, The University of North Carolina Press. Chapel Hill, NC, 1994.

Rubec, P. and B.W. Coad, First record of the margined madtom (*Noturus insignis*) from Canada, *J. Fish. Res. Board Canada*, 31:1430–1431. 1974.

Stedman, R.M. and C.A. Bowen, Introduction and spread of the threespine stickleback (*Gasterosteus aculeatus*) in Lakes Huron and Michigan, *J. Great Lakes Res.*, 11:508–511, 1985.

Summerfelt, R.C. and M.C. Warner., Incidence and intensity of infection of *Plistophora ovariae*, a microsporidian parasite of golden shiner, *Notemigonus crysoleucas*. In S.F. Snieszkop, Ed., *A Symposium on Diseases of Fishes and Shellfishes*, Am. Fish. Soc. Spec. Publ. 5. 1970, pp 142–160.

Terrell, J.W. and T.T. Terrell, Macrophyte control and food habits of the grass carp in Georgia ponds, *Verh. Internat. Verein. Limnol.*, 19:2515–2520, 1975.

Thomsen, H.P. and A.D. Hasler, The minnow problem in Wisconsin, *Wisconsin Conserv. Bull.*, 9(12):6–8, 1944.

Trautman, M.B., *The Fishes of Ohio*, Ohio State University Press. 1981.

Wurtsbaugh, W.A. and R.A. Tapia, Mass mortality of fishes in Lake Titicaca (Peru-Bolivia) associated with the protozoan parasite *Ichthyophthirius multifiliis*, *Trans. Amer. Fish. Soc.*, 117:213–217, 1988.

12 Ecological Impacts of Introductions Associated With the Use of Live Baitfish

Cheryl D. Goodchild

INTRODUCTION

The use of live baitfish is a major component of recreational fishing activities throughout much of Canada and the U.S. The extent of this practice is demonstrated by the large size of the industry (Litvak and Mandrak in previous chapter) that has evolved in response to the huge demand for live baitfish by anglers. While any activity of this magnitude would be expected to have a considerable impact, until recently there has been insufficient recognition of the extent of impacts associated with baitfish use in North America.

This chapter primarily documents ecological impacts of introductions of nonindigenous species linked to continued use of live baitfish, expanding on some of the topics introduced in the previous chapter by Litvak and Mandrak. Further, it documents the impacts of nontarget species transported with baitfish (smaller organisms whose presence in the transport medium or attachment to the target species may be overlooked or considered unimportant — invertebrates, algae, plants, and pathogens) which have received very little attention to date.

BAITFISH INTRODUCTIONS

The collection, transport, and release of small fish and invertebrates for use as bait has probably resulted in the introduction and establishment of many aquatic species (Carlton 1992). There are a number of factors that influence the ability of species to invade new environments, most of which have been discussed in detail in other chapters of this publication. For baitfish introductions, certain areas may be more susceptible to invasions through this pathway due to their geographic location or regulatory practices.

Most jurisdictions now restrict baitfish use to avoid the introduction of nuisance species through release of live bait. The risk of bait-related introductions may only be partly related to differences in baitfish regulations among jurisdictions. Information on the relative level of restrictions on baitfish importation (Figure 12.1) and use (Figure 12.2) in Canada and a number of selected American states was obtained in a telephone survey (Goodchild 1997). Although jurisdictions with more liberal baitfish regulations may be more susceptible to introductions from this pathway, neighboring jurisdictions may still be at risk despite stiffer baitfish restrictions. Introduced baitfishes can disperse rapidly into new areas. Anglers probably introduced the Arkansas River shiner (*Notropis girardi*) into the Pecos River, New Mexico, and within 3 years the species moved 260 km downstream. Similarly, the banded darter (*Etheostoma zonale*) dispersed 400 km in the Susquehanna River, Pennsylvania after probably being released as unused bait (LoVullo and Stauffer 1993).

Proximity to the Great Lakes combined with extensive baitfish harvest and use in the area may have a profound effect on the risk of introduction of a wide variety of aquatic species. Many

baitfish importation restrictions

- no import permitted
- no known import
- import permitted
- Not surveyed

FIGURE 12.1   Baitfish importation restrictions in Canada and selected American states. Information obtained during a telephone survey conducted in 1995.

introduced species are now established in the Great Lakes. As baitfish regulations are relatively liberal in Great Lakes bordering jurisdictions, continued harvest, transportation, and use of live bait in the area greatly intensifies the risk of introducing many species further inland.

While there is a greater chance of baitfish introductions in the Great Lakes Basin, other jurisdictions are not exempt. The use of live baitfish has been prohibited in Alberta since 1963; however, in the past considerable numbers of fishes were transported from one waterbody to another through the indiscriminate movement of baitfish. The current distribution of fish in Alberta reflects the extent to which species were transferred before implementation of the ban on live baitfish (D. Berry, Alberta Environmental Protection, Edmonton, Alberta, pers. comm.). In New Brunswick, internal transfers of baitfish have resulted in the widespread almost ubiquitous distribution of most native baitfish species (M. Campbell, Department of Fisheries and Oceans, Moncton, New Brunswick, pers. comm.). It is extremely unlikely that populations of golden shiners (*Notemigonus crysoleucas*) in two small inland lakes of Prince Edward Island are native. Their presence is probably due to deliberate stocking as forage or possibly accidental release of baitfish. Attempts to eradicate golden shiners using rotenone have not been successful (A. Smith, Department of Environmental Resources, Charlottetown, P.E.I., pers. comm.).

baitfish use restrictions

- no use permitted
- many/extensive restrictions
- some/few/no restrictions
- not surveyed

FIGURE 12.2   Baitfish use restrictions in Canada and selected American states. Information obtained during a telephone survey conducted in 1995.

THE ROLE OF NONTARGET ORGANISMS

The threat from bait related introductions extends beyond the target baitfish species to "hitchhiking" or nontarget organisms that can also be transferred through the disposal of the contents of bait-buckets and live wells. Since the larval stages of zebra mussels (*Dreissena polymorpha*) and spiny water fleas (*Bythotrephes cederstroemi*) are nearly invisible to the naked eye, these exotic species, recently established in the Great Lakes, might easily be spread through the harvest and transportation of live baitfish, particularly in bordering jurisdictions such as Ontario. The zebra mussel has been introduced in many waterways in eastern North America, although most of this spread has probably been through recreational and commercial boating traffic. Spiny water fleas are now present in 16 inland lakes in Ontario, possibly through unintentional angler introductions (Yan et al. 1992; Krishka et al. 1996). Since some aquatic species such as spiny water fleas reproduce parthenogenically, the release of a single gravid female could start an invasion.

Diseases and parasites either infecting baitfish or contaminating the containers in which they are transported can also be readily transferred through baitfish use. Although many diseases associated with North American baitfishes may be ubiquitous, there is the potential for exotic diseases to be inadvertently introduced along with transferred bait.

BAITFISH INDUSTRY AS A SYNERGISTIC PATHWAY

The role of bait transfers in the dispersal of aquatic species introduced through other pathways (stocking, immigration through canals and shipping corridors, disposal of ballast water from ships, transport of species used in aquaculture and the aquarium trade) is a continuing concern.

Synergistic transfers of exotic aquatic species involving baitfish have been reported throughout North America as illustrated by the following examples. The blue tilapia (*Tilapia aurea*), a popular exotic species originally imported for stocking and aquaculture, is currently being further dispersed in the lower Colorado River by the use of juveniles as baitfish (Courtenay et al. 1984). Although exotic rudd (*Scardinius erythrophthalmus*) were imported on many occassions, the species did not become established in the U.S. until it became popular as a cultured baitfish in Arkansas and was subsequently transported to 14 states. Continued dispersal is probably a result of the release of live baitfish (Easton et al. 1993). In Canada, the first record of the rudd in the province of Ontario was from a bait-bucket in eastern Ontario (E. Holm, Royal Ontario Museum, Toronto, pers. comm.). The species is now established in the St. Lawrence River (Crossman et al. 1992). Also, the current distribution of the Asian freshwater clam (*Corbicula fluminea*) is partly due to transfer by fishermen using it as a bait organism (Carlton 1992), acting in concert with accidental introductions associated with aquaculture (Counts 1986), and intentional introductions by people purchasing it as a food item. The Asian freshwater clam may have originally been introduced to the west coast of the U.S. as a food item by Chinese immigrants (Foster and Fuller 1997).

EXAMPLES OF ECOLOGICAL IMPACTS OF INTRODUCTIONS RELATED TO THE USE OF LIVE BAITFISH

The impacts of introduced species have been observed as functions of competition, trophic and spatial alteration, predation, environmental changes, genetic effects and hybridization, and transmission of pathogens. Reported ecological impacts attributed to species introduced through the use of live baitfish are discussed within these categories and summarized in Table 12.1.

COMPETITION

Competitive interactions between introduced and native species that have similar requirements for a limited supply of food and space may negatively impact native populations. Two species of gobies introduced to the Great Lakes very recently could potentially be further dispersed through the use of live baitfish and compete with native fishes. They are reportedly already favored by anglers as a bait for largemouth bass (*Micropterus salmoides*) in Ohio portions of Lake Erie (E. Crossman, pers. comm.). Small gobies resemble native sculpins (*Cottus* sp.), although gobies have a distinctive suctorial disk (formed by fused pelvic fins) that is visible on closer examination. The tubenose goby (*Proterorhinus marmoratus*) has remained uncommon, but the round goby (*Neogobius melanostomus*) has dispersed rapidly (Marsden and Jude 1995). Gobies have successfully colonized many other areas throughout the world (Jude et al. 1992). In North America, the round goby may compete with other native benthic species such as sculpins and darters. Reductions in sculpin populations (*Cottus bairdi* and *C. cognatus*) have been reported from areas where gobies have become established (Marsden and Jude 1995).

The ruffe (*Gymnocephalus cernuus*), native to northern Europe, has a history of invading and displacing native species. Ruffe were first discovered in western Lake Superior, in Duluth Harbor, Minnesota, in 1989 (Pratt et al. 1992) and subsequently captured in Ontario from the Kaministiquia River at Thunder Bay, Lake Superior in 1991 (MacCallum 1994). Ruffe have become established in both Duluth Harbor and Thunder Bay and appear to be expanding their range at a rapid rate. The species has recently been found in Lake Huron (Krishka et al. 1996). Although the major mechanism for the initial introduction and transfer of this species within the Great Lakes appears to have been through the disposal of ballast water, there is great risk that the species may be further transported in live wells and bait buckets despite being an illegal baitfish species in many jurisdictions, including Ontario. Adult ruffe are generally less than 20 cm in length and therefore have no value as commercial or sport fish. Fisheries biologists throughout the Great Lakes area, consider the ruffe to be a serious threat to native species. In the Duluth area, they have already become the most abundant species, and subsequently numbers of many native species have declined.

Before 1980, the threespine stickleback (*Gasterosteus aculeatus*) was unknown from the Great Lakes above Niagara Falls, but has since become established in the upper Great Lakes. Specimens have been collected from small bays and tributaries to lakes Huron and Michigan (Stedman and Bowen 1985), and also from Thunder Bay, Lake Superior, Ontario (Ball and Tost 1992). The most likely pathway for these introductions was through transportation by bait dealers and subsequent release by anglers, although they may also have been transported in ballast water. Competition for food between introduced threespine sticklebacks and native ninespine sticklebacks (*Pungitius pungitius*) could negatively impact ninespine stickleback populations. Both species have similar dietary preferences, yet the two species appear to coexist successfully in Lake Ontario (Stedman and Bowen 1985).

In Hasse Lake, Alberta, self-sustaining populations of introduced threespine sticklebacks supplanted native brook sticklebacks (*Culaea inconstans*) and fathead minnows (*Pimephales promelas*). There are indications that threespine sticklebacks may also have gained entry to adjacent lakes through dispersal or further bait-bucket introductions. Rehabilitation is being considered (D.Berry, Alberta Environmental Protection, Edmonton, Alberta, pers. comm.). Conversely, introduced brown bullhead (*Ameiurus nebulosus*) and pumpkinseed (*Lepomis gibbosus*) have caused the extirpation of native threespine sticklebacks in many small lakes on Vancouver Island, British Columbia. There are serious concerns that these introduced species will cause the extirpation of other unique sticklebacks such as the Enos Lake species pair, two sympatric, genetically distinct, and reproductively isolated forms (McPhail 1989).

Extremely costly rehabilitation efforts involving the use of rotenone were required to eliminate the redside shiner (*Richardsonius balteatus*) from Lees Lake, Alberta, a valuable trout lake (D. Berry, pers. comm.). The redside shiner was probably transported as bait from southern British Columbia and unintentionally introduced. The species was eliminated due to concerns that it would negatively affect trout populations. Redside shiners had a serious impact on a trout fishery in Paul Lake, near Kamloops in British Columbia (Larkin and Smith 1954). Introduced redside shiners fed on invertebrates at sizes too small for rainbow trout (*Oncorhynchus mykiss*) and also behaviorally prevented small trout access to invertebrates.

The release and subsequent introduction of yellow perch (*Perca flavescens*) into stocked trout lakes has posed a serious problem in Alberta. Yellow perch compete for food and space with other species. Also, biting perch may frustrate anglers in search of trout. The introduction of yellow perch through bait-bucket release has similarly negatively impacted brook trout (*Salvelinus fontinalis*) populations in many small Ontario lakes, despite being prohibited for use as bait in the province. Such introductions have often resulted in costly reclamation efforts with rotenone, with subsequent live bait bans. Similarly, the recent discovery of yellow perch and white perch (*Morone americana*) in Skiff Lake, New Brunswick, has caused considerable concern that through competition the perch will have adverse effects on both brook trout and landlocked salmon fisheries (C. Ayer, New Brunswick Department of Natural Resources and Energy, Fredericton, NB, pers. comm.).

The first occurrence of the red shiner (*Cyprinella lutrensis*) in the Colorado River, California may be attributed to escape from fish ponds or release of live bait. Although the species was subsequently stocked as forage in the northern part of the state, further dispersal is probably due to the release of bait. Red shiners have recently and rapidly invaded the San Joaquin Valley, California (Jennings and Saiki 1990). The native California roach (*Hesperoleucus symmetricus*) may be vulnerable to displacement by the red shiner. Although recent studies failed to collect roach from parts of the San Joaquin Valley, there are no definite links to competition with red shiners as roach may be absent or rare in this area (Jennings and Saiki 1990). Red shiners are aggressive and highly adaptable, which may provide a competitive advantage over certain native cyprinids. Red shiners have reportedly caused depletion of endemic fishes particularly in Arizona (A. Cordone, c/o California Department of Fish and Game, Sacramento, CA, unpublished data) and are considered a great threat to indigenous fishes of the southwestern U.S. Specifically, red shiners have been discovered in Aravaipa Creek, Arizona, probably the result of another bait-bucket introduction. Aravaipa Creek is one of the few remaining perennial streams in the arid southwest that still contains native and endemic fishes, including the federally threatened spikedace (*Meda fulgida*). In the Moapa River, Nevada, red shiner populations have also been implicated in the decline of native fish, including spikedace, woundfin (*Plagopterus argentissiumus*), and Virgin River chubs (*Gila seminuda*) (Fuller and Nico 1996). Attempts to eradicate this species have so far been unsuccessful.

The fathead minnow is another popular baitfish introduced to a large number of American states. Both red shiners and fathead minnows may compete with and adversely affect young Colorado squawfish (*Ptychocheilus lucius*) (Fuller and Nico 1996). The Colorado squawfish is an endangered species that is found only in large rivers of the Colorado basin (Lee et al. 1980).

Golden shiners were probably introduced to California with stocked game fishes and through importation of live bait. Although they remain one of the most commonly used baitfishes in California, they are believed to negatively impact trout fisheries through competition for food (A. Cordone, unpublished data). Introduction of golden shiners into California lakes usually results in decreased growth and reproduction in trout. Golden shiners may also compete with young centrarchids for food (Fuller and Nico 1996). Consequently, golden shiners have often been targeted in chemical rehabilitation programs in California.

Many nontarget aquatic organisms that may have a competitive advantage over native species can also potentially be transferred through discarding the contents of baitfish buckets. The zebra mussel was introduced to North America in the late 1980s, most probably through the discharge of ballast water containing veligers originating from a European port (Griffiths et al. 1991). The species has spread throughout the Great Lakes, and has invaded inland waters in Ontario, particularly the Kawartha Lakes, Lake Simcoe, and the Muskoka Lakes (O'Neill and Dextrase 1994). Recently, zebra mussels have also been reported from the Rideau Canal system and eastern Ontario Lakes, despite public education on the dangers of transporting the species through movements of boats. Adult zebra mussels, which attach themselves to many hard surfaces including boats and boat trailers, can survive out of the water for weeks and be transported long distances. There is also considerable potential for the planktonic veliger larvae of zebra mussels, which are invisible to the human eye, to be further dispersed through the disposal of bait-bucket, live-well, and bilge water. Zebra mussels appear to have a competitive advantage over native mussel species and readily colonize the shells of native mussels. Dramatic declines in abundance and species richness of native unionid mussels have been observed in areas where zebra mussels have proliferated, leading to the extirpation of native mussels in many areas (Garton et al. 1993). Similarly, Asian clams (*Corbicula fluminea*) appear to dominate benthic communities (Sickel 1986), displacing native species through competition.

TROPHIC AND SPATIAL ALTERATION

Successful introduced species may cause changes to the food web or force native species to occupy other habitats in the aquatic environment. Native species may compensate for competition from introduced species through changes in food and habitat selection. Magnan and Fitzgerald (1982, 1984) studied the impact of introduced creek chub (*Semotilus atromaculatus*) on brook trout (*Salvelinus fontinalis*) populations in Québec. Brook trout switched from feeding on benthos to feeding on zooplankton after creek chub became established. This dietary change apparently caused reduced growth rates in trout from 30 to 70%. These and other studies of the impacts of introduced baitfish species in Québec indicate that substantial losses in revenue from sport fisheries likely resulted from reductions in brook trout growth (Poirier 1992).

The zebra mussel may also be a powerful food web modifier causing considerable trophic shifts from a pelagic system to one dominated by benthic organisms as discussed by Nalepa and colleagues in Chapter 16 of this volume.

PREDATION

Baitfishes normally provide forage for other fishes and therefore there are few examples of introduced baitfishes acting as predators. Some baitfish species, however, may be piscivorous during certain periods and feed on the eggs and larval stages of other fishes.

Rainbow smelt can occupy the role of prey, predator or competitor (Franzin et al. 1994). As predators they are opportunistic feeders that may feed on eggs or young of other species. Lake whitefish (*Coregonus clupeaformis*) populations have been affected by the introduction of rainbow smelt and declines in lake whitefish recruitment have been observed after rainbow smelt invasions. Extensive predation on larval lake whitefish by rainbow smelt has been documented (Loftus and Hulsman 1986). Rainbow smelt may also prey on lake trout (*Salvelinus namaycush*) eggs and larvae, however this does not apparently have any negative impacts on lake trout populations (Loftus and Hulsman 1986). Although rainbow smelt have rapid dispersal potential, human actions have been the most important agent in the spread of the species (Franzin et al. 1994). Despite being prohibited for use as live bait in both Ontario and Manitoba, most road-accessible lake trout lakes in both provinces are at great risk of rainbow smelt introductions (Franzin et al. 1994). Rainbow smelt were introduced into Quetico Park, Ontario, by anglers using them for lake trout bait (E. Crossman, pers. comm.). The timing of rainbow smelt spawning and spring recreational fishing may be a factor in its spread by anglers. Viable, fertilized eggs may survive in a bait-bucket for many hours (Franzin et al. 1994), and could be transported in this manner to nearby lakes. The rainbow smelt has been widely introduced in the Great Lakes and inland waters and has often established large populations and affected indigenous fishes through both predation and competition (Evans and Loftus 1987).

The use of live baitfish is suspected to have contributed to the dispersal of both rainbow smelt and alewife (*Alosa pseudoharengus*) in New York (P. Festa, NY Dept. Environmental Conservation, Albany, pers. comm.). Alewife introductions have probably affected walleye (*Stizostedion vitreum*), yellow perch, and whitefish (*Coregonus* sp.) populations through increased competition and heavy predation by alewives on the fry and eggs of other species. The decline and in some instances extirpation of lake herring (*Coregonus artedi*) populations have been attributed to introductions of alewives (Smith 1985). Alewives have also been implicated in the collapse of walleye populations in Conesus Lake and Schohari Reservoir, New York, within 5 years of introduction (P. Festa, pers. comm.) and blamed for declines in fish populations, particularly whitefishes, in the Great Lakes.

In Québec, small coarse fishes, such as yellow perch and sunfishes (Centrarchidae) that had probably been introduced by anglers releasing bait, were observed eating lake trout eggs (Wilson 1958). These fish may have been responsible for lake trout decline. Predation on lake trout eggs by introduced gobies has also been observed in laboratory experiments (Marsden and Jude 1995).

The endangered razorback sucker (*Xyrauchen texanus*) may be unable to reproduce successfully due to predation by introduced red shiners and other species on its spawn (Miller et al. 1982). Predation on young fish by mosquitofish (*Gambusia affinis*), combined with inhibition of reproduction, contributed to the decline of the Gila topminnow (*Poeciliopsis occidentalis*) in Arizona (Schoenherr 1981).

As mentioned, the spiny water flea successfully established populations in many areas of the Great Lakes, after suspected introduction through ballast water discharge (Sprules et al. 1990). Recently, they have become established in many inland areas of Ontario. There is a continued risk of further dispersal in bait buckets as well as other pathways associated with recreational activities. Although few adverse impacts have been identified to date, the species has been identified in the gut contents of many fish species and the impact on native zooplankton and fish communities have been debated in the literature. Declines in native *Daphnia* sp. may possibly be attributed to predation by spiny water flea. However, some studies indicate that the reproductive rate of *Daphnia* exceeds the predatory ability of *Bythotrephes* and observed declines in *Daphnia* populations may have been caused by increased predation by planktivorous fish (Sprules et al. 1990).

Environmental Changes

Many introduced species have been associated with environmental changes in North America, including removal of aquatic vegetation, increased turbidity, and eutrophication. Most jurisdictions have banned the use of the common carp and goldfish for use as bait because these species have been associated with increased turbidity and harmful alteration of fish habitat. Taylor et al. (1984) summarized the effects of introductions of common carp as follows. The foraging and spawning behaviour of common carp generally results in reductions in macrophytes that can precipitate increases in turbidity and siltation. Observations of slower growth and lower recruitment in centrarchids and behavioral changes in other species have been observed after increases in turbidity. Effects of increased turbidity on native fishes include disruption or elimination of reproductive activities and interference with normal respiratory functions with a resultant decline in population numbers and species diversity. Destruction of aquatic vegetation may also be a factor in declining waterfowl populations.

Conversely, increased water clarity has been associated with the formation of zebra mussel colonies in the Great Lakes, apparently a result of the efficiency by which zebra mussels feed on phytoplankton (Leach 1993). In zebra mussel infested areas, a proliferation of aquatic macrophytes has been observed in many areas because of increased light penetration. Concern over introductions of the Asian clam have primarily centered around its biofouling effects on industrial water systems (Isom et al. 1986). Benthic substrate alterations have also been observed after establishment of Asian clams (Sickel 1986). Impacts may be less severe in areas with low winter temperatures due to a higher mortality rate and possibly delayed sexual maturity (Mills et al. 1993).

Largely through their use as live bait, rusty crayfish (*Orconectes rusticus*) have been introduced into Minnesota, Wisconsin, and Ontario (Gunderson 1995). The rusty crayfish is probably native to the Ohio River basin and the states of Ohio, Kentucky, Tennessee, Indiana, and Illinois but is now found in many other American states (Michigan, Massachusetts, Missouri, Iowa, New Mexico, New York, New Jersey, and Pennsylvania). Before the late 1960s, rusty crayfish were extremely rare in Ontario, found only at six locations in south central Ontario and Lake of the Woods (Crocker and Barr 1968). A decade later it was collected at 44 sites in the Kawarthas, was the only crayfish species found at 15 sites, and was predominant at 13 other sites (Berrill 1978). Berrill predicted accelerated dispersal of crayfishes through fishermen transporting bait. Rusty crayfish are now one of the two most common Orconectes species found in southern Ontario. It has also recently been collected from several new localities in northwestern Ontario (Momot 1992). Rusty crayfish introductions are also associated with reductions in invertebrate fauna, particularly snails (Lodge and Lorman 1987; Olsen et al. 1991).

Although displacement of native species by rusty crayfish is a concern, the most serious impact appears to be destruction of aquatic vegetation, particularly in unproductive northwestern Ontario shield lakes (Olsen et al. 1991; Momot 1992). Submerged aquatic plants provide habitat for invertebrates, shelter for fishes, spawning habitat, and resistance to wave action that can cause erosion (Gunderson 1995). The species has created problems through reduction of aquatic vegetation in lakes and has negatively impacted sport fisheries (R. Johannes, Minnesota DNR, St. Paul, Minnesota, pers. comm.).

Eurasian watermilfoil (*Myriophyllum spicatum*) has spread rapidly throughout North America and has become a major pest in British Columbia. Anglers are warned against the transfer of live bait in British Columbia in order to limit distribution of watermilfoil (Fisheries and Oceans 1992).

Rapid dispersal is possible because the species reproduces by vegetative fragmentation and dispersal is assisted by unintentional transportation by boats or bait buckets. Dense stands of Eurasian watermilfoil may displace native vegetation or invade unvegetated areas. The loss of blackchin shiner (*Notropis heterodon*) from several Wisconsin lakes appears to be the result of introduced Eurasian watermilfoil (Lyons 1989). Several other exotic plant species potentially could also be transferred by bait buckets and may have negative impacts on fish and wildlife habitats. These include flowering rush (*Butomus umbellatus*), purple loosestrife (*Lythrum salicaria*), and European frog-bit (*Hydrocharis morsus-ranae*). Invasions of exotic aquatic plants have caused the loss of both fish and wildlife habitats in Minnesota (R. Johannes, pers. comm.).

GENETIC EFFECTS AND HYBRIDIZATION

The preservation of biological diversity is crucial to ensuring long-term ecosystem sustainability and ultimately to the well-being of mankind, yet the worldwide loss of biodiversity is primarily associated with human activities (Olver et al. 1995; Winter and Hughes 1995).

The potential for introductions of baitfish to have negative impacts on the genetic integrity of native species is still largely unrecognized. However, there is growing concern that rudd may endanger the genetic integrity of the native golden shiner through hybridization. The two species have been successfully hybridized experimentally. Burkhead and Williams (1991) suggest that this intergeneric cross will probably occur in nature based on the relative ease with which the two species hybridize in the laboratory. Apparently only hybrids of male golden shiners and female rudd survive, and preliminary investigations suggest these hybrids are sterile. However, sterile hybrids can reduce the number of individuals available for spawning, resulting in lower population numbers. The possibility also exists for the hybridization of these two species to result in fertile hybrids that could compromise the genetic integrity of indigenous golden shiner populations and possibly lower their genetic fitness. Because it hybridizes with native golden shiners, the rudd should be considered an undesirable species based on the protocol developed by Kohler and Stanley (1984). Since rudd are morphologically similar, small specimens may be difficult to differentiate from native golden shiners.

Many closely related species readily hybridize when isolating mechanisms are removed. A probable bait-bucket release is thought to be responsible for the introduction of cutlips minnows (*Exoglossum maxillingua*) to the New drainage of Virginia and West Virginia. Cutlips minnow are now hybridizing with native tonguetied minnows (*E. laurae*), a closely related species (Fuller and Nico 1996). Tonguetied minnows have a very limited and disjunct distribution in the Ohio River basin (Lee et al. 1980). Hybrids between tessellated darter (*Etheostoma olmstedi*) and banded darter (*E. zonale*) were also documented following the introduction and subsequent rapid dispersal of banded darter into the Susquehanna River (LoVullo and Stauffer 1993). The Lahontan tui chub (*Gila bicolor obesa*) became established in Crowley Lake, California after being introduced by bait bucket release. There is concern that this introduced subspecies will hybridize with and deplete populations of indigenous Owens tui chubs (*G. b. snyderi*), an endangered species (Moyle 1976) — possibly leading to its extirpation. The Mohave tui chub (*G. b. mohavensis*) was nearly extirpated through hybridization with the introduced arroyo chub (*G. orcutti*) (Fuller and Nico 1996).

The Banff longnose dace (*Rhinichthys cataractae smithi*) was endemic to a marsh created by hotsprings southwest of Banff, Alberta. The species was quite abundant before the early 1900s based on collection records. Initial decline of the Banff longnose dace can be attributed to introductions of many exotic species, including mosquitofish stocked in 1924. The final extinction of the subspecies, however, has been attributed to introgression with the eastern subspecies of longnose dace (*R. c. cataractae*) (Miller et al. 1989). Longnose dace are not commonly used as baitfish in Canada. However, they are widely used in central and western parts of the U.S. (Scott and Crossman 1973).

The long-term genetic effects of the continued transfer and introduction of billions of locally adapted baitfishes may have profound effects on the fitness and continued survival of many forage species.

TRANSMISSION OF PATHOGENS

The dispersal of disease agents is considered by the American Fisheries Society to be one of the most significant threats that introduced species may pose to native fish communities (Ganzhorn et al. 1992). Diseases caused by bacteria, viruses, and parasites may be transported along with introduced aquatic species (Kohler and Courtenay 1986). Many parasites have been spread through the transfer of fish. At least 50 were reported by the early 1970s (Hoffman 1970). Although the extent of transfer of diseases or parasites through baitfish is unknown, diseases are at least as widespread in baitfishes as in other fish species (Fisheries and Oceans 1992). There is a high risk of transmission of pathogens through continued transportation of baitfishes, particularly as baitfishes are commonly shipped without strict disease control inspections. Highly stressed populations, such as crowded baitfish in tanks, are more susceptible to infection (Fisheries and Oceans 1992). Infectious septicemia and infectious pancreatic necrosis virus (IPNV) are both commonly found in baitfish (Fisheries and Oceans 1992). The transfer of disease agents to hatcheries through the use of baitfishes as forage has been observed in Illinois (R. Horner, Illinois DNR, Topeka, Illinois, pers. comm.), and suspected in many other jurisdictions.

Besides the potential for diseases associated with baitfishes to be released into new environments and infect other species, the tanks used to transport baitfishes may also be contaminated. Resistant stages of many parasites can persist outside the host and survive on containers or in the water used to transport bait (Ganzhorn et al. 1992). Viral pathogens may also be transferred through shipments of fish and may survive for extended periods outside the host. Fish viral diseases are untreatable and may have severe impacts on populations of fishes (Ganzhorn et al. 1992).

Recreational fishes may be susceptible to infection from diseases transmitted by baitfishes. Many baitfishes are hardy and may be resistant to certain diseases and parasites. Even dead or frozen baitfishes may communicate diseases to recreational fishes (Wilson 1958). Transmission of the fungi (*Saprolegnia* sp.) does not require fish-to-fish contact. The infection, which causes death through osmoregulatory failure, is associated with significant baitfish mortality and is also widespread in salmonids (Fisheries and Oceans 1992). Since the rate of infection increases with maturity, larger salmonids are at greater risk and the disease is probably affecting recreational fisheries (Fisheries and Oceans 1992). Currently in Canada, there is considerable concern over transmission of diseases such as furunculosis, caused by a bacterial parasite (*Aeromonas salmonicida salmonicida*), and IPNV. Outbreaks of IPNV have been recorded throughout most of Canada, but furunculosis outbreaks have only been reported from New Brunswick, Québec, Ontario, and British Columbia. Both pathogens survive well in Canadian waters, are widespread in baitfish, and infect other susceptible fish taxa including salmonids (Fisheries and Oceans 1992). Under experimental conditions, *A. s. salmonicida* caused furunculosis in four species of freshwater baitfish (Ostland et al. 1987). Although direct transmission in the wild has not been proven, in experimental studies furunculosis was transmitted from common shiners (*Luxilus cornutus*) to coho salmon (*Oncorhynchus kisutch*) and brook trout (Ostland et al. 1987).

Of great concern in Alberta is the potential for yellow perch to act as a vector for the transmission of diseases and parasites (D. Berry, pers. comm.). Both intentional and unintentional releases of yellow perch have resulted in the transfer of many diseases and parasites. Currently, there are fears that the tapeworm (*Triaenophorus crassus*), which infects cisco and lake whitefish populations in northern Alberta, will be transferred to populations in south and central areas of the province (D. Berry, pers. comm.). Although not harmful, the parasite is aesthetically unappealing, making fish less desirable for consumption and not acceptable for export. This could result in considerable economic losses to commercial fisheries.

Five cyprinid fishes imported from New York and sold in Pennsylvania for bait were discovered to be infested with glochidia (LoVullo and Stauffer 1993). Glochidia, the larval stage of freshwater mussels (Unionidae), parasitizes fishes. Apart from the obvious danger of transporting diseases, this is also an example of the potential for the transportation of exotic mussel species between drainages (LoVullo and Stauffer 1993).

The protozoan parasite (*Plistophora ovariae*) has been reported in native stocks of golden shiners (Litvak and Mandrak 1993), where it slowly destroys the ovaries of female shiners (Anon. Undated). Golden shiners are one of the most popular baitfish species and are widely transported in many jurisdictions. Restricting the use and transport of baitfishes to common species such as golden shiners has been recommended (Litvak and Mandrak 1993; Moyle and Li 1994), but this approach does not consider the potentially deleterious effects of the introduction of associated pathogens to native species.

An introduced Eurasian trematode (*Bolbophorus confusus*) was recently discovered in minnows shipped from South Dakota to Arkansas. Infected native snails in Montana apparently passed on the parasite to trout (Hoffman and Schubert 1984). Golden shiners and fathead minnows are just two of the many cyprinid species that can be infected by the Asian tapeworm (*Bothriocephalus opsarichthydis*). The species reportedly has also infested channel catfish (*Ictalurus punctatus*). According to Ganzhorn et al. (1992), the Asian tapeworm probably originated in China and was transported with grass carp (*Ctenopharyngodon idella*) introduced to North America. The life cycle of this parasite involves a secondary host, a planktonic crustacean, which is eaten by fish. Asian tapeworms now infect many indigenous wild fish populations in many North American locations. Golden shiners in Arkansas and Missouri have been heavily contaminated (Anon. Undated). Red shiners, a popular baitfish species in the southwestern United States, have also become hosts to Asian tapeworms in their native range, probably transmitted from introduced grass carp. When red shiners were subsequently introduced to Utah, they caused heavy infestations of the tapeworm in the woundfin (*Plagopterus argentissimus*), an endangered cyprinid (Fuller and Nico 1996). Recently, the parasite has also been found in the endangered Colorado squawfish from New Mexico (Hoffman and Schubert 1984).

The potential for disease agents to be introduced through the use of other types of live bait has received little attention but presents another substantial risk to indigenous populations. The introduction of the North American signal crayfish (*Pacifasticus leniusculus*) into Europe is one example of such a deleterious introduction. Thompson (1990) discusses the spread of crayfish plague (a fungus) in Europe following the introduction of American specimens into rivers in Italy in the 1860s and the subsequent dispersal throughout Europe and Great Britain. The results have been devastating. In Sweden the annual harvest of native crayfish was reduced to zero in only 10 years and the native species was extirpated from many areas. North American crayfish are generally resistant to the plague. Introductions of other exotic species into North American waters, however, could have similar impacts.

DISCUSSION

From the preceding discussion, it is clear that even relatively local transfers of baitfish can have as profound an effect as invaders from other parts of the globe. Although it is often difficult to definitely attribute specific introductions to the use of baitfish, there is widespread belief that baitfish use is associated with many negative impacts. The continued use and transportation of live baitfish substantially increases the risk of introducing many organisms, including nontarget organisms and pathogens. It also enhances the risk of further dispersing unwanted species originally introduced through other pathways. Once established, introduced species are very difficult, if not impossible, to eradicate.

With increased recognition of the many risks associated with baitfish use (particularly those associated with introductions) most agencies have responded with changes to their management practices. These have occurred primarily through the imposition of greater restrictions (in some cases complete bans) on use of live baitfish. Jurisdictions where baitfish use is less common and where economic impacts would be less severe, have been more expedient in implementing these changes. Other jurisdictions, where baitfish are frequently used and a large industry has developed, have been slower in imposing restrictions because industry and angler opposition to proposed changes has often been a stumbling block.

Live bait was used extensively before an awareness of the intrinsic risks developed and therefore many effects were not recorded. These have since been obscured by the passage of time and disguised by other anthropogenically induced changes. Rigorous documentation and scientific proof are unfortunately lacking. Without valid experimental analysis, comparative methods are used to elucidate impacts. The effects of introduced species have been discerned largely through a comparison of before and after habitat or population changes. For example, a reduction in unwanted introductions was observed in Saskatchewan after the use of live baitfish was banned (E. Dean, Saskatchewan Environment and Resource Management, Regina, Saskatchewan, pers. comm.), suggesting that the use of live baitfish had contributed to the introduction and dispersal of nuisance species.

The degree of risk of introductions from continued baitfish use in different jurisdictions is related to a combination of factors including angler preference, size of industry, geographic location, and extent of restrictions on importation, transportation, and use both within the jurisdiction and in neighboring jurisdictions. Although many jurisdictions prohibit the importation and release of live baitfish, widespread use is generally allowed. The risk may be intensified in some geographic areas despite specific regulations designed to reduce risk. For example, Ontario and several American states border the Great Lakes where introductions from a variety of sources have had considerable environmental impact. Clearly, multijurisdictional cooperation between Great Lakes bordering states and the province of Ontario are necessary to establish effective management policies designed to reduce baitfish related risks. Successful partnerships such as the Great Lakes Fishery Commission have established precedents and furnish models that could provide the basis for cooperative baitfish management efforts in the entire Great Lakes watershed. For these efforts to be effective, they would require implementation of consistent regulations and effective enforcement efforts on the part of all cooperating jurisdictions. Similar approaches could be used to mitigate risk from baitfish use in other parts of Canada and the U.S.

Management in the past has been largely reactive and consequently restrictions and regulations have often been implemented as needed and not with a view to the future. Regulations are merely tools that can be used to achieve certain management objectives. It is imperative to tailor regulations to specific situations and objectives. The efficacy of current baitfish restrictions in realizing specific management goals to reduce previous and potential environmental damage has not been proven. Certainly, confusion over the various regulations and restrictions among states and provinces hinders the ability of jurisdictions to achieve angler compliance and reduces the credibility of management agencies. Therefore, agencies are increasingly turning to public education rather than adding to the already overwhelming regulatory maze, as a means of ensuring adherence to conservation measures.

Resource management agencies are increasingly cognizant of their role as resource stewards and now have greater regard for protecting native species and their critical habitats. Contemporary values on the part of the general public are also changing and recognition of environmental issues growing. Ultimately, enhanced understanding will lead to more informed decision-making.

ACKNOWLEDGMENTS

The Ontario Ministry of Natural Resources provided partial funding for this project. Sincere thanks to the many individuals in Canadian and American natural resource agencies who generously provided information and reference materials. Finally, I am particularly grateful for the support and assistance of Alan Dextrase, Ontario Ministry of Natural Resources.

REFERENCES

Anon. *Baitfish importation: the position of the Maine Department of Inland Fisheries and Wildlife*. Unpublished report, undated.

Ball, H.E. and J. Tost. (1992). *Summary of small fish surveys conducted in the rivers entering Thunder Bay Harbour, 1990*. North Shore of Lake Superior Remedial Action Plan Technical Report No. 11.

Berrill, M., Distribution and ecology of crayfish in the Kawartha Lakes region of southern Ontario. *Canadian Journal of Zoology*, 56, pp. 166–177, 1978.

Burkhead, N.M. and J.E. Williams, An intergeneric hybrid of a native minnow, the golden shiner and an exotic minnow, the rudd. *Transactions of the American Fisheries Society*, 120, pp. 781–795, 1991.

Carlton, J.T., Dispersal of living organisms into aquatic ecosystems as mediated by aquaculture and fisheries activities, in *Dispersal of living organisms into aquatic ecosystems*, Rosenfield, A. and R. Mann, Eds., Maryland Sea Grant Publication, College Park, MD, 1992.

Counts, C.L., III., The zoogeography and history of the invasion of the United States by *Corbicula fluminea* (Bivalvia: Corbiculidae). *American Malacological Bulletin*, Special Edition No. 2, pp. 7–39, 1986.

Courtenay, W.R., Jr., D.A. Hensley, J.N. Taylor, and J.A. McCann, Distribution of exotic fishes in the continental United States, ch. 4, in *Distribution, biology, and management of exotic species*, Courtenay, W.R., Jr., and J.R. Stauffer, Jr., Johns Hopkins University Press, Baltimore, MD, 1984.

Crocker, D.W. and D.W. Barr, *Handbook of the crayfishes of Ontario*. ROM Life Sci. Miscellaneous Publication. University of Toronto Press, 1968.

Crossman, E.J., E. Holm, R. Cholmondeley, and K. Tuininga, First record for Canada of the rudd, *Scardinius erythrophthalmus*, and notes on the introduced round goby, *Neogobius melanostomus. Canadian Field-Naturalist*, 106, pp. 206–209, 1992.

Easton, R.S., D.J. Orth and N.M. Burkhead, The first collection of rudd, *Scardinius erythrophthalmus* (Cyprinidae), in the New River, West Virginia. *Journal of Freshwater Ecology*, 8, pp. 263–264, 1993.

Evans, D.O. and D.H. Loftus, Colonization of inland lakes in the Great Lakes region by rainbow smelt, *Osmerus mordax*: their freshwater niche and effects on indigenous fishes. *Canadian Journal of Fisheries and Aquatic Sciences*, 44(suppl. 2), pp. 249–266, 1987.

Fisheries and Oceans Canada. An assessment of the risks of introducing pathogens with the importation of baitfish. *Canada Department of Fisheries and Oceans, Aquaculture and Resource Development Branch Report*, 1992.

Foster, A.M. and P. Fuller, *Corbicula fluminea* Muller 1774. [www.nas@nfrcg.gov]. U.S. Department of the Interior, U.S. Geological Survey, Biological Resources Division, Florida Caribbean Science Center, Gainesville, FL, 1998.

Franzin, W.G., B.A. Barton, R.A. Remmant, D.B. Wain, and S.J. Page, Range extension, present and potential distribution, and possible effects of rainbow smelt in Hudson Bay drainage waters of northwestern Ontario, Manitoba, and Minnesota. *North American Journal of Fisheries Management*, 14, pp. 65–76, 1994.

Fuller, P. and L. Nico, *Cyprinidae*. [www.nas@nfrcg.gov]. U.S. Fish and Wildlife Service, National Biological Service, Southeastern Biological Science Centre, Gainesville, FL, 1996.

Ganzhorn, J., J.S. Rohovec, and J.L. Fryer, Dissemination of microbial pathogens through introductions and transfers of finfish, in *Dispersal of living organisms into aquatic ecosystems*. Rosenfield, A. and R. Mann, Eds., Maryland Sea Grant Publications, College Park, MD, 1992.

Garton, D.W., D.J. Berg, A.M. Stoeckmann, and W.R. Haag, Biology of recent invertebrate invading species in the Great Lakes: the spiny water flea *Bythotrephes cederstroemi* and the zebra mussel, *Dreissena polymorpha*, in *Biological Pollution: the control and impact of invasive exotic species*. B.N. McKnight, Ed., Proceedings of a symposium held at the Universtiy Place Conf. Center, Indiana University-Purdue University at Indianapolis, October 25 & 26, pp. 63–84, 1991.

Goodchild, C.D., *Live baitfish: an analysis of the value of the industry, ecological risks, and current management strategies in Canada and selected American states, with emphasis on Ontario*. Ontario Ministry of Natural Resources, 1997.

Griffiths, R.W., D.W. Schloesser, J.H. Leach, and W.P. Kovalak, Distribution and dispersal of the zebra mussel (*Dreissena polymorpha*) in the Great Lakes Region. *Canadian Journal of Fisheries and Aquatic Sciences*, 48, pp. 1381–1388, 1991.

Gunderson, J., Rusty crayfish: a nasty invader. Biology, identification and impacts of the rusty crayfish. *Minnesota Sea Grant*, 1995.

Hoffman, G.L., Intercontinental and transcontinental dissemination and transfaunation of fish parasites with emphasis on whirling disease (*Myxosoma cerebralis*), in *A symposium on diseases of fishes and shellfishes*, S.F. Snieszko, Ed., American Fisheries Society, Washington, DC, p. 69–81, 1970.

Hoffman, G.L. and G. Schubert, Some parasites of exotic fishes, ch. 12, in *Distribution, biology, and management of exotic species*, Courtenay, W.R., Jr. and J.R. Stauffer, Jr., Johns Hopkins University Press, Baltimore, MD, 1984.

Isom, B.G., Historical review of Asiatic clam (*Corbicula*) invasion and biofouling of waters and industries in the Americas. *American Malacological Bulletin*, Special Edition No. 2, pp. 1–5, 1986

Jennings, M.R. and M.K. Saiki, Establishment of red shiner, *Notropis lutrensis*, in the San Joaquin Valley, California. *California Fish and Game*, 76, pp. 46–57, 1990.

Jude, D.J., R.H. Reider, and G.R. Smith, Establishment of Gobiidae in the Great Lakes basin. *Canadian Journal of Fisheries and Aquatic Sciences*, 49, pp. 416–421, 1992.

Kohler, C.C. and W.R. Courtenay, Jr., American Fisheries Society position on introductions of aquatic species. *Fisheries*, 11, pp. 39–42, 1986.

Kohler, C.C. and J.G. Stanley, A suggested protocol for evaluating proposed exotic fish introductions in the Unites States, Chapter 18, in *Distribution, biology and management of exotic fishes*, W.R. Courtenay, Jr. and J.R. Stauffer, Jr., Eds., The Johns Hopkins Universtiy Press, Baltimore, MD, 1984.

Krishka, B.A., R.F. Cholmondeley, A.J. Dextrase, and P.J. Colby, *Impacts of introductions and removals on Ontario percid communities*. Percid Synthesis. Introductions and Removals Working Group Report, Ontario Ministry of Natural Resources, 1996.

Larkin, P.A. and S.B. Smith, Some effects of introduction of the redside shiner on the Kamloops trout in Paul Lake, British Columbia. *Transactions of the American Fisheries Society*, 83(1953), pp. 161–175, 1954.

Leach, J.H., Impacts of the zebra mussel (*Dreissena polymorpha*) on water quality and fish spawning reefs in western lake Erie, in *Zebra mussels: biology, impacts and control*, Nalepa, T.F. and D.W. Schloesser, Eds., Lewis Publishers, Boca Raton, FL, 1993.

Lee, D.S., C.R. Gilbert, C.H. Hocutt, R.E. Jenkins, D.E. McAllister, J.R. Stauffer, Jr., *Atlas of North American freshwater fishes*. North Carolina Biological Survey Publication No. 1980–12, 1980.

Litvak, M.K. and N.E. Mandrak, Ecology of freshwater baitfish use in Canada and the United States. *Fisheries*, 18, pp. 6–13, 1993.

Lodge, D.M. and J. G. Lorman, Reductions in submersed macrophyte biomass and species richness by the crayfish, *Orconectes rusticus*. *Canadian Journal of Fisheries and Aquatic Sciences*, 44, pp. 591–597, 1987.

Loftus, D.H. and P.F. Hulsman, Predation on boreal lake whitefish *Coregonus clupeaformis* and lake herring *C. artedi* by rainbow smelt *Osmerus mordax*. *Canadian Journal of Fisheries and Aquatic Sciences*, 43, pp. 812–818, 1986.

LoVullo, T.J. and J.R. Stauffer, Jr., The retail bait-fish industry in Pennsylvania — source of introduced species. *Journal of the Pennsylvania Academy of Science*, 67, pp. 13–15, 1993.

Lyons, J., Changes in the abundance of small littoral zone fishes in Lake Mendota, Wisconsin. *Canadian Journal of Zoology*, 67, pp. 2910–2916, 1989.

MacCallum, W.R., *Ruffe surveillance program, 1991-1993, Thunder Bay Harbour, Lake Superior*. Lake Superior Management Unit, Ontario Ministry of Natural Resources, Thunder Bay, Ontario, 1994.

Magnan, P. and G.J. Fitzgerald, Resource partitioning between brook trout, *Salvelinus fontinalis* Mitchill, in selected oligotrophic lakes of southern Quebec. *Canadian Journal of Zoology*, 60, pp. 1612–1617, 1982.

Magnan, P. and G.J. Fitzgerald, Mechanisms responsible for the niche shift of brook charr, *Salvelinus fontinalis* Mitchill, when living sympatrically with creek chub, *Semotilus atromaculatus* Mitchill. *Canadian Journal of Fisheries and Aquatic Sciences*, 62, pp. 1548–1555, 1984.

Marsden, J.E. and D.J. Jude, Round gobies invade North America. Illinois-Indiana Sea Grant. *Ohio Sea Grant Communications*, 1995.

McPhail, J.D., Status of the Enos Lake Stickleback species pair, *Gasterosteus* spp.* *Canadian Field-Naturalist*, 103, pp. 216–219, 1989.

Miller, W.H., H.M. Tyus and C.A. Carlson, Eds., *Fishes of the upper Colorado River System: present and future*. Western Division, American Fisheries Society, Bethesda, MD, 1982.

Miller, R.R., J.D. Williams and J.E. Williams, Extinctions of North American fishes during the past century. *Fisheries*, 14, pp. 22–38, 1989.

Mills, E.L., J.L. Leach, J.T. Carlton, and C.L., Secor. Exotic species in the Great Lakes: a hitory of biotic crises and anthropogenic introductions. *Journal of Great Lakes Research*, 19, pp. 1–54, 1993.

Momot, W.T., Further range extensions of the crayfish *Orconectes rusticus* in the Lake Superior basin of northwestern Ontario. *Canadian Field-Naturalist*, 106, pp. 397–399, 1992.

Moyle, P.B., Fish introductions in California: history and impact on native fishes. *Biological Conservation*, 9, pp. 101–118, 1976.

Moyle, P.B. and H. W. Li., Good report but should go much farther. *Fisheries*, 19, pp. 22–23, 1994.

Olsen, T.M., D.M. Lodge, G.M. Capelli, and R.J. Houlihan, Mechanisms of impact of an introduced crayfish (*Orconectes rusticus*) on littoral congeners, snails, and macrophytes. *Canadian Journal of Fisheries and Aquatic Sciences*, 48, pp. 1853–1861, 1991.

Olver, C.H., B.J. Shuter, and C.K. Minns, Toward a definition of conservation principles for fisheries management. *Canadian Journal of Fisheries and Aquatic Sciences*, 52, pp. 1584–1594, 1995.

O'Neill, C.R., Jr., and A.J. Dextrase, The introduction and spread of the zebra mussel in North America, in *Proceedings: 4th International Zebra Mussel Conference*, Wisconsin Sea Grant Institute, Madison, Wisconsin, pp. 433–446, 1994.

Ostland, V.E., B.D. Hicks and J.G. Daly, Furunculosis in baitfish and its transmission to salmonids. *Diseases of Aquatic Organisms*, 2, pp. 163–166, 1987.

Poirier, P., *Problématique des introductions des poissons-appâts dans la région administrative Mauricie*-Bois-Francs. G.D.G. Environment Ltée for Québec Ministère du Loisir de la Chasse et de la Pêche, 1992.

Pratt, D.M., W.H. Blust and J. H. Selgeby, Ruffe, *Gymnocephalus cernuus*: newly introduced in North America. *Canadian Journal of Fisheries and Aquatic Sciences*, 49, pp. 1616–1618, 1992.

Schoenherr, A.A., The role of competition in the replacement of native species by introduced species, in *Fishes in North American Deserts*, Naiman, R.J. and D.L. Soltz, Eds., John Wiley, New York, 1981.

Scott, W.B. and E.J. Crossman, *Freshwater Fishes of Canada*. Fisheries Research. Board of Canada Bulletin, 184, 1973.

Sickel, J.B., *Corbicula* population mortalities: factors influencing population control. *American Malacological Bulletin*, Special Edition No. 2, pp. 89–94, 1986.

Smith, C.L., *The inland fishes of New York State*. New York State Department of Environmental Conservation, Albany, NY, 1985.

Sprules, W.G., H.P. Riessen, and E.H. Jin, Dynamics of the *Bythotrephes* invasion of the St. Lawrence Great Lakes. *Journal of Great Lakes Research*, 16, pp. 346–351, 1990.

Stedman, R.M. and C. A. Bowen II., Introduction and spread of the threespine stickleback (*Gasterosteus aculeatus*) in Lakes Huron and Michigan. *Journal of Great Lakes Research*, 11, pp. 508–511, 1985.

Taylor, J.N., W.R. Courtenay, Jr., and J.A. McCann, Known impacts of exotic fishes in the continental United States, ch. 16, in *Distribution, biology, and management of exotic fishes*, Courtenay, W.R., Jr. and J.R. Stauffer, Eds., The Johns Hopkins University Press, Baltimore, MD, 1984.

Thompson, A.G., The danger of aquatic species. *World Aquaculture*, 21, pp. 25–32, 1990.

Wilson, L., Minnows are monsters. *Family Herald*, pp. 45-48, 1958.

Winter, B.D. and R.M. Hughes, AFS Draft Position Statement on Biodiversity. *Fisheries*, 20, pp. 20–26, 1995.

Yan, N.D., W.I. Dunlop, T.W. Pawson, and L.E. MacKay, *Bythotrephes cederstroemi* (Schoedler) in Muskoka Lakes: first records of the European invader in inland lakes in Canada. *Canadian Journal of Fisheries and Aquatic Sciences* , 49, pp. 422–426, 1992.

APPENDIX 12.1
Summary of Reported Impacts of Nonindigenous Species Introduced to the United States and Canada through the Use of Live Baitfish, With an Indication of Origin, Pathway, and Status

Taxon	Scientific Name	Common Name	Origin[a]	Pathway[b]	Status[c]		Ecological Impacts
			Bacteria				
—	*Aeromonas s. salmonicida*	(bacterial parasite)	—	B	NT	•	Causes furunculosis in fishes, potentially transmitted from baitfish to sport fish (has been transmitted experimentally from common shiners to coho salmon and brook trout)
			Fungi				
Saprolegniales	*Saprolegnia sp.*		—	B	NT	•	Causes death of fishes through osmoregulatory failure, may be transmitted to salmonids by infected baitfish; larger salmonids are at greater risk of infection
			Submerged Plants				
Haloragaceae	*Myriophyllum spicatum*	Eurasian watermilfoil	E	OP+B	NT	•	Disperses rapidly and may displace native vegetation or invade unvegetated areas; has become a major pest in British Columbia; implicated in the extirpation of blackchin shiner in several Wisconsin lakes
			Protozoans				
—	*Plistophora ovariae*		—	B	NT	•	Parasite destroys ovaries of golden shiners
			Trematodes				
—	*Bolbophorus confusus*	Eurasian trematode	E	OP+B	NT	•	Discovered in shipments of baitfish; native snails infected with this parasite may have transferred it to trout in Montana
			Tapeworms (Cestodes)				
Cestodaria	*Bothriocephalus opsarichthydis*	Asian tapeworm	E	OP+B	NT	•	Many native cyprinids heavily contaminated with the Asian tapeworm; probably transmitted from infected red shiners to woundfin, an endangered cyprinid, in Utah; recently found in endangered Colorado squawfish from New Mexico; reportedly infected channel catfish and may infect many other indigenous fish

Family	Species	Common name				Impacts
	Triaenophorus crassus	tapeworm	—	B	NT	• Infects cisco and lake whitefish populations in northern Alberta; infections not harmful but aesthetically unappealing, potentially causing economic losses to commercial fisheries
Molluscs						
Corbiculidae	*Corbicula fluminea*	Asiatic clam	E	OP+B	NT	• May displace native clams, in southern areas may damage and clog water intake systems, alterations in benthic substrate observed.
Dreissenidae	*Dreissena polymorpha*	zebra mussel	E	OP+B	NT	• Dramatic declines or extirpation of native unionid mussels; causes trophic alterations, increased water clarity; proliferation of aquatic macrophytes associated with zebra mussel colonies
Unionidae		glochidia larvae	—	B	NT	• Parasitizes fishes; found in shipments of baitfishes
Crustaceans						
Cladocera	*Bythotrephes cederstroemi*	spiny water flea	E	OP+B	NT	• Declines of *Daphnia* sp. possibly due to predation by spiny water flea; impacts on native zooplankton and fish have been debated
Decapoda						
Astacidae	*Orconectes rusticus*	rusty crayfish	TP	B	T	• May displace native crayfish species; associated with destruction of aquatic vegetation, which negatively affects sport fish populations; particularly of concern in northwestern Ontario shield lakes
Fish						
Clupeidae	*Alosa pseudoharengus*	alewife	TP	B	T	• Probably negatively impacted walleye, yellow perch, and whitefish populations through competition and predation on fry and eggs; implicated in collapse of walleye populations in two lakes in New York; alewife may have contributed to decline of whitefish in the Great Lakes

APPENDIX 12.1 (CONTINUED)
Summary of Reported Impacts of Nonindigenous Species Introduced to the United States and Canada through the Use of Live Baitfish, With an Indication of Origin, Pathway, and Status

Taxon	Scientific Name	Common Name	Origin[a]	Pathway[b]	Status[c]	Ecological Impacts
Cyprinidae	Carassius auratus	goldfish	E	OP+B	T	• Associated with negative alterations in fish habitat, including increased turbidity
	Cyprinella lutrensis	red shiner	TP	B	T	• Invasive species may displace California roach and other endemic fishes in western United States; implicated in decline of federally threatened spikedace, woundfin, and Virgin River chubs; may displace young Colorado squawfish
	Cyprinus carpio	common carp	E	OP+B	T	• Associated with negative alterations in fish habitat, including increased turbidity and reduction in macrophytes
	Exoglossum maxillingua	cutlips minnow	TP	B	T	• Hybridizes with native tonguetied minnows where introduced; potentially decreasing genetic fitness
	Gila bicolor obesa	Lahontan tui chub	TP	B	T	• Concern that this species may hybridize with and deplete populations of indigenous Owens tui chub, an endangered species in California
	Gila orcutti	arroyo chub	TP	B	T	• Mohave tui chub was nearly extirpated through hybridization with the introduced arroyo chub
	Notemigonus crysoluecas	golden shiner	TP	B	T	• Negatively impacted trout in lakes in California; competes with young sunfish and bass for food
	Pimephales promelas	fathead minnow	TP	B	T	• May displace young Colorado squawfish
	Rhinichthys cataractae cataractae	eastern longnose dace	TP	B	T	• Decline of Banff longnose dace can be attributed to introductions of many exotic species but the final extinction of the subspecies was probably due to introgression with this subspecies.
	Richardsonius balteatus	redside shiner	TP	B	T	• May negatively affect rainbow trout

Family	Scientific name	Common name					Impact
	Scardinius erythrophthalmus	rudd	E	OP+B	T	•	May hybridize with native golden shiners in the wild leading to a reduction of numbers available for spawning due to possibility of sterile hybrids; possibly could endanger genetic integrity of native golden shiners
	Semotilus atromaculatus	creek chub	TP	B	T	•	Reductions in brook trout growth; brook trout switched from feeding on benthos to zooplankton after introduction of creek chub
Gasterosteidae	*Gasterosteus aculeatus*	threespine stickleback	TP	B	T	•	Compete for food with native ninespine sticklebacks, brook sticklebacks, and fathead minnows
Centrarchidae	*Lepomis gibbosus*	pumpkinseed	TP	B	T	•	Threespine sticklebacks extirpated after introductions of this species; concern that it will cause extirpation of other stickleback species in British Columbia
Percidae	*Etheostoma zonale*	banded darter	TP	B	T	•	Hybridized with tessellated darters after being introduced to the Susquehanna River, possibly could affect genetic integrity of tessellated darters
	Gymnocephalus cernuus	ruffe	E	OP+B	T	•	Invasive species, a number of native species have reportedly declined in areas where introduced
	Perca flavescens	yellow perch	TP	B	T	•	Negatively impacted brook trout lakes in Alberta; may have adverse effects on brook trout and landlocked salmon fisheries; have been observed feeding on lake trout eggs and may have been responsible for observed decline in lake trout populations
Gobiidae	*Neogobius melanostomus*	round goby	E	OP+B	T	•	Competes with and may displace native benthic species such as sculpins and darters; predation on lake trout eggs may occur in the wild
	Proterorhinus marmoratus	tubenose goby	E	OP+B	T	•	Competes with native benthic species such as sculpins and darters; predation on lake trout eggs may occur in the wild
Ictaluridae	*Ameiurus nebulosus*	brown bullhead	TP	OP+B	T	•	Threespine sticklebacks extirpated after introductions of this species; concern that will extirpate other stickleback species in British Columbia

APPENDIX 12.1 (CONTINUED)
Summary of Reported Impacts of Nonindigenous Species Introduced to the United States and Canada through the Use of Live Baitfish, With an Indication of Origin, Pathway, and Status

Taxon	Scientific Name	Common Name	Origin[a]	Pathway[b]	Status[c]	Ecological Impacts
Osmeridae	*Osmerus mordax*	rainbow smelt	TP	B	T	• Invasive species that may feed on eggs of sport fishes, including lake trout; declines in lake whitefish populations observed after rainbow smelt introductions
Percichthyidae	*Morone americana*	white perch	TP	B	T	• May have adverse effects on brook trout and landlocked salmon fisheries

[a] TP = transplanted species indigenous to part of the U.S. and or Canada; E = exotic from another continent.

[b] B = introduced by use of baitfish only; OP+B = introduced by other pathway and further dispersed through use of bait.

[c] T = target — small fish or invertebrates harvested for bait (does not imply a nonrestricted or specifically targetted species); NT = nontarget — plants, invertebrates, pathogens found in transport medium or attached to bait species.

Section IV

Ballast Water Vector

13 The Role of Ships as a Vector of Introduction for Nonindigenous Freshwater Organisms, with Focus on the Great Lakes

C.J. Wiley and R. Claudi

INTRODUCTION

There exists an ever increasing body of information about the introduction of nonindigenous organisms into different parts of the world and impact they have on the receiving environment. Ships have been identified as one of the primary vectors of introduction. This is not surprising, as waterborne trade prior to the industrial revolution was, in practical terms, the only means of transport to different parts of the world. Certainly from the European perspective, ships were the only way to the New World. Ships carried the goods needed to expand empires, trade with natives, and make war. With the ships many organisms came to the New World. The Spanish brought non indigenous horses to North America in 1519 and, subsequently, cows during their exploration.

As exploration gave way to colonization, the cargoes carried reflected the desire to bring the benefits of civilization to the New World. Cargoes offloaded were considered needed or beneficial and were intentionally brought to the New World to make the lives of travelers and immigrants more comfortable. A great variety of plants, seeds, and livestock survived the rigors of an ocean crossing from Europe in the dank holds of sailing ships, to be successfully introduced into a new environment. Often plants and flowers were brought by colonists to remind them of homes that in many cases they might not see again. Lilacs, for example, were introduced to North America in this way.

Other nonintentional introductions arrived by ship as well. The black death or plague was associated with ships. Certainly, rats were transported to a number of medieval cities around the Mediterranean by the sailing ships of the day. Convict ships brought many uninvited guests to Australia, including the housefly. Slave ships were often associated with not only their unfortunate cargo but also the "pestilence" aboard. Innumerable stories of suffering and death were associated with ships transporting human cargo — both voluntary and otherwise — to various parts of the world. Quarantine stations were set up as a response.

TRANSPORT OF ORGANISMS VIA HULL FOULING

In the last century, the most common form of inadvertent transfer of any aquatic species was on the hulls of wooden sailing ships. Sailing ships often spent days or weeks at anchor waiting for cargoes or orders. Not only did hulls become fouled with organisms, but anchors and rigging as well. The slow speed of sailing ships, coupled with trade routes that often passed through tropical waters to take advantage of prevailing wind patterns, assisted in this fouling. Many records exist of great quantities of living material fouling the bottoms of sailing ships. Wealthy shipowners and naval vessels often "coppered" the underwater portion of ships hulls to prevent this fouling. The average shipowner, however, could not afford such a luxury, and the vast majority of ships collected a mass of living organism from whichever parts of the world the ships traded. It was common practice to have to careen ships on beaches to scrape off this mass of seaweed, hydroids, seasquirts, bacteria, and fungi. Organisms were, as a result, deposited many thousands of miles from where they had become attached to the ships hull.

Similarly, boring organisms such as ship worms and insects found homes in the planking of wooden ships and were transported to all areas of the globe. With the rate of shipwrecks prior to the dawn of the 20th century, it was virtually a guarantee that these organisms were soon found in places far from home.

From a Great Lakes perspective, hull fouling was the predominant vector of aquatic introductions in the last century. The advent of canal building fever in North America in the early 1800s resulted in Lake Erie being accessible from the Hudson river by 1825 and Lake Ontario from the St. Lawrence by 1847 (Mance 1988). The opportunity was provided for organisms to be transported upstream on vessels using these canals. There was some direct trade via the canals from deep sea ports as early as 1847. However, because of the time required to transit the myriad of locks between the sea and the inland ports and the limitations on length and draft of vessels using the locks, a whole secondary transportation system developed. Height restrictions forced sailing ships to literally be dismasted to enter some of the lock systems. Most ocean-going vessels stopped at Montreal or New York and transshipped cargo into "canallers" or Erie canal barges (LeStrang 1981). It is postulated that the hulls of these secondary vessels offered the sea lamprey and several species of algae rides into the Great Lakes (Mills et al. 1994). These introductions were essentially an assisted upstream expansion of range of organisms that already existed close to the Great Lakes.

TRANSPORT OF ORGANISMS VIA BALLAST

It is ballast that is currently most frequently associated with the transference of nonindigenous organisms worldwide. What then is ballast? *Ballast* is the material that is used to maintain stability and assist in allowing the vessel to be steered when the ship has little or no cargo on board. Ships have always required some form of ballast for safe and efficient operation. Among other things, ballast provides the ship with stability by adjusting the center of gravity. In this manner, rolling, pitching, and slamming of the vessel are minimized. It also improves maneuverability of lightly loaded ships by assuring adequate draft for immersion of the propeller and allows the captain to reduce the drift from the effects of wind. From a safety point of view, ballast minimizes bending and shear forces on the hull, preventing structural damage to the hull in heavy seas, and counteracts nonuniform cargo loading. Ballast can be permanent, actually included in the structure of the ship, or nonpermanent, only added or discharged as the needs of the ship dictate.

SOLID BALLAST

Originally, stone, sand, lead ingots or practically anything that was available near shore, that was sufficiently dense, and that could be manually shoveled into an empty cargo hold was used as nonpermanent ballast for sail or early steam vessels. At the cargo loading port this material was again manually shoveled out and deposited either into the harbor (although this was frowned upon, as the harbor would then have to be dredged) or onto the shore. Larger ports such as New York and Philadelphia had dedicated ballast dumping grounds. Coastal ports received many species of nonindigenous plants, insects, clams, and snails via this route (Mills et al. 1994).The ports of the Great Lakes basin were not immune to solid ballast introductions. The appearance of certain organisms, most notably species of snails, have been attributed to this vector (Mills et al. 1993). Luckily, the large variety and scale of introductions experienced in the coastal ports were not mirrored in freshwater ports. A major factor that prevented greater number of introductions in the freshwater ports was the trading pattern of ships. The trade into the Great Lakes historically consisted of loaded ships carrying supplies, baggage, and people and returning with resources. There was little economic incentive to make a ballast voyage to these areas, unlike the major seaports where a cargo could always be found. It must be remembered that the main entrance port to the Great Lakes at Montreal is about 1,650 km from the Atlantic Ocean. Depending upon the weather, just getting to Montreal could add a week or more to a sailing ship's passage, not to mention the additional time to actually get into a Great Lakes port.

Within the lakes the use of disposable solid ballast was not nearly as prevalent as in the coastal ports. Voyages were short, there were many harbors for shelter against weather, and cargoes tended to be contracted. Indeed many Great Lakes schooners (and some early wooden bulk carriers) utilized "centerboards" in preference to the use of solid ballast. Literally, a large wooden board that could be raised or lowered through the bottom of the hull was used. This center board gave many of the advantages offered by loading solid ballast — i.e., stability, increased steerage way, etc. — without the time-consuming offloading of a nonpaying cargo of solid ballast (Devendorf 1996).

WATER BALLAST

The opportunity for transport and introduction of aquatic organisms increased greatly when ships started carrying ballast water instead of dry ballast. The switch from dry ballast to water was the result of several concurrent events. As the demand for shipping of bulk cargoes (especially iron ore) increased in the last quarter of the last century, ships became larger. Steam ships replaced schooners in the bulk trades and iron or steel replaced wood for construction. Concurrently, regulations were put in place in certain coastal ports prohibiting dumping of solid ballast.

The typical schooner of the 1850s seldom exceeded 135 feet, and could carry a cargo of 100–200 tons. To put this in perspective, no truck on our highways today can carry 200 tons. By 1869 the iron ore trade was so profitable in the Great Lakes that even larger ships were required. The R.J. Hackett, the first ship in the world built expressly for carrying bulk cargoes, was the result. Built of wood, this ship carried an unheard-of cargo of 1013 tons on a length of 200' and a draft of 11'6". By 1882 a true revolution took place with the construction of the first modern design iron bulk carrier — the Onoko. Not only was this vessel built of iron, it was almost a hundred feet longer and carried three times the cargo of the Hackett (3024 tons on a draft of 16'6"). However, most importantly from the point of view of the introduction of nonindigenous species, iron vessels like her introduced the use of water ballast. Steam technology had advanced sufficiently to allow the efficient pumping of liquids, and iron construction allowed the formation of tanks within the ships' structure (Greenwood 1995). As a footnote, a number of wooden-hulled vessels were built to carry water ballast, the first being the Fred Pabst of the brewery of the same name (Devendorf 1996). If it had hauled beer instead of water, history might have been made; it predated the first oil tanker by 5 years.

To understand the ballast water mechanism of introduction of aquatic species, one has to examine how ships take on and carry ballast water. Ships carry ballast water in internal water tanks located in various regions of the ship (Figure 13.1). For a typical Great Lakes vessel, these tanks surround the cargo holds on the bottom and sides. However, each type of vessel utilizes ballast tanks in a way that maximizes cargo capacity and assists in structural stability. The tanks are filled or emptied according to the amount of cargo the ship carries, providing balance and buoyancy when required. The water is taken in through underwater intake ports, usually located underneath the stern of the ship. It is pumped forward to fill the various tanks. The tanks may be interconnected. The process is generally controlled from the engine room of the ship.

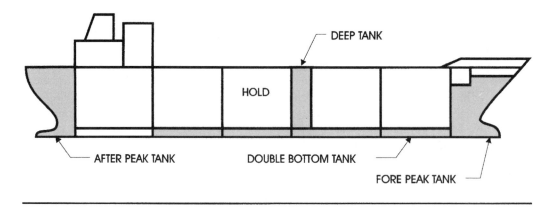

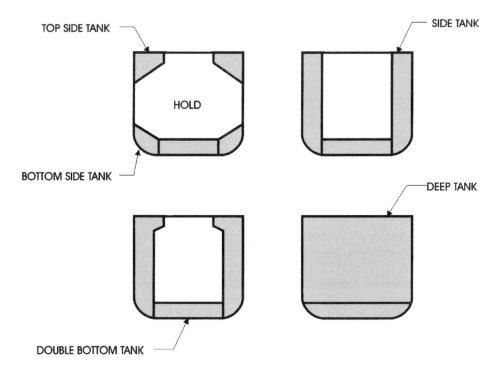

FIGURE 13.1   Typical ballast water tank locations.

The ballast water that is pumped in can be either salty or fresh, depending on the port from which it was taken. Some ports may contain salt water most of the time, but contain fresh water at certain times of the year. For example, St. Petersburg, Russia is considered a saltwater port. However, during spring runoff it becomes a freshwater port. Fresh or salt, the harbor water is likely to be teaming with living organisms.

When this water is discharged at the final destination, all of the organisms taken up are released into the receiving waters. If the ballast water is salty and the harbor fresh or vice versa, the survival chances of the organisms are likely to be diminished. If however both the ballast water and the receiving waters are of approximately the same salinity, the probability of survival of individual organisms increases.

From the point of view of introductions of freshwater organisms through ballast water discharge, the Great Lakes were relatively well protected until 1959. The canal systems with access to external waters had expanded in size a number of times. However, as late as 1959, the largest vessels with access to the Great Lakes from the St. Lawrence could not exceed 261 ft in length, 44 ft in breadth and a 25 ft draft. This represents a vessel of approximately 3000 tons and maximum ballast capacity of about 1000 tons of water. Therefore, the locks, while improved from the 1850s, were still a significant limiting factor economically for any vessel entering without a cargo. A large efficient system of canallers existed for most bulk cargoes. Transshipment at Montreal was the norm. With the advent of the railroads as a competing, faster, and more efficient method of transport for finished and manufactured goods, access to the Great Lakes from the Hudson River and the Erie Canal diminished in importance as both a trade route and as a source for the introduction of organisms during this time period.

The Mississippi River at Chicago is another possible conduit for organism transfer via the Ship and Sanitary Canal. Fortunately, the draft limitations and the fact that the majority of trade from the Great Lakes region goes down this waterway, rather than into the lakes, limits the threat potential. It must be noted, however, that the spread of the zebra mussel down the Mississippi was via this route, and there is sufficient concern of the threat aspects of this route, both ways, that the U.S. Army Corps of Engineers is investigating a mechanism to prevent species transfer.

After 1959 the game changed. The economic barriers to entry of ships in ballast were brought down with the opening of the present St. Lawrence Seaway. Not only were there fewer and larger locks allowing access to bigger foreign ships, at less economic penalty, but the Great Lakes region aggressively marketed the business. Saltwater ships were welcomed with open arms, and cargoes were ready for them to take away. The grain ports of Duluth, Fort William, and Port Arthur had ample grain, often paid for as aid cargo to developing nations. The number of ships in ballast increased exponentially at these ports. Further, the types of vessels courted were exactly the kinds of vessels that potentially carried ballast. Bulk carriers have a much higher volume of ballast than general cargo ships. The general cargo ships, while a significant part of the trade, had to compete with the railways and trucks. Today, general cargo is containerized and travels from coastal cities to Great Lakes ports on trains. By far the majority of ships entering the system are bulk carriers.

Fortunately for the Great Lakes, deep-sea vessels are still limited by the size of the locks of even the new, improved St. Lawrence Seaway. Vessels greater than 78 ft wide × 735 ft long with a draft of 25 ft 6 in. simply don't fit. This, in practical terms, allows a vessel of a maximum size of approximately 40,000 tons to enter. For those vessels that actually enter in a fully ballasted condition and no cargo, this translates into approximately 25,000 tons of water ballast. In comparison, the largest deep-sea vessels exceed 600,000 tons and have an approximate ballast capability of 250,000 tons, or ten times the average possible for vessels entering the Seaway.

Even so, from the point of view of the possibility of introduction, a combination of an exponential increase in the number of ships, a 25 times increase in the amount of ballast carried, and a transit time cut by a factor of 10 results in a significant increase in risk. This is reflected by the increased rate of aquatic introductions to the Great Lakes since 1959. There were 90 introductions between 1810 and 1959 and 43 introductions between 1960 and 1990 (Mills et al. 1993).

Almost 30% of all the introductions are considered likely ship source introductions. The majority of this percentage are considered to be ballast water introductions. In other words, freshwater ballast originating offshore was released in the Great Lakes and contained organisms that were able to successfully colonize the new environment. Surprisingly perhaps, the other 70% were introduced by other vectors.

Real interest in this problem by society at large was sparked primarily by the arrival of the zebra mussel (*Dreissena polymorpha*) in Lake St.Clair in 1985–86. The zebra mussel was one of the few introductions causing not only environmental upheaval (detailed by Nalepa and colleagues in Chapter 16 of this book), but also inflicting substantial economic penalties on many different users of the Great Lakes water.

Other recent invaders thought to have originated in ballast water include the following:

- The spiny waterflea (*Bythotrephes cederstroemi*), which is a tiny crustacean with a long, sharp, barbed tail spine. Native of Great Britain and northern Europe, it was first found in Lake Huron in 1984. Since then it has spread throughout the Great Lakes. It feeds on smaller water fleas, probably competing with young perch and other small fish for food.
- The ruffe (*Gymnocephalus cernuus*), a perch-like fish, maximum of 12.5 cm in length. It is native to central and eastern Europe, where it is considered a pest species. It was introduced to Duluth harbor around 1985 and is spreading to other rivers and bays around Lake Superior. Where it is present, populations of yellow perch and emerald shiner have declined.
- The tube-nosed goby (*Proterorhinus marmoratus*), and the round goby (*Neogobius melanostomus*) are small fish, first found in the St.Clair River. There is a concern that they may displace the native sculpins.

Paradoxically, the pollution-control efforts worldwide have increased the likelihood of successful introductions. As the water quality in many of the major ports around the world improves, more organisms are able to survive, thus increasing the chances of being taken up with the ballast water. Similarly, when the ballast water is discharged in a foreign port, the water quality is possibly good enough to allow the organisms contained in the ballast water to survive. This may in part explain why the invasion of zebra mussels, anticipated by Kew in 1893, Johnson in 1921, and Sinclair in 1964 (Carlton 1993), did not in fact occur until 1986.

GUIDELINES AND REGULATIONS ON BALLAST WATER DISCHARGE

As ballast water became an acknowledged vector of aquatic introductions to the Great Lakes, the Canadian and U.S. government came under increased pressure to regulate its discharge in the Great Lakes. In 1989, the Canadian Coast Guard set in place guidelines (Canada 1993) for the voluntary open ocean exchange of ballast water of ocean-going vessels traveling upstream into the fresh waters of the St. Lawrence. Under these guidelines, these ships could only carry ballast water taken in open ocean where the depth was greater than 2,000 m. The exception to this rule are those ships that had not left the North American continental shelf. The intent of these guidelines was to get ships in ballast to flush out freshwater taken in foreign ports and replace it with high-salinity water, thereby flushing out most of the freshwater organisms present and exposing any remaining biota to water of high salinity. As most freshwater species do not survive in salinity above 8 ppt, salinity around 35 ppt was surmised to provide an effective means of control for any freshwater organisms that may be present.

In the United States, the Aquatic Nuisance Prevention and Control Act of 1990 (United States 1990) required the U.S. Coast Guard to issue voluntary guidelines for ballast water exchange within 6 months, followed by mandatory controls. These went into effect on May 10, 1993. The regulations require that ballast water has salinity of at least 30 ppt if it is to be discharged in the Great Lakes. U.S. Coast Guard officials now check the salinity of the ballast water of every ship entering the Great Lakes at Massena. If the ballast water is of insufficient salinity, it either has to be treated in a manner approved by the Coast Guard, or retained on board for the duration of the journey in the Great Lakes.

A good example of how effective the guidelines and regulations can be was observed in December of 1994. The Liberian flagged, Peruvian owned, freighter Pal Wind arrived in Montreal on December 4th. The master reported that he performed the required ballast exchange at sea. His claim could not be verified due to sampling difficulties, and rather than take time to remove the access hatches to allow testing, the vessel's operator and agents opted to retain their ballast water on board. The vessel then proceeded to Thunder Bay. On December 13, the vessel's agents requested permission to deballast so that the ship could take on cargo in Hamilton, Ontario before leaving the Great Lakes. Before granting permission, the Captain of the Port in Buffalo ordered that the ballast water be tested by an independent laboratory. The test revealed that the water was nearly completely salt-free Congo River water containing live organisms. The request to discharge ballast was denied, and the master was warned of his criminal liability by a joint Canadian/U.S. Coast Guard team. The vessel was subsequently convicted and fined.

The experience of the Great Lakes, as well as other areas in the world affected by introduced non indigenous aquatic species such as Australia, has brought the whole subject to the attention of the international marine community. The International Maritime Organization (IMO) has convened a permanent working group examining the issues of ballast water under its Marine Environmental Protection Committee (MEPC). Indeed, IMO resolution A774 (Reid and Carlton 1997) suggests methods by which ships in the international trade may decrease the possibility of spreading nonindigenous organisms. In the future it is likely that some variant of this will become binding as an annex to international marine pollution legislation.

OUTSTANDING ISSUES IN BALLAST WATER CONTROL

The current ballast water exchange requirement is a big step toward minimizing introductions of nonindigenous aquatic organisms into the Great Lakes by not allowing ships in ballast originating on another continent to discharge the unexchanged ballast into the Great Lakes. However, opportunities for introduction of unwanted organisms still exist. The following are some outstanding questions being addressed by ongoing research.

- How effective is the exchange of fresh for salt water? Is there sufficient mixing in the ballast water tanks to expose all freshwater organisms present to high salinity or is there a formation of layers of water of different salinity? Could freshwater organisms or their dormant/encysted forms survive in such a layer? Are there salinity-tolerant species that would survive the exchange regardless? Studies done for Environment Canada (Locke et al. 1991, 1993) and Transport Canada (Pollutech 1992; Aquatic Sciences 1996) in fact indicate that exchange at best is only partially effective. It is a stop-gap solution to the problem, and freshwater organisms do indeed survive in the exchanged ballast water, although in diminished quantities.
- What happens if a ship reports no ballast on board? In most cases that actually means that there is unpumpable layer of water and sediment in the ballast water tanks. A recent Transport Canada Study (Aquatic Sciences 1996) found live organisms present in these remnants in virtually all ships that reported no ballast on board. Furthermore, the water present was usually fresh water. If such a ship takes on ballast water in the Great Lakes, it is mixed with the "unpumpable ballast," and it can be released without any regulation in another part of the Great Lakes. This scenario in fact represents the majority of ships entering the Great Lakes. Only a small percentage of the ships entering the system actually enter in ballast. Basic economics dictate that to enter a freshwater system some distance from the sea costs money, and therefore very few ships enter without cargo and full ballast water tanks. For ships that report No Ballast On Board (NOBOB) there is no effective regulatory control at this time. Fortunately, the recent NISA legislation (20) in the U.S. directs funds to examine the problem.
- What can be done if a ship does not or cannot exchange ballast on the high seas? The North Atlantic is known to be one of the most dangerous waters on earth. There is considerable concern at this point that the process of exchanging ballast is not safe in terms of structural stresses it places on the ship's hull (Melville Shipping 1995). Further, for ships entering the Seaway that have a 10 to 1 length to breadth ratio, there is a history of structural cracking regardless of the possibility of ballast exchange induced stresses

(Aquatic Sciences 1995). All present and proposed regulations are likely to ensure that the master of the ship is not required to put the ship in danger. Currently, there is a possibility of a secondary ballast exchange site in the Laurentian channel, which theoretically is less likely to have the rough weather of the Atlantic. If this site is not utilized, the ballast tanks are sealed for the duration of the ship's stay in the Great Lakes This imposes an economic penalty equal to the amount of cargo that cannot be carried. If, however, the ship required deballasting, at this point in time, there are no methods of ballast water treatment clearly identified as acceptable.

- Are there possible unintended negative effects from the regulatory regime? A direct result of all the publicity of the oil spill caused by the Exxon Valdez was a focus on the tanker industry worldwide. The U.S. brought in legislation that in part governed the construction of tankers. Other countries followed suit. Without going into details of the various legislation, the intent was to promote the phasing out of single hull tankers and replace them with those constructed of double hull design. Additionally, the new tankers built were to have ballast tanks that were segregated from the cargo tanks. From the point of view of the Great Lakes and the introductions of nonindigenous species, there have been two unintended negative results. First, tankers that up to this point had pumped all their ballast ashore, to be treated in refineries, now pump any ballast directly into the port. Secondly, because of the decreasing need brought about by segregated ballast, the few remaining ballast reception facilities are being shut down. Paradoxically, refineries that do accept ballast are requiring tankers to ensure any ballast pumped ashore is "zebra mussel veliger free."

- Ballast water may also provide means of transfer for various microorganisms, including pathogens. This would be true for trans-Atlantic trips, as well as those along the continental coast lines and through inland navigational waterways. As an example, the cholera bacterium *Vibrio cholerae* 01 was found in oysters from Mobile Bay, Alabama during routine seafood sampling in 1991. The same strain was found to be present in the ballast water of three ships in the area. These ships have previously visited South America, where an epidemic of cholera was taking place (McCarthy 1994). An in-depth study of both ocean-going and domestic trade was undertaken in the Welland Canal during the 1995 shipping season. One hundred and five vessels were boarded that yielded samples that could be analyzed for the presence of live organisms, including microorganisms. Most vessels sampled had water that contained live freshwater organisms — regardless of salinity. This included phytoplankton, zooplankton, and bacteria. Cholera-causing bacteria was initially suspected in three of the samples collected. This later turned out to be another, closely related organism. However, other bacteria, including some potentially dangerous to humans, were found (see Chapter 17 of this book).

INTRABASIN TRANSFER IN THE GREAT LAKES

Realistically, the success of introductions is dependent upon the amount of water being discharged. The more untreated ballast water enters the lakes, the greater the chance of an unwelcome introduction. A number of studies have characterized the amount of ballast water entering the Great Lakes during a shipping season. Despite different mechanisms of calculation, they all agreed that actual water being deballasted from ships coming into the Great Lakes was approximately 5 – 7 million tons. This included both ships entering in ballast condition and those entering as NOBOB and discharging only after taking Great Lakes water on board. However, these studies also indicated that domestic shipping within the Great Lakes discharged volumes in excess of 10 times that amount. What this means is that once a species is introduced to the Great Lakes, the chances of it being transported throughout the system are excellent, if not by the great quantities of water distributed around the system by the domestic fleet, then by other vectors such as pleasure boats and the bait or aquarium trade.

The spread of the ruffe (*Gymnocephalus cernuus*) out of Duluth Harbor is a good example of this mechanism. Some fishery managers on the Great Lakes view the ruffe as a serious threat to the fisheries of the Great Lakes. To prevent or at least minimize the potential interbasin transfer of the ruffe, a voluntary ballast water management scheme has been implemented for Duluth Harbor and other ruffe-infested areas by the ship operators themselves (Canadian Shipowners 1997). Commercial ships are not to take on ballast water from ruffe-inhabited ports from May to July when small ruffe could be entrained in the ballast water. Ballast is exchanged in the middle of Lake Superior to prevent any expansion of the ruffe by this vector. Despite these precautions, ruffe have been found in Alpena, Michigan on Lake Huron and are progressing eastward along the shores of Lake Superior.

OTHER FRESHWATER ECOSYSTEMS AT RISK FROM BALLAST WATER

The Great Lakes are an excellent example of an ecosystem that has been thrown out of balance, in part, by the introduction of nonindigenous species. Once the natural protection offered by the relative isolation of the system was broken by the ability of large vessels to go to all parts of the system, the introduction of nonindigenous species increased considerably. This example unfortunately may be a model for other freshwater systems.

The Mississippi River is already a known conduit for the spread of zebra mussels from the Great Lakes. The considerable barge trade going to all parts of the system has no doubt acted as a vector to spread many organisms throughout its tributaries. Fortunately for the northern parts of this system, from the point of view of introduction by ships, the primary transfer medium is likely the hulls of tugs and barges. Because of the limitations of the system, ballast water, while used, is primarily employed to increase the drafts for barges that travel outside the system — either down to the Gulf or out into the Great Lakes. These represent a small portion of the total traffic.

Of more concern is the fact that up to the port of Baton Rouge in the south, large deep-sea vessels are regular visitors to the system. Many are tankers subject to the same regulations with regard to segregated ballast. Others enter in ballast looking to take cargoes of grain. The well-known currents of the Mississippi probably afford some protection through a flushing of any introduced organism, but like the Great Lakes, the opportunities for intrabasin transfer via secondary vectors are plentiful.

The model is mirrored in all freshwater rivers exiting at coastal ports. The opportunities for freshwater organisms to survive and be transported to new and exotic locations are myriad and are repeated many times daily throughout North America. The wonder is not that such organisms are introduced, but rather that more of them are not.

LONG-TERM SOLUTIONS?

Efforts are underway on many fronts to examine technology that might be appropriately used to treat ballast water so that it can be discharged without any threat to freshwater ecosystems. The Great Lakes have an advantage over many other freshwater ecosystems in that the ships entering the lakes must first pass through the St. Lawrence Seaway. The Seaway system has the potential to serve as both shoreside support for available treatment options and a choke point for regulators to ensure the effectiveness of the treatment. In both cases major slowdown of commercial activities can be avoided. Similarly, geography dictates that the majority of vessels entering the Great Lakes are economically required to carry cargo both in and out of the system. Therefore, rather than having to treat large volumes of water in fully ballasted vessels, the majority of vessels need treatment of the one to two inches of unpumpable slop that remains in the bottom of ballast tanks.

Shipboard treatment options considered to date include physical measures such as filtration, ultraviolet sterilization, acoustics, heat treatment in various forms, and redesign of ballast tanks to better promote effective exchange. Chemical treatment options examined have included many compounds found to be effective in industrial mitigation strategies such as chlorine, hydrogen peroxide, organic acids, sodium metabisulphite, and gluteraldehyde. However, none of these options, chemical or physical, have yet been demonstrated to be both effective and economical in commercial application. The safety implications, both to the ship itself and to shipboard personnel, have yet to be fully researched for many of proposed treatments. Similarly, the cost implications of many treatment options make implementation problematic.

CONCLUSION

That the ecosystem of the Great Lakes has been detrimentally affected by nonindigenous freshwater organisms is not in doubt. Ships in pursuit of international commerce have been identified as a significant vector. Research continues in order to find a range of options with which to effectively mitigate the threat posed. However, until such time as a regime is in place, the Great Lakes are still potentially susceptible to invasion by the next " zebra mussel." Next time, the threat might be to human health as well as to the environment.

REFERENCES

Aquatic Sciences Inc., St. Catharines, Ont., Examination of Aquatic Nuisance Species Introductions to the Great Lakes through Commercial Shipping Ballast Water and Assessment of Control Options, Phase 1 & Phase II, *Transport Canada*, Aug 1996.

Canada. Voluntary Guidelines for the Control of Ballast Water Discharges from Ships Proceeding to the St. Lawrence River and the Great Lakes. Amendment #4, March 1993

Canadian Shipowners Association, Lake Carriers Association, The Thunder Bay Harbour Commission, Seaway Port Authority of Duluth, Shipping Federation of Canada, U.S. Great Lakes Shipping Association, "Great Lakes Maritime Industry Voluntary Ballast Water Management Plan for the Control of Ruffe in Lake Superior Ports." 1997.

Carleton, J.T., "Dispersal Mechanism of the Zebra Mussel (*Dreissena polymorpha*)", in *Zebra Mussels: Biology, Impact and Control*, T.F.Nalepa and D.W.Schloesser, Eds., Lewis Publishers, Boca Raton, FL, 1993.

Devendorf, J.F., *Great Lakes Bulk Carriers 1869 – 1985*. John F. Devedorf, Niles, MI, 1996.

Farley, B.R., Analysis of Overseas Vessel Transits into the Great Lakes and Resultant Distribution of Ballast Water, unpub. paper, no. 331, Department of Naval Architecture and Marine Engineering, University of Michigan, Ann Arbor, MI, October 1996.

Gauthier, D. and Steel D.A., "A Synopsis of the Situation Regarding the Introduction of Non Indigenous Species by Ship Transported Ballast Water in Canada and Selected Countries. *Fisheries and Oceans Canada*, 1996.

Greenwood, J.O., *The Fleet Histories — Volume III*, Freshwater Press, 1995.

LeStrang, J., *Cargo Carriers of the Great Lakes*, American Legacy Press, New York, 1981.

Locke, A., Reid D.M., van Leeuwen H.C., Sprules W.G., and Carlton J.T., "Ballast Water Exchange as a Means of Controlling Dispersal of Freshwater Organisms by Ships," *Canadian Journal of Fishery and Aquatic Science*, 50, pp 2086–2093, 1993.

Locke, A., Reid D.M., Sprules W.G., Carlton J.T., and van Leeuwen H.C., "Effectiveness of Mid Ocean Exchange in Controlling Freshwater and Coastal Zooplankton in Ballast Water. *Canadian Fisheries and Aquatic Sciences*, no. 1822, 1991.

Mance, T., Know Your Ships — The Seaway Issue, 29th ed., Marine Publishing Company, 1988.

McCarthy, A.S. and Khambaty F.M., "International Dissemination of Epidemic *Vibrio Cholerae* by Cargo Ship Ballast Water and Other Non Potable Waters." *Applied and Environmental Microbiology*, 60, no. 7 pp. 2597–2601, July 1994.

Melville Shipping Ltd., Ottawa Ont., Ballast Water Exchange Study: Phase 1, *Transport Canada,* March 1995.

Mills, L.E., Leach H.J., Carlton T.J., and Secor L.C., "Exotic Species and the Integrity of the Great Lakes: Lessons from the Past," *Bioscience*, 44, no. 10, 1994,

Mills, L.E., Leach H.J., Carlton T.J., and Secor L.C., "Exotic Species in the Great Lakes: A History of Biotic Crises and Anthropogenic Introductions" *J. Great Lakes Res.*, 19(1) 1–54, 1993.

Parsons, G.M., Moll R., Mackay P.T., and Farley B.R., Great Lakes Ballast Demonstration Project — Phase I. Cooperative Institute for Limnology and Ecosystem Research, University of Michigan, Ann Arbor, May 1997.

Pollutech Environmental Limited, "A Review and Evaluation of Ballast Water Management and Treatment Options to Reduce the Potential for the Indroduction of Non Native Species to the Great Lakes." *Transport Canada*, March 31 1992.

Reid, M.D. and Carlton T.J., A Study of the Introduction of Aquatic Nuisance Species by Vessels Entering the Great Lakes and Canadian Waters Adjacent to the United States, Shipping Study I-A, USCG Report o. CG-D-17-97. National Technical Information Service, Springfield, VA, March 1997.

United States, Nonindigenous Aquatic Nuisance Prevention and Control Act of 1990 (NANPCA), Public Law 101-646 (29 November 1990).

United States, National Invasive Species Act of 1996 (NISA), Public Law 104-332 (26 October 1996).

14 Summary of North American Fish Introductions through the Ballast Water Vector

E.J. Crossman and B.C. Cudmore

We have listed 12 species (Table 14.1) of fishes introduced by the ballast water vector. The reproductive status of each form is included in the table. Most of the species involved are foreign, and therefore usually stand out when encountered in North American waters. The largest number of these introductions have resulted from trans-Atlantic shipping arriving in North America in water ballast and penetrating one or more of the Great Lakes. Others may have arrived in boats entering the Mississippi. It is interesting that although a large number of foreign marine species have been introduced by discharge of ballast water in the salt and brackish waters on the West Coast of North America, only two species introduced into freshwaters there are listed in the NAS database as attributable to ballast water. These are the yellowfin goby, *Acanthogobius flavimanus*, and the shimufuri goby, *Tridentiger bifasciatus* (a new species resurrected from chameleon goby, *Tridentiger trigonocephalus*).

Two species that arrived in ballast water are, of all introduced fishes, represented by the greatest amount of publicity in the scientific literature and popular press (see Ogle 1995 and Marsden et al. 1996). As indicated above, this results from the fact that those species (ruffe and round goby) have rapidly established reproductive populations, expanded beyond the original location, and have resulted in significant negative impacts on the waters in which they now occur. The ruffe, originally introduced into the Duluth–Superior area of the western tip of Lake Superior, has now appeared near Thunder Bay, Ontario, and in Lake Huron. Both these expansions have been attributed to secondary transfer in ballast water. The round goby was reported first in the St.Clair River near Sarnia, Ontario. Since then it has spread to southern Lake Huron, southern Lake Michigan, the western tip of Lake Superior, throughout Lake Erie, and to extreme eastern Lake Ontario. Some of the new locations on the lower lakes may be attributable to bait bucket transfer, since the abundant and easily caught species was quickly adopted as a bait species. The more distant extensions in the upper lakes and eastern Ontario are most likely secondary ballast water transfers.

Two other ballast water fishes are a strong contrast to those two. The tubenose goby, *Proterorhinus marmoratus*, like the round goby, was introduced by this vector into the same area of the St. Clair River but has not expanded its range to any extent. The European flounder, *Platichthys flesus*, was first recorded in North America in Lake Erie in 1974 (Emery and Teleki 1978), and individuals have been captured at least 12 times from the Great Lakes other than Huron and Ontario. Each of these records of the flounder is judged to be a separate ballast water introduction, and no individuals less than young of the year of this brackish-water spawner have, as yet, been found in North American freshwaters.

Almost all of the concern for species introduced in ballast water h

TABLE 13.1
List of Fishes Introduced in Ecological North America (Excluding Mexico) Outside Their Native Range Through Ballast Water Introductions and Their Reproductive Status in the Nonindigenous Area(s)

Scientific Name	Common Name	Status[a]
Anguillidae — freshwater eels		
Anguilla anguilla	European eel	Not established
Cyprinidae — carps and minnows		
Scardinius erythropthalmus	rudd	Established in most areas
Fundulidae — topminnows and killifishes		
Lucania parva	rainwater killifish	Established in most areas
Gasterosteidae — sticklebacks		
Apeltes quadracus	fourspine stickleback	Established
Gasterosteus aculeatus	threespine stickleback	Established in most areas
Percidae — perches		
Gymnocephalus cernuus	ruffe	Established
Sciaenidae — drums		
Aplodinotus grunniens	freshwater drum	Established
Gobiidae — gobies		
Acanthogobius flavimanus	yellowfin goby	Established
Neogobius melanostomus	round goby	Established
Proterorhinus marmoratus	tubenose goby	Established
Tridentiger bifasciatus	shimufuri goby	Established
Pleuronectidae — righteye flounders		
Platichthys flesus	European flounder	Not established

[a] Status refers to the reproductive nature of "populations" of that fish in areas where it has been introduced, regardless of vector(s) used. Many statements on status were derived from information in the NAS database. Some notations may be open to criticism (e.g., hybrids).

as been for European species from European waters. Two species deemed to have been introduced into the Great Lakes by this vector, threespine stickleback, *Gasterosteus aculeatus*, and fourspine stickleback, *Apeltes quadracus*, were probably transferred in ballast water from nearshore North American locations.

While not a record from ballast water, two other species possibly associated with ships were the sea lamprey and the American eel. The access of the sea lamprey to the Upper Great Lakes via the canals and invasive nature may have been aided by its attachment to the hulls of ships passing through the canals of the St. Lawrence Seaway. Proposals to replace two marine railways with navigable canals on the Severn portion of the Trent–Severn Waterway in Ontario were negated, in part, by the desire to prevent the sea lamprey from entering Lake Simcoe and the Kawartha Lakes.

In addition, crew members of ships passing up the lakes apparently (Emery, 1985) took live American eels from Lake Ontario along as food and oftern haphazardly released them alive into the other lakes.

REFERENCES

Emery, L. Review of Fish Species Introduced into the Great Lakes, 1819-1974. *Great Lakes Fishery Commission, Technical Report 45*, Great Lakes Fishery Commission, Ann Arbor, MI, 1985.

Emery, A.R. and G. Teleki. European Flounder (*Platichthys flessus*) captured in Lake Erie, Ontario. *Canadian Field-Naturalist*, 92, pp. 89–91, 1978.

Marsden, J.E., P. Charlebois, K. Wolfe, D. Jude, and S. Rudnicka, The Round Goby (*Neogobius melanostomus*): A Review of European and North American Literature with Notes from the Round Goby Conference, Chicago, 1996. *Illinois Natural History Survey, Centre for Aquatic Ecology, Aquatic Ecology, Technical Report*, 96/10, 1996.

Ogle, D.H., Ruffe (*Gymnopcephalus cernuus*): A Review of Published Literature. Wisconsin Department of Natural Resources, Administative Report No. 38, 1995.

15 Ballast Water Introductions of Mollusca

Gerald L. Mackie

INTRODUCTION

The taxonomy of freshwater gastropods and bivalves is still somewhat in disarray, but on the basis of species described in Burch (1975a,b, 1989) and Rosenburg and Ludyanskiy (1994), there are 755 species of freshwater Mollusca in North America. Of these, 485 species are gastropods and 270 species are bivalves. About 3% (15) of the gastropods and 3% (7) of the bivalves are exotic. Of the 22 species of exotic molluscs, 10 (about 45%) have been introduced via ballast water. Four of the species are gastropods and six are bivalves. Most of the species have had no documented ecological impacts. Two, both dreissenids, have had severe ecological and socioeconomic impacts, as detailed in Chapter 16 of this book.

Each species introduced by ballast water was examined for the following ten attributes, when known, to establish what common characteristics, if any, these invaders share and what threat they may pose to the host environment:

1. *Species characteristics* — Only those features diagnostic of each species are summarized. Most of the characteristics are taken from taxonomic keys.
2. *Mode of life and habitat* — This section describes how and where the animal lives. Native sphaeriids occur in a wide variety of aquatic habitats and conditions. However, the exotic species, while tolerant of a wide variety of water quality, including some organic enrichment (e.g., mesotrophy to eutrophy), are more common in lentic waters than in lotic ones, except large rivers (e.g., St. Lawrence River). Unlike native species of *Musculium* and one species of *Sphaerium* (*occidentale*), which have adaptations for life in ephemeral habitats (McKee and Mackie 1980, 1981, 1983), the exotic species will occur mainly in permanent aquatic habitats. None of the exotic species of sphaeriids are known to occur in small streams, but large densities are common in rivers with depositional substrates (Mackie and Qadri 1973). Like all bivalves, sphaeriids are filter feeders but are capable of a form of deposit feeding where cilia on the foot direct food particles through the pedal gape to the labial palps where initial food sorting occurs. Pennak (1989) summarizes the modes of life and habitats of gastropods and bivalves in general.
3. *Dispersal mechanisms* — Other than ballast water of ships, only the most common mechanisms are summarized. Kew (1893), Boycott (1936), Baker (1945), Malone (1965, 1966), and Rees (1965) have treated the subject thoroughly, and all agree that waterfowl and shorebirds are primarily responsible for dispersing molluscs overland to isolated bodies of water. Gastropods attach to feet and feathers of birds, and once removed from water, retain their viability for a sufficient period of time to effect overland dispersal for great distances (Malone 1965). Large insects are effective dispersal agents for small snails and clams over short distances (Rees 1965). In general, external transport (e.g., feet and feathers) is a more effective dispersal mechanism than internal transport via the digestive tract (Mackie 1979). Hanna (1966) ascribes the introduction of many marine

0-15667-0449-9/99/$0.00+$.50

and terrestrial molluscs in western North America to ports of entry in baggage and mail, but this mechanism is not commonly attributed to freshwater molluscs.

4. *Adaptations* — Any morphological, behavioral, and/or physiological adaptations that explain, at least in part, the specie's success in a particular habitat are summarized. Most of the discussions examine physiological adaptations, because the physiological ecology of molluscs has received more attention than morphological or behavioral adaptations. Burky (1983), McMahon (1983), and Aldridge (1983) give detailed accounts, including feeding, respiration, reproduction, life cycles, etc., in bivalves, pulmonates, and prosobranchs, respectively. Although few references are made to exotic species, the descriptions for native species probably apply (with some exceptions) to exotic species.

   In general, molluscs have shown adaptations for (1) avoiding desiccation or surviving prolonged periods of exposure; (2) living an infaunal existence in the soft sediments; (3) living an epifaunal existence on firm substrates; (4) tolerating high turbidities characteristic of high order streams/rivers; (5) being eurythermous over their normal temperature range (e.g., tropical eurytherms and temperate eurytherms); and (6) having the physiological adaptations to deal with short periods of anoxia or low oxygen tensions or else live only in the epilimnion, which may at times be supersaturated with oxygen. The prerequisites for a specie's ability to adapt to any of these six scenarios can be summarized in its tolerances or requirements for surviving periods of desiccation, siltation, or high turbidity, variable temperatures or extreme temperatures, and anoxia. There are distinct differences in the abilities of exotic species of molluscs to tolerate anoxia (Aldridge 1983; Burky 1983; McMahon 1983; Holopainen 1987; Matthews and McMahon 1994), but none can survive prolonged anoxic conditions. The relatively poor tolerance of all exotic species to prolonged anoxia will restrict most species to the shallow, well-oxygenated surface waters of reservoirs.

5. *Predators* — Fish and waterfowl, and occasionally aquatic mammals (muskrat, otters), are the main predators of freshwater molluscs, including exotic species. Many waterfowl seek invertebrates high in protein and calcium contents, like Mollusca, during egg production periods (Rogers and Korschgen 1966; Krapu and Swanson 1975). Some turtles (e.g., *Graptemys pseudographica kohnii*) also feed on Asian clams and small gastropods (Lindeman 1995).

6. *Parasites* — The parasites of freshwater molluscs fall under five groups: Protista, especially ciliates; Digenea; Aspidogastrea; Nematoda; and Arthropoda, especially hydrachnids (parasitic mites). The aspidogastrids and parasitic mites parasitize unionid and corbiculid bivalves as their definitive hosts, but the other parasitic groups use most gastropods and bivalves as intermediate hosts.

7. *Reproductive potential and life cycle* — Most gastropods are oviparous, laying eggs in masses on firm substrates such as macrophytes, rocks, and logs. Development is direct and tiny snails emerge from the eggs. The exceptions to this are the Viviparidae, which are ovoviviparous, and the Ampullariidae, which oviposit calcareous-shelled eggs under fairly dry conditions (Aldridge 1983). The numbers of eggs oviposited by oviparous forms varies considerably within and among species (Aldridge 1983), but rarely are more than 100 produced by one female (ampullariids may produce several hundred per female). The ovoviviparous forms are less fecund and usually produce less than 10 larvae per parent, but may range up to 50 per parent.

   In general ovoviviparous forms are less fecund than oviparous forms because brooding space for the developing larvae within parents is limiting. For example, the internal shell volume in sphaeriid clams remains relatively constant so that the reproductive output cannot be significantly altered and will always be relatively low (Kilgour and Mackie

1990). Although the numbers are highly variable, ovoviviparous forms are about an order of magnitude less fecund than oviparous forms (e.g., 10:100). Even though many ovoviviparous forms are parthenogenetic, reducing the risk of having to find a mate, the fecundities are still relatively low. Even ovipositing, oviparous forms have low natalities relative to planktonic oviparous forms. Ovipositing oviparous forms are about three to four orders of magnitude less fecund than planktonic oviparous forms (e.g., 100:100,000 to 100:1,000,000).

8. *Food and feeding habits* — All gastropods, including the introduced species, are grazers. They have a radula that is used like a rasp to grind attached bacteria, protists, and/or algae from hard surfaces. Some feed on algae attached to sand or mixed in with sand (epipsammic feeders). Some feed on detritus and take in sand particles as triturating agents to grind the detritus (Aldridge 1983; McMahon 1983). Epipsammic feeding involves taking in one or a few large sediment particles into the buccal cavity, where organic material is ingested and sand particles are regurgitated (Aldridge 1983). Other gastropods are deposit feeders, usually on sediments that are fine-grained, relative to the body size of the animal. Some viviparids are adapted for filter-feeding using cilia on their ctenidial gill (Aldridge 1983), although not as efficiently as occurs in bivalves.
In bivalves, food selection is performed by a variety of cilia, including those in the mantle cavity (gills, labial palps, and foot) and in the stomach and midgut. In some, cilia in the mantle cavity and stomach select particles of 15–40 µm for food but can filter out particles as small as 0.7–1.0 µm in diameter from the water (Mackie et al. 1989). Most sphaeriaceans (Sphaeriidae and Corbiculidae) are also capable of deposit feeding on bacteria, algae, and fine organic matter in interstitial pore water of the sediments.

9. *Impact potential* — All ecological and/or socioeconomic impact(s) ascribed to a species are summarized. The potential impacts are of three general types: (a) Little or no potential ecological or socioeconomic impact; (b) potential beneficial ecological and/or socioeconomic impacts (uses); (c) potential ecological and/or socioeconomic harm. The extent of the impacts will depend not only on the species that invade a habitat but the quality of the habitat visa viz. the tolerances and requirements of the invading species.

10. *Control measures* — Control measures are described only for biofouling species.

DETAILED REVIEW OF MOLLUSCA INTRODUCED BY BALLAST WATER

CLASS: GASTROPODA

Much of the information on life cycle, food and feeding, biotic associations and the tolerances and requirements of many of the gastropods and bivalves discussed below were taken from Clarke (1973), Harman (1974), Fuller (1974), Jokinen (1983), and Pennak (1989).

Subclass: Prosobranchia

Shells small to large; dextral; operculum present with either concentric, paucispiral, or multispiral growth lines; respiration by gills; mostly dioecious, except for Valvatidae.

Family: Hydrobiidae.
Shells small (<8 mm high usually), dextral, umbilicate, or nonumbilicate; operculum is paucispiral (= subspiral); mostly dioecious, except for a few parthenogenetic species; radula with seven teeth in each row.

Potamopyrgus antipodarum (New Zealand hydrobiid)

The species was first reported from Lake Ontario and the Saint Lawrence River, although it was first found (but not reported) in North America in 1990, from Green River, Colorado (Zaranko et al. 1997). It was apparently introduced to the Great Lakes in 1991 (Zaranko et al. 1997). It is widespread throughout Europe, including Britain (Wallace 1978; 1985), Denmark (Lassen 1978), Switzerland (Ribi 1986), Poland, Hungary and Slovakia (Cejka 1994). It is also widely distributed in southeastern Australia and Tasmania, where it was introduced in the middle of the last century (Ponder 1988). Zaranko et al. (1997) consider the species to be native to New Zealand, but Ponder (1988) has examined the taxonomy of the species and gives convincing evidence, along with Winterbourn (1970a,b; 1972), that *P. antipodarum* of Australia and New Zealand is the same species as *Potamopyrgus jenkinsi* (Smith) of Britain; he also proposes that the species was introduced to Australia by Europeans in the early 1800s.

Species characteristics: Shell 3–12 mm high; shape variable, slender and elongate to ventricose; 4–8 whorls; in many specimens whorls have a hirsute carina; sutures deeply impressed; operculum ovate.

Mode of life and habitat: On silty sand in Lake Ontario but in its native country, New Zealand, it occurs in lowland rivers, stony streams, creeks, ponds, lakes, springs, and estuaries where salinities can range up to 26% (Zaranko et al. 1997). Found only in permanent waters with a wide range in calcium content, on soft and hard substrates and among vegetation (Winterbourn 1970). In Australia, the species appears to live almost entirely in, or close to, disturbed habitats, such as those that have been altered by urban development or by agricultural or forestry activities (Ponder 1988). They cannot tolerate water temperatures greater than 28°C (Winterbourn 1969; 1972).

Dispersal mechanisms: Haynes and Taylor (1984) state that the species spreads passively by river and current transport and by attaching to extremities on birds and fish; they also attribute internal transport in the digestive tract of fish as a mechanism. Hubendick (1950) attributes part of its effectiveness at passive dispersal to its ability to survive liberation into estuarine areas, such as the River Thames in England. The snails have been known to breed in freshwater tanks and reservoirs in the Sydney area of Australia and have been distributed through water pipes to emerge from domestic taps (Ponder 1988). Ponder (1988) suggests that it was first introduced to Tasmania by way of drinking water supplies on ships and probably entered Europe at about the same time in the same way. Zaranko et al. (1997) suggest that the species may have been introduced to Lake Ontario by the aquarium trade, but the species is so small and of no socioeconomic value or appeal that the most likely vector was ballast water. However, the aquarium trade has been attributed in part to the wide distribution in Europe and Australia (Winterbourn 1972). Once introduced, the species is capable of rapid dispersal (see references in Zaranko et al. 1997).

Adaptations: The species is capable of tolerating a wide range of salinities (up to 26%) and habitats, which may account for its variable shell shape and its wide distribution. It is also parthenogenetic, which White (1954), Mayr (1963), and Tomlinson (1966) suggest is advantageous to organisms inhabiting marginal habitats, because it eliminates the need for finding a mate.

Predators: Probably fish and waterfowl, perhaps some amphibians.

Parasites: None reported for North American specimens, but in Europe, infestations by the trematode, *Microphallus* sp., have been attributed to inducing or maintaining sexual reproduction (the species is normally parthenogenic, see below).

Reproductive potential and life cycle: The species is ovoviviparous, but populations may be entirely parthenogenic or contain varying proportions of sexually functional males. Sexual reproduction probably occurs to some extent in many populations because males constitute up to 20% in some European populations (Wallace 1985), 9% in some Australian populations (Wallace 1978), but usually 0 to <3% in most populations (Ponder 1988). In Lake Ontario only females were found by Zaranko et al. (1997).

Food and feeding habits: The species is apparently a detrivore and/or a herbivore and prefers to feed on plant and animal detritus, but it also ingests algae and diatoms (Calow and Calow 1975).

Impact potential: The species has not been present long enough to make any evaluations on its positive or negative impact potential. Zaranko et al. (1997) suggest that the biofouling potential is low, its most serious threat being competition with native gastropods. However, Cotton (1942) has recorded the species (as *Bythinella pattisoni*, a synonym of *P. antipodarum* (Ponder 1988)) as blocking water pipes and meters in Australia.

Control: None, since there is no evidence to suggest that the species requires immediate control.

Family: Valvatidae.
Shells small (usually <5 mm high), dextral, umbilicate; radula with seven teeth in each row; operculum multispiral; hermaphroditic.

Valvata piscinalis (European valve snail)

The species was introduced from Europe into the lower Great Lakes where it is currently very abundant. The species was first recorded from Lake Ontario in 1913 (Clarke 1980) and has since spread to lakes and large rivers in all states and provinces (Mackie et al. 1980) bordering the Great Lakes.

Species characteristics: The shells of most valvatids are almost discoidal, but in *V. piscinalis* the shell has a spire with very distinct, thread-like striae. It has a multi-spiral operculum like other valvatids.

Mode of life and habitat: Most commonly found on or near the surface of the substrate or on submersed macrophytes in water less than 2 m deep. Found commonly in oligotrophic to mesotrophic lakes.

Dispersal mechanisms: Usually attach to the feet or feathers of waterfowl.

Adaptations: A euthermous species but anoxia intolerant.

Predators: Mainly fish and waterfowl.

Parasites: Probably similar to most other gastropods (see introduction).

Reproductive potential and life cycle: Like all valvatids, it is hermaphroditic and oviparous, reproduction occurring in late spring to late summer/early autumn. In *V. piscinalis*, eggs are deposited on macrophytes after three separate periods of migration into shallow waters (Heard 1963).

Food and feeding habits: The species grazes mostly on epiphytic algae.

Impact potential: The species appears to be relatively benign in its impact potential and competition with native gastropods is probably its most serious threat.

Control: None, since there is no evidence to suggest that the species requires immediate control.

Family: Bithyniidae.
Shell medium-size (up to ~1.5 cm), dextral, with elevated spire; operculum typically with concentric lines of growth (sometimes paucispiral); radula with seven teeth in each row; dioecious.

Bithynia tentaculata tentaculata (Faucet snail)

The species is native to Europe and was first introduced into the Great Lakes. It was first recorded in Lake Michigan in 1871 (Robertson and Blakeslee 1948) and has since spread throughout most states and provinces bordering the Great Lakes. It also occurs in the mid-Atlantic U.S. The species is the only representative of its genus in North America.

Species characteristics: Shell conical, up to 1.5 cm high, moderately inflated, nonumbilicate; sutures of spire impressed; operculum usually at opening of aperture when relaxed (withdrawn in most other prosobranch species).

Mode of life and habitat: The species forms large populations in richly eutrophic and moderately polluted waters of large lakes, large rivers, and canals (Harman 1968; 1974). It is one of the most abundant molluscs in the Great Lakes (Berry 1943) but apparently cannot live in streams with turbulent water (Harman 1968). It can tolerate low levels of oxygen, down to about 4 mg/l (Harman 1974) which partly explains its abundance in eutrophic waters.

Dispersal mechanisms: Probably waterfowl, on feathers and feet, and aquatic mammals.

Adaptations: Prolific breeding habits and iteroparous reproduction (Calow 1978) allow it to increase in population numbers rapidly.

Predators: Mainly fish and waterfowl. Some leeches (e.g., *Glossiphonia complanata*) also prey on the species (Sawyer 1974).

Parasites: Bythinia is a first intermediate host to three digean trematodes, *Stichorchis subtriquetrus*, whose final host is the beaver, *Plagiorchis muris* whose definitive host is the herring gull, and *Prosthogonimus macrorchis* whose definitive hosts are ducks, chickens, and pheasants, English sparrows, and crows that live near or on lakes of the Great Lakes region (Olsen 1974). The snail is not host to any parasites of quarantine significance.

Reproductive potential and life cycle: It is a prolific and opportunistic species, with breeding periods in July and August. It attaches its egg capsules mostly on the shells of other individuals or of other species. The animal takes about 2 years to reach maximum size and they die shortly after. Eggs are laid in the spring by adults born the previous year and hatch in June and July. Growth is rather slow with increments of 3.5–4 mm in shell height each year. When the animal reaches 7 mm the females begin laying eggs (Mattice 1970; Vincent et al. 1981). The life span is two (Pinel-Alloul and Magnin 1971) to 3 years (Vincent et al. 1981).

Food and feeding habits: The species has two feeding modes, grazing on epiphytic algae on rocks and plants and suspension feeding (Tashiro and Coleman 1982; Brendelberger and Jurgens 1993; Brendelberger 1995). Apparently suspension feeding is energetically more advantageous than grazing (Hunter 1975) and is used whenever suspended particles are available (Tashiro and Coleman

1982). However, the suspension feeding mode appears not to be able to provide sufficient energy to support reproduction. More weight is gained by grazing than by suspension feeding (Brendel-berger 1995).

Impact potential: The species appears to be relatively benign in its impact potential and competition with native gastropods is probably its most serious threat. *Bythinia* is thought to be responsible for the eradication of pleurocerids in the Erie Canal (Baker 1916) and Oneida Lake (New York), presumably because of *Bythinia*'s unique ability to utilize suspension feeding in addition to grazing (Harman 1968).

Control: None, since there is no evidence to suggest that the species requires immediate control.

Family: Lymnaeidae.

Shell small to large, dextral, thin, spire typically elevated; operculum absent; respiration by a "lung" (vascularized mantle cavity); hermaphroditic.

Radix auricularia (European ear snail)

"Widely introduced but of spotty occurrence in North America" (Burch 1989). The species is native to Europe and Asia and was first recorded from Montana by Baker (1913). It is known to occur as far west as British Columbia and Alberta; most common in states and provinces bordering the Great Lakes.

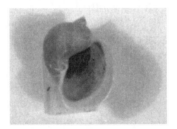

FIGURE 15.1   *Radix auricularia.*

Species characteristics: Shell up to 30 mm high, with five whorls, the body whorl being greatly inflated and accounting for more than 90% of shell volume; aperture greatly dilated, somewhat ear-shaped.

Mode of life and habitat: Clarke (1980) states that it is especially common in muddy substrates of lakes, ponds, and slow-moving rivers near major cities. Like native pulmonates, the exotic species, including *R. auricularia,* are most abundant in the littoral and sublittoral regions of lakes. The species has a thin, fragile shell and cannot maintain itself in strong currents characteristic of low order streams. Moreover, as the river continuum concept predicts, grazers, like gastropods, are not abundant until epiphytic algae begins to accumulate in stream orders above 3.

Dispersal mechanisms: Probably waterfowl, on feathers and feet.

Adaptations: Facultative air-breather and able to avoid anoxia, allowing it to live in a great variety of habitats.

Predators: Mainly fish and waterfowl and some glossophoniid leeches.

Parasites: Many lymnaeids are common intermediate hosts of digean trematodes, and *R. auricularia* is probably no exception.

Reproductive potential and life cycle: Hermaphroditic and oviparous.

Food and feeding habits: Feeds mainly on epiphytic and filamentous algae.

Impact potential: Its ecological and socioeconomic significance is unknown, but it is probably an efficient competitor with some native lymnaeids.

Control: None, since there is no evidence to suggest that the species requires immediate control.

CLASS: BIVALVIA

Subclass: Heterodonta

Order: Corbiculacea

Family: Sphaeriidae.
Tiny (<1.5 mm) to medium-sized (2 cm) clams; hinge with cardinal teeth, two in the left valve, one in the right; lateral teeth paired in the right valve, single in the left; periostracum yellowish or brownish; shell with little or no ornamentation; hermaphroditic and ovoviviparous, producing 1–5 (usually) larvae in brood sacs on each inner gill.

Four species of exotic sphaeriids have been introduced from Europe into the Great Lakes via ballast water discharge. All four species are common in the Great Lakes and/or its drainages (Mackie et al. 1980) and have shown little dispersal beyond the Great Lakes drainage basin. All sphaeriids are easily dispersed by flying insects and birds (Mackie 1979), as well as by mammals. The ecological and socioeconomic impacts have not been determined for any of the introduced species. Most exotic species are common intermediate hosts of trematode parasites, but so are the native species (Mackie 1976).

McMahon in another chapter of this book discusses the Corbiculidae in detail. It is briefly discussed here to put its reproductive biology into perspective with the closely related Sphaeriidae. The Corbiculacea are monoecious, the Sphaeriidae being simultaneous hermaphrodites (Mackie 1984) and *C. fluminea* apparently being dioecious or monoecious in lentic waters (Morton 1983) and female or monoecious in lotic waters (Morton 1986). Internal fertilization is used in all native species of bivalves and in *Corbicula*. However, in *Corbicula* the fertilized eggs are brooded in gill chambers of the parent, where they develop through the trochophore, veliger, pediveliger, and straight-hinge stages before being released as umbonal juveniles to the plankton (Figure 15.1b). The umbonal juveniles may swim for 1 to 3 days before they settle on the substrate to begin their benthic existence.

The Sphaeriidae are ovoviviparous (Mackie 1978) and brood their young in brood sacs on the inner gills of the parent. Four marsupial stages are recognized: (1) Embryos first appear as gastrula in single-walled primary sacs; (2) the embryos develop most of the organ systems and mature through the fetal larvae stage to the shelled prodissoconch stage within brood sacs; (3) the prodissoconch larvae grow in size to become extra-marsupial larvae which outgrow their brood sacs and tear the sac wall to eventually lie free in the inner gill space; (4) the extra-marsupial larvae are born through the excurrent siphon and begin a benthic existence as newborn. Complete larval development requires 1 to 3 months, depending on species.

The best reproductive adaptations for life in streams are an ovoviviparous habit, where the young are brooded inside the parent, or an oviparous one where the eggs are attached to the substrate.

These adaptations allow the animals to maintain their position in the streams. Many can migrate further upstream by crawling against the current, especially during low-flow periods.

Mackie (1986) has reviewed the adaptations of sphaeriid clams to life in a variety of habitats. The adaptations include: shell structure is of complex-crossed lamellar type, which is considered an adaptation to high resistance to abrasion; ability to change internal shell volume to compensate for increased natalities during optimal conditions; ability to change shell allometry so that thickness in the protective calcium carbonate layer can be maintained in habitats with low calcium supplies; physiological adaptations for oxygen uptake during hypoxia; and life history traits that can be adapted to life in either temporary or permanent aquatic habitats.

Sphaerium corneum (European fingernail clam)

Apparently introduced from Europe into the Great Lakes (it is most common in Lake Erie and Lake Ontario) (Herrington 1962) in the early 1900s, the species has dispersed slowly into lakes and large rivers of states bordering the Great Lakes. It is the most common *Sphaerium* in Britain (Ellis 1978), and adapts well to almost any kind of habitat (e.g., ponds, ditches, swamps). However, in North America it is relatively uncommon. The earliest record for the species in North America appears to be 1924 (Adamstone 1924) from Lake Ontario.

Species characteristics: Shell medium-sized, up to 9 mm long, thin; moderately inflated, somewhat ovate in outline; periostracum yellowish in juveniles, brownish in adults, somewhat glossy; dorsal margin evenly curved, sometimes with a rather straight and equal slope anterior and posterior to umbone; hinge plate long with narrow cardinal and lateral teeth (Figure 15.3)

Mode of life and habitat: In Europe it is a common infaunal species of eutrophic lakes, stagnant and slowly running waters (Zhadin 1952; Adam 1960), and in North America it occurs in a variety of substrates, from organic ooze to fine sand. The species may occur in impoundments, but it will not be found in the drawdown zone because it does not like drying ponds and streams" (Boycott 1936).

Dispersal mechanisms: Its slow rate of dispersal may be an indication of its poor ability to compete with native species of *Sphaerium*, although it is moderately common in Lake Erie (Mackie et al. 1980).

Adaptations: Has physiological adaptations for survival in water with low oxygen tensions and is common in eutrophic lakes. The ovoviviparous habits of all sphaeriids, including *S. corneum*, is considered an adaptation for survival for short-term aerial dispersal (Mackie 1979). See also discussion in introduction.

Predators: Fish and waterfowl.

Parasites: Mostly digean trematodes, especially *Crepidostomum transmarinum*, *Bunodera luciper- cae*, and *Phyllodistomum simile* (Mackie 1976). The definitive hosts of the parasites are fish, including salmon and brown trout.

Reproductive potential and life cycle: Hermaphroditic and ovoviviparous, with 1–10 larvae developing on each inner gill.

Food and feeding habits: Filter feeders of planktonic algae, especially diatoms.

Impact potential: Probably none, because it does not seem to have displaced any native species in the Great Lakes, where it is only moderately common in Lake Ontario and Lake Erie, absent or

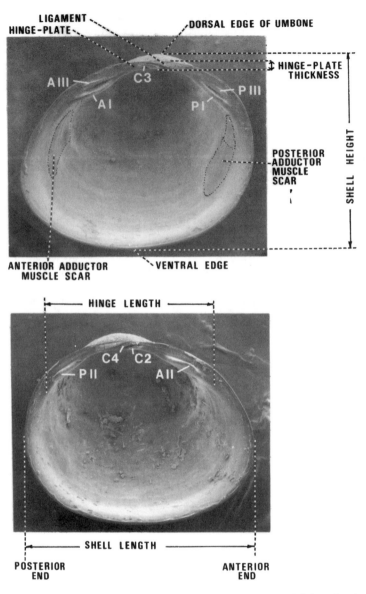

FIGURE 15.2 Valve characteristics of sphaeriid clams with inside view of right valve (top) and lift valve (bottom).

rare in other Great Lakes. There is no apparent negative ecological or socioeconomic impact of the species in its native range in Europe.

Control: None, since there is no evidence to suggest that the species requires immediate control.

Pisidium supinum (Hump-backed pea clam)

This species has been considered an introduced species (Burch 1975b; Clarke 1973), but several fossil records from Idaho and Alberta (Herrington 1962) suggest that it may be indigenous to North America. Most common in Great Lakes and St. Lawrence drainage basins.

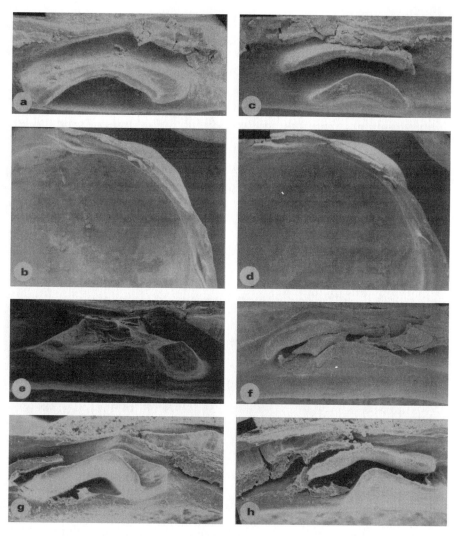

FIGURE 15.3 Hinge features of Sphaerium corneum: a = C3, b = hinge, plate of right valve, c = C2 (inner tooth) and C4, d = hinge plate of left valve; e–f = different variations of C3 and C2/C4 teeth.

Species characteristics: Shell medium-sized, up to 4.5 mm in length, somewhat triangular; beaks with an oblique ridge; dorsal margin short, strongly curved; anterior margin elongated, straight, roundly pointed ventrally; posterior margin somewhat truncated; hinge teeth thick and heavy.

Mode of life and habitat: More common in large rivers (e.g., Ottawa River, St. Lawrence River) than in lakes and occurs sporadically in the Great Lakes. Infaunal in fine sand or mud

Dispersal mechanisms: Probably the same mechanisms as for sphaeriids in general — flying insects and birds, mammals (otters, muskrat, etc.).

Adaptations: No adaptations appear to be peculiar to this species' ability to grow and reproduce in North American waters, which may explain why it does not appear to have displaced any native species of pisidia.

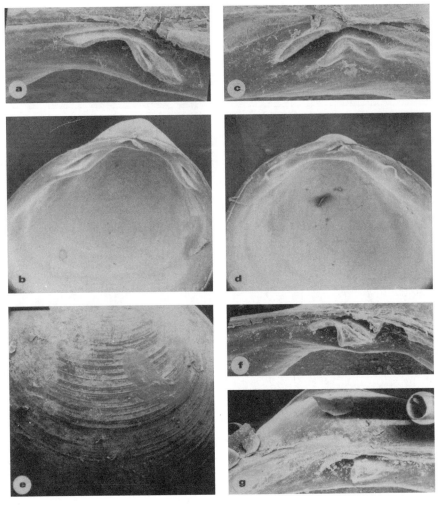

FIGURE 15.4 Hinge features of *Pisidium supinum*, form supinum. a = C3, b = hinge plate of right valve, c = C2 (lower tooth) and C4, d = left hinge plate, e = distinct raised striae, f and g = variations of C3 and C2/C4, respectively.

Predators: Fish and waterfowl and some glossiphoniid leeches.

Parasites: Probably digenean trematodes.

Reproductive potential and life cycle: As implied in *"Adaptations"* listing above, the species does not appear to have a reproductive potential that exceeds our native species.

Food and feeding habits: Filter feeds on planktonic algae, especially diatoms.

Impact potential: Probably none, because it does not seem to have displaced any native species in most of the Great Lakes, where it is only moderately common, or in the Ottawa River and other tributaries of the St. Lawrence River and the Great Lakes.

Control: None, since there is no evidence to suggest that the species requires immediate control.

Pisidium amnicum (greater European pea clam)

Found in the drainage systems of St. Lawrence River and the Great Lakes (Burch 1975b), the species occurs throughout Europe as far north as Naples, eastward through Siberia to Lake Baikal, and in Algiers (Woodward 1913). Both *P. amnicum* and *S. corneum* were probably introduced in the mid to late 1800s, the earliest record being 1899 in Hamilton Bay of Lake Ontario (Sterki 1899).

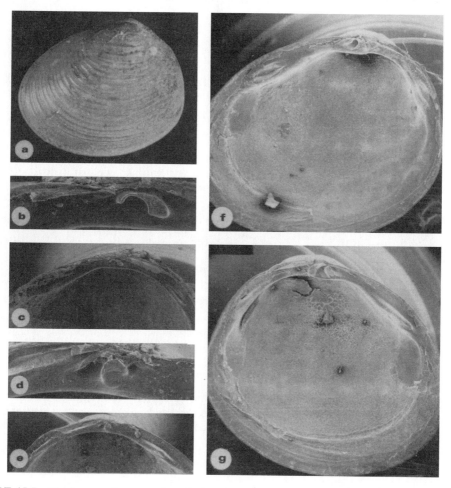

FIGURE 15.5   Pisicium amnicum. a = lateral view, b = C3, c = right hinge plate, d = C2 and variation of C4, e = variation of left hinge plate, f and g = internal views of right and left valves, respectively.

Species characteristics: Among the largest of the *Pisidium* species, being up to 1 cm in shell length; the shell is rather elliptical in outline, compressed, and surface has heavy striae, right up over the beaks; the periostracum is glossy and yellow to brown; C3 cardinal tooth of hinge is U-shaped and is diagnostic of the species.

Mode of life and habitat: In Europe the species lives in rather clean waters, large and small rivers, brooks, ditches and lakes among vegetation (Ellis 1978). It has been found as deep as 30 m in Europe. In North America it occurs in large (e.g., St. Lawrence, Delaware) and small (e.g., Ottawa) rivers and large lakes (e.g., Great Lakes). It is not common, or perhaps has not been dispersed to smaller habitats. It has very contagious distribution patterns, in mud, sand, or gravel, and is only moderately abundant when it does occur.

Dispersal mechanisms: Fish, waterfowl, and insects.

Adaptations: None apparently unique to the species, but it appears to possess most of the adaptations common to all sphaeriids, including native species, as discussed in the introduction.

Predators: Fish and waterfowl and some glossophoniid leeches.

Parasites: Mostly digean trematodes, especially *Crepidostomum farionis* (Mackie 1976). Other intermediate hosts are mayflies and amphipods. The definitive host of the parasite is salmon.

Reproductive potential and life cycle: Vincent et al. (1981) have examined growth and reproduction of the species in the St. Lawrence River, but its reproductive potential falls within that of native species (described in introduction).

Food and feeding habits: Same as other sphaeriids.

Impact potential: Its ecological and socioeconomic impacts are unknown, but its low densities in most populations suggest it is not a good competitor with most native sphaeriids.

Control: None, since there is no evidence to suggest that the species requires immediate control.

Pisidium henslowanum (Henslow's pea clam)

This species is common and generally distributed in Scandinavia. France, Germany, Belgium, former U.S.S.R., and Britain (Woodward 1913; Ellis 1978). It is relatively uncommon in North America, but is generally distributed in the Laurentian Great Lakes, where it was probably introduced. Most records of its occurrence are in rivers and lakes in states bordering the Great Lakes. Harris (1973) gives an anomalous record of this species in western Canada, but it is probably *Pisidium supinum*, because he based the identification on Herrington's (1962) description for *Pisidium henslowanum* form *supinum*.

Species characteristics: Shell small, attaining 4 mm in shell length maximum; beak in most specimens has a short, concentric ridge; dorsal margin short and curved, somewhat alate at anterior and posterior margins; shell surface silky to glossy, with rather coarse striae.

Mode of life and habitat: In Europe the species inhabits streams, rivers, canals, lakes, and ponds, usually in fine sediments (Ellis 1978).

Dispersal mechanisms: Insects, fish, waterfowl, and aquatic mammals.

Adaptations: Probably possesses all or most of the features described in the introduction to the Sphaeriidae family.

Predators: Fish and waterfowl and some glossiphoniid leeches.

Parasites: None reported to date, but the species is probably an intermediate host of digean trematodes, as are most sphaeriids.

Reproductive potential and life cycle: The reproductive potential of the species has not been examined for North American populations, but in Europe it has average sphaeriid reproductive features, with 1–10 larvae per parent and one to two reproductive periods per year.

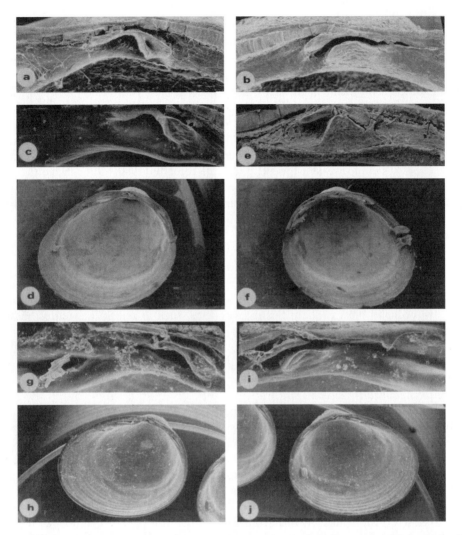

FIGURE 15.6   Hinge features of *Pisidium henslowanum*. a = typical C3, b = typical C2 (lower tooth) and C4, c and d = C3 and right hinge plate, e and f = C2 and C4 and left hinge plate of a variant; same sequence in g–j of another variation in form.

Food and feeding habits: Same as other sphaeriids — bacteria, diatoms and other algae, and organic detritus.

Impact potential: It appears to have little ecological significance other than competing with other sphaeriids and increasing the diversity of the macroinvertebrate community.

Control: None, since there is no evidence to suggest that the species requires immediate control.

Family: Dreissenidae.
Shell mytiliform, shell solid in color or with yellow, brown, and/or black stripes; byssate, with byssal opening on ventral side in anterior half of shell; ventral side flat, convex or concave; hinge teeth reduced to absent; heteromyarian (posterior adductor muscle much larger than anterior adductor muscle); oviparous, with free-swimming veliger larvae.

Dreissena polymorpha (zebra mussel)

The species was found and described for the first time in the northern part of the Caspian Sea and in the Ural River by Pallas in 1769 (Mackie et al. 1989). In the early 19th century the geographical range of *D. polymorpha* was dramatically extended: to Hungary in 1794; then a rapid invasion of Britain, first in Cambridgeshire in the 1820s, London in 1824, Yorkshire 1831–33, Forth and Clyde Canals in 1833, and in the Union Canal near Edinburgh in 1834, with numerous other records from 1835 onward (Morton 1993); the invasion was just as rapid in Germany, to Rotterdam in 1827, Hamburg in 1830, and Copenhagen in 1840 (Morton 1979). The invasion of the zebra mussel throughout the USSR was just as startling; from the Davina River Basin in 1845, through the Mariinsk Canal system to western Europe, and more recently (1960s) through the Oginski and Moscow Canals (Mackie et al. 1989). Extension of its range in the USSR is still occurring at a rapid rate. The species appeared in the Scandinavias in the 1940s and has since appeared in numerous Swiss lakes (Mackie et al. 1989). It is now in Italy, Finland, Ireland, and the Iranian coast of the Caspian Sea (Mackie et al. 1989).

The zebra mussel was first introduced into Lake St. Clair in 1985 or 1986 via freshwater ballast from an oceangoing vessel (Hebert et al. 1989) and spread to all the Great Lakes within 2 years. By 1991 it had spread outside the Great Lakes drainage systems to the Mississippi River and within 2 years the species had migrated downstream to its current southern limit at New Orleans. The species now occupies most of the major river systems, such as the St. Lawrence, Mississippi, Missouri, Ohio, Illinois, Tennessee, Cumberland, and Hudson Rivers.

Species characteristics: Shell typically with yellow, brown, and black stripes, occasionally with few or no stripes; byssal opening on ventral side in anterior third of shell; ventral side flat to concave (compare to *Dreissena bugensis*).

Mode of life and habitat: Large freshwater lakes and rivers are the favorite habitats of dreissenids (Strayer, 1991), but they will also do well in cooling ponds, quarries, and irrigation ponds of golf courses (Mackie and Schloesser 1996). Zebra mussels are capable of living in brackish water or estuaries where the salinity does not exceed 8 to 12 ppt (Strayer and Smith 1993, Kilgour et al. 1994; Strayer et al. 1996). The incipient lethal salinity for zebra mussel post-veligers is near 2 ppt and for adults (5–15 mm) it is between 2 and 4 ppt (Kilgour et al. 1994). These studies are supported by Strayer et al. (1996), who show that zebra mussels occur in estuarine portions of the Hudson River where salinities range between 3 and 5 ppt. Stanczykowska (1964, 1976, 1977) and Stanczykowska et al. (1976) relate biomass and production of zebra mussels to several environmental variables and in numerous Polish lakes.

Dispersal mechanisms: Carlton (1993) describes 23 different potential mechanisms by which the species has been or can be dispersed in the larval and adult stages. These are divided into three natural (e.g., water currents, birds, and other animals) and 20 human-mediated mechanisms (e.g., ballast water, hulls of sailing vessels, etc.). Strayer (1991), Mellina and Rasmussen (1994), and Johnson and Carlton (1996) describe various mechanisms and conditions needed for the wide-scale spread of the zebra mussel. The great diversity of potential dispersal mechanisms has impressed upon us the near impossibility of preventing the spread of zebra mussels once they have been introduced. "It is not a question of IF it will get here, it is a question of WHEN it will get here" is now a common cliche for describing the dispersal powers of biofouling dreissenids.

While currents are an effective mechanism for dispersing planktonic larval stages, they are not effective for sustaining populations in streams. The byssal apparatus will help to maintain the position of the adults in the streams, but the flow of water will carry the larvae well downstream of the parent population, which will ultimately disappear unless an upstream population can rejuvenate the colony. The distance that the larvae are carried depends upon the velocity and the

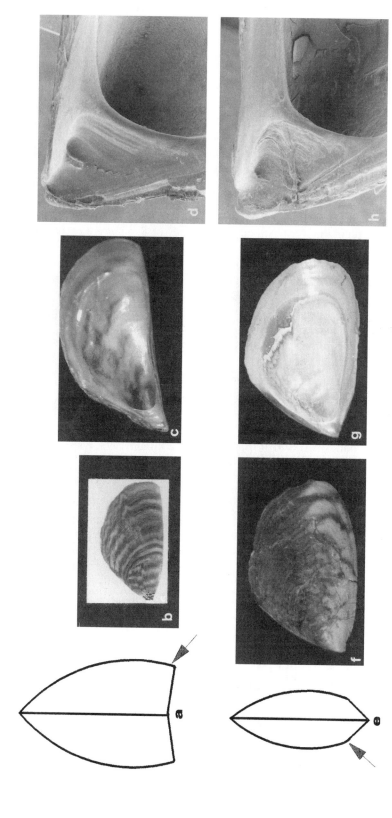

FIGURE 15.7 *Dreissena polymorpha* (a–d). a = end view showing flat-to-concave ventral surface with angulate lateral/ventral edge (arrow). *Dreissena bagensis* (e–h): e = end view showing convex ventral surface with rounded lateral/ventral edge (arrow).

duration of the planktonic stage. Since the planktonic stage of most dreissenids lasts 10 to 30 days (average 21 days) (Baldwin 1995), the larvae will be carried long distances before they are able to settle and attach to the substrate. Even in streams with low velocities (0.1 m/s) larvae are not able to swim against the current and will be carried downstream. As an example, and selecting a stream with a low current velocity (0.1 m/s) and larvae that develop and settle quickly in 10 days, the larvae will be carried 86.4 km downstream! (0.1 m/s × 1 km/1000 m × 10 days × 60 s/min × 60 min/h × 24 h/day = 86.4 km, or 8.64 km/day). Hence, dreissenids introduced to a stream will survive one life span at best, unless they are dispersed again to the same site, which is not likely if there are no adults upstream, or there is an impoundment to retain the planktonic larvae. Indeed, reservoirs will serve to provide breeding habitats for establishing and maintaining populations downstream. Once rivers become slow enough and currents are such that the position of the developing larvae can be maintained up to and including settlement, the populations can be self sustained by the adults attached to the bottom. Indeed, without mainstream reservoirs, dreissenids would not succeed (or at least would not be as great a pest) in most rivers in North America.

Adaptations: McMahon (1996) and Dietz et al. (1996) provide exhaustive reviews of the physiological ecology and ionic regulation of the North American zebra mussel. Readers are encouraged to review those papers for much more detail than can be provided here.

One adaptation often overlooked is the zebra mussel's shell morphology, which allows the species to exploit an epifaunal existence in fresh waters. The ventral surface is flat, and with help from the byssal apparatus, allows the species to be pulled tightly against the surface of the substrate. With a triangular shape (in end view), it makes it very difficult for a predator to grasp and pull the mussel from the surface. The zebra mussel is the only freshwater bivalve that has this adaptation.

Other adaptations include abilities to survive periods of desiccation, turbidity and, to a lesser extent, hypoxia. Quagga and zebra mussels can survive only 5 and 13 days, respectively, in low humidities, while Asian clams can survive up to 2 to 4 weeks in low humidities (Ussery and McMahon 1994). The dreissenid mussels succumb to accumulations of toxic anaeorobic end-products during desiccation (Ussery and McMahon 1994). The time to death of zebra mussels is even more rapid at freezing temperatures, occurring in less than 24 h at $-3°C$ (= 27°F) (Clarke et al. 1993).

Of all the physiological parameters of importance to exotic molluscs, tolerances to turbidity is the least understood. Associated with increasing loads of suspended particles is increasing sedimentation rates, especially in lakes or streams with depositional current velocities. One effect of turbidity is a depression of respiratory rate, with normal levels of turbidity having little or no effect on respiration rate (Aldridge et al. 1987). However, in *D. polymorpha*, the respiration rate is relatively depressed at low turbidities (5 NTU Alexander et al. 1994). But Summers et al. (1996) showed that both zebra and quagga mussels can partially acclimate to turbid water conditions by adjusting their metabolic rate. This occurs when mussels are allowed to acclimate at a relatively high turbidity (80 NTU) for about 4 weeks. Payne et al. (1995) showed that there are ecophenotypic differences within populations of *D. polymorpha* in which populations in habitats characterized by high suspended loads showed a marked increase in labial palp to gill area ratios.

Also, increased turbidities reduce light penetration resulting in lower primary production. Therefore, nongrazing and infaunal forms of exotic molluscs will be favored over grazing and epifaunal forms. As a result, gastropods will be replaced by filter-feeding bivalves. Although filter-feeding rates may be depressed in waters with extremely high silt loads (Sprung and Rose 1988), much of the inorganic material (i.e., silt) is captured by gill cilia and mucous and is rejected as pseudofaeces. The onset of pseudofaeces production indicates overload of the filtration apparatus (Lei et al. 1996). Horgan and Mills (1997) found that the population filtration rate depends on individual clearance rates (volume of water cleared of particles per unit time), the percentage of mussels filtering and the time of day, with filtering activity being about 6%–9% higher at night than during the day; however, they observed no diel change in clearance rates. Clearance rate also depends on mussel size but filtering activity does not (Horgan and Mills 1997).

Studies in European lakes and canals indicate that *Dreissena* plays a significant role in processes of biological self-purification and improvement in water quality in aquatic systems (Mackie et al. 1989). This is not a consequence of a high individual filtration capacity, but rather to the enormous numbers of mussels that usually prevail in aquatic systems. As bivalves, the zebra mussel's filtration rate (10–100 ml/ind./h) is intermediate between Unionidae (60–490 ml/ind./h) and Sphaeriidae (0.6–8.3 ml/ind./h) (Hinz and Schiel 1972; Stanczykowska et al. 1976). More recently, Summers et al. (1996) have shown that size, temperature, acclimation turbidity, and ambient turbidity significantly affect oxygen consumption rates and that zebra mussels adjust their metabolic rate in response to chronic exposure turbidity.

In areas heavily populated by zebra mussels, large biomasses of pseudofaeces accumulate on the bottom, often causing a shift in energy from the pelagic zone to the benthic zone (Griffiths 1993). This has implications on altering concepts in reservoir ecology, especially pulse stability and successional changes. Dreissenids may add to the duration and/or the magnitude of pulse stability because their population densities exhibit their own pulse stabilities; the filtering activities of enormous numbers of dreissenids is bound to affect the pulse stability of a reservoir. Similarly with succession, dreissenids dramatically accelerate the sedimentation rate of suspended particles and the rates of trophic changes that normally occur in new impoundments could be dramatically altered.

The thermal tolerances of biofouling dreissenids are well described. For zebra mussels the chronic lethal temperature is 34–37°C (McMahon et al. 1993), while the acute lethal temperature ranges from about 33°C (Jenner and Janssen-Mommen 1993) to 42.3°C (Neuhauser et al. 1993). However, previous acclimation temperature greatly affects both the acute and chronic lethal temperatures (Iwanyzki and McCauley 1993; McMahon et al. 1993). The tolerance times at different temperatures increases with increasing acclimation temperature and decreasing shell size and decreases with increasing exposure temperature (McMahon et al. 1993).

Nearly all the thermal tolerance studies were conducted to determine the feasibility of using thermal stress to control biofoulers. Epilimnial temperatures in lakes and reservoirs are usually higher than temperatures in the parent streams, but the hypolimnial temperatures are colder than parent stream temperatures. In general, molluscs, like all other aquatic organisms in impoundments, adjust their upper thermal tolerance limits by acclimating to increasing summer temperatures as the summer progresses. Most molluscs have the ability to seek a preferred temperature and can usually find it in thermally stratified reservoirs. This includes even the byssate forms because they can translocate by releasing themselves from their byssal attachment and resettle in a more suitable thermal regime. Although their new position will depend on current patterns in the reservoir, most will resettle in a more downstream position.

Dreissenids cannot tolerate even short periods of anoxia but can survive short periods of hypoxia, down to 2 mg/l for a few days (Mackie et al. 1989).

Predators: The dreissenids have planktonic larval stages that are preyed upon mainly by crustacean zooplankton (e.g., cyclops) and larval fish. The relative importance of these prey groups to the total mortality of larval stages is unknown (Mackie et al. 1989). Conn and Conn (1993) reported predation of zebra mussel larvae by the cnidarian, *Hydra americana*. These and other hydrozoans are often present in large numbers attached to mussels and capture food in currents generated by the siphons of the mussels.

Adult zebra mussels have a high nutritional value of the tissues, with 60.7% protein, 19.0% carbohydrate, 12.0% lipid, and 5.9% ash (Cleven and Frenzel 1992) and are consumed in large quantities by crayfish, fish, and waterfowl (Mackie et al. 1989). The nutritional values of these food items change seasonally (Walz, 1978) so that prey selection also varies seasonally. In North America, predation by crayfish has been documented by Love and Savino (1993), MacIssac (1994) and Perry et al. (1997). Perry et al. (1997) suggest that streams containing *Orconectes* crayfish either will be less susceptible to invasion by dreissenids or at least the mussels will be maintained at lower densities than streams without crayfish. However, as discussed below, zebra and quagga mussels will not likely succeed in streams unless there is an impoundment upstream, as discussed above. The extent of predation by crayfish on zebra mussels appears to depend upon at least the size classes of mussels present and on colonization of mussels by microorganisms and a learning period for the crayfish (Hazlett 1994; MacIssac 1994).

The relative contribution of the fish species to total mortality by predation of zebra mussels is unknown. Zebra mussels have been found in the stomachs of freshwater drum, white suckers, walleye, yellow perch, bass, and a few others (French 1993), some of which are exotic species themselves, such as the round (*Neogobius melanostomus*) and tubenose gobies (*Proterorhinus marmoratus*) (Jude et al. 1995). The freshwater drum, *Aplodinotus grunniens*, has molariform pharyngeal teeth for crushing shells of the dreissenids (French and Bur 1993). The species is most common in large lakes and rivers, and its feeding habits has been investigated in detail by French and Bur (1993); they reported that predation on zebra mussels increases as drum size increases, with large drum feeding almost exclusively on zebra mussels. Nagelkerke and Sibbing (1996) showed that prey (zebra mussels) size is limited by the crushing power of the pharyngeal apparatus of common bream and white bream, and such species have a distinct advantage in competing for molluscs. Similar observations were made by Morrison et al. (1997), who observed size-selective predation in both young freshwater drum and yellow perch *(Perca flavescens)*, but they concluded that predation intensity by large fish was related more to the availability of mussels.

In North America, dreissenids form a major part of the diet of greater (*Aythya marila*) and lesser (*A. affinis*) scaup, buffleheads (*Bucephala albeola*), oldsquaws (*Clangula hyemalis*), and white-winged scoters (*Melanitta deglandi*) (Hamilton 1992; Wormington and Leach 1993; Hamilton et al. 1994); the distribution of greater and lesser scaup at Long Point, Lake Erie appears to be influenced by the distribution of dreissenid beds (Hamilton et al. 1994). The impact of waterfowl on zebra mussel densities is variable, from little or no effect to significant decreases, up to 97% in zebra mussel biomass in some lakes in the Netherlands, especially in the winter months (Mackie et al. 1989).

There is a high probability that crayfish, fish, reptiles, waterfowl, and mammals will be important predators contributing to the total mortality of many species of exotic molluscs in North American lakes and streams. While some predators may limit the population sizes of some species of exotic gastropods, it is unlikely that any will limit the population sizes of the byssate biofoulers, or of *Corbicula*, because of their high reproductive potential.

Parasites: Dreissenids are not common vectors of parasites (Stanczykowska 1977). Protists and digeans are the most common parasites, with Nematoda observed sporadically in zebra mussels. Many protists are common parasites but they do not seem to "affect the numbers" of the zebra mussel (Stanczykowska 1977). Trematodes are less common parasites of zebra mussels than are protists, the greatest infestation rate observed being 10%. The most dangerous protists are ciliates of the family Ophryoglenidae which parasitize the digestive gland and may kill the mussel (Stanczykowska 1977). The only other parasitic ciliate reported in dreissenids is *Concophthirus*, but adverse effects on the bivalve host are unknown (Molloy et al. 1995).

Species of *Phyllodistomum* and *Bucephalus* are important parasites of dreissenids in lakes of the Netherlands (Davids and Kraak 1993); the infestation prevalence is usually about 1% and may go as high as 10%. The effects of parasites on *D. polymorpha* appear to be minimal, at least until high emissions of cercariae of *B. polymorphus* occur (Mackie et al. 1989). Intensity of parasitism by *P. folium* is directly correlated to shell size of *D. polymorpha*, the maximum number recorded being 200 at a shell size of 24–28 mm.

Toews et al. (1993) reported a 2.9% prevalence of plagiorchiid metacercariae in mussels and 2.7% prevalence of adults and juvenile aspidogastrids in mussels from two sites in Lake Erie. The ciliate, *Ophryoglena*, occurred with 1.3% prevalence at a site in Lake St. Clair and 2.7% to 4.3% prevalence at two sites in Lake Erie. Toews *et al.* (1993) concluded that mass development of zebra mussels may increase the infection rate in definitive hosts, especially fish and waterfowl. The oligochaete, *Chaetogaster limnaei*, and the chironomid, *Paratanytarsus*, have also been reported as commensals in zebra and quagga mussels (Conn et al. 1994, Ricciardi 1994).

The only other "parasitic group" reported is sponges. Ricciardi et al. (1995) recently reported lethal overgrowths of three sponge species, *Eunapius fragilis*, *Ephydatia muelleri*, and *Spongilla lacustris* on zebra mussels. The sponge colonies spread over the entire shell of the zebra mussel and smother their siphons. The sponges also inhibit settlement of dreissenids and out compete them for hard substrate (Miner et al. 1995).

Reproductive potential and life cycle **(Figure 15.8):** The most prolific mollusc species in freshwater are the introduced dreissenids, with over one million eggs per female each year in its 2 to 3 year life span (Nalepa and Schloesser 1993; Nichols 1996). As in European populations, there are one or two spawning seasons per year. The first lasts about three months (early to mid-May to early to mid-August) and is comprised of several spawning events; the second occurs in August or September (Claudi and Mackie 1994; Jantz 1996; Nichols 1996). If only one spawning season occurs, spawning peaks about the end of August, but several small spawning events may occur even into October in the Great Lakes.

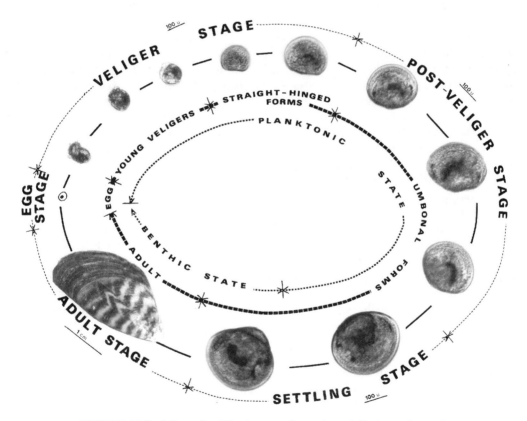

FIGURE 15.8   Life cycle of *Dreissena polymorpha* and *Dreissena bagensis*.

The larvae pass through several developmental stages, including trochophore, D-shape, veliconcha, pediveliger, and plantigrade, as described by Ackerman et al. (1994). However, up to 99% mortality of the larval stages may occur (Nalepa and Schloesser 1993). Most of the mortality may occur during the settling event if the plantigrade form does not find a suitable substrate on which to attach. In spite of this high mortality, the dreissenids are the most productive of all the exotic molluscs. Even *Corbicula* displays a lower fecundity than *Dreissena*, with 25,000 to 75,000 veligers produced in the lifetime of a single clam (Aldridge and McMahon 1978), compared to more than one million eggs produced annually by a single female zebra mussel (Nalepa and Schloesser 1993). Native species of Sphaeriidae have very low fecundities (e.g., 1–40 eggs per adult) because they are brooders (Mackie 1984). Since only a small number of larvae can be brooded by any one parent, the number of larvae that are produced is rather small. Hence, external fertilization and development partly explains why *Dreissena* is much more prolific than native species of bivalves in North American surface waters. The physiological aspects of zebra mussel maturation, spawning, and fertilization are reviewed by Ram et al. (1996). Jantz (1996) and Nichols (1996) review variations in the reproductive cycle of *D. polymorpha* in Europe, Russia, and North America.

Food and feeding habits: Adult zebra mussels feed on (filter) phytoplankton, protists, bacteria and other microscopic plankton and zooplankton (MacIssac et al. 1992; Bruner et al. 1994; Fanslow et al. 1995), including their own veligers (MacIssac et al. 1991). Mitchell et al. (1996) used stable isotopes of nitrogen and carbon to demonstrate that zebra mussels utilize the entire seston resource, whereas other filter feeders, such as *Daphnia*, use distinct sources of organic carbon. The mussels can filter particles less than 1 μm and are capable of filtering a wide range of bacteria ranging in size from 1–4 μm. The importance of bacteria in the zebra mussel's diet has been demonstrated recently by Silverman et al. (1996), who showed that zebra mussels have a more efficient gill ciliature than many native unionids and *Corbicula*.

The filtration rate of dreissenids is affected by their shell size, turbidity, temperature, and concentrations of certain sizes and kinds of algal cells (e.g., *Chlamydomonas*) and bacterial cells (see reviews in Neumann and Jenner 1992 and Lei et al. 1996). The filtration capability of *D. polymorpha* in relation to its role as a clarifier of water in an entire lake, epilimnion, or littoral zone has been reviewed by Neumann and Jenner (1992) for some European systems; they also describe applications of zebra mussels in water quality management, particularly as water clarifiers. Mackie and Wright (1993) have demonstrated that zebra mussels are not only capable of removing suspended solids but also are capable of significantly reducing biochemical oxygen demand and phosphorous levels in dilute (15%) activated sewage sludge.

Since the arrival of the zebra mussel, numerous studies have related increased water clarity to the removal of nutrients, algae, and seston by the filtering activities of zebra mussels (Makerewicz and Bertram 1991; Holland 1993; Leach 1993; Nicholls and Hopkins 1993; Fahnensteil et al. 1995a,b; Fanslow et al. 1995; Holland et al. 1995; MacIssac et al. 1995; Madenjian 1995; Klerks et al. 1996; MacIssac 1996). Although mussels are efficient clarifiers, the suspended materials accumulate on the bottom as faeces and pseudofaeces. The size of faecal and pseudofaecal pellets varies with mussel size and the settling velocities and rate of accumulation greatly exceed normal sedimentation processes (Dean 1994).

Impact potential: The ecological and socioeconomic impacts are the severest of all species introductions to date. Mackie (1995), Mackie et al. (1989), Beeton (1995), and MacIssac (1996) describe several different impacts. These can be divided into Industrial/Utility Impacts; Fisheries and Wildlife Impacts and other Biological Impacts; Navigational and Vessel Impacts; and Private Property Impacts.

Industrial/Utility Impacts

(1) *Industrial and domestic pipelines*: Accumulations cause (1) reductions in the bore of pipes; (2) reduced flow through the pipe due to friction loss (turbulent flow instead of laminar); (3) electro-corrosion of steel or cast iron pipes; (4) deposition of empty mussel shells at the pipe outlet; and (5) tainting and possible contamination of the water upon death (especially when killed as part of a massive control program). All have been documented in North American facilities.

(2) *Underground irrigation systems*: Some golf courses that draw water from the Great Lakes have reported this impact.

Fisheries and Wildlife and Other Biological Impacts

(3) *Fouling of fishing gear:* Trap nets and pound nets that are left in the water for extended periods of time, are especially commonly reported in Lake Erie.

(4) *Potential loss of commercial fisheries:* Fish species that feed on plankton will be affected through their filtering feeding activities of the zebra mussels. This has not yet been proved for Great Lakes fisheries, but conversations with commercial fishermen on Lake Erie and Lake St. Clair suggest that impacts are only now being manifested in smaller catches of perch, smelt, and walleye. This may be related, in part, to declines in some deepwater benthos. Dermott and Kerec (1997) suggest that recent declines in smelt populations are being enhanced by diversion of energy to large *D. bugensis* populations in eastern Lake Erie, both through induced changes to the biomass and composition of the plankton as well as reduced *Diporeia hoyi* populations. With the introduction of the zebra mussel, there has been a shift from a pelagic-dominated food web to a benthic-dominated food web. In some lakes (e.g., Lake St. Clair, Lake Ontario), there has been an increase in benthic biomass, especially in annelids, gastropods, amphipods, and crayfish (Griffiths 1993; Dermott et al. 1993; Stewart and Haynes 1994).

(5) *Elevated levels of contaminants:* Some *Dreissena* have elevated levels of organic and inorganic contaminants (Bruner et al 1994; Mazak 1995; Dobson and Mackie 1997) that are passed up the food chain. Biomagnification has resulted in reproductive problems such as reduced clutch sizes and high embryo mortality in ducks (de Kock and Bowmer 1993).

(6) *Reduced seston levels/increased water transparency:* The dense populations of zebra mussels in North American waters has resulted in the filtration of suspended materials (seston) from the water which has resulted in reduced phytoplankton biomass (Bunt et al. 1993; MacIssac 1996; Caraco et al. 1997) and increased water clarity and nutrient levels (Holland et al. 1995; MacIssac; Caraco et al. 1997). MacIssac (1996) reviews the potential negative impact of reduced levels of phytoplankton on planktivorous fish. However, there have been some beneficial impacts. Macrophyte and benthic algae biomass and diversity has increased (Griffiths 1993; Lowe and Pilsbury 1995; Skubinna et al. 1995; Stuckey and Moore 1995) and SCUBA divers have benefited from the increased water clarity.

(7) *Loss of endemic species of Unionidae:* Gillis and Mackie (1994) have shown that all native unionids on the Ontario shores of Lake St. Clair, where the mussel was first introduced, have been eliminated by the zebra mussel. Similar results have been documented for other North American lakes (mostly Great Lakes) and rivers by Nalepa et al. (1991), Hebert et al. (1991), Schloesser and Kovalak (1991), Hunter and Bailey (1992), Haag et al. (1993), Nalepa (1994), Schloesser and Nalepa (1994), Strayer et al. (1994), Tucker (1994), Ricciardi et al. (1995; 1996) and Schloesser et al. (1996). Ricciardi et al. (1996) suggest that significant declines in unionid densities occur when infestation levels exceed 10/unionid. Baker and Hornbach (1997), however, provide data to show that the decline in unionid diversity since the introduction of dreissenids is due, in part at least, to species-specific rates of starvation in unionids. Mastellar et al. (1993) were unable to show any impact of the zebra mussel on unionids, in Presque Isle Bay, Lake Erie; however, they considered the zebra mussel a threat to the continued existence of Unionidae in this unique habitat of Lake Erie.

(8) *Alteration of benthic community structure and abundance:* Changes in biomass of benthic algae and macrophytes (see (5) above) are probably followed by changes in the benthic invertebrate community, with grazers and herbivores showing a greater contribution to the total diversity of functional feeding groups. Numerous studies document changes in community structure and abundance in benthic invertebrates, in part due to increased loadings of faeces, pseudofaeces, and seston placed on the bottom by zebra mussel filtering activities (Griffiths 1993; Stewart and Haynes 1994; Botts et al. 1996; Howell et al. 1996; Ricciardi et al. 1997; Dermott and Kerec 1997) which ultimately affect fish community structure and abundance (MacIssac 1996).

(9) *Increased potential as vectors of parasites whose definitive hosts are commercially important species of fish and/or waterfowl:* This has not yet been demonstrated in North American waters.

Navigational and Vessel Impacts

(10) *Encrusting the hulls of boating and sailing vessels:* This is a common sight in the Great Lakes.

(11) *Encrusting of navigation buoys to the point that the buoys sink deeper in the water than normal:* This has also been demonstrated in the Great Lakes, especially for fishing markers/buoys.

Private Property Impacts

(12) *Formation of shoals of shell debris on beaches that will detract from the beach's recreational and aesthetic value:* This was a common sight on some large beaches on Lake Erie in 1990–1992.

(13) *Fouling of cottage plumbing and intake structures:* There are increasing reports that foot valves at the end of pipes in lakes and/or screens protecting the plumbing lines are being fouled by zebra mussels. In fact, some entrepreneurs in Canada and the U.S. have established a small market in sand filters and other devices for preventing biofouling of cottage plumbing structures.

Control: Claudi and Mackie (1994) review numerous methods that are categorized as either chemical control; physical control; mechanical control or biological control. Table 15.1 summarizes the various methods tested or being tried in the Great Lakes or used in Europe.

Dreissena bugensis (quagga mussel) (Figure 15.7)

The species was first found in Lake Ontario in 1989 and may have been introduced at the same time as the zebra mussel but, because it can live in deeper water than the zebra mussel, the species may not have been discovered until later. Its ecological and socioeconomic impacts are similar to the zebra mussel. Mackie and Schloesser (1996) and Marsden et al. (1996) compare the genetic, biological and ecological characteristics of both dreissenid species and Mills et al. (1996) compare the ecological and biological characteristics of the species in Europe and North America.

Species characteristics: Shell ornamentations very similar to the zebra mussel, but ventral surface is convex and the byssal opening is more distal in position.

Mode of life and habitat: The mode of life and habitat of quagga mussels is similar to zebra mussels, with two major exceptions. It can live as an infaunal species in littoral and profundal sediments (Claxton and Mackie 1995) and it can live as an epifaunal or infaunal species in either the littoral or profundal zones of lakes.

Dreissena polymorpha has an exclusively epifaunal habit, but *D. bugensis* has an ability for an infaunal habit in deeper waters and epifaunal habit in shallower waters (Mackie 1995). In the Great Lakes the ratio of quagga to zebra mussels increases with increasing depth down to about 50 m (Mackie 1995). The greatest densities of both species in Lake Erie occur at about 12 m. Only quagga mussels occur in Ekman grab samples taken at these depths (Dermott 1993). It appears that quagga mussels are able to alter their shell allometry for an infaunal existence; they are thinner and lighter in weight in soft profundal sediments than on firm inshore substrates, whereas zebra mussels exhibit little or no changes in shell allometry on firm or soft substrates (Claxton and Mackie 1995).

With the adaptability of quagga mussels to life in soft sediments at great depths, provided oxygen is near saturation, dreissenids have the potential to invade more northern lakes than can the zebra mussel. Although dreissenids tend to develop greater biomasses in flowing waters (e.g., pipelines) of moderate velocities (<1.5 m/s), where their filter feeding efficiency can be enhanced by a constant delivery of food to the siphons, the planktonic larvae will prevent them from establishing large and permanent populations in natural streams and rivers that lack impoundments.

Dispersal mechanisms: Same as the zebra mussel.

Adaptations: The species can grow and reproduce at lower temperatures (<9°C) than the zebra mussel (8–10°C for growth, 12–15°C for reproduction) and therefore has a greater potential to infest more northern lakes than the zebra mussel (Claxton et al. 1997). Mitchell et al. (1996) found that quagga mussels were more abundant at thermally enriched sites than at sites unaffected by thermal discharges and dispute claims (e.g., Domm et al. 1993) that zebra mussels can tolerate higher temperatures than quagga mussels. Mitchell et al. (1996) suggest that quagga mussels prefer the relatively constant temperatures characteristic of deep water (e.g., 1–4°C) than the highly variable temperatures found in shallow waters. However, their studies were performed in waters <12 m where temperatures are just as variable as in shallow waters.

Apparently quagga mussels have a lower upper thermal limit and a greater instantaneous mortality rate across acclimation temperatures than do zebra mussels (Domm et al. 1993). However, McMahon et al. (1994) found no difference in instantaneous temperatures required to cause 100% mortality in each species. Quagga mussels also have the same functional response as zebra mussels to turbidity, acclimation turbidity, and ambient turbidity, and both respire normally in turbidities as high as 80 NTU when previously acclimated to high suspended loads (Summers et al. 1996).

Predators: Most, or all, of the discussion for the zebra mussel applies for the quagga mussel.

Parasites: Similar to the zebra mussel, with the naidid oligochaete, *Chaetogaster limnaei*, and various nematode species constituting the majority of parasites. Only *C. limnaei* appeared to cause any pathology where some worms feed on oocytes of females (Conn et al. 1994).

Reproductive potential and life cycle: Much greater reproductive potential than the zebra mussel because it can reproduce at lower temperatures, as low as 4°C (Claxton, University of Guelph, pers. comm.) and grow at lower temperatures (4°C). It is also capable of an infaunal existence, which means it can occupy a greater percentage of the bottom of lakes than can the zebra mussel.

Food and feeding habits: See discussion for the zebra mussel.

Impact potential: It has the same impact potential as the zebra mussel, but because it can live as an infaunal species in profundal sediments (Claxton and Mackie 1995), its ecological impacts will be extended to deeper water and to lakes further north than will the zebra mussel.

Control: It has the same control strategies as the zebra mussel. However, *D. bugensis* exhibits greater mortality than *D. polymorpha* in desiccation experiments when both species are exposed to 95% relative humidities (Ussery and McMahon 1995).

SUMMARY OF IMPACT POTENTIAL OF SPECIES INTRODUCED VIA BALLAST WATER

SPECIES WITH LITTLE OR NO ECOLOGICAL IMPACT

Most introduced species will have little or no obvious impacts on the ecology of the ecosystem. Although most freshwater molluscs are intermediate hosts of numerous species of trematodes whose definitive hosts are fish or waterfowl, there are few studies that have assessed the added impact of introduced molluscs with their trematodes on definitive hosts. It is not anticipated that the exotic species of molluscs will add significantly to the infestation of definitive hosts that are already infected by trematodes of native species of molluscs. All exotic species will have to compete with their native counterparts for food and space but very few (*Bythinia tentaculata* may be an exception) are known to have replaced native species; rather they seem to have added to the diversity of molluscan fauna. Eight of the ten species are included in this group:

Potamopyrgus antipodarum
Valvata piscinalis
Bythinia tentaculata tentaculata (with some exceptions)
Radix auricularia
Sphaerium corneum
Pisidium amnicum
Pisidium supinum
Pisidium henslowanum

SPECIES OF POTENTIAL SOCIOECONOMIC BENEFIT

None of the species introduced via ballast water are known (documented) to have any beneficial socioeconomic uses.

SPECIES OF POTENTIAL ECOLOGICAL AND/OR SOCIOECONOMIC HARM

Several types of potentially harmful socioeconomic impacts are recognized here. These include Vectors of Human Parasites; Agricultural Impacts; Industrial and/or Utility Impacts; Navigational Impacts; and Recreational Impacts. The ecological impacts can be divided into Biological Impacts and Limnological Impacts.

The only species introduced via ballast water with demonstrated socioeconomic and ecological impacts are *D. Polymorpha* and *D. Bugensis*. These impacts include the following:

1. Industrial and Utility Impacts: Accumulations of *Dreissena* species are known to cause (i) reductions in the bore of pipes; (ii) reduced flow through the pipe due to friction loss (turbulent flow instead of laminar); (iii) electro-corrosion of steel or cast iron pipes; (iv) deposition of empty mussel shells at the pipe outlet; and (v) tainting and possible contamination of the water upon death (especially when killed as part of a massive control program).
2. Navigational Impacts: If the reservoir is on a major shipping route or is used by barges, *Dreissena* can encrust the hulls of boats and sailing vessels and encrust navigation buoys to the point that the buoys will sink deeper in the water than normal.
3. Recreational Impacts: Most reservoirs have some recreational value, including swimming, sailing, and fishing. *Dreissena* is known to cause formations of shoals of shell debris on beaches that will detract from the beach's recreational and aesthetic value. For example, broken shells are sharp and cut the feet of swimmers. Many beaches on Lake Erie have signs posted recommending the use of sandals to avoid cuts on feet. The impacts of dreissenids on sailing and fishing are described above under Navigational Impacts and below under Biological Impacts.
4. Ecological Impacts: Included among the most severe of biological impacts are losses of native bivalves. The Mississippi River and Illinois River, in particular, are valuable resources for Unionidae, which are harvested for the Pacific cultured pearl industry. Dreissenids have been shown to cause losses of endemic species of Unionidae through a variety of mechanisms (Mackie 1991; Gillis and Mackie 1994). *Dreissena*, occurring in enormous numbers, also increases the potential of parasites finding an intermediate

host, thus increasing the risk for parasitizing commercially important species of fish and/or waterfowl. Losses of commercial fisheries, especially fish species that feed on plankton removed by the filtering feeding activities of zebra mussels, can also be anticipated. These losses can also occur through fouling of fishing gear, especially trap nets and pound nets that are left in the water for extended periods of time. Increases in dreissenid biomasses may be accompanied by decreases in biomasses of sphaeriids and some gastropods and chironomids (Dermott et al. 1993; Dermott and Kerec 1995) and increases in biomasses of amphipods, flatworms, oligochaetes, and some gastropods and chironomids (Griffiths 1993). However, recent studies by Dermott and Kerec (1995) show that increases in abundance of *D. bugensis* have been followed by decreases in the abundance of the burrowing amphipod, *Diporeia*, which in turn has been accompanied by declines of smelt which feed on the profundal amphipod. The increases in light transmission (see below) has been accompanied by increases in biomasses of sublittoral macrophytes.

Of the *Limnological Impacts,* because of their efficient filtering activities, large numbers of *Dreissena* can cause rapid changes in the limnological features of lakes and impoundments. This includes (a) increases in water clarity, where Secchi depths doubled within three or four years of invasion, with a concomitant lowering of phosphorous and chlorophyll *a* levels (Leach 1993); (b) increases in levels of soluble reactive phosphorous, silica, nitrate nitrogen and ammonia nitrogen (Beeton 1995); and (c) accelerated sedimentation rates of particulates and biodeposits and inorganic and organic contaminants bound to the particulates (Dobson 1994; Dean 1994; Roditi et al. 1997). Nalepa and colleagues discuss in detail these and other limnological impacts of dreissenid mussels in Chapter 16.

TABLE 15.1
Summary of Mollusc Species Introduced by Ballast Water Discharge and Their Origin, Date of Entry, First Documented Location in North America, and Reported Impact(s)

Species	Origin	Date	First Location	Impact(s)
Potamopyrgus antipodarum	New Zealand or Britain	1991	Lake Ontario	None reported to date
Valvata piscinalis	Europe	1913	Lake Ontario	None reported to date
Bithynia t. tentaculata	Europe	1871	Lake Michigan	Perhaps a competitor with pleurocerids
Radix auricularia	Europe	1913	Montana	None reported to date
Sphaerium corneum	Europe	1924	Lake Ontario	None reported to date
Pisidium supinum	Europe	?	Great Lakes	None reported to date
Pisidium amnicum	Europe	mid-1800s	Hamilton Bay, Lake Ontario	None reported to date
Pisidium henslowanum	Europe	mid 1900s	Great Lakes	None reported to date
Dreissena polymorpha	Aral/Caspian Sea area	1985	Lake St. Clair	Numerous impacts to industries and utilities, irrigation systems, navigation, private property, ecology and limnology of lakes
Dreissena bugensis	Aral/Caspian Sea area	1990	Lake Ontario	

SUMMARY

Table 15.1 summarizes the country of origin, date of entry, first documented location in North America, and their reported impact(s), when known, for molluscs introduced via ballast water discharge. As indicated in the text, *Pisidium supinum* may not be an exotic species because there is some fossil evidence that the species is native to Canada.

REFERENCES

Ackerman, J.D., B. Sim, S.J. Nichols, and R. Claudi. 1994. A review of the early life history of zebra mussels (*Dreissena polymorpha*): comparisons with marine bivalves. *Canad. J. Zool.* 72: 1169–1179.

Adam, W. 1960. Faune de Belgique. Mollusques. Tome 1. Mollusques terrestres et dulcicoles. L'Institue Royal des Sciences Naturelles de Belgique, Bruxelles. pp. 337–363.

Adamstone, F.B. 1924. The distribution and economic importance of the bottom fauna of Lake Nipigon with an appendix on the bottom fauna of Lake Ontario. Toronto University Studies, Biology Series No. 25, Ontario Fisheries Research Laboratory Publications 24: 34-100.

Aldridge, D.W. 1983. Physiological ecology of freshwater prosobranchs. In W.D. Russel-Hunter (Ed.), *The Mollusca*, vol. 6, *Ecology*. Academic Press, New York, pp. 330–359.

Aldridge, D.W. and R.F. McMahon. 1978. Growth, fecundity and bioenergetics in a natural population of the Asiatic clam, *Corbicula manilensis* Philippi, from north central Texas. *J. Moll. Stud.* 44: 49–70.

Alexander, J.E., J.H. Thorp, and R.D. Fell. 1994. Turbidity and temperature effects on oxygen consumption in the zebra mussel (*Dreissena polymorpha*). *Canad. J. Fisher. Aquat. Sci.* 51: 179–184.

Baker, F.C. 1913. A new *Lymnaea* from Montana. *Nautilus* 26: 115–116.

Baker, F.C. 1916. The freshwater Mollusca of Oneida Lake, New York. *Nautilus* 30: 5–9.

Baker, F.C. 1945. *The Molluscan Family* Planorbidae. The University of Illinois Press, Urbana.

Baker, S.M. and D.J. Hornbach. 1997. Acute physiological effects of zebra mussel *(Dreissena polymorpha)* infestation on two unionid mussels, *Actinonais ligamentina* and *Amblema plicata. Canad. J. Fisher. Aquat. Sci.* 54: 512–519.

Baldwin, B. 1995. Settlement and metamorophosis of larval zebra and quagga mussels: implications for their colonization and spread. Abstract, *The Fifth International Zebra Mussel & Other Aquatic Nuisance Organisms Conference*, Toronto, Ontario, Feb. 21–24, 1995, p. 69.

Beeton, A.M. 1995. Ecosystem impacts of the zebra mussel, *Dreissena polymorpha*. Abstract, *The Fifth International Zebra Mussel & Other Aquatic Nuisance Organisms Conference*, Toronto, Ontario, Feb. 21–24, 1995, p. 3.

Berry, E.G. 1943. The Amnicolidae of Michigan: Distribution, ecology, and taxonomy. *Misc. Publ. Mus. Zool. Univ. Mich.* 57: 1–68.

Botts, P.S., B.A. Patterson and D.W. Schloesser. 1996. Zebra mussel effects on benthic invertebrates: physical or abiotic? *J. N. Amer. Benthol. Soc.* 15: 179–184.

Boycott, A.E. 1936. The habitats of freshwater Mollusca in Britain. *J. Anim. Ecol.* 5: 116–186.

Brendelberger, H. 1995. Growth of juvenile *Bythinia tentaculata* (Prosobranchia, Bythiniidae) under different food regimes: A long-term laboratory study. *J. Moll. Stud.* 61: 89–95.

Brendelberger, H. and S. Jurgens. 1993. Suspension feeding in *Bythinia tentaculata* (Prosobranchia, Bythiniidae), as affected by body size, food and temperature. *Oecologia* 94: 36-42.

Bruner, K.A., S.W. Fisher, and P.F. Landrum. 1994. The role of the zebra mussel *Dreissena polymorpha*, in contaminant cycling: II. Zebra mussel contaminant accumulation from algae and suspended particles, and transfer to the benthic invertebrate, *Gammarus fasciatus. J. Great Lakes Res.* 20: 735–750.

Bunt, C.M., H.J. MacIssac and W.G. Sprules. 1993. Pumping rates and projected filtering impacts of juvenile zebra mussels (*Dreissena polymorpha*) in western Lake Erie. *Canad. J. Fisher. Aquat. Sci.*

Burch, J.B. 1975a. *Freshwater Unionacean Clams (Mollusca: Pelecypoda) of North America*. Malacological Publ., Hamburg, MI.

Burch, J.B. 1975b. *Freshwater Sphaeriacean Clams (Mollusca: Pelecypoda) of North America*. Malacological Publ., Hamburg, MI.

Burch, J.B. 1989. *North American Freshwater Snails*. Malacological Publ., Hamburg, MI.

Burky, A.J. 1983. Physiological ecology of freshwater bivalves. In: W.D. Russel-Hunter (Ed.), *The Mollusca*, vol. 6, *Ecology*. Academic Press, New York. pp. 281–329.

Calow, P. 1978. The evolution of life cycle strategies in freshwater gastropods. *Malacologia* 17: 351–364.

Calow, P. and L.J. Calow. 1975. Cellulase activity and niche separation in freshwater gastropods. *Nature* 255: 478–480.

Caraco, N.F., J.J. Raymond, D.L. Strayer, M.L. Pace, S.E.G. Findlay, and D.T. Fischer. 1997. Zebra mussel invasion in a large, turbid river: Phytoplankton response to increased grazing. *Ecology* 78: 588–602.

Carlton, J.T. 1993. Dispersal mechanisms of the zebra mussel. pp. 677–697. In: T.F. Nalepa and D.W. Schoesser (Eds.). *Zebra Mussels: Biology, Impacts, and Control*. Lewis/CRC Press, Boca Raton, FL.

Cejka, T. 1994. First record of the New Zealand mollusc Potamopyrgus antipodarum (Gray 1843) (Gastropoda, Hydrobiidae) from the Slovak section of the Dunaj River. Biologia, *Bratislava* 49: 657–658.

Claudi, R. and G. L. Mackie. 1994. *Practical Manual for Zebra Mussel Monitoring and Control*. Lewis, Boca Raton, FL. 227 p.

Claxton, T. and G. L. Mackie. 1995. Determining the colonization success of zebra and quagga mussels at different depths in Long Point Bay, Lake Erie. Abstract, *The Fifth International Zebra Mussel & Other Aquatic Nuisance Organisms Conference*, Toronto, Ontario, Feb. 21–24, 1995, p. 115.

Clarke, A.H. 1973. The freshwater molluscs of the Canadian interior basin. *Malacologia* 13: 1–509.

Clarke, A.H. 1980. *The freshwater molluscs of Canada*. National Museums of Canada, Ottawa, Ontario.

Clarke, M., R.F. McMahon, A.C. Miller, and B.S. Payne. 1993. Tissue freezing points and time for complete mortality on exposure to freezing air temperatures in the zebra mussel (*Dreissena polymorpha*) with special reference to dewatering during freezing conditions as a mitigation strategy. In: J.L. Tsou and Y.G. Mussalli (Eds.), *Proceedings: Third International Zebra Mussel Conference*, 1993, Toronto, Ontario. Electric Power Research Institute, Palo Alto, CA, Publ. No. TR-102077, pp. 4-119 – 4-145.

Cleven, E. and P. Frenzel. 1992. Population dynamics and production of *Dreissena polymorpha* in the River Seerhein, the outlet of Lake Constance. In: D. Neumann and H.A. Jenner (Eds.), *The Zebra Mussel Dreissena Polymorpha*. Gustav Fischer Verlag, New York. pp. 45–47.

Conn, D.B., M.N. Babapulle, K.A. Klein, and D.A. Rosen. 1994. Invading the invaders: Infestation of zebra mussels by native parasites in the St. Lawrence River. In: *4th International Zebra Mussel Conference '94 Proceedings*, Madison, Wisconsin. pp. 515–523.

Conn, D.B. and D.A. Conn. 1993. Parasitism, predation, and other associations between dreissenid mussels and native animals in the St. Lawrence River. In: J.L. Tsou and Y.G. Mussalli (Eds.), *Proceedings: Third International Zebra Mussel Conference*, 1993, Toronto, Ontario, Electric Power Research Institute, Palo Alto, CA, Publ. No. TR-102077, pp. 2-25–2-34.

Cotton, B.C. 1942. Some Australian freshwater Gastropoda. *Trans. Royal Soc. South Australia* 66: 75–82.

Davids, C. and H.S. Kraak. 1993. Trematode parasites of the zebra mussel (*Dreissena polymorpha*). In: T. F. Nalepa and D. W. Schloesser (Eds.). *Zebra Mussels: Biology, Impacts, and Control*. Lewis/CRC Press, Boca Raton, FL, pp. 749–760

de Kock, W.C. and C.T. Bowmer. 1993. Bioaccumulation, biological effects, and food chain transfer of contaminants in the zebra mussel (*Dreissena polymorpha*). In: T.F. Nalepa and D.W. Schloesser (Eds.). *Zebra Mussels: Biology, Impacts, and Control*. Lewis/CRC Press, Boca Raton, FL, pp. 503–533.

Dean, D.M. 1994. Investigations of biodeposition by *Dreissena polymorpha* and settling velocities of faeces and pseudofaeces. M.Sc. Dissertation, University of Guelph, Guelph, Ontario.

Dermott, R.M. 1993. Distribution and ecological impact of "quagga" mussels in the lower Great Lakes. In: J.L. Tsou and Y G. Mussalli (Eds.), *Proceedings: Third International Zebra Mussel Conference*, 1993, Toronto, Ontario. Published by Electric Power Research Institute, Palo Alto, CA, Publ. No. TR-102077, pp. 2-1 – 2-21.

Dermott, R.N. and D. Kerec. 1995. Changes in the deep-water benthos of eastern Lake Erie between 1979 and 1993. In: *The Fifth International Zebra Mussel & Other Aquatic Nuisance Organisms Conference*, Toronto, Ontario, Feb. 14–21, 1995, pp. 57–64.

Dermott, R.N. and D. Kerec. 1997. Changes in the deep-water benthos of eastern Lake Erie since the invasion of *Dreissena*. 1979–1993. *Canad. J. Fisher. Aquat. Sci.* 54: 922–930.

Dermott, R.N., J. Mitchell, I. Murray and E. Fear. 1993. Biomass and production of zebra mussels (*Dreissena polymorpha*) in shallow waters of northeastern Lake Erie. In: T.F. Nalepa and D.W. Schloesser (Eds.), *Zebra Mussels: Biology, Impacts, and Control*. Lewis, Boca Raton, FL, pp. 399–414.

Dietz, T.H., S.J. Wilcox, R.A. Byrne, J.W. Lynn, and H. Silverman. 1996. Osmotic and ionic regulation of North American zebra mussels (*Dreissena polymorpha*). *Amer. Zool.* 36: 364–372.

Dobson, E. 1994. Biodeposition and uptake of polychlorinated biphenyls and cadmium by the zebra mussel (*Dreissena polymorpha*). M.Sc. dissertation, University of Guelph, Guelph, Ontario.

Domm, S., R.W. McCauley, E. Kott, and J.D. Ackerman. 1993. Physiological and taxonomic separation of two dreissenid mussels in the Laurentian Great Lakes. *Canad. J. Fisher. Aquat. Sci.* 50: 2294–2298.

Ellis, A.E. 1978. *British Freshwater Bivalve Mollusca. Keys and Notes for the Identification of the Species.* Academic Press, London.

Fahnenstiel, G.L., G.A. Lang, T.F. Nalepa and T.H. Johengen. 1995a. Effects of the zebra mussel (*Dreissena polymorpha*) colonization on water quality parameters in Saginaw Bay, Lake Huron. *J. Great Lakes Res.* 21: 435–448.

Fahnenstiel, G.L., T.B. Bridgeman, G.A. Lang, M.J. McCormick, and T.F. Nalepa. 1995b. Phytoplankton productivity in Saginaw Bay, Lake Huron: effects of zebra mussel (*Dreissena polymorpha*) on natural seston from Saginaw Bay, Lake Huron. *J. Great Lakes Res.* 21: 465–475.

Fanslow, D.L., T.F. Nalepa and G.A. Lang. 1995. Filtration rates of the zebra mussel (*Dreissena polymorpha*) on natural seston from Saginaw Bay, Lake Huron. *J. Great Lakes Res.* 21: 489–500.

French III, J.R.P. 1993. How ell can fishes prey on zebra mussels in North America. *Fisheries* 18: 13–19.

French III, J.R.P. and M.T. Bur. 1993. Predation of the zebra mussel (*Dreissena polymorpha*) by freshwater drum in western Lake Erie. In: T.F. Nalepa and D.W. Schoesser (Eds.), *Zebra mussels: Biology, impacts and control.* Lewis/CRC Press, Boca Raton, FL, pp. 453–464.

Fuller, S.L.H. 1974. Clams and mussels (Mollusca: Bivalvia). In: Hart, C.W. and L.H. Fuller (Eds.), *Pollution Ecology of Freshwater Invertebrates.* pp. 215–274.

Gillis, P.L. and G.L. Mackie. 1994. The impact of *Dreissena polymorpha* on populations of Unionidae in Lake St. Clair. *Canad. J. Zool.* 72: 1260–1271.

Griffiths, R.W. 1993. Effects of zebra mussels (*Dreissena polymorpha*) on benthic fauna of Lake St. Clair. In: T.F. Nalepa and D.W. Schloesser (Eds.), *Zebra Mussels: Biology, Impacts, and Control.* Lewis, Boca Raton, FL, pp. 415–438.

Haag, W.R., D.J. Berg, D.W. Garton, and J.L. Farris. 1993. Reduced survival and fitness in native bivalves in response to fouling by the introduced zebra mussel (*Dreissena polymorpha*) in western Lake Erie. *Canad. J. Fisher. Aquat. Sci.* 50: 13–19.

Hanna, G.D. 1966. Introduced molluscs of western North America. *Occ. Papers Calif. Acad. Sci.* No. 48.

Harman, W.N. 1968. Replacement of pleurocerids by *Bythinia* in polluted waters of central New York. *Nautilus* 81: 77–83.

Harman, W. N. 1974. Snails (Mollusca: Gastropoda). In: Hart, C.W. and L.H. Fuller (Eds.), *Pollution Ecology of Freshwater Invertebrates.* pp. 275–313.

Harris, S.A. 1973. *Pisidium henslowanum* (Sheppard) in western Canada. *Nautilus* 87: 86–87.

Haynes, A. and B.J.R. Taylor. 1984. Food finding and food preference in *Potomopyrgus jenkinsi* (Smith E.A.) (Gastropoda: Prosobranchia). *Arch. Hydrobiol.* 100: 479–491.

Hazlett, B.A. 1994. Crayfish feeding responses to zebra mussels depend on micro-organisms and learning. *J. Chem. Ecol.* 20: 2623–2630.

Hebert, P.D.N., B.W. Muncaster, and G.L. Mackie. 1989. Ecological and genetic studies on *Dreissena polymorpha* (Pallas): A new mollusc in the Great Lakes. *Canad. J. Fisher. Aquat. Sci.* 46: 1587–1591.

Hebert, P.D.N., C.C. Wilson, M.H. Murdoch, and R. Lazar. 1991. Demography and ecological impacts of the invading mollusc *Dreissena polymorpha*. *Canad. J. Zool.* 59: 405–409.

Herrington, H.B. 1962. A revision of the Sphaeriidae of North America (Mollusca: Pelecypoda). *Misc. Publ. Mus. Zool.*, Univ. Michigan 118: 1–74.

Hinz W. and H.G. Scheil. 1972. Filtration rate of *Dreissena*, *Sphaerium* and *Pisidium* (Eulamellibranchiata). Oecologia 11, 45–54.

Holland, R.E. 1993. Changes in planktonic diatoms and water transparency in Hatchery Bay, Bass Island, western Lake Erie since the establishment of the zebra mussel. *J. Great Lakes Res.* 19: 717–724.

Holland, R.E., T.H. Johengen, and T.M. Beeton. 1995. Trends in nutrient concentration in Hatchery Bay, western Lake Erie, before and after *Dreissena polymorpha*. *Canad. J. Fisher. Aquat. Sci.* 52: 1202–1209.

Holopainen, I.J. 1987. Seasonal variation in survival time in anoxic water and glycogen content of *Sphaerium corneum* and *Pisidium amnicum* (Bivalvia: Pisidiidae). *Am. Malacol. Bull.* 5: 41–48.

Horgan, M.J. and E.L. Mills. 1997. Clearance rates and filtering activity of zebra mussel *(Dreissena polymorpha)*: implications for freshwater lakes. *Canad. J. Fisher. Aquat. Sci.* 54: 249–255.

Howell, E.T., C.H. Marvin, R.W. Bilyea, P.B. Kauss, and K. Somers. 1996. Changes in environmental conditions during *Dreissena* colonization of a monitoring station in eastern Lake Erie. *J. Great Lakes Res.* 22: 744–756.

Hubendick, B. 1950. The effectiveness of passive dispersal in *Hydrobia jenkinsi. Zool. Bidrag Uppsala* 28: 493–504. (Not seen, cited by Ponder 1988).

Hunter, R.D. 1975. Growth, fecundity and bioenergetics in three populations of *Lymnaea palustris* in upstate New York. *Ecology* 56: 50–63.

Hunter, R.D. and R.C. Bailey. 1992. *Dreissena polymorpha* (zebra mussel): Colonization of soft substrata and some effects on unionid bivalves. *Nautilus* 106: 60–70.

Iwanyzki, S. and R.W. McCauley. 1993. Upper lethal temperatures of adult zebra mussels. In: T.F. Nalepa and D.W. Schloesser (Eds.), *Zebra Mussels: Biology, Impacts, and Control.* Lewis, Boca Raton, FL, pp. 667–673.

Jantz, B. 1996. Wachstum, reproduktion, populationsentwicklung und beeintrachtigung der zebramuschel *(Dreissena polymorpha)* in einem groËen fieËgewasser, dem Rhein. Ph.D. dissertation (with some English chapters) Mathematisch-Naturwissenschaftlichen, Falkutat der Universitat zu Koln.

Jenner, H.A. and J.P.M. Janssen-Mommen. 1993. Monitoring and control of *Dreissena polymorpha* and other macrofouling bivalves in the Netherlands. In: T.F. Nalepa and D.W. Schloesser (Eds.), *Zebra Mussels: Biology, Impacts, and Control.* Lewis, Boca Raton, FL, pp. 537–554.

Johnson, R.K. and J.T. Carlton. 1996. Post-establishment spread in large-scale invasions: dispersal mechanisms of the zebra mussel *Dreissena polymorpha. Ecology* 77: 1686–1690.

Jokinen, E.H. 1983. The freshwater snails of Connecticut. *St. Geol. Nat. Hist. Surv. Connecticut*, Bull. No. 109.

Jude, D.J., J. Janssen, and G. Crawford. 1995. Ecology, distribution and impact of the newly introduced round and tubenose gobies on the biota of the St. Clair and Detroit Rivers. In: M. Munawar, T. Edsall, and J. Leach (Eds.), *The Lake Huron Ecosystem: Ecology, Fisheries and Management.* SPB Academic, Amsterdam, The Netherlands.

Kajak, Z. 1988. Considerations on benthos abundance in freshwaters, its factors and mechanisms. *Int. Rev. Ges. Hydrobiol.* 73: 5–19.

Kew, H.W. 1893. The dispersal of shells. Kegan Paul, French, French Trubner & Co., London.

Kilgour, B.W. and G.L. Mackie. 1990. Relationships between reproductive output and shell morphometrics of the pill clam, *Pisidium casertanum* (Bivalvia: Sphaeriidae). *Canad. J. Zool.* 68: 1568–1571.

Kilgour, B.W., G.L. Mackie, M.A. Baker, and R Keppel. 1994. Effects of Salinity on the condition and survival of zebra mussels *(Dreissena polymorpha). Estuaries* 17: 385–393.

Klerks, P.L., J.K. Holen, S.G. Monismith, and J.E. Cloern. 1993. Effects of zebra mussels *(Dreissena polymorpha)* on seston levels and sediment deposition in western Lake Erie. *Canad. J. Fisher. Aquat. Sci.* 53: 2284–2291.

Klerks, P.L., P.C. Fraleigh, and J.E. Lawniczak. 1996. Effects of zebra mussels *(Dreissena polymorpha)* on seston levels and sediment deposition in western Lake Erie. *Canad. J. Fisher. Aquat. Sci.* 53: 2281–2291.

Krapu, G.L. and G.A. Swanson. 1966. Some nutritional aspects of reproduction in prairie nesting pintails. *J. Wildl. Mgmt* 39: 156–162.

Lassen, H.H. 1979. The migration potential of freshwater snails exemplified by the dispersal of *Potamopyrgus jenkinsi. Natura Jut.* 20: 24.

Leach, J.H. 1993. Impacts of the zebra mussel *(Dreissena polymorpha)* on water quality and fish spawning reefs in western Lake Erie. In: T.F. Nalepa and D.W. Schloesser (Eds.), *Zebra mussels: Biology, impact and control.* Lewis, Boca Raton, FL, pp. 381–398.

Lei, J., B.S. Payne, and S.Y. Wang. 1996. Filtration dynamics of the zebra mussel, *Dreissena polymorpha. Canad. J. Fisher. Aquat. Sci.* 53: 29–37.

Lindeman, P.V. 1995. Comparative ecology of two map turtles, *Graptemys ouachitensis and G. pseudogeographica*, in Kentucky Lake. Abstract, *Contributed Papers at the Sixth Symposium on the Natural History of Lower Tennessee and Cumberland River Valleys.* Brandon Spring Group Camp, Land Between the Lakes, Mar. 3-4, 1995. (Available from The Center for Field Biology, Austin Peay State University, Clarkesville, TN 37044).

Love, J. and J.F. Savino. 1993. Crayfish *(Orconectes virilis)* predation on zebra mussels *(Dreissena polymorpha). J. Freshwat. Ecol.* 8: 253–259.

Lowe, R.L. and R.W. Pillsbury. 1995. Shifts in benthic algal community structure and function following the appearance of zebra mussels (*Dreissena polymorpha*) in Saginaw Bay, Lake Huron. *J. Great Lakes. Res.* 21: 558–566.

MacIssac, H. J. 1994. Size-selective predation on zebra mussels (*Dreissena polymorpha*) by crayfish (*Orconectes propinquus*). *J. N. Amer. Benthol. Soc.* 13: 206–216.

MacIssac, H. J. 1996. Potential abiotic and biotic effects of zebra mussels on inland waters of *North America Amer. Zool.* 36: 287–299.

MacIssac, H.J., W. G. Sprules, O.E. Johannsson, and J.H. Leach. 1992. Filtering impacts of larval and sessile zebra mussels (*Dreissena polymorpha*) in western Lake Erie. *Oecologia* 92: 30–39.

MacIssac, H.J., C.J. Lonnee, and J.H. Leach. 1995. Suppresssion of microzooplankton by zebra mussels: Importance of mussel size. *Freshwat. Biol.* 34: 379–387.

Mackie, G.L. 1976. Trematode parasitism in the Sphaeriidae clams, and the effects in three Ottawa River species. *Nautilus* 90: 36–41.

Mackie, G.L. 1978. Are sphaeriid clams ovoviviparous or viviparous? *Nautilus* 92: 145–147.

Mackie, G.L. 1979. Dispersal mechanisms in Sphaeriidae. *Bull. Am. Malacol. Union* 1979: 17–21.

Mackie, G.L. 1984. 5. Bivalves. In: K.M. Wilbur (Ed.). *The mollusca*. Vol. 7. *Reproduction* (Eds. A.S. Tompa, N.H. Verdonk, and J.A.M. van de Biggelaar), Academic Press, New York, pp. 351–418.

Mackie, G.L. 1993. Biology of the zebra mussel (*Dreissena polymorpha*) and observations of mussel colonization on unionid bivalves in Lake St. Clair. In: T.F. Nalepa and D. W. Schoesser (Eds.). *Zebra mussels: Biology, impacts, and control*. Lewis/CRC Press, Boca Raton, FL, pp. 153–166.

Mackie, G.L. 1991. Biology of the exotic zebra mussel, *Dreissena polymorpha*, in relation to native bivalves and its potential impact in Lake St. Clair. In M. Munawar and T. Edsall (Eds.), *Environmental assessment and habitat evaluation of the upper Great Lakes connecting channels*. Hydrobiologia 219: 251–268.

Mackie, G.L., W.N. Gibbons, B.W. Muncaster, and I.M. Gray. 1989. The zebra mussel, *Dreissena polymorpha*: a synthesis of European experiences and a preview for North America. Report prepared for Water Resources Branch, Great Lakes Section. Available from Queen's Printer for Ontario, ISBN 0-7729-5647-2.

Mackie, G.L. and S.U. Qadri. 1973. Abandance and diversity of Mollusca in an industrialized portion of the Ottawa River near Ottawa-Hull, Canada. *J. Fish. Res. Bd.* Canada 30: 167–172.

Mackie, G.L. and D.W. Schloesser. 1996. Comparative biology of zebra mussels in Europe and North America: An overview. *Am. Zool.* 36: 244-258.

Mackie, G.L., D.S. White, and T.W. Zdeba. 1980. A guide to the freshwater molusks of the Laurentian Great Lakes with special reference to the *Pisidium*. U.S. Environ. Protect. Agency, EPA-600/3-80-068.

Mackie, G.L. and C. Wright. 1994. Biodeposition of seston and removal of phosphorous and biochemical oxygen demand from activated sewage sludge by *Dreissena polymorpha* (Bivalvia: Dreissenidae). *Wat. Res.* 28: 1123–1130.

Madenjian, C.P. 1995. Removal of algae by the zebra mussel (*Dreissena polymorpha*) population in western Lake Erie: A bioenergetics approach. *Canad. J. Fisher. Aquat. Sci.* 52: 381–390.

Makarewicz, J.C. and P. Bertram. Evidence for the restoration of the Lake Erie ecosystem. *Bioscience* 41: 216–223.

Malone, C.R. 1965. Dispersal of aquatic gastropods via the intestinal tract of water birds. *Nautilus* 78: 135–139.

Malone, C.R. 1966. Regurgitation of food by mallard ducks. *Wilson Bull.* 78: 227–228.

Marsden, J.E., A.P. Spidle, and B. May. 1996. Review of genetic studies of *Dreissena* spp. *Amer. Zool.* 36: 259–270.

Mastellar, E.C., K.R. Maleski, and D.W. Schloesser. 1993. Unionid bivalves (Mollusca: Bivalvia: Unionidae) of Presque Isle Bay, Erie, Pennsylvania. *J. Pennsylvania Acad. of Sci.* 67(3): 120–126

Matthews, M.A. and R.F. McMahon. 1994. The survival of zebra mussels (*Dreissena polymorpha*) and Asian clams (*Corbicula fluminea*) under extreme hypoxia. In: *Proceedings, 4th International Zebra Mussel Conference '94*, Madison, WI, Mar 7-10, 1994. pp. 231–250.

Mattice, J.S. 1970. Trophic biology of a natural population of *Bythinia tentaculata* in terms of ecological energetics. PhD dissertation, Syracuse University, Syracuse N.Y.

Mayr, E. 1963. *Animal species and evolution*. Harvard Univ. Press, Boston.

McKee, P.M. and G.L. Mackie. 1980. Desiccation resistance in *Sphaerium occidentale* and *Musculium securis* from a temporary pond. *Canad. J. Zool.* 58: 1693–1696.

McKee, P.M. and G.L. Mackie. 1981. Life history adaptations of the fingernail clams *Sphaerium occidentale* and *Musculium securis* to ephemeral habitats. *Canad. J. Zool.* 59: 2219–2229.

McKee, P.M. and G.L. Mackie. 1983. Respiratory adaptations of the fingernail clams *Sphaerium occidentale* and *Musculium securis* to ephemeral habitats. *Canad. J. Fisher. Aquat. Sci.* 40: 783–791.

McMahon, R.F. 1983. Physiological ecology of freshwater pulmonates. In: W.D. Russell-Hunter (Ed.), *The mollusca*, vol. 6, *Ecology*. Academic Press, New York, pp. 360–430.

McMahon, R.F. 1996. The physiological ecology of the zebra mussel, *Dreissena polymorpha*, in North America and Europe. *Amer. Zool.* 36: 339–365.

McMahon, R.F., T.A. Ussery, A.C. Miller, and B. S. Payne. 1993. Thermal tolerance in zebra mussels (*Dreissena polymorpha*) relative to rate of temperature increase and acclimation temperature. In: J.L. Tsou and Y.G. Mussalli (Eds.), Proceedings: Third International Zebra Mussel Conference, 1993, Toronto, Ontario. Electric Power Research Institute, Palo Alto, CA, Publ. No. TR-102077, pp. 4-98 – 4-118.

McMillan, N. F. 1968. British shells. Frederick Warne & Co. Ltd. London, England.

Mellina, E. and J.B. Rasmussen. 1994. Patterns in the distribution and abundance of zebra mussels *(Dreissena polymorpha)* in rivers and lakes in relation to substrate and other physiochemical factors. *Canad. J. Fisher. Aquat. Sci.* 51: 1024–1036.

Mills, E.L., G. Rosenburg, A.P. Spidle, M. Ludyanski, Y. Pligin, and B. May. 1996. A review of the biology and ecology of the quagga mussel (*Dreissena bugensis*), a new species of freshwater dreissenid introduced into North America. *Am. Zool.* 36: 271–287.

Miner, J.G., R. Lowe, T. Stewart, F. Snyder, and D. Kelch. 1995. Distribution, abundance and growth of freshwater sponge (Spongillidae) in western Lake Erie: Projecting impacts on zebra mussels. Abstract, *The Fifth International Zebra Mussel & Other Aquatic Nuisance Organisms Conference*, Toronto, Ontario, Feb. 21–24, 1995, p. 94.

Mitchell, J.S., R.C. Bailey, and R.W. Knapton. 1996. Abundance of *Dreissena polymorpha* and *Dreissena bugensis* in a warmwater plume: effects of depth and temperature. *Canad. J. Fisher. Aquat. Sci.* 53: 1705–1712.

Mitchell, M.J., E.L. Mills, N. Idrisi, and R. Michener. 1996. Stable isotopes of nitrogen and carbon in an aquatic food web recently invaded by *Dreissena polymorpha*. (Pallas). *Canad. J. Fisher. Aquat. Sci.* 53: 1445–1450.

Molloy, D.P., A. Karatayev, L. Burlakova, D. Kurandina, and S. Fokin. 1995. Ciliate parasites of European zebra mussels. Abstract, *The Fifth International Zebra Mussel & Other Aquatic Nuisance Organisms Conference*, Toronto, Ontario, Feb. 21–24, 1995, p. 92.

Morrison, T.W., W.E. Lynch, Jr., and K. Dabrowski. 1997. Predation on zebra mussels by freshwater drum and yellow perch in western Lake Erie. *J. Gr. Lakes Res.* 23: 177–189.

Morton, B. 1979. Freshwater biofouling bivalves. In J.C. Britton (Ed.), *Proceedings, First International Corbicula Symposium*, Texas Christian Univ., Research Foundation, Fort Worth TX, pp. 1–14.

Morton, B. 1983. The sexuality of *Corbicula fluminea* in lentic and lotic waters in Hong Kong. *J. Moll. Stud.* 49: 81–83.

Morton, B. 1986. *Corbicula* in Asia — an updated synthesis. *Am. Malacol. Bull.*, spec. ed. No. 2. 2: 113–124.

Morton, B. 1993. The anatomy of *Dreissena polymorpha* and the evolution and success of the heteromyarian form in the Dreissenoidea. In: T.F. Nalepa and D.W. Schloesser (Eds.), *Zebra Mussels: Biology, Impacts, and Control*. Lewis/CRC Press, Boca Raton, FL, pp. 185–216.

Nagelkerke, L.A.J. and F.A. Sibbing. 1996. Efficiency of feeding on zebra mussel *(Dreissena polymorpha)* by common bream (*Abramis brama*), white bream (*Blicca bjoerkna*), and roach (*Rutilus rutilus*): the effects of morphology and behavior. *Canad. J. Fisher. Aquat. Sci.* 53: 2847–2861.

Nalepa, T.F. 1994. Decline of native bivalves (Unionidae: Bivalvia) in Lake St. Clair after infestation by the zebra mussels *Dreissena polymorpha*. *Canad. J. Fisher. Aquat. Sci.* 51: 2227–2233.

Nalepa, T.F and D.W. Schloesser (Eds.). 1993. *Zebra Mussels: Biology, Impacts, and Control*. Lewis/CRC Press, Boca Raton, FL.

Neuhauser, E.F., J.J. Knowlton, D.P. Lewis, and G.L. Mackie. 1993. Thermal treatment to control zebra mussels at the Dunkirk Steam Station. In: J.L. Tsou and Y.G. Mussalli (Eds.), *Proceedings: Third International Zebra Mussel Conference*, 1993, Toronto, Ontario. Electric Power Research Institute, Palo Alto, CA, Publ. No. TR-102077, pp. 4-71–4-94.

Neumann, D. and H.A. Jenner. 1992. *The ZebraMussel* Dreissena polymorpha: *Ecology, Biological Monitoring and First Applications in Water Quality Management*. VCH Ppublishers, Deerfield Beach, FL.

Nichols, S.J. 1996. Variations in the reproductive cycle of *Dreissena polymorpha* in Europe, Russia, and North America. *Amer. Zool.* 311–325.

Nicholls, K.H. and G.J. Hopkins. 1993. Recent changes in Lake Erie (north shore) phytoplankton: Cumulative impacts of phosphorous loading reductions and the zebra mussel introduction. *J. Great Lakes Res.* 19: 637–647.

Olsen, O.W. 1974. *Animal parasites: Their life cycles and ecology.* University Park Press, London.

Pathy, D.A. 1994. The life history and demography of zebra mussel, *Dreissena polymorpha*, populations in Lake St. Clair, Lake Erie, and Lake Ontario. M.Sc. thesis, Univ. Guelph, Guelph, Ontario.

Payne, B.S., J. Lei, A.C. Miller, and E.D. Hubertz. 1995. Adaptive variation in palp and gill size of the zebra mussel (*Dreissena polymorpha*) and Asian clam *(Corbicula fluminea). Canad. J. Fisher. Aquat. Sci.* 52: 11130–1134.

Pennak, R.W. 1989. *Fresh water invertebrates of the United States.* John Wiley, New York. 628 p.

Perry, W.L., D.M. Lodge, and G.A. Lamberti. 1997. Impact of crayfish predation on exotic zebra mussels and native invertebrates in a lake-outlet stream. *Canad. J. Fisher. Aquat. Sci.* 54: 120–125.

Pinel-Alloul, B. and E. Magnin. 1971. Cycle vital et croissance de *Bythinia tentaculata* L. (Mollusca; Gastropoda; Prosobranchia) du Lac St. Louis pres de Montreal. *Ca. J. Zool.* 49: 759–766.

Ponder, W.F. 1988. *Potamopyrgus antipodarum* — a molluscan coloniser of Europe and Australia. *J. Moll. Stud.* 54: 271–285.

Ram, J.L., P.P. Fong, and D.W. Garton. 1996. Physiological aspects of zebra mussel reproduction: Maturation, spawning, and fertilization. *Amer. Zool.* 36: 326–338.

Rees, W.J. 1965. The aerial dispersal of Mollusca. *Proc. Malacol. Soc. London* 36: 269–282.

Ribi, G. 1986. Within-lake dispersal of the prosobranch snails, *Viviparus ater* and *Potamopyrgus jenkinsi. Oecologia* 69: 60–63,

Ricciardi, A. 1994. Occurrence of chironomid larvae (*Paratanytarsus*) as commensals of dreissenid mussels (*Dreisssena polymorpha* and *D. bugensis*). *Canad. J. Zool.*

Ricciardi, A., H.M. Reiswig, F.L. Snyder, and D.O. Kelch. 1995. Lethal overgrowth of dreissenid mussels by freshwater sponges: Potential biological control? Abstract, *The Fifth International Zebra Mussel & Other Aquatic Nuisance Organisms Conference*, Toronto, Ontario, Feb. 21–24, 1995, p. 93.

Ricciardi, A., F.G. Whoriskey, and J.B. Rasmussen. 1995. Predicting the intensity and impact of *Dreissena* infestation on native unionid bivalves from *Dreissena* field density. *Canad. J. Fisher. Aquat. Sci.* 52: 1449–1461.

Ricciardi, A., F. G. Whoriskey and J. B. Rasmussen. 1996. Impact of the *Dreissena* invasion on native unionid bivalves in the upper St. Lawrence River. *Canad. J. Fisher. Aquat. Sci.* 53: 1434–1444.

Ricciardi, A., F.G. Whoriskey, and J.B. Rasmussen. 1997. The role of the zebra mussel (*Dreissena polymorpha*) in structuring macroinvertebrate communities on hard substrata. *Canad. J. Zool.*

Robertson, I.C.S. and C.L. Blakeslee. 1948. The Mollusca of the Niagara frontier region. *Bull. Buffalo Soc. Nat. Sci.* 19: 1–191.

Roditi, H.A., D.L. Strayer, and S.E.G. Findlay. 1997. Characteristics of zebra mussel (*Dreissena polymorpha*) biodeposits in a tidal freshwater estuary. *Arch. Hydrobiol.* In press.

Rogers, J.P. and L.J. Korschgen. 1966. Foods of lesser scaups on breeding, migration, and wintering areas. *J. Wildl. Mgmt.* 30: 258–264.

Rosenburg, G. and M. Ludyanskiy. 1994. A nomenclatural review of *Dreissena* (Bivalvia: Dreissenidae), with identification of the quagga mussel as *Dreissena bugensis. Canad. J. Fisher. Aquat. Sci.* 51: 1474–1484.

Sawyer, R.T. Leeches (Annelida: Hirudinea). In: Hart, C.W. and L.H. Fuller (Ed.), *Pollution ecology of freshwater invertebrates*, pp. 82–142.

Schloesser, D.W. and W. Kovalak. 1991. Infestation of unionids *by Dreissena polymorpha* in a power plant canal in Lake Erie. *J. Shellfish Res.* 10: 355–359.

Schloesser, D.W. and T.F. Nalepa. 1994. Infestation of unionids in offshore waters of western Lake Erie after the invasion by the zebra mussel, *Dreissena polymorpha. Canad. J. Fisher. Aquat. Sci.* 51:2234–2242.

Schloesser, D.W., T.F. Nalepa, and G.L. Mackie. 1996. Zebra mussel infestation of unionid bivalves (Unionidae) in North America. *Amer. Zool.* 36: 300–310.

Silverman, H., J.W. Lynn, E.C. Achberger, and T.H. Dietz. 1996. Gill structure in zebra mussels: Bacterial-sized particle filtration. *Amer. Zool.* 36: 373–383.

Skubinna, J.P., T.G. Coon, and T.R. Batterson. 1995. Increased abundance and depth of submersed macrophytes in response to decreased turbidity in Saginaw Bay, Lake Huron. *J. Great Lakes Res.* 21: 476–488.

Sprung, M. and U. Rose. 1988. Influence of food size and food quantity on the feeding of the mussel *Dreissena polymorpha. Oecologia* 77, 526–532.

Stanczykowska, A. 1964. On the relationship between abundance, aggregations and "condition" of *Dreissena polymorpha* Pall. in 36 Masurian lakes. *Ekol. Pol.* A12: 653–690.

Stanczykowska, A. 1976. Biomass and production of *Dreissena polymorpha* (Pall.) in some Masurian lakes. *Ekol. Pol.* 24: 103–112.

Stanczykowska, A. 1977. Ecology of *Dreissena polymorpha* (Pallas) (Bivalvia) in lakes. *Pol. Arch. Hydrobiol.* 24(4):461–530.

Stanczykowska A., L. Lodzimierz, J. Mattice and S. K. Lewandowski. 1976. Bivalves as a factor effecting circulation of matter in Lake Mikolajskie (Poland). *Limnologia* 19, 347–352.

Sterki, V. 1899. Pisidia new to our country, and new species. *Nautilus* 13: 9–12.

Stewart, T.W. and J.M. Haynes. 1994. Benthic macroinvertebrate communities of southwestern Lake Ontario following invasion of *Dreissena*. *J. Great Lakes Res.* 20: 479–493.

Strayer, D.L. 1991. Projected distribution of the zebra mussel, *Dreissena polymorpha*, in North America. *Canad. J. Fisher. Aquat. Sci.* 48: 1389–1395.

Strayer, D.L., D.C. Hunter, L.C. Smith, and C.K. Borg. 1994. Distribution, abundance and roles of freshwater clams (Bivalvia: Unionidae) in the freshwater tidal Hudson River. *Freshwat. Biol.* 31: 239–248.

Strayer, D.L., J. Powell, P. Ambrose, L.C. Smith, M.L. Pace, and D.T. Fischer. 1996. Arrival, spread, and early dynamics of a zebra mussel (*Dreissena polymorpha*) population in the Hudson River estuary. *Canad. J. Fisher. Aquat. Sci.* 53: 1143–1149.

Strayer, D.L. and L.C. Smith. 1993. Distribution of the zebra mussel (*Dreissena polymorpha*) in estuaries and brackish waters. In: T.F. Nalepa and D.W. Schloesser (Eds.), *Zebra Mussels: Biology, Impacts, and Control*. Lewis/CRC Press, Boca Raton, FL, pp. 715–728.

Strayer, D.L. and L.C. Smith. 1996. Relationship between zebra mussels *(Dreissena polymorpha)* and unionid clams during the early stages of zebra mussel invasion of the Hudson River. *Freshwat. Biol.* 36: 771–779.

Summers, R.B., J.H. Thorp, J.E. Alexander, and R.D. Fell. 1996. Respiratory adjustment of dreissenid mussels (*Dreissena polymorpha* and *Dreissena bugensis*) in response to chronic turbidity. *Canad. J. Fisher. Aquat. Sci.* 53: 1626–1631.

Tashiro, J.S. and S.D. Coleman. 1982. Filter feeding in the freshwater prosobranch snail *Bythinia tentaculata*: Bioenergetic partitioning of ingested carbon and nitrogen. *Am. Midl. Nat.* 107: 114–132.

Thomlinson, J. 1966. The advantage of hermaphroditism and parthenogenesis. *J. Theoret. Biol.* 11: 54–58.

Toews, S., M. Beverley-Burton, and T. Lawrimore. 1993. Helminth and protist parasites of zebra mussels, *Dreissena polymorpha* (Pallas, 1771), in the Great Lakes region of southwestern Ontario, with comments on associated bacteria. *Canad. J. Zool.* 71: 1763–1766.

Tucker, J.K. 1994. Colonization of unionid bivalves by the zebra mussel, *Dreissena polymorpha*, in pool 26 of the Mississippi River. *J. Freshwat. Ecol.* 9: 129–134.

Ussery, T.A. and R.F. McMahon. 1994. Comparative study of the desiccation resistance of zebra mussels (*Dreissena polymorpha*) and quagga mussels (*Dreissena bugensis*). In: *Proceedings, Fourth International Zebra Mussel Conference '94*, Madison, Wisconsin, Mar 7-10, 1994. pp. 351–369.

Vincent, B., G. Vaillaincourt, and M. Harvey. 1981. Cycle de developpment, croissance effectifs, biomasse et production de *Bythinia tentaculata* L. (Gastropoda: Prosobranchia) dans le Saint-Laurent (Quebec). *Can. J. Zool.* 59: 1237–1250.

Wallace, C. 1978. Notes on the distribution of sex and shell characteristics in some Australian populations of *Potamopyrgus* (Gastropoda: Hydrobiidae). *J. Malacol. Soc. Australia* 4: 71–76.

Wallace, C. 1985. On the distribution of the sexes of *Potamopyrgus jenkinsi* (Smith). *J. Moll. Stud.* 51: 290–296.

Walz, N. 1978. The energy balance of the freshwater mussel *Dreissena polymorpha* Pallas in laboratory experiments and in Lake Constance. I. Pattern of activity, feeding, and assimilation efficiency. *Arch. Hydrobiol. Suppl.* 55: 83–105.

White, M.J.D. 1954. *Animal cytology and evolution*. Cambridge Univ. Press.

Winterbourne, M. 1970a. The New Zealand species of *Potamopyrgus* (Gastropoda: Hydrobiidae). *Malacologia* 10: 283–321.

Winterbourn, M. 1970b. Population studies on the New Zealand species of *Potamopyrgus antipodarum* (Gray). *Proc. Malacol. Soc. London* 39: 139–149.

Winterbourn, M. 1972. Morphological variation of *Potamopyrgus jenkinsi* (Smith) from England and a comparison with the New Zealand species, *Potamopyrgus antipodarum* (Gray). *Proc. Malacol. Soc. London* 40: 133–145.

Woodward, B.B. 1913. *Catalogue of the British species of Pisidium (recent and fossil) in the collections of the British Museum (Natural History), with notes on those of western Europe.* British Museum (Natural History), London.

Zaranko, D.T., D.G. Farara, and F.G. Thompson. 1997. Another exotic mollusc in the Great Lakes: the New Zealand native, *Potamopyrgus antipodarum* (Gray 1843) (Gastropoda: Hydrobiidae). *Canad. J. Fisher. Aquat. Sci.* 54: 809–814.

Zhadin, V.I. 1952. Mollusks of fresh and brackish waters in the U.S.S.R. Zoological Institute of the Academy of Sciences of the U.S.S.R. Publication No. 46. (English translation 1965, Israel Program for Scientific Translation, Jerusalem. IPST Cat. No. 10001.)

16 Impacts of the Zebra Mussel (*Dreissena polymorpha*) on Water Quality: A Case Study in Saginaw Bay, Lake Huron

Thomas F. Nalepa, Gary L. Fahnenstiel, Thomas H. Johengen

INTRODUCTION

Impacts of benthic, suspension-feeders on pelagic measures of water quality (i.e., phytoplankton, water clarity, nutrients) have been well-documented in both freshwater and marine environments (Cloern 1982; Officer et al. 1982; Wright et al. 1982; Cohen et al. 1984; Dame et al. 1991). These organisms filter particles from the water and ingest material that is either assimilated and incorporated into biomass, or rejected and deposited as feces and pseudofeces. As a result, energy is shifted from the pelagic region to the benthic region, and changes occur in the normal pathways by which nutrients are utilized and cycled. Impacts of these feeding activities depend on the characteristics of the particular system, and on the density of the suspension-feeding population. Greatest impacts generally occur in productive, shallow water systems with high population densities. Under these conditions, the population can be capable of filtering water at a time rate constant that is much greater than the water residence time within the system, and at a rate greater than, or comparable to, phytoplankton growth.

After the suspension-feeding bivalve *Dreissena polymorpha* (zebra mussel) became established in the Great Lakes, changes in water quality parameters immediately became apparent in bays and nearshore regions where this species was most abundant. Water clarity increased (Hebert et al. 1991; Marsden et al. 1993; Leach 1993), chlorophyll and phytoplankton abundances declined (Leach 1993; Nicholls and Hopkins 1993; Holland 1993), and nutrient cycles were altered (Holland et al. 1995; Arnott and Vanni 1996; Mellina et al. 1995). While suspension feeders were present in the Great Lakes prior to *D. polymorpha* (i.e., bivalves of the families Unionidae, Sphaeriidae, and Corbiculidae), populations were too low and filtration capacity too limited to have any significant impact on water quality. Because of recent changes resulting from the filtering activities of *D. polymorpha*, responses of water quality variables to nutrient abatement programs are no longer predictable, and management approaches to water quality issues must be completely reevaluated.

In this paper, we examine impacts of *D. polymorpha* on the Saginaw Bay, Lake Huron ecosystem, emphasizing changes in pelagic measures of water quality. We focus primarily on summarizing changes during the early years of the invasion (1991–93), but also include preliminary results of water quality changes observed in 1994 and 1995. Specific details of changes during the 1991–93 period are given in a series of papers published in Volume 21 (4) of the *Journal of Great Lakes Research*. For sake of brevity and purpose, results given herein will emphasize changes occurring within the eutrophic inner portion of the bay. Changes in the outer bay were minimal and/or more local in nature and details can be found in the journal volume.

JUSTIFICATION OF STUDY SITE

Soon after the zebra mussel was discovered in the Great Lakes in 1988, we identified Saginaw Bay as an ideal location to assess ecological changes that might result from the filtering activities of this organism. Specific considerations that led to the decision to initiate a monitoring program in the bay were as follows: (1) at the time, zebra mussels were not yet established in the bay, thus baseline conditions immediately prior to the mussel's invasion could be documented; further, previous surveys of water quality parameters in 1974–80 (Smith et al., 1977; Bierman et al., 1984) could provide a longer term perspective to assess potential changes; (2) the bay had extensive areas of hard bottom, along with ideal temperature and food regimes, and thus large populations of mussels were expected to develop; (3) there existed an important commercial and sport fishery that could be affected; (4) the natural gradient between the eutrophic inner bay and the more oligotrophic outer bay provided an opportunity to assess impacts over a wide range of trophic conditions; and (5) the bay is an Area of Concern as designated by the International Joint Commission and the subject of remedial action to reduce nutrient inputs (Richardson and Kreis 1987). After a decade of little or no monitoring in the bay, surveys of water quality parameters initiated as part of this study not only provide information to assess impacts of the zebra mussel, but also provide information to assess the bay's response to continued efforts to improve water quality.

DESCRIPTION OF STUDY SITE

Saginaw Bay is a shallow, well-mixed extension of the western shoreline of Lake Huron (Figure 16.1). The bay is 21–42 km wide, about 82 km long , and has a drainage basin of about 21,000 km^2. Total area of the bay is 2.77×10^9 m^2, and total water volume is 24.54×10^9 m^3 (Table 16.1). The bay can be functionally divided into an inner and outer region by a line extending along its narrowest width (21 km) from Sand Point to Point Lookout (Figure 16.1). A broad shoal and several islands (Charity Islands) along this line provide a natural demarcation between the two regions. Differences in physical and chemical features of the inner and outer bay regions are distinct (Beeton et al. 1967; Smith et al. 1977). The inner bay has a mean depth of 5.1 m, is nutrient-rich, and is heavily influenced by input from the Saginaw River, which accounts for over 70% of the total tributary flow into the bay. The outer bay has a mean depth of 13.7 m and is more influenced by the colder, nutrient-poor waters of Lake Huron.

Circulation within the inner bay is generally weak; currents average about 7 cm s^{-1} (Danek and Saylor 1977). Exchange and flushing of water in the inner bay occurs when winds blow along the long axis of the bay (southwest/northeast). Dominant winds in the summer are from the southwest. Little exchange occurs when winds are perpendicular to the long axis (west/east). Most water exchange/flushing between the inner and outer bay occurs on the northern side of the bay within the deep channel located between Point Lookout and Charity Island and that continues into the inner bay. Although some water may exit the inner bay along the southern shoreline, it is of minor significance because of the shallowness of the region (Danek and Saylor 1977). Furthermore, preliminary results of Lagrangian current measurements in the outer bay during the summers of 1992 and 1993 suggest that the flushing of inner bay waters into Lake Huron is episodic in nature (M. McCormick, unpublished data). Water residence times are about 120 days for the inner bay and 60 days for the outer bay (Bratzel et al. 1977).

Bottom substrates in Saginaw Bay range from silt to mostly cobble and rock. The inner bay has a wide sand-gravel bar that extends along the eastern side of the bay from the Saginaw River to the Charity Islands. Another sand-gravel bar extends along the western shoreline to Point Au Gres. Both sand bars have irregular areas of cobble along with patches of sand, gravel, and pebbles. The bars extend into the shorelines as extensive flats grade into marshes. Between the two sand bars is an area of maximum depth where sediment deposition occurs; the substrate in this region consists of fine-grained sediments (silt/mud). Based on areal estimates of substrate type by Wood

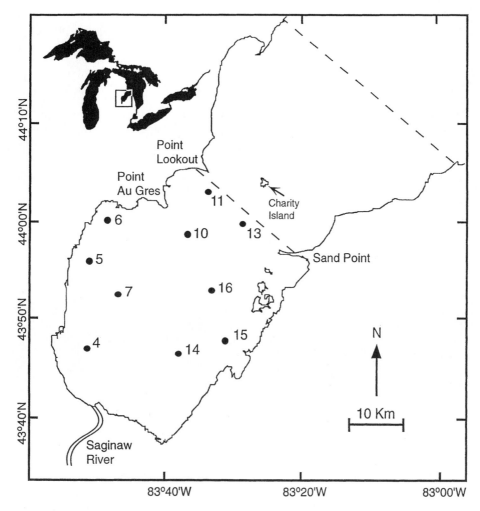

FIGURE 16.1 Location of sampling sites in inner Saginaw Bay, 1990–95. Dashed lines differentiate the inner bay from the outer bay, and the outer bay from Lake Huron.

TABLE 16.1
Mean Depth, Surface Area, and Water Volume of the Inner Bay, the Outer Bay, and the Bay as a Whole

	Mean Depth (m)	Surface Area (m²)	Volume (m³)
Inner Bay	5.09	1.55×10^9	7.91×10^9
Outer Bay	13.66	1.22×10^9	16.63×10^9
Whole Bay	8.86	2.77×10^9	24.54×10^9

Note: Values were computed from digitized NOAA chart no. 14863. A 0.66 m offset was used to account for low water datum.

(1964) and extensive benthic sampling in the late 1980s (T. Nalepa, unpublished data), we estimated that 70% of the bottom in the inner bay consists of sand, gravel, and cobble, and 30% consists of silt/mud.

SAMPLING DESIGN AND METHODS

Over the period 1991–95, samples were generally collected monthly from May to October at eight sites in the inner bay (Figure 16.1). In some years, samples were collected in April and/or November. Additionally, there were two sampling dates in May in every year except 1993. Water samples were collected with a Niskin bottle at a depth of 1 m at every site, and at the mid-water column depth at Station 10 (6 m) and Station 11 (5 m). Water depth ranged from 3 m to 11 m (Table 16.2).

TABLE 16.2
Sampling Depth and Substrate Type at Sampling Sites in Inner Saginaw Bay

Station	Depth (m)	Substrate
4	7.0	Silt/mud
5	3.5	Cobble, sand, gravel
6	4.0	Sand, gravel, some cobble
7	7.0	Silt/mud
10	11.0	Silt/mud
11	9.0	Silty sand
13	3.0	Sand, gravel, some cobble
14	3.5	Sand, gravel, some cobble
15	3.0	Sand, some cobble
16	3.5	Sand, gravel, some cobble

Chlorophyll was measured in triplicate using the method of Strickland and Parsons (1972), and nutrients were determined using standard automated colorimetric techniques (APHA 1990) on a Technicon Auto Analyzer II as detailed in Davis and Simmons (1979). Water clarity was measured with a 25-cm secchi disk. Analytical techniques, sampling dates, site locations, quality control, and values of all measured variables for each date and site can be found in Nalepa et al. (1996a).

Densities of zebra mussels were estimated in the fall of each year at each of the eight sampling sites and at two additional sites (Stations 6 and 15; Figure 16.1). The collection method depended upon substrate type. At sites with a hard substrate (sand, gravel, cobble; Table 16.2), divers randomly placed a 0.25 or 0.5 m² frame on the bottom and hand-collected all material within the frame area. Triplicate samples were collected at each site with divers moving about 2–3 m between replicates. At sites where the bottom consisted of silt, samples were collected using a Ponar grab. Triplicate samples were washed into an elutriation device fitted with a Nitex sleeve having 0.5-mm openings (Nalepa 1987). Details of counting and sizing procedures, as well as methods to estimate ash-free dry weight (AFDW) biomass, are given in Nalepa et al. (1995).

RESULTS

ZEBRA MUSSEL POPULATION TRENDS

Zebra mussels were first discovered in Saginaw Bay in 1990, but the population did not become widespread and abundant until 1991 (Nalepa et al. 1995). Yearly trends in densities in 1991–95 at sites with hard substrates are given in Table 16.3. Mean densities in the inner bay increased between

TABLE 16.3
Mean Density (Individuals m⁻²) of *D.*
***polymorpha* at Each Hard-Substrate Site**
Sampled in Inner Saginaw Bay, 1991–95

Station	Year				
	1991	1992	1993	1994	1995
5	28,244	75,296	237	2,959	1,018
6	4,453	3,620	3,557	10,724	2,291
13		8,956	376	854	211
14	208	63,242	7,506	3,900	2,564
15	43,117	5,556	7,341	9,725	6,728
16	26	46,360	4,830	1,727	60
Mean	10,130	33,838	3,975	4,982	2,145

1991 and 1992 to reach a peak of 33,800 m⁻². Densities declined after 1992; mean yearly densities varied between 2,000–5,000 m⁻² in 1993–95. These results indicate that the population apparently reached an equilibrium on hard substrates by fall 1993, just a few years after the first major recruitment. Factors that likely limited population growth were lack of suitable substrate, adults filtering the larvae before settling occurred, and a decline in food availability (Nalepa et al. 1995). Such a dramatic increase and decline within just a few years of the initial colonization is not unusual. A similar trend was noted when zebra mussels colonized the freshwater portion of the Hudson River estuary (Strayer et al. 1996).

Variation in densities between individual sites was large for any given year, with densities often differing by an order of magnitude (Table 16.3). However, no individual site had densities that were consistently higher or lower than the other sites over the entire 5-year sampling period. Site variation was likely related to the nature of "hard" substrate within the inner bay. As noted, the substrate at stations designated as "hard-substrate sites" consisted of a patchy mixture of cobble, sand, and gravel. Most mussels were found on cobble, and the amount of cobble at some sites varied from 5% to 50% between years. Thus, while the navigation system (Loran C) provided accurate positioning, even slight variation in sampling location between years could lead to large differences in density estimates.

To examine the extent of spatial variation at individual sites, we conducted a high-frequency sampling program at two sites (Stations 5 and 14) in spring 1994. Nine replicate quadrat samples were collected at the designated site location, and at locations that were 0.4 and 0.8 km from the designated site on north, south, east, and west transects (nine replicates per nine sampling sites). Mean densities at the nine sampling locations varied from 410 to 7,690 m⁻² at Station 5, and from 840 to 4,760 m⁻² at Station 14 (Nalepa et al. 1995). Coefficients of variation of the mean densities at the nine sampling locations were 64% and 63% at the two sites, respectively. In comparison, coefficients of variation for the means for all six sites with hard substrate in 1993, 1994, and 1995 were 81%, 84%, and 115%. Thus, variation within 1.6 km of an individual site was only slightly lower than variation between all sites within a given year. This further indicates that densities on hard substrates were generally similar within the bay by 1993, and that year-to-year differences thereafter were primarily a function of substrate variability. Densities at "soft" substrate sites (mud/silt) were minimal and insignificant over the entire sampling period (Nalepa et al. 1995; Nalepa, unpublished data).

Yearly trends in AFDW biomass at the hard-substrate sites were similar to trends in densities. Mean biomass peaked in 1992 at 61.9 g m⁻² and then declined; mean biomass in 1993–1995 varied from 3.1 to 4.5 g m⁻² (Table 16.4).

TABLE 16.4
Mean Biomass (g AFDW m^{-2}) of *D. polymorpha* at Each Hard-Substrate Site in Inner Saginaw Bay, 1991–95

			Year		
Station	1991	1992	1993	1994	1995
5	10.5	106.9	0.2	5.6	2.1
6	4.4	8.9	6.7	6.9	6.3
13		24.7	0.8	1.2	0.7
14	0.1	144.0	11.6	1.7	6.1
15	34.1	8.6	3.4	1.9	9.4
16	<0.1	78.3	4.4	1.1	0.1
Mean	9.8	61.9	4.5	3.1	4.1

IMPACTS ON CHLOROPHYLL, TOTAL PHOSPHORUS, AND WATER CLARITY

The impact of mussel filtering activities were apparent soon after the first large population recruitment occurred in summer 1991. For example, mean chlorophyll in late summer/fall 1991 was 51% lower than mean chlorophyll in late summer/fall 1990 (Table 16.5). To put subsequent changes into a long-term perspective, mean seasonal values of chlorophyll, total phosphorus, and secchi-disk transparency were determined for spring (April–June) and fall (August–October) of each year over

TABLE 16.5
Mean Values of Total Phosphorus, Chlorophyll, and Secchi Depth in Inner Saginaw Bay

	Total Phosphorus (µg/L)		Chlorophyll a (µg/L)		Secchi Depth (m)	
Year	Spring	Late Summer/Fall	Spring	Late Summer/Fall	Spring	Late Summer/Fall
1974	32.5	27.0	21.5	29.7	1.20	0.93
1975	33.5	33.5	17.3	21.3	1.50	1.20
1976	46.7	39.3	20.0	27.2	0.90	1.04
1977	—	—	—	—	1.53	0.93
1978	47.5	33.1	18.9	14.7	1.31	1.09
1979	39.6	30.2	9.8	13.6	1.20	0.92
1980	26.1	24.5	10.2	11.1	1.52	1.39
1990	—	—	8.2	10.9	—	—
1991	24.9	21.7	13.3	5.3	1.29	1.61
1992	14.1	17.2	3.7	7.6	2.48	1.73
1993	13.2	17.4	3.1	4.9	2.92	2.18
1994	14.7	25.8	4.9	11.2	2.73	1.70
1995	7.4	21.1	3.1	6.8	2.69	1.33

Note: Values of total phosphorus and chlorophyll are from the 1-m depth interval. 1974–80 values were calculated from data taken from STORET (U.S. EPA) as defined by Bierman et al. (1984) and 1990–95 values are from this study. Monthly data were aggregated into spring (April–June) and late summer/fall (August–October) (Bierman 1984).

the sampling period and compared to corresponding seasonal means in 1974–80. The 1974–80 data were taken from STORET (U.S. EPA), and means were calculated from values at sites that were in close proximity to our sampling locations. Rational for these seasonal categories is given in Bierman et al. (1984).

As noted by Fahnenstiel et al. (1995a), three distinct periods are readily distinguished based on chlorophyll values: the period before phosphorus control (prior to 1976), the period after phosphorus control but before zebra mussel (1978 to spring 1991), and the period after zebra mussel (late summer/fall 1991 to present). Prior to 1976, mean chlorophyll levels varied between 17–30 μg l⁻¹ (Table 5); such values were considered among the highest within the Great Lakes (Vollenweider et al. 1974). During the mid-1970s, municipal treatment plants were upgraded and phosphorus was banned from detergents by the State of Michigan. As a result of these remedial efforts, annual phosphorus loads declined 55% between 1974–76 and 1978–80 and, over the same period, chlorophyll declined 53% in spring and 61% in late summer/fall (Bierman et al. 1984). In the period immediately after phosphorus control (1978–80), mean chlorophyll levels declined to 10–19 μg l⁻¹ (Table 16.5). After 1980, there was no monitoring of water quality parameters in the bay until our study was initiated in 1990. Chlorophyll levels in 1990 and in spring 1991 were very similar to values found in 1978–80 (Table 16.5). However, after the zebra mussel became established, mean chlorophyll declined further, varying from 3–11 μg/l in late summer/fall 1991 through 1995. Chlorophyll levels declined 54% between 1978 and spring 1991 and late summer/fall 1991–95. Thus, the decline in chlorophyll after mussels became established was comparable to the decline observed after phosphorus control measures were initiated.

A comparison of mean values of total phosphorus and secchi-disk transparency between the pre-phosphorus control and post-phosphorus control periods (1974–76 vs. 1978 to spring 1991) showed that total phosphorus tended to decline, but the decline was far less than the decline in chlorophyll over the period (Table 16.5). Also, there was no apparent change in secchi-disk transparency. Bierman et al. (1984) suggested that the relatively minor change in total phosphorus was a result of sediment resuspension. Phosphorus associated with resuspended material can contribute to concentrations in the water column, but not in a form readily available for use by phytoplankton. However, after zebra mussel became established, both total phosphorus and secchi-disk transparency changed dramatically. A comparison of means between the post-phosphorus control period and the post-zebra mussel period (1978 to spring 1991 vs. late summer/fall 1991–95) showed that total phosphorus declined 47% and secchi-disk transparency increased 72%.

IMPACTS ON PRIMARY PRODUCTION

While chlorophyll is an important measure of phytoplankton biomass, primary production is a measure of the photosynthetic rate of carbon fixation. Primary production thus more accurately reflects a system's trophic status. Water column primary production is a function of both phytoplankton biomass (e.g.,chlorophyll) and the amount of available light (e.g., underwater extinction and surface irradiance). As shown, chlorophyll levels in the bay declined after zebra mussel became established, which would lead to a decline in production. On the other hand, water clarity increased, which would lead to greater light penetration in the water column, a condition more favorable to production. Fahnenstiel et al. (1995b) measured primary production in 1990–93 using the C¹⁴ method within a photosynthetron. Photosynthetic rates were used to construct a photosynthesis-irradiance (P-I) curve and resulting output was then modeled to determine the areal and volumetric rate of primary production. Mean areal production was 942 mgC m⁻² d⁻¹ in 1990 (pre-zebra mussel year) and 912 mgC m⁻² d⁻¹ in 1991 (transition year). Areal production was 483 mgC m⁻² d⁻¹ and 536 mgC m⁻² d⁻¹ in 1992 and 1993 (post-zebra mussel years), respectively. Thus, excluding the transitional year 1991, production declined 38% after zebra mussel became established. Based on input data from Canale et al. (1976), areal production in 1974/75 was calculated to be 753 mgC m⁻² d⁻¹. A sensitivity analysis indicated that the decline in production was solely a result of the decline in chlorophyll as other components of model input, such as the underwater light extinction coefficient (kPar), the maximum photosynthetic rate at light saturation (P_{max}), and the initial linear slope at low irradiances (α), actually increased production by 32-35%.

In a light-limited system such as pre-zebra mussel Saginaw Bay, any increase in water clarity would be favorable to benthic primary production since more light would reach the bottom. Fahnenstiel et al. (1995b) made the assumption that the 1% light level was the lower limit for photosynthesis and then compared the ratio of the 1% light level in the water to the bottom depth. Prior to the zebra mussel infestation (1975 and 1990), the ratio in the inner bay was 0.6–0.8. After infestation (1992 and 1993), the ratio increased to 1.1–1.3, indicating a shift to light conditions that favor benthic production. Indeed, although the extent of the light increase was variable depending upon the site location, the overall abundance, depth of colonization, and areal coverage of both submersed vascular macrophytes and benthic macrophytic algae were greater in 1992 and 1993 than in 1991 (Skubinna et al. 1995; Lowe and Pillsbury 1995). Taxa that increased to the greatest extent were the vascular hydrophyte *Chara gobularis* and the following benthic macrophytic algae: *Cladophora, Spirogyra, Zygnema, Hydrodictyon,* and *Oedogonium* (Figure 16.2). Macrophytic algae have high growth rates, high nutrient absorption efficiency, and low light adaptations, making them especially well-adapted for a rapid response to any changes in light availability. To put changes in system productivity in perspective, primary production estimates of mostly benthic macrophytic algae (Lowe and Pillsbury 1995) were compared to estimates of primary production of phytoplankton for the years after the zebra mussel infestation (Fahnenstiel et al. 1995b). The decrease in pelagic primary production was nearly compensated for by the increase in benthic production, without even considering production associated with vascular macrophytes. This finding indicates that zebra mussel did not change the overall productivity of the bay, but only changed how this productivity was distributed between the pelagic and benthic regions.

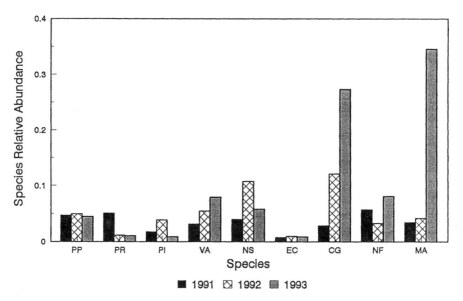

FIGURE 16.2 Relative abundance of the most abundant taxa in inner Saginaw Bay in July 1991–93. Relative abundance is defined as the proportion of the total number of samples containing the particular taxa. PP = *Potamogeton pectinatus,* PR = *Potamogeton richardsoni,* PI = *Potamogeton illinoensis,* VA = *Vallisneria americana,* NS = *Najas sp.,* EC = *Elodea canadiensis,* CG = *Chara globularis,* NF = *Nitella flexilis*; MA = macrophytic algae. From Skubinna et al., *J. Great Lakes Res.* 21, 476–488, 1995. With permission.

DID ZEBRA MUSSELS CAUSE THESE CHANGES?

The evidence clearly suggests that zebra mussels were the cause of changes in water clarity, chlorophyll, and total phosphorus in the inner bay beginning in late 1991. Yet to further establish mussels as the cause, it must be shown that the filtering capacity of the mussel population was high enough to initiate the observed changes, and also that other potential causes were insignificant. Filtration capacity can be determined from estimates of population biomass and filtration rates. Population biomass in the inner bay was estimated by first multiplying mean biomass on the two dominant substrates (sand/cobble, silt) by the total bottom area covered by the two substrates, and then summing the two values. Seasonal filtration rates on natural seston from the inner bay were measured in both 1992 and 1993 (Fanslow et al. 1995). Mean filtration rates were 8.6 and 15.6 ml mg^{-1} h^{-1} in the 2 years, respectively. Assuming mussels filter 17 h per day (Walz 1978), the population filtered 6.4 m^3 m^{-2} d^{-1} and 0.9 m^3 m^{-2} d^{-1} in the two years. With an inner bay water volume of 7.9×10^9 m^3, the population theoretically filtered the entire volume of the inner bay at a rate of 0.8 d^{-1} in 1992, and 0.2 d^{-1} in 1993. For the mussel population to cause a decline in chlorophyll levels, filtration turnover times must equal or exceed algal growth rates. Algal growth rates in the inner bay were 0. 25 d^{-1} in 1992 and 0.20 d^{-1} in 1993 (Fahnenstiel et al. 1995a). Thus, the filtering activities of the mussel population could certainly account for the changes observed.

Other variables that could potentially cause changes in water quality variables are phosphorus loading and zooplankton grazing. While there was a decline in phosphorus loads after control measures were initiated in the mid-1970s (Bierman et al. 1984), there was no significant trend in loadings between 1978 and 1994 (Limno-Tech 1995; Limno-Tech, unpublished data). The mean ($\pm$ SD) yearly phosphorus load for the 1978–89 period was 781 ± 464 t (metric tons). In the 1990–93 period, when dramatic changes occurred in chlorophyll, total phosphorus, and water clarity, phosphorus loads were 506, 1150, 611, and 724 t in each of the four years, respectively. Note that, over this period, the phosphorus load was highest in 1991, the year chlorophyll declines were first observed. Also, note that the phosphorus load in 1990 was somewhat lower than loads in 1992 and 1993, but mean chlorophyll levles were higher (Table 16.5). Since it has been previously shown that chlorophyll is directly correlated with phosphorus loads in the inner bay (Bierman et al. 1984), it seems improbable that variation in annual loads were the cause of the observed changes over this period.

During the early years of the zebra mussel invasion in the western basin of Lake Erie, it was argued that zooplankton grazing could account for declines in phytoplankton and increases in water clarity (Wu and Culver 1991). While these findings have since been discounted for the western basin (MacIsaac et al. 1992; Nicholls and Hopkins 1993), zooplankton grazing in Saginaw Bay was examined in 1991 and 1992 to determine if zooplankton played a role in the observed changes in water quality (Bridgeman et al. 1995). For the seasonal period when zooplankton biomass was the greatest (May/June), biomass-specific grazing rates were similar in both years. However, because of a decrease in biomass in 1992 compared to 1991, total community grazing rates were actually 58% lower in the later year. It is not clear why biomass was lower in 1992, but certainly the decrease in grazing pressure would indicate zooplankton did not play a major role in the decline in chlorophyll and increase in water clarity in that year. Based on maximum community grazing estimates in May/June, zooplankton could theoretically filter the volume of the inner bay in 17 days in 1991 and 37 days in 1992 (Bridgeman et al. 1995). Given these theoretical rates, which were far lower than those for the zebra mussel population, grazing by zooplankton was not likely a cause of the changes observed.

IMPACTS ON NUTRIENT DYNAMICS

There are several different ways in which the zebra mussel affects nutrient concentrations and cycling. First, by its filtering activities, the zebra mussel removes particles from the water column, thereby reducing the pool of nutrients associated with these particles. Second, the zebra mussel excretes dissolved nutrients (phosphate and ammonia) as part of the digestive process and thus changes supply rates to phytoplankton. In a phosphorus-limited system such as Saginaw Bay (Heath et al. 1995), phosphate excretion increases phosphorus availability for growth in species (algae or bacteria) that are not heavily grazed by the mussels. Also, since less phytoplankton are present because of mussel filtering, there is less demand for dissolved nutrients. Third, the zebra mussel affects nutrients directly by accumulating nutrients in soft-tissue biomass.

In Saginaw Bay, all measured particulate nutrients (total suspended solids, particulate organic carbon, particulate phosphorus, and particulate silica) declined significantly after the zebra mussel became established (Johengen et al. 1995). These particulate nutrients were likely either incorporated into mussel biomass, or tied up in biodeposits of feces and pseudofeces. The relative significance of phosphorus incorporated into mussel biomass was examined by comparing phosphorus in soft-tissue biomass to phosphorus loads and to changes in amounts in the water column (Johengen et al. 1995). Phosphorus in biomass was determined by multiplying the mean annual biomass in the inner bay by a measured phosphorus content of 1.0%. Thus determined, the mass of phosphorus contained in the soft-tissue of the zebra mussel population in the inner bay was 108, 682, and 52 t in 1991, 1992, and 1993, respectively. From annual loads given previously, phosphorus in mussel biomass accounted for 9%, 111%, and 7% of the load in each of the three years. Additional data for biomass and loads in 1994 and 1995 (loads of 941 t in 1994 and 578 t in 1995; Limnotech, Inc., unpublished data) indicated that phosphorus in soft-tissue biomass accounted for 4% and 8% of the loads in these two years. The extremely high percentage in 1992 might be considered atypical since it was directly related to the unsustainably high biomass found in that year.

In theory, mussels can increase dissolved nutrients both by direct excretion and by decreasing algal demand. In the inner bay, nitrate (NO_3), ammonia (NH_4), and silica (SiO_2) increased between 1991 and 1993, but soluble reactive phosphorus (SRP) decreased (Johengen et al. 1995). Reasons for the contrary trend in SRP are not immediately apparent, and explanations are particularly difficult given that all observed concentrations over the period were very low and near the detection limit. However, a potential explanation may be found in results of controlled mesocosm experiments in the bay (Heath et al. 1995). Zebra mussels were placed at two different densities in enclosures containing 1600 l of water, while other enclosures without mussels served as controls. Over the short time period of the experiments (6 days), SRP concentrations in the mesocosms with zebra mussels initially increased but then declined. In the same mesocosms, phytoplankton abundance declined but growth rates increased. These results were interpreted in terms of the dynamic relationship between nutrient supply rates and phytoplankton uptake kinetics. As P-limited phytoplankton were grazed by zebra mussels, the concentration of SRP initially increased because of lower demand by phytoplankton and excretion by the mussels. Over time, the remaining, ungrazed phytoplankton rapidly adjusted to the increase in available phosphorus with increased rates of growth, leading to a decrease in SRP levels. Growth rates of phytoplankton in mesocosms with mussels were over two times greater than control mesocosms without mussels. Extending these mesocosm results to *in situ* changes in the bay, the decrease in SRP in the inner bay between 1991 and 1993 may indeed be related to increased uptake by phytoplankton that were ungrazed by the mussels. However, phytoplankton growth rates did not change between 1992 and 1993, which may indicate that other factors were also involved. Most likely, the decrease in SRP was related to increased uptake by the more extensive benthic algae community. Also, both bacteria and protozoa can influence the dynamics of nutrient supply rates to phytoplankton; the dynamics and community structure of both of these groups were affected by mussel grazing and excretion activities (Cotner et al. 1995; Lavrentyev et al. 1995).

Besides phosphate, mussels also excrete ammonia. In the above-mentioned mesocosm experiments (Heath et al. 1995), and also in short-term bottle experiments using bay water (Gardner et al. 1995), concentrations of ammonia increased in treatments with mussels as compared to controls without mussels, which is consistent with increases in both nitrate and ammonia in the inner bay between 1991 and 1993 (Johengen et al. 1995).

IMPACT ON PHYTOPLANKTON COMPOSITION

While detailed accounts of changes in phytoplankton community composition are not yet available, there is enough evidence to suggest that zebra mussel filtering and/or excretion activities have likely contributed to major changes in relative abundances of phytoplankton species. The most apparent change is the occurrence of a summer bloom of the cyanophyte (blue-green algae) *Microcystis*; intense blooms have occurred in the inner bay in late summer/fall 1994 and 1995 (Lavrentyev et al. 1995; Vanderploeg et al. 1996). Cyanophyte blooms have not been reported in the bay since phosphorus control measures were initiated in the mid-1970s (Richardson and Kreis 1987). The exact role of zebra mussels in initiating and sustaining these blooms is not clear. One likely theory is that mussels create favorable conditions for mass blooms of cyanophytes by selectively removing more desirable algae species and rejecting the less-palatable cyanophytes. Such activities would favor cyanophytes by decreasing competition for nutrients and allowing cyanophytes to rapidly grow when conditions are favorable (i.e., warm water temperature in late summer). Evidence for selective filtration/rejection by mussels is derived from both experimental results and direct observation. In the previously mentioned mesocosm experiments, not all algal classes were equally affected by the zebra mussel — diatoms and chlorophytes declined, but biovolumes of cyanophytes (*Microcystis*) and chryosphytes *(Synura)* remained unchanged (Heath et al. 1995). Further, during the intense bloom in 1994, zebra mussels selectively removed *Cyclotella* and *Cryptomonas*, but had no affect on abundances of *Microcystis* (Lavrentyev et al. 1995). Other studies using bay water showed that zebra mussels had little effect on chlorophyll when cyanophytes were abundant (Fanslow et al. 1995; Gardner et al. 1995). Recent observations using microcinematography showed that zebra mussels do not stop pumping water in the presence of *Microcystis*, but rather continued pumping while rejecting this species as unconsolidated pseudofeces (Vanderploeg et al. 1996).

Another theory is that zebra mussels may be affecting phytoplankton composition by altering N:P molar ratios. Many phytoplankton species growing at maximum rates will maintain a cell ratio of 16:1 (Redfield 1958); however, some species grow best at other ratios, and changes in species composition will occur depending upon how this ratio changes over time. Preliminary results indicate that mussels in the bay excrete nutrients at a ratio of greater than 40:1 (Johengen et al. 1995). This high ratio may explain the increase in nitrogen compared to phosphorus in the bay since the mussels became established. Yet, since *Microcystis* grows best at low N:P ratios (Rhee and Gotham 1980), it is unlikely that changes in the N:P ratio within the bay contributed to conditions that favor this species.

DISCUSSION

The impact that a suspension-feeding bivalve such as *D. polymorpha* will have on a given ecosystem will depend on a number of factors, including population densities, phytoplankton composition and growth rates, nutrient levels and dynamics, and physical factors such as water residence times and water column mixing. Since there have been a number of studies assessing the impacts of zebra mussels in North American waters over the past several years, particularly in western Lake Erie, changes observed in Saginaw Bay offer some interesting comparisons and contrasts.

Mean densities of the mussel population within inner Saginaw Bay are similar to, or lower than densities reported for other distinctly defined systems (Table 16.6). Relatively modest densities in the bay are likely a result of the limited amount of suitable hard substrate. Mussels are mostly found on cobble, and the proportion of cobble in a given area is highly variable and generally less than the total bottom area (Nalepa et al. 1995). Also, few mussels are found on soft sediments in the inner bay. In contrast, mussels are quite abundant in soft bottom regions of other systems such

TABLE 16.6

Densities of _Dreissena polymorpha_ and Corresponding Declines in Chlorophyll and Total Phosphorus in Different Water Bodies after Mussels Became Established

Water Body	Density (individuals m[-2])	Decline in Chlorophyll (%)	Decline in Total Phosphorus (%)	Filtration Turnover Rate (d[-1])
Saginaw Bay	1,200–3,400[a]	54	47	0.2–0.8[e]
Western Lake Erie	10,500[b]	67[c]	10[c]	2.2–9.8[c]
Oneida Lake	36,800[c]	34[c]	13[c]	0.5–1.4[c]
Lake St. Clair	3,000[d]	68[f]	0[c]	0.25[d]

Note: The filtration turnover rate is the theoretical time (days) it would take for the mussel population to filter the entire water volume. Densities are derived from whole-system surveys or from calculations based on the proportion of suitable substrate.

[a] from Nalepa et al. (1995) and Nalepa (unpublished).

[b] from Arnott and Vanni (1996).

[c] from Mellina et al. (1995).

[d] from Nalepa et al. (1996b).

[e] from Fanslow et al. (1995).

[f] From Nalepa et al. (1993).

as Lake St. Clair and western Lake Erie (Dermott and Munawar 1993; Nalepa et al. 1996b). In these two systems, unionids (both live and dead) are found in soft sediments and the shells provide a substrate for mussel attachment, and a means for the population to further expand (Hunter and Bailey 1992). Unionid shells are not found in regions with soft sediment in Saginaw Bay.

Despite marked differences in the spatial distribution of mussels as related to substrate type, spatial differences in the extent of water quality changes based on mussel densities could not be readily discerned (Fahnenstiel et al. 1995a). This finding might be expected since water circulation patterns respond rapidly to wind changes in the inner bay, resulting in a well-mixed water column (Danek and Saylor 1977). Also, vertical profiles of algal fluorescence (a function of chlorophyll) showed that near-bottom values were not lower than values in the upper water column (Nalepa et al. 1996a). If the water column was not well-mixed, near-bottom depletion would have occurred. Overall, impacts of suspension feeders are most prominent in well-mixed systems (Officer et al. 1982).

A key consideration in assessing the impact of the zebra mussel population on water quality parameters is the relationship between filtration capacity and phytoplankton growth rates. In 1992, when the population in Saginaw Bay was at a peak, filtration turnover rate exceeded algal growth rates by threefold (0.8 d[-1] vs. 0.25 d[-1]), but mean densities decreased in 1993, and filtration turnover rates were lower and similar to algal growth rates (0.2 d[-1] vs. 0.2 d[-1]). Interestingly, despite the decline in filtration turnover rates between 1992 and 1993, corresponding mean values of chlorophyll, total phosphorus, and water clarity were similar for the two years. This suggests that the impact of mussel filtering activity on water quality parameters is not linear, and that impacts will be observed as long as filtration turnover rates are at, or above, some threshold level compared to algal growth rates. A comparison between changes observed in western Lake Erie and Saginaw Bay illustrate this point. In the western basin of Lake Erie, filtration turnover rates were estimated to be between 2.2 and 9.8 d[-1] (Mellina et al. 1995). While algal growth rates in the western basin have not been measured, they were not likely greater than the general mean growth rate of 1.0 d[-1] (Reynolds 1984). Thus, although filtration turnover rates were much lower relative to algal growth rates in Saginaw Bay than in western Lake Erie, changes in chlorophyll were generally similar, and changes in total phosphorus were greater (Table 16.6).

The decline in total phosphorus in Saginaw Bay was greater than observed in other lake systems invaded by zebra mussels (Table 16.6). Exact reasons for the greater relative decline in total phosphorus are not clear, but the magnitude of the decline appears consistent with other changes within the bay itself. Given that the mean steady-state biomass of the zebra mussel population in the inner bay is about 3.9 g m^{-2} (Table 16.4), and assuming phosphorus in soft tissue is about 1.0% (Johengen et al. 1995), the amount of phosphorus bound in mussel biomass is 39 mg m^{-2}. With a relative decline of total phosphorus of about 15 µg l^{-1} (Table 16.5), and with a mean depth of 5 m in the inner bay, the decline in total phosphorus amounts to about 75 mg m^{-2}. The difference of 36 mg phosphorus between water column loss and that bound in mussel biomass can certainly be attributed to increased biomass of benthic algae and macrophytes (Skubinna et al. 1995), or perhaps to incorporation into biodeposited material such as feces and pseudofeces. However, it is unlikely that biodeposits significantly contributed to losses of total phosphorus in the water column. This material is readily resuspended (Haven and Morales-Alamo 1966), and mass balance models imply that this material does not represent permanent removal from the water column (Mellina et al. 1995). The lower decline in water column total phosphorus in western Lake Erie (from 39 µg l^{-1} before mussels to 35 µg l^{-1} after mussels) observed by Holland et al. (1995) was attributed to sediment resuspension. Yet sediment resuspension of phosphorus is also significant in Saginaw Bay (Bierman et al. 1984). Conceivably, differences in relative declines of total phosphorus between the various systems were a result of differences in phosphorus inputs and throughput rates in relation to population biomass. For example, total phosphorus apparently did not decline in Lake St. Clair after mussels became established (Table 16.6). Mean biomass of the zebra mussel population in Lake St. Clair in 1994 was about 3.1 g m^{-2} (Nalepa et al. 1996b). Assuming a phosphorus tissue content of 1.0%, the amount of phosphorus bound in soft tissue is 31 mg m^{-2}, or 34 t for the entire lake (lake area = 1.11×10^9 m^2). Since the annual phosphorus load of Lake St. Clair is 3,100 t (mean of 1975–80 period; Lang et al. 1988), phosphorus bound in mussel tissue is only 1% of the annual load. Also, phosphorus throughput time is very rapid as water residence time in Lake St. Clair is only 9 days. Accurate basin-wide estimates of population biomass in western Lake Erie are not available. However, if overall densities are about four times greater in western Lake Erie than in Saginaw Bay (Table 16.6), and assuming all other factors are equal (size-frequency, length-weight), then the amount of phosphorus bound in mussel tissue in western Lake Erie would be four times that of Saginaw Bay, or 156 mg m^{-2}. This amount would be equivalent to 574 t (western basin area = 3.68×10^9 m^2) or 12% of the annual load of 6,693 t (mean of 1984–91 period; Mellina et al. 1995). While the proportion of the phosphorus load bound in mussel tissue is less in Saginaw Bay (7% for 1993) than in western Lake Erie, throughput time is less since water residence time is longer (120 days for inner Saginaw Bay vs. 51 days for the western basin of Lake Erie).

Evidence that Saginaw Bay is more severely P-limited than western Lake Erie can be derived from relative concentrations of SRP in the water column. Values of SRP are three to four times lower in Saginaw Bay (Johengen et al. 1995) and, as such, any dissolved phosphate excreted by mussels would less likely accumulate in the water column. Hence, SRP levels were generally lower in Saginaw Bay after mussels became established, while SRP in western Lake Erie increased by 11% (Holland et al. 1995). Also, zebra mussels in Saginaw Bay excreted nutrients at a N:P ratio of greater than 40:1, while this ratio was less than 20:1 for mussels in western Lake Erie (Arnott and Vanni 1996), indicating that phosphorus in Saginaw Bay is more efficiently retained by mussels and not as rapidly recycled into the water column. Interestingly, despite different phosphorus levels and N:P ratios in the two systems, blooms of *Microcystis* have also recently appeared in western Lake Erie (Taylor 1995), which gives credence to the selection/rejection hypothesis of Vanderploeg et al. (1996).

Obviously, the invasion of *D. polymorpha* into North American waters has altered fundamental relationships between phosphorus loads, phosphorus cycling, and measures of water quality. For Saginaw Bay, reductions in loads may now be more reflected in corresponding reductions in benthic algae, a group far more difficult to measure and quantify. Also, there is no longer a simple, direct relationship between nutrient reductions and diminished probability of cyanophyte blooms. With the likely role of zebra mussel in creating conditions favorable for these blooms, efforts to reduce nutrient loads below target levels must be reassessed if blooms are to be prevented.

Eutrophication models are valuable tools in assessing and predicting responses of water quality variables to nutrient abatement efforts. Because of the zebra mussel, new assumptions will need to be validated with experimental data, and models will need to be recalibrated with long term data sets. Submodels of zebra mussel population dynamics need to be incorporated into such modeling efforts to not only evaluate mussel impacts, but also to assess interactive responses of the population to created changes. Such models are being developed (Meisner et al. 1993; Limno-Tech 1995), and early results have led to some interesting predictions. For example, the model of Meisner et al. (1993) for the Bay of Quinte, Ontario, predicted that declines in total phosphorus in the water column will be restricted to the short term. After full colonization (steady state), total phosphorus would return to preinvasion levels because of sediment resuspension and net flux out of the sediments. In Saginaw Bay, total phosphorus has remained lower for the entire post-zebra mussel period despite the importance of resuspension in influencing total phosphorus levels in the water (Bierman et al. 1984).

Models of estuarine systems have shown that suspension-feeding bivalves that form pseudofeces are the most important determinant of stability in water quality variables (Herman and Scholten 1990; Gerritsen et al. 1994). Suspension feeders contribute to functional stability by exerting continuous pressure on phytoplankton. These models predict that feeding influences are so great that even major increases in nutrient loads have little impact on phytoplankton biomass; on the other hand, only a minor decline in suspension-feeder biomass could lead to major increases in phytoplankton at the same nutrient level (Herman and Scholten 1990). The role of zebra mussel as a stabilizing influence on phytoplankton in Saginaw Bay remains questionable. Certainly, chlorophyll levels in both spring and late summer/fall are lower now than before mussels became established, but seasonal differences are more pronounced. Chlorophyll levels in late summer/fall averaged 1.15 times greater than spring levels in years prior to mussels (1978–90), but average 2.02 times greater in years after mussels (1992–95).

Have water quality parameters and zebra mussel populations reached a steady state in Saginaw Bay? Density and biomass of the mussel population have remained relatively constant over the past few years (1993–95), and distinct trends in chlorophyll have not been apparent over the same period. In addition, seasonal changes in 1994 and 1995 have followed a characteristic pattern — an intense clear-water phase in the spring followed by a late summer bloom of the cyanophyte *Microcystis*. Additional yearly data, particularly in years when phosphorus loadings are atypical, should provide valuable information in predicting the response of water quality parameters to nutrient abatement efforts in the post-zebra mussel era.

ACKNOWLEDGMENTS

We thank the following for their assistance during this project: Tom Bridgeman, Jenny Buchanan, Bill Burns, Joann Cavaletto, David Fanslow, Mark Ford, Wendy Gordon, Gerry Gostenik, David Hartson, Ann Krause, Greg Lang, Ann Vielmetti, Bruce Wagner, MaryJo Wimmer, and Jim Wojcik. We especially thank Mike McCormick for his technical assistance and thought-provoking ideas, Cathy Darnell for her editorial comments, and Al Beeton, Dave Reid, and Hank Vanderploeg for their support and encouragement. This is GLERL Contribution Number 1037.

REFERENCES

APHA. *Standard Methods for the Examination of Water and Wastewater.* 17th ed., American Public Health Association, Washington, DC, 1990.

Arnott, D.L. and M.J. Vanni. Nitrogen and Phosphorus Cycling by the Zebra Mussel (*Dreissena polymorpha*) in the Western Basin of Lake Erie. *Can. J. Fish. Aquat. Sci.*, 53, pp. 646–659, 1996.

Beeton, A.M., S.H. Smith, and F.H. Hooper. *Physical Limnology of Saginaw Bay, Lake Huron.* Technical Report Number 12, Great Lakes Fishery Commission, Ann Arbor, MI, p. 56, 1967.

Bierman, V.J., D.M. Dolan, and R. Kasprzyk. Retrospective Analysis of the Response of Saginaw Bay, Lake Huron, to Reductions in Phosphorus Loadings. *Environ. Sci. Technol.*, 18, pp. 23–31, 1984.

Bratzel, M.P., M.E. Thompson, R.J. Bowden (Eds.). *The Water of Lake Huron and Lake Superior, Vol. II (Part A): Lake Huron, Georgian Bay, and the North Channel.* Report to the International Joint Commission by the Upper Great Lakes Reference Group, Windsor, Ontario, 1977.

Bridgeman, T.B., G.L. Fahnenstiel, G.A. Lang, and T.F. Nalepa. Zooplankton Grazing During the Zebra Mussel (*Dreissena polymorpha*) Colonization of Saginaw Bay, Lake Huron. *J. Great Lakes Res.*, 21, pp. 567–573, 1995.

Canale, R.P., P.L. Freedman, M.T. Auer, and J.J. Sygo. *Saginaw Bay Limnological Data.* Tech. Rep. #54. MICHU-SG-76-207, Michigan Sea Grant Program, Ann Arbor, MI, 1976.

Cloern, J.E. Does the Benthos Control Phytoplankton Biomass in South San Francisco Bay? *Mar. Ecol. Prog. Ser.*, 9, pp. 191–202, 1982.

Cohen, R.R.H., P.V. Dresler, E.J.P. Phillips, and R.L. Cory. The Effect of the Asiatic Clam, *Corbicula fluminea*, on Phytoplankton of the Potomac River, Maryland. *Limnol. Oceanogr.*, 29, pp. 170–180, 1984.

Cotner, J.B., W.S. Gardner, J.R. Johnson, R.H. Sada, J.F. Cavaletto, and R.T. Heath. Effects of Zebra Mussels (*Dreissena polymorpha*) on Bacterioplankton: Evidence for Both Size-Selective Consumption and Growth Stimulation. *J. Great Lakes Res.*, 21, pp. 517–528, 1995.

Dame, R., N. Dankers, T. Prines, H. Jongsma, and A. Smaal. The Influence of Mussel Beds in Nutrients in the Western Wadden Sea and Eastern Scheldt Estuaries. *Estuaries*, 14, pp. 130–138, 1991.

Danek, L.J., and J.H. Saylor. Measurements of the Summer Currents in Saginaw Bay, Michigan. *J. Great Lakes Res.*, 3, pp. 65–71, 1977.

Davis, C. O. and Simmons, M. S. 1979. *Water Chemistry and Phytoplankton Field and Laboratory Procedures.* Special Report No. 70, Great Lakes Research Division, University of Michigan, Ann Arbor, MI, 1979.

Dermott, R. and M. Munawar. Invasion of Lake Erie Offshore Sediments by *Dreissena*, and Its Ecological Implications. *Can. J. Fish. Aquat. Sci.*, 50, pp. 2298–2304, 1993.

Fahnenstiel, G.L., G.A. Lang, T.F. Nalepa, and T.J. Johengen. Effects of Zebra Mussel (*Dreissena polymorpha*) Colonization on Water Quality Parameters in Saginaw Bay, Lake Huron. *J. Great Lakes Res.*, 21, pp. 435–448, 1995a.

Fahnenstiel, G.L., T.B. Bridgeman, G.A. Lang, M.J. McCormick, and T.F. Nalepa. Phytoplankton Productivity in Saginaw Bay, Lake Huron: Effects of Zebra Mussel (*Dreissena polymorpha*) Colonization. *J. Great Lakes Res.*, 21, pp. 465–475, 1995b.

Fanslow, D.L., T.F. Nalepa, and G.L. Lang. Filtration Rates of Zebra Mussels (*Dreissena polymorpha*) on Natural Seston From Saginaw Bay, Lake Huron. *J. Great Lakes Res.*, 21, pp. 489–500, 1995.

Gardner, W.S., J.F. Cavaletto, T.H. Johengen, J.R. Johnson, R.T Heath, and J.B. Cotner, Jr. Effects of the Zebra Mussel (*Dreissena polymorpha*) on Nitrogen Dynamics in Saginaw Bay, Lake Huron. *J. Great Lakes Res.*, 21, pp. 529–544, 1995.

Gerritsen, J., A.F. Holland, and D.E. Irvine. Suspension-Feeding Bivalves and the Fate of Primary Production: An Estuarine Model Applied to Chesapeake Bay. *Estuaries*, 17, pp. 403–416, 1994.

Haven, D.S. and R. Morales-Alamo. Aspects of Biodeposition by Oysters and Other Invertebrate Filter Feeders. *Limnol. Oceanogr.*, 11, pp. 487–498, 1966.

Heath, R.T., G.L. Fahnenstiel, W.S. Gardner, J.F. Cavaletto, and S.J. Hwang. Ecosystem-Level Effects of Zebra Mussels (*Dreissena polymorpha*): An Enclosure Experiment in Saginaw Bay, Lake Huron. *J. Great Lakes Res.*, 21, pp. 501–516, 1995.

Hebert, P.D.N., C.C. Wilson, M.H. Murdoch, and R. Lazar. Demography and Ecological Impacts of the Invading Mollusc, *Dreissena polymorpha. Can. J. Zool.*, 69, pp. 405–409, 1991.

Herman, P.M.J. and H. Scholten. Can Suspension Feeders Stabilise Ecosystems?, in *Trophic Relationships in the Marine Environment*, Barnes, M. and Gibsen R.N., Eds., Aberdeen University Press, Aberdeen, U.K., 1990.

Holland, R.E. Changes in Planktonic Diatoms and Water Transparency in Hatchery Bay, Bass Island area, Western Lake Erie Since the Establishment of the Zebra Mussel. *J. Great Lakes Res.*, 19, pp. 617–624, 1993.

Holland, R.E., T.H. Johengen, and A.M. Beeton. 1995. Trends in Nutrient Concentrations in Hatchery Bay, Western Lake Erie, Before and After *Dreissena polymorpha*. *Can. J. Fish. Aquat. Sci.*, 52, pp. 1202–1209, 1995.

Hunter, R.D., and J.F. Bailey.. *Dreissena polymorpha* (Zebra Mussel): Colonization of Soft Substrata and Some Effects of Unionid Bivalves. *Nautilus*, 106, pp. 60–67, 1992.

Johengen, T.H., T.F. Nalepa, G.L. Fahnenstiel, and G. Goudy. Nutrient Changes in Saginaw Bay, Lake Huron after the Establishment of the Zebra Mussel *Dreissena polymorpha*. *J. Great Lakes Res.*, 21, pp. 449–464, 1995.

Lang, G.A., J.A. Morton, and T.D. Fontaine, III. Total phosphorus budget for Lake St. Clair: 1975-1980. *J. Great Lakes Res.*, 14, pp. 257–266, 1988.

Lavrentyev, P.J., W.S. Gardner, J.F. Cavaletto, and J.R. Beaver. Effects of the Zebra Mussel (*Dreissena polymorpha*) on Protozoa and Phytoplankton in Saginaw Bay, Lake Huron. *J. Great Lakes Res.*, 21, pp. 545–557, 1995.

Leach, J.H. Impacts of the Zebra Mussel (*Dreissena polymorpha*) on Water Quality and Fish Spawning Reefs in Western Lake Erie, in *Zebra Mussels: Biology, Impacts, and Control*, Nalepa, T.F. and Schloesser, D.W., Eds., Lewis Publishers/CRC Press, Boca Raton, FL, 1993.

Limnotech, Inc. *A Preliminary Ecosystem Modeling Study of Zebra Mussels (Dreissena polymorpha) in Saginaw Bay, Lake Huron*. LTI-Limnotech, Ann Arbor, MI, p. 120, 1995.

Lowe, R.L. and R.W. Pillsbury. Shifts in Benthic Algal Community Structure and Function Following the Appearance of Zebra Mussels (*Dreissena polymorpha*) in Saginaw Bay, Lake Huron. *J. Great Lakes Res.*, 21, pp. 558–566, 1995.

MacIsaac, H.J., W.G. Sprules, O.E. Johannsson, and J.H. Leach. Filtering Impacts of Larval and Sessile Zebra Mussels (*Dreissena polymorpha*) in Western Lake Erie. *Oecologia*, 92, pp. 30–39, 1992.

Marsden, J.E., N. Trudeau, and T. Keniry. *Zebra Mussel Study of Lake Michigan*. Aquatic Ecology Technical Report 93/14, Illinois Natural History Survey, Zion, IL, p. 51, 1993.

Meisner, J.D., M. Kuc, and C.H.R. Wedeles. *Effects of Zebra Mussels, Dreissena polymorpha, on Summer Phosphorus and Algal Biomass of the Upper Bay of Quinte: Implications for Remedial Action Scenarios.* Report Prepared for the Bay of Quinte Remedial Action Plan Coordinating Committee, Great Lakes Laboratory for Fisheries and Aquatic Sciences, Fisheries and Oceans Canada, Burlington, ON, 1993.

Mellina, E., J.B. Rasmussen, and E.L. Mills. Impact of Zebra Mussel (*Dreissena polymorpha)* on Phosphorus Cycling and Chlorophyll in Lakes. *Can. J. Fish. Aquat. Sci.*, 52, pp. 2553–2573, 1995.

Nalepa, T.F. Long term Changes in the Macrobenthos of Southern Lake Michigan.. *Can. J. Fish. Aquat. Sci.*, 44, pp. 515–524, 1987.

Nalepa, T.F., J.F. Cavaletto, M. Ford, W.M. Gordon, and M. Wimmer. Seasonal and Annual Variation in Weight and Biochemical Content of the Zebra Mussel, *Dreissena polymorpha*, in Lake Huron. *J. Great Lakes Res.*, 19, pp. 541–552, 1993.

Nalepa, T.F., J.A. Wojcik, D.L. Fanslow, and G.A. Lang. Initial Colonization of the Zebra Mussel (*Dreissena polymorpha*) in Saginaw Bay, Lake Huron: Population Recruitment, Density and Size Structure. *J. Great Lakes Res.*, 21, pp. 417–434, 1995.

Nalepa, T.F., G.L. Fahnenstiel, M.J. McCormick, T.H. Johengen, G.L. Lang, J.F. Cavaletto, and G. Goudy. *Physical and Chemical Variables of Saginaw Bay, Lake Huron in 1991-93*. NOAA Technical Memorandum ERL GLERL-91, Great Lakes Environmental Research Laboratory, Ann Arbor, MI, p.78, 1996a.

Nalepa, T.F., D.J. Hartson, G.W. Gostenik, D.L. Fanslow, and G.A. Lang. Changes in the Freshwater Mussel Community of Lake St. Clair: From Unionidae to *Dreissena polymorpha* in Eight Years. *J. Great Lakes Res.*, 22, pp. 354–369, 1996b.

Nicholls, K.H. and G.J. Hopkins. Recent Changes in Lake Erie (North Shore) Phytoplankton: Cumulative Impacts of Phosphorus Loading Reductions and the Zebra Mussel Introduction. *J. Great Lakes Res.*, 19, pp. 637–647, 1993.

Officer, C.B., T.J. Smayda, and R. Mann. Benthic Filter Feeding: A Natural Eutrophication Control. *Mar. Ecol. Prog. Ser.*, 9, pp. 203-210, 1982.

Redfield, A.C. The Biological Control of Chemical Factors in the Environment. *Am. Sci.*, 46, pp. 205–221, 1958.

Reynolds, C.S. *The Ecology of Freshwater Phytoplankton*. Cambridge University Press, Cambridge, UK, pp. 193–199, 1984.

Rhee, G.Y. and I.J. Gotham. Optimum N:P Ratios and Co-Existence of Planktonic Algae. *J. Phycol.*, 16, pp. 486–489, 1980.

Richardson, W.L. and R.G. Kreis, Jr. Historical Perspectives of Water Quality in Saginaw Bay, in *Proceedings: A New Way for the Bay, A Workshop for the Future of Saginaw Bay*, East Central Michigan Planning and Development Region, Saginaw, MI, 1987.

Skubinna, J.P., T.G. Coon, and T.R. Batterson. Increased Abundance and Depth of Submersed Macrophytes in Response to Decreased Turbidity in Saginaw Bay, Lake Huron. *J. Great Lakes Res.*, 21, pp. 476–488, 1995.

Smith, V.E., K.W. Lee, J.C. Filkins, K.W. Hartwell, K.R. Rygwelski, and J.M. Townsend. *Survey of Chemical Factors in Saginaw Bay (Lake Huron)*. Ecological Research Series. EPA-600/3-77-125, Environmental Protection Agency, Duluth, MN, p. 159, 1977.

Strayer, D.L., J. Powell, P. Ambrose, L.C. Smith, M.L. Pace, and D.T. Fischer. Arrival, Spread, and Early Dynamics of a Zebra Mussel *(Dreissena polymorpha)* Population in the Hudson River Estuary. *Can. J. Fish. Aquat. Sci.*, 53, pp. 1143–1149, 1996.

Strickland, J.D.H. and Parsons, T.R. *A Practical Handbook of Seawater Analysis*, 2nd ed., Bull. Fish. Res. Bd. Can. No. 167, 1972.

Taylor, R. Bloom of Blue-Green Alga Returns to Lake Erie, in *Twineline*, Ohio Sea Grant Program, 15, pp. 1,14, 1995.

Vanderploeg, H.A., T.H. Johengen, J.R. Strickler, J.R. Liebig, and T.F. Nalepa. Zebra Mussels May Be Promoting *Microcystis* Blooms in Saginaw Bay and Lake Erie. *Abstract, 44th Annual Meeting North American Benthological Society*, Kalispell, MT, 1996.

Vollenweider, R.A., Munawar, M., and Stadelman, P.A. Comparative Review of Phytoplankton and Primary Production in the Laurentian Great Lakes. *J. Fish. Res. Bd. Can.*, 31, pp. 739–762, 1974.

Walz, N. The Energy Balance of the Freshwater Mussel *Dreissena polymorpha* Pallas in Laboratory Experiments and in Lake Constance. I. Pattern of Activity, Feeding and Assimilation Efficiency. *Arch. Hydrobio./Suppl.*, 55, pp. 83–105, 1978.

Wood, L.E. Bottom Sediments of Saginaw Bay, Michigan. *J. Sediment. Petrol.*, 34, pp. 173–184, 1964.

Wright, R.T., R.B. Coffin, C.P. Ersing, and D. Pearson. Field and Laboratory Measurements of Bivalve Filtration of Natural Marine Bacterioplankton. *Limnol. Oceanogr.*, 27, pp. 91–98, 1982.

Wu, L. and D.A. Culver. 1991. Zooplankton Grazing and Phytoplankton Abundance: An Assessment Before and After Invasion of *Dreissena polymorpha*. *J. Great Lakes Res.*, 17, pp. 425–436, 1964.

17 A Microbiological, Chemical, and Physical Survey of Ballast Water on Ships in the Great Lakes

G.E. Whitby, D.P. Lewis, M. Shafer, and C.J. Wiley

INTRODUCTION

Ballast water from cargo vessels is considered to be one of the major factors in the dissemination of aquatic organisms worldwide. While there have been a number of reported incidences of microbiological movements or introductions in the marine environment through ballast water, to date little attention has been paid to freshwater ecosystems and, in particular, activities in the Great Lakes. Ballast water from commercial shipping was found to be the vehicle for the introduction of *Vibrio cholerae* to Mexico from 1991 to 1992 (McCarthy and Khambaty, 1994). In addition, three fish diseases are also known to have been introduced to the Great Lakes; a bacterium and two protozoans (Mills et al., 1993). The ocean can serve as a reservoir for the caliciviruses and these can emerge and infect terrestrial hosts, including humans. The best-documented example of ocean caliciviruses causing disease in terrestrial species is the animal disease vesicular exanthema of swine (VES) (Smith et al., 1998).

The potential danger to the ecosystem, economy, and human health posed by ballast water introductions has prompted international initiatives for preventing ballast water-mediated invasion by aquatic nonindigenous organisms. The most recent activities were prompted by the invasion of the zebra mussel (*Dreisenna polymorpha*) into the ecosystem of the Great Lakes.

In the fall of 1994, the Canadian Coast Guard commissioned a study to investigate the presence of aquatic nuisance species in the ballast water of commercial ships. A part of this study looked at the potential threat to the ecosystem and human health of microorganisms in the ballast water of commercial ships entering the Great Lakes.

The majority of vessels that come into the Great Lakes from foreign ports report that they have No Ballast On Board (NOBOB). These vessels often pick up ballast in the Great Lakes after they unload their initial cargo. They may then proceed to a second port to take on new cargo where they will dump the mixture of water from the Great Lakes and foreign ballast water remnants. This mixture may contain aquatic nuisance organisms and microbial pathogens from foreign freshwater sources or from secondary staging areas in the Great Lakes.

To date, very little attention has been given to the introduction of pathogenic microorganisms into the Great Lakes via this vector. In this study, ballast waters, and in the cases where no ballast was reported on board, ballast water remnants and associated sediment, were examined for fecal indicator organisms and *Vibrio anguillarum*. *V. anguillarum* was picked because it is pathogenic to fish and found only in salt water. It is not been found in the Great Lakes to date. Therefore, its presence would indicate that picking up ballast water at sea could introduce new fish pathogens.

In addition to microbiological sampling, some chemical and physical parameters were also measured that may be of significance when considering methods of monitoring or controlling microbial pathogens in ballast waters.

MATERIALS AND METHODS

SAMPLE COLLECTION AND ANALYSIS

The field program was completed over a 2-year period. The microbiological study was included in the second year of the work. More than 100 vessels in the Welland Canal and nearby ports were boarded during the study period, and ballast water or ballast water remnants were sampled. Samples from 60 of these vessels were submitted for microbial analysis.

Samples were collected from double bottom, stern or aft ballast tanks. Each tank was sampled through sounding pipes and access ways by the pumping action of a modified inertial hand pump. Samples of 1 liter were collected in sterile bottles and refrigerated until processed within 24 hours. Prior to each sample collection, 20 liters of water was passed through the sampler to prevent contamination by the previous sample.

The majority of the vessels sampled had reported No Ballast On Board (NOBOB) and so they were not subject to ballast exchange regulations. Nevertheless, there was ample water associated with the sediment in the bottom of these tanks to collect adequate sample volumes.

PHYSICAL AND CHEMICAL MEASUREMENTS

The following physicochemical properties of the samples of ballast water were measured: pH, salinity, total and dissolved iron, hardness, total suspended solids, and UV transmittance at a wavelength of 254 nm.

MICROBIOLOGICAL EXAMINATION

The initial study was not designed to isolate any human or aquatic pathogens except for *V. anguillarum*. The presence of considerable numbers of *Escherichia coli* and vibrios prompted further work to determine if there were any pathogenic *E. coli* or *V. cholerae*.

The microbial examination included the heterotrophic plate count, total coliforms, fecal coliforms, *E. coli* (enteropathogenic and nonpathogenic), as well as species of *Vibrio* pathogenic to humans and/or fish (i.e., *V. anguillarum*). Pathogenic *E. coli* and *V. cholerae* identifications were sent to the National Laboratory for Enteric Pathogens, LCDC, Health Canada to be confirmed.

RESULTS AND DISCUSSION

CHEMICAL AND PHYSICAL RESULTS AND DISCUSSION

The investigation took place between November 12 and December 12, 1995, and included inland or coastal and ocean-going vessels. A total of 71 water samples were collected from the multiple ballast water tanks of 60 cargo ships. The majority of these vessels were in transit through the Welland Canal that connects Lake Ontario and Lake Erie in Canada. Several vessels were also boarded at the ports of Toronto and Hamilton in Ontario, Canada.

The pH values of the 71 ballast water samples ranged between 6.5 and 8.2 with a mean of 7.5. The pH of water in the Great Lakes and the ocean usually ranges between 7.5 and 8.5 (Mackie and Kilgour, 1995; Austin, 1993). The pH values of the ballast waters investigated in this study were close to the values reported for the Great Lakes and presented no stress for the microorganisms.

Salinity measurements were performed on 71 water samples with values ranging from 0 to 40 ppt. The mean salinity of ballast waters from the ocean going ships was 14 ppt versus the U.S.

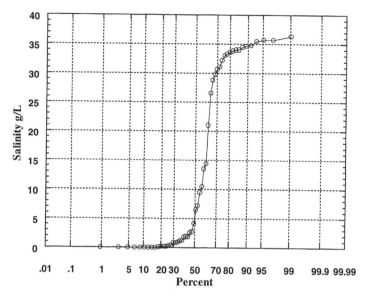

FIGURE 17.1   Distribution of salinities in the samples from the ballast water and ballast water remnants of the ships (includes vessels reporting No Ballast On Board).

Coast Guard regulatory regime of 30 ppt, which is used as an indicator of satisfactory ballast water exchange (Wiley, 1996). Figure 17.1 shows that 75% of all samples exhibited salinities below 30 ppt, and in fact, 50% were less than 8 ppt. Samples collected from ocean going vessels varied significantly in salinity between ships (Figure 17.1) and even from tank to tank on a given ship (Table 17.1).

TABLE 17.1
The Level of Salinity in the Various Double Bottom Ballast Tanks of a Single Ocean-Going Vessel

Ballast Tank	Salinity (ppt)
5ST	13.5
6ST	0.9
3ST	0.2
8ST	31.1
3P	34
5P	33.8

Hardness and iron can effect chemical and physical disinfectants such as chlorine and ultraviolet light, respectively. The determination of total hardness revealed varied results among the 71 water samples tested. The total hardness of the water samples varied from 20 to 11,000 mg/l with a mean of 2,300 mg/l. The iron content of the ballast water samples varied from zero to 600 mg/l with a mean of 17 mg/l.

Sixty-eight ballast water and ballast water remnant samples were analyzed for their total suspended solids content, and the distribution of values is shown in Figure 17.2. The total suspended

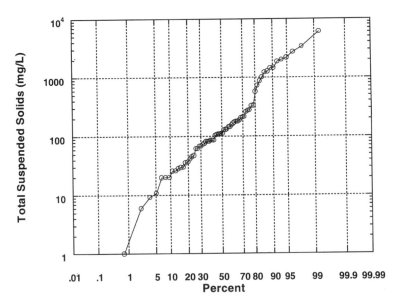

FIGURE 17.2   Distribution of suspended solids in the samples from the ballast water of the ships.

solids ranged from 1 to 6,024 mg/l with a mean of 478 mg/l. Suspended solids can harbor and protect microbial pathogens, making them very difficult to kill with chemical or physical disinfectants (Bitton, 1994).

The percent UV transmittance was measured at a wavelength of 254 nm to determine whether UV light would be a suitable disinfectant. The majority of UV lamps that are used for disinfecting water produce almost all of their UV light at a wavelength of 254 nm. Figure 17.3 shows the results for the 71 samples before and after filtration through a 0.45 µm filter. The UV transmittance ranged from 0.04% to 94% with a mean of 57% for the unfiltered samples and 90% for the filtered samples with a range of 67–100%. Over 70% of the samples had UV transmissions of 55% or better before filtration.

MICROBIOLOGICAL RESULTS AND DISCUSSION

The results of the bacteriological sampling of the ballast water for total and fecal coliforms, *E. coli*, enterococci and heterotrophic plate count are shown in Table 17.2.

The presence of any fecal indicator organisms in ballast water indicates the potential for the presence of microbial pathogens. In this study, 48% of the samples were positive for fecal contamination. Fecal coliforms and *E. coli* were found in 31 (45%) of the 69 samples collected from vessels in the study. When the results are based on the presence of enterococci, 74% of the samples showed fecal contamination. Enterococci may be considered a better indicator of fecal contamination in studies of ballast water from ocean going vessels because of their longer survival in salt water (Bitton, 1994).

It has been recommended that the heterotrophic plate count of drinking water should be less than 100/ml or 500/ml, especially if it is reclaimed water (Grabow, 1990). It is reasonable to argue that since the ballast waters are of foreign origin and not from a drinking water source they may contain exotic pathogens. These exotic organisms may pose a greater risk to the local environment or human health, and therefore these waters should meet North American drinking water standards to minimize this risk. A heterotrophic plate count over 500 per ml was found in 43 samples, and 57 were over 100 per ml.

The known human pathogen enterohemorrhagic *E. coli* serotype O111 was isolated from one ship. *Pseudomonas aeruginosa*, an opportunistic pathogen known to cause pink eye in infants and

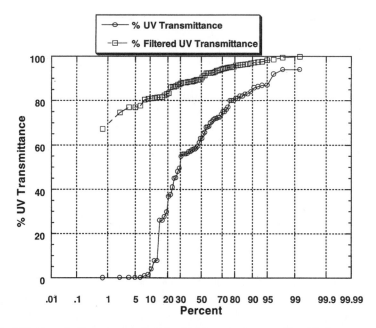

FIGURE 17.3   UV transmission at a wavelength of 254 nm in the samples of ballast water from the ships before and after filtration through a 0.45 µm filter.

TABLE 17.2
Results of the Bacteriological Sampling of the Ballast Water for Total and Fecal Coliforms, *E. coli*, Enterococci, and Heterotrophic Plate Count

Indicator Organisms	Number of Samples	Mean Colonies per 100 ml	Standard Deviation	Range	Positive Samples	Percent Positive Samples
Total coliforms	69	700	2300	0–17,600	45	65
Fecal coliforms	69	29	70	0–400	33	48
E. coli	69	16	43	0-240	31	45
Enterococci	70	51	94	0–480	52	74
Heterotrophic plate count per ml	70	3500	10,600	2–86,000		

implicated in some forms of pneumonia (Benenson, 1990), was recovered from the ballast water of another ship. *Provodencia rettgeri*, implicated in nosocomial infections, was also isolated from a separate ballast water sample.

The media used for the initial isolation of the vibrios was specific for *V. anguillarum*. As a follow up, all of the vibrios that had been isolated from this media were tested to determine if they were *V. cholerae* as well as *V. anguillarum*. Since *V. anguillarum* is only found in seawater, its presence could indicate that other pathogens to fish or other aquatic species are in the ballast water. These pathogens could have been picked up during the ballast water exchange at sea. *Vibrio* spp. were isolated from the ballast waters of 17 vessels. One isolate was initially identified as *V. cholerae* by limited biochemical and serological testing, but Health Canada (Laboratory Centre for Disease Control) confirmed it as the closely related and opportunistic pathogen, *Aeromonas sobria*. The other isolates were identified as *V. fluvialis, V. alginoliticus*, and two related species, *A. hydrophila* and *A. caviae. A. hydrophila* is an opportunistic pathogen.

SUMMARY AND CONCLUSION

The open ocean exchange of ballast water will not prevent the spread of pathogenic organisms by ships as it is presently practiced, and may result in the picking up of additional aquatic pathogens of saltwater origin that may survive in freshwater.

The pH values of the ballast water investigated in this study were close to pH 8, which is similar to the values reported for the Great Lakes and presented no stress for the organisms.

Present methods for exchanging ballast water appear not to ensure adequate exchange in all ballast tanks, did not produce consistently high levels of salinity that would be required to kill aquatic nuisance organisms, and did not remove all the freshwater microorganisms, or result in their total mortality.

The isolation of enteropathogenic *E. coli* serovar O111, *Aeromonas* sp., and *Vibrio* sp. shows the pathogenic potential of the ballast water entering the Great Lakes ecosystem and the potential for pathogens to be discharged in ports in the vicinity of heavily populated areas, swimming beaches, and drinking water intakes. The potential threat these introductions represent to human health has yet to be determined. It is apparent, however, that any method of treating ballast water to control nuisance species will have to include a disinfection component.

Ballast water should be tested for viruses and protozoan cysts that are much more resistant to adverse conditions in the environment than the bacterial indicators. In this work, bacteriophages may be useful to test for as surrogates for enteric viruses (Lee et al., 1997).

It is apparent from the present study that domestic and foreign fleets operating in the North American Great Lakes system may contain significant numbers of human pathogens. Little detailed work has been completed with regard to the microbiology of ballast water as it relates to the introduction of nonindigenous pathogens into North American freshwaters, although some work has been done in other jurisdictions. *Clostridium botulinum* C was found in the sediment of a Norwegian vessel docked in Australia after having visited Singapore (Anderson, 1992).

Our study did not confirm the presence of *V. cholerae*; however, it has been documented that *V. cholerae* can enter a viable and nonculturable state. It may be necessary to carry out further tests for this organism under conditions more conducive to their detection (Colwell, 1996).

The microbiological sampling took place in the late fall and early winter when the temperature of the ballast water would depress the proliferation of pathogens (Geldreich, 1996). A study in the summer when the ballast water is much higher in temperature may show greater numbers of fecal indicator organisms and pathogens.

The ecology of the Great Lakes is presently in a state of flux. As stresses such as the success of new invading species and continued global warming keep the system reeling out of balance, new opportunities for further introductions exist (Colwell, 1996). The opportunities for pathogens are presently no less than for other invaders. It is well known that shellfish can concentrate viruses and other enteric pathogens from the water column. The presence of extremely high densities of zebra mussels may prove to be an ideal reservoir and vector for pathogenic microorganisms. In addition many industrial ports in the Great Lakes are in close proximity to heavily populated areas, which use the lakes for recreation and as a source of potable water. It is clear that the issue of pathogen introductions to the Great Lakes and other freshwater systems, via ballast water, needs further investigation, and that monitoring ballast discharges in these areas may be an important consideration for the protection of public health.

REFERENCES

Anderson I., End of the Line for Deadly Stowaways? *New Scientist* 1992.
Austin, B., *Marine Microbiology*. Cambridge University Press, Cambridge UK, 1993, p. 4.
Benenson A.S., *Control of Communicable Diseases in Man*. American Public Health Association, 15th ed., 1990.

Bitton, G., *Wastewater Microbiology.* Wiley-Liss, New York, 1994, pp. 116 and 363.

Colwell, R.R. Global Climate and Infectious Disease: The Cholera Paradigm. *Sci.* 274, pp. 2025–2031, 1996.

Geldreich, E.E. Pathogenic Agents in Freshwater Resources. *Hydrological Processes* 10, pp. 315–333, 1996.

Grabow, W.O.K., Microbiology of Drinking Water Treatment: Reclaimed Wastewater. In: *Drinking Water Microbiology,* McFeters, G.A., Ed., Springer-Verlag, New York, 1990.

Lee, J.V., S.R. Dawson, S. Ward, S.B. Surman, and K.R. Neal. Bacteriophages are a Better Indicator of Illness Rates Than Bacteria Amongst Users of a White Water Course Fed by a Lowland River. *Wat. Sci. Tech.* 35 (11–12), pp. 165–170, 1997.

Mackie, G.L. and B.W. Kilgour. Efficacy and role of alum in removal of zebra mussel veliger larvae from raw water supplies. *Water Research* 29 (2), pp. 731–744, 1995.

McCarthy, S.A. and F.M. Khambaty. International dissemination of epidemic *Vibrio cholerae* by cargo ship ballast and other nonpotable waters. *Appl. Environ. Microbiol.,* 60(7), pp. 2597–2601, 1994.

Mills, E.L., J.H. Leach, C.L. Secor, and J.T. Carlton. *Whats New? The Prediction and Management of Exotic Species in the Great Lakes.* Great Lakes Fisheries Commission, 1993.

Smith, A.W., D.E. Skilling, N. Cherry, J. H. Mead, and D.O. Matson. Calicivirus Emergence from Ocean Reservoirs: Zoonotic and Interspecies Movements. *Emerging Infectious Diseases* 4 (1), 13–20,1998.

Wiley, C.J., Aquatic Nuisance Species: Nature, Transport, and Regulation. In *Zebra Mussels and Aquatic Nuisance Species*, D'Itri F.M., Ed., Ann Arbor Press, Chelsea, MI, 1997.

Section V

Aquaculture Vector

18 The Blue Revolution and Sustainability: At a Crossroads

Paul H. Patrick

SECTION INTRODUCTION

Similar to how changes in production methods created the Green Revolution in agriculture, the Blue Revolution is expected to begin in aquaculture (Weber, 1996). Everywhere, including the Great Lakes region, the aquaculture industry is poised for further expansion. As the industry grows, there is an increased awareness of the various environmental consequences of this development. We are now at a crossroads on whether industrial expansion of aquaculture can meet sustainable development objectives.

Aquaculture can be broadly defined as the farming of aquatic organisms in freshwater, estuarine, or marine waters. This includes fish, molluscs, crustaceans (shellfish), and aquatic plants. Although many people think of aquaculture as a marine activity, it is noteworthy that much of the aquaculture production today takes place in freshwater or estaurine systems. During the past several decades aquaculture has emerged as a major food producing sector growing at a rate of about 10% annually. Recently, this annual rate has increased more dramatically. In 1994 world production totalled 12.1 million metric tonnes. By 1995, total production of finfish, shellfish, and aquatic plants more than doubled the 1994 figures reaching a record 27.8 million tonnes, valued at approximately \$42 billion. As world capture fishery landings level off, and the production of value-added species are on the decline, aquaculture and stock enhancement are viewed as the only means of meeting the demand to feed the expanding human populations in the next millennium.

Over 90% of the aquaculture products are farmed in Asia, primarily in China and India. In Canada, commercial aquaculture has grown to an industry that generated \$300 million in 1995, and contributed more than 17% of the total landed value of the Canadian fisheries sector (Moccia et al., 1997). In the U.S., the industry is substantially greater, with a total value of \$5.6 billion. Aquaculture has already provided 5200 jobs in Canada (Moccia et al., 1997) and 181,000 in the United States, mainly in the production, supply, and service sectors. In the Great Lakes there are over 1000 producers involved in a wide range of activities including food fish, baitfish, fishing for stocking, ornamental culture, fee-fishing operations, as well as aquatic plants for food (e.g., wild rice). In Ontario alone, over 3300 tonnes of rainbow trout (*Oncorhynchus gairdneri*) are produced from some 200 licensed farms (Moccia et al., 1997). This represents more than 95% of the aquaculture industry production output. This situation is likely to change based on amendments to the Ontario Game and Fish Act, which now allow the culture of 38 different aquatic species including baitfish, tilapia (various genera), and arctic charr (*Salvelinus alpinus*) (the latter two are not endemic to the Great Lakes). The precedent for this change is the Great Lake states that have allowed the culture of a number of different species, some of which are not native to the Great Lakes basin area, i.e., triploid grass carp, hybrid striped bass, and, of course, tilapia.

PROBLEMS WITH SUSTAINABILITY OF THE BLUE REVOLUTION

Any industry that uses water in the focal part of its operation has to have some effect on the environment. Aquaculture is no exception. Several problems already exist and others are likely to emerge as the industry grows:

- The use of nonendemic species for culture and the possibility of causing new introductions to the ecosystem through escape
- Potential for escape of such fish species during live shipments
- Geographic transfer of nonendemic organisms other than the target fish species as well as pathogenic organism through aquaculture stock transfer
- Escapes of domesticated endemic fish from the fish farm and their impact on the integrity of the wild fish gene pool
- Nutrient enrichment, potential for disease magnification, and the presence of chemicals (such as antibiotics) in water discharged by fish farms

These problems are considered significant by many people involved in ecosystem health. Unfortunately, much of the earlier research focused on production rather dealing with such problems. More recently, research has been more focused on sustainable development issues. Academic institutions, state and provincial agencies, and technological companies are now more active in dealing with aquaculture issues such as those described above.

FISH ESCAPE, NONENDEMIC SPECIES, AND THE GENE POOL

An introduced fish species is defined as one that is not native to a particular ecosystem. For this reason, fish escape is definitely a concern, especially where aquaculture involves cage culture facilities. Cage breakage, vandals, poaching, and/or predator breaching of cages can all result in escaped fish. In the case of nonendemic species, the potential hazards are obvious. Another nonnative organism could be introduced to the receiving ecosystem. There are concerns about species, such as tilapia, that could escape and become resident in the Great Lakes. Still, there is no scientific evidence of tilapia living and reproducing in the warm discharges of power plants on the Great Lakes. However, in southern California, some of the rivers now have resident tilapia populations (J. Rounds, pers. commun.). Cold-water species such as arctic charr are also a concern, since they are not endemic to the Great Lakes but might survive in the Great Lakes environment.

In the case of fish escape involving endemic species, the hazards are less obvious. The experience from Scotland and especially Norway have indicated the negative impact that a large number of escaped cultured Atlantic salmon can have on wild populations. The negative impact is the result of the genetic selection done by the aquaculture industry. The most desirable farmed animals are those that breed fast and have progeny that exhibit rapid growth, disease resistance, and can grow under high-density situations. Further, docile fish that accept handling and readily take feed are favored. The progeny of such fish does not necessarily fare well when introduced into the natural environment where genetic traits such as good foraging and fear of predators provide for better survival. There are reports that there are more escaped Atlantic salmon in certain Norwegian rivers than there are wild fish. This immediately raises concern about the dilution of the gene pool and possible lower breeding success. In the Great Lakes, fish have already escaped from aquaculture facilities located in Georgian Bay, although these escapes are likely very minimal relative to the lake populations. Yet concerns exist especially if this trend continues in the future. For example, investigators have already questioned the ability of large numbers of escaped hatchery reared coho salmon in the Great Lakes to rehabilitate wild populations based on data on breeding success (Fleming and Gross, 1993). Aquaculturists claim that escape from their industry is insignificant compared to the stocking of large numbers of salmonid fingerlings by provincial and state agencies.

This debate will continue for a long time. To mitigate this situation in California, several groups are considering the mixing of wild stock at specific time intervals with the existing domesticated brood stock to maintain some genetic integrity of the populations (J. Palmer, pers. commun.). This would always maintain some "wild" trait in the domesticated stock and would minimize the impact if large numbers of fish escaped.

There must be sound controls to reduce the probability of escapes. In Ontario, the Ministry of Natural Resources has, in part, dealt with this issue by developing a risk assessment protocol for using nonendemic species for aquaculture. Even with this legislation, escapes are likely to occur. Research is ongoing to address escapes in cage culture situations. In Norway, submergible cages are being investigated as an alternative to floating cages to minimize escape of fish, especially during adverse winter conditions (Reinertsen et al., 1993). Another system being considered involves a floating fish farm with bag pens rather than nets (Reinertsen et al., 1993). A similar technology is currently being tried in Ontario. Experimental techniques such as the use of sound conditioning technology to recover escaped "programmed" fish perhaps offer another option for cage-based systems (Patrick et al., 1992). Obviously, land based operations have less risk of escape, especially recirculation-based systems. Considerable research and development is ongoing with recirculation systems, primarily in the U.S. (Libey and Timmons, 1996). Although these systems at present are expensive to establish, the probability of escape would be minimal. Partial recirculation systems are being considered for some applications in the Great Lakes region by several aquaculture groups.

Another concern is that fish introduced into the Great Lakes may actually have originated outside the basin area. For instance, grass carp (*Ctenopharyngodon idella*) are now found in many of the Great Lakes. The big-head carp (*Hypophthalmichthys nobilis*) have been reported in Lake Erie (E. Crossman, pers. commun.), although they are not endemic to the area. These species are raised for aquaculture mainly in the southern U.S. Escape may have occurred through the southern water system into Lake Ontario or possibly from the shipments of live fish brought into the Great Lake region for domestic consumption by so-called "live haulers." Currently, there are over a dozen haulers who bring live fish into Ontario for human consumption to meet the demand for live animals created by the large Oriental market in Toronto.

Live haulers as vectors of unwanted introductions are not restricted to fish as food. At present, live catches of species such as pumpkinseed caught by commercial fisherman in the Port Dover area of Lake Erie are sold to haulers, who bring these fish into neighboring states for fee-fishing operations in small lakes or ponds. As part of this transport, zebra mussels and pathogens may also be transported. Recently, in Ireland, a shipment of live eels from offshore contained zebra mussel, which have now been introduced into the region previously mussel-free (P. Cullen, pers. commun.).

DISEASE AND CONTROL

Pathogens are always present in the environment. Some are endemic and some may be introduced with the aquaculture stock. Disease outbreaks generally occur when high densities of a monocultured species experience environmental stress as well as unfavourable environmental conditions such as rapid temperature changes. Disease outbreaks may begin in the aquaculture facility and then spread to the wild stocks as a result of waste discharge from aquaculture farms.

Antibiotics and prophylactic treatments are continually being used to treat diseased fish. The use of any antibiotic for any food production is not advised and should be minimized where possible. A more sustainable option would be reducing fish densities to create more optimal conditions. The lower "farm gate" revenue resulting from lower densities of fish stocks would be compensated by lower feed costs, labour and less medication costs. Many of the fish operations that I have seen appear to have extremely high densities of fish in their operations. Disease problems were also prevalent.

An integrated fish management plan is required as a preventative measure for disease control. This includes improvements in design, operation and inspection of fish culture establishments;

inspection, quarantine, and treatment of sick animals, as well as adoption and enforcement of adequate and comprehensive legislation.

Further, ongoing research is focusing on developing methods for a genetic test to diagnose certain gram bacterial diseases. Pathologists and microbiologists are also developing rapid diagnostic testing methodology for key diseases as well as management and treatment protocols for diseases such as bacterial gill disease and tail rot (Moccia, 1992).

NUTRIENT ENRICHMENT

There is concern about nutrient enrichment resulting from fish metabolites, faeces, and unused feed. Part of the problem relates to a lower than expected feed conversion ratio. Feed conversion ratio (FCR), defined as a ratio of fish feed weight to fish biomass (wet weight), is typically 1.5 or greater, which results in feed waste contributing a large proportion to the total waste output in most operations. Although the FCR for species such as rainbow trout has gradually decreased, largely because of better diets and animal domestication (including improved genetics), this ratio will likely continue to be high for valued-added species such as walleye and yellow perch. There are still technological constraints for alternative species such as walleye and yellow perch that must be addressed before the aquaculture of these species is deemed feasible (Stickney, 1993; Kestemont and Dabrowski, 1995; Summerfelt, 1996). Behavioral technologies such as the use of light and sound to optimize feeding to improve conversion rate are currently being investigated (Reinertsen et al., 1993; Patrick et al., 1992). Other advanced automated feeding systems with adaptive feedback mechanisms to assess fish feeding response are also being developed.

Nutrient enrichment issues are more relevant for some aquaculture practices than others. For example, it is likely more pertinent for net pen or cage culture operations than land-based ones, since it is more difficult to collect wastes. Traditionally, monitoring of water chemistry has been at the discretion of the aquaculturist. Some have developed limited monitoring protocols and have managed to meet specific regulatory guidelines using "dilution is the solution" concepts. Yet, estimates of total phosphorus output from a single well-run salmonid operation producing 100 tonnes of fish annually can be the equivalent of the raw sewage effluent from a community of over 850 people (Folke et al., 1994). From an ecological perspective, nutrients from fish farming can be considered similar to nutrients added from municipal sewage, as they have the same potential to cause eutrophication problems (Persson, 1992). Still, aquaculturists argue that nutrient waste from their industry is minor compared to the other industries such as agriculture.

The feed companies have responded to this nutrient issue by manufacturing feed with less phosphorus content. Considerable thought and research are also being focused on the use of plant protein as a substitute for fish meal (major ingredient of pelleted fish food). Furthermore, considerable research is ongoing by many institutions such as the University of Guelph on the nutritional requirements of fish so that less expensive and more sustainable diets can be developed. Also, new biological and nutritional approaches that include formulation of high nutrient-dense, low pollution diets and software designed to predict aquaculture wastes are being developed as alternatives to the more conventional chemical approaches (Cho et al., 1994).

Other recent technological developments are occurring that could allow a more sustainable solution to the nutrient problem especially for net pen operations. For example, "in-situ" bag cages with a means of collecting wastes from the cage bottom are currently being used in Ontario as an alternative to the more traditional mesh cages. For shore-based operations, fine meshed rotary drum filters and similar technologies are also being considered in the Great Lakes region as a means of collecting fine particulates from shore-based operations. These filters have already been used successfully on the east coast in significantly reducing total phosphorus in large hatchery operations (D. Guest, pers. commun.). Other research focuses on nutrient recycling involving recirculation systems, which are being considered for many land-based operations (Libey and Timmons 1996).

SUMMARY

The Blue Revolution will continue to provide much-needed fresh food to consumers as well as providing necessary job creation and opportunities in the Great Lakes area. Yet there are environmental concerns that currently exist and others that are expected to have more of an impact in future years. These issues include escape of domesticated fish, integrity of the wild gene pool, the use of nonendemic species for culture, introductions of new pathogens and other organisms with live aquaculture shipments, nutrient enrichment, disease magnification, and the use of chemicals (prophylactic treatment) and antibiotics to treat fish. We are at a crossroads where various environmental groups are at times in heated debate with many small business aquaculture operations.

These problems are not insurmountable, and it is my belief that science and technology can be used successfully to resolve many of these issues and keep the Blue Revolution operating in a sustainable manner. These technologies must be environmentally benign and focus on eco-efficiency and recycling. At the same time, better communication and government tax incentives will also be required for the aquaculturist to take advantage of these new developments. Most aquaculture in the Great Lakes region consists of small business operators, and cash flow is always a significant issue with this industry. For this reason, financial incentives should be created by government to make such a technological transition possible. Better and consistent regulatory controls for the Great Lakes region are also required especially in the transportation of live fish, and the culture of nonnative species.

There are still criticisms by some environmentalists that solutions such as screening technology are "end of the pipe-type solutions" and not really sustainable in the long run. They argue that few attempts have been made to restructure the production itself such as move away from a high-density monoculture philosophy. Future aquaculture may eventually turn away from large-scale monoculture methods to more traditional methods based on polyculture and integrated farming that promote waste recycling. The Chinese have done this for generations, although generally on small-scale applications. Furthermore, much of the research and development efforts that compare the ecological advantages of a polyculture approach over a monoculture one deal with warm-water systems (Milstein, 1997). There appears to be little information on polyculture within temperate waters, as found in the Great Lakes. Whether large-scale integrated systems in the Great Lakes basin are economic and achievable remains to be seen.

REFERENCES

Cho, C.Y., J.D. Hynes, K.R. Wood, and H.K. Yoshida. Development of high-nutrient-dense, low-pollution diets and prediction of aquaculture wastes using biological approaches. *Aquaculture* 124, pp. 293–305, 1994.

Fleming, I.A. and M.R. Gross. Breeding success of hatchery and wild coho salmon (*Oncorhynchus kisutch*) in competition. *Ecological Applications* 3, pp. 230–245, 1993.

Folke, C., N. Kautsky, and M. Troell. The costs of eutrophication from salmon farming: Implications for policy. *Journal of Environmental Management* 40, pp. 173–182, 1994.

Kestemont, P. and K. Dabrowski. Recent advances in percid fish. *Journal of Applied Ichthyology* 12, pp. 137–200, 1995.

Libey, G.S. and M.B. Timmons, Successes and Failures in Commercial, in *Commercial Recirculating Aquaculture: Proceedings of Recirculating Aquaculture Conference, Roanoke, VA, 1996*, p. 651.

Milstein, A. Do management procedures affect the ecology of polyculture ponds? *World Aquaculture* 28, pp.12–19, 1997.

Moccia, R.D. Aquaculture's growing in Ontario and World-Wide. *Highlights Magazine*, Ontario Ministry of Agriculture and Food 15, pp. 24–27, 1992.

Moccia, R.D., S. Naylor, and G. Reid. An overview of aquaculture in Ontario. *University of Guelph Publication* No. 96-003, 1997.

Patrick, P.H., A.E. McLeod, and J. Kowalewski. Advanced aquaculture techniques. Sound conditioning of fish. *Ontario Hydro Research Division Report* No. 92-64-K, 1992.

Persson, G. Eutrophication resulting from salmonid fish culture in freshwater and saltwater: Scandinavian experiences, in *Nutritional Strategies & Aquaculture Waste,* Cowey, G.B. and Cho. C.Y., Eds., University of Guelph Press, Guelph, ON, 1992.

Reinertsen, H., L. Dahle, L. Jorgensen, and K. Tvinneriem. Fish Farming Technology, in *Proceedings of the First International Conference on Fish Farming Technology,* Trondheim, Norway, August 9-12, 1993. A.A. Balkema, Rotterdam, p. 482.

Stickney, R.R. *Culture on Non-Salmonid Freshwater Fishes. Advances in Fisheries Science.* CRC Press, Ann Arbor, MI, 1993, p. 331.

Summerfelt, R.C. *Walleye culture manual.* NCRAC Culture Series 101. p. 415, 1996.

Weber, M.L. 1996. So you say you want a blue revolution? *Amicus Journal,* Fall, pp. 39–42, 1996.

19 The American Bullfrog in British Columbia: The Frog Who Came to Dinner

Stan A. Orchard

INTRODUCTION

Today, in many parts of southwestern British Columbia, the American bullfrog (*Rana catesbeiana*) is abundant, and its population numbers and geographical distribution seem to be expanding rapidly (Figure 19.1). The novelty of this fact is that American bullfrogs are not native to British Columbia, but were brought here within the past few decades. There are many aspects to this story:

- Why were they transported to places so distant from their native lakes and ponds?
- What has been the impact of this transplantation on the native frogs?
- Have the bullfrogs created other ecological problems? If so, should they be eradicated?
- Why have the problems been ignored for so long? Are there solutions?

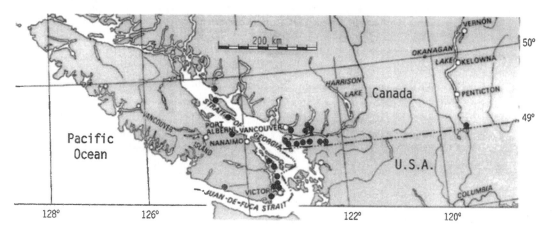

FIGURE 19.1 Locality records for American bullfrog populations in British Columbia, Canada. No comprehensive survey has been conducted, but their actual distribution, particularly on the southwest coast, is probably much wider than is currently documented.

The situation in British Columbia has not developed suddenly and is certainly not unique. Nevertheless, it has recently become topical because of the scientific interest in the phenomenon of declining amphibian populations. This is another parable of the discordance that exists between "human nature" and the rest of the natural world, but this is one problem that we could probably come to grips with if we cared enough to act.

BACKGROUND ON THE SPECIES

The American bullfrog is a robust, brilliant green amphibian that occurs in nature only in North America, from southern Québec and Ontario, throughout the Mississippi drainage, south to the Isthmus of Tehuantepec in southern Mexico. Its natural range does not extend west of the Rocky Mountains.

It is large, by frog standards, with an adult body length in excess of 7 in., easily twice the size of any native frog. It is muscular, with very long powerful legs for leaping and swimming. It is also a productive species, annually generating as many as 20,000 or more eggs in a single jelly mass. The tadpoles are extraordinarily large as well, looking somewhat like olive-brown golf balls with high, laterally compressed tails. In the southern quarter of its distribution it can transform from a tadpole to a frog in a single season. Further north, however, the development rate becomes retarded by lower mean temperatures and a much briefer active season. Thus, in British Columbia, the larval period, or tadpole stage, can last 3 to 4 years before reaching metamorphosis. Another 2 to 3 years is required for the young frogs to become sexually mature.

This heat-seeking amphibian is the last frog species to emerge from winter torpor in British Columbia, and it is the first species to become inactive in the fall. Consequently, at this latitude, bullfrogs remain inactive for about half of the year. While conditions are favorably warm they must feed very aggressively to build up the fat reserves that will sustain them over winter months and provide the energy and nutrients that they will need to reproduce the following spring. They are voracious predators and will swallow any creature that will fit into their large, gaping mouths, including earthworms, leeches, insects, centipedes, millipedes, spiders, crayfish, snails, salamanders, tadpoles, frogs (including other bullfrogs), fish, small alligators, small turtles, lizards, snakes, shrews, moles, mice, bats, and birds up to the size of robins and orioles (Bury and Whelan 1984).

The loud *basso profundo* mating call of the male has been variously described as *jugo-rum, more rum, blood'n'ouns, br-wum, be drowned, knee deep*, and *bottle-o-rum* (Wright and Wright 1949). However, my personal impression of the call is more in accord with the description by Mark Catesby, the American naturalist in whose honor the species is scientifically named, who said ". . . the noise they make has caused their name, for at a few yards distance their bellowing sounds very much like that of a bull a quarter of a mile off" (Holbrook 1842).

Bullfrogs are highly aquatic and thrive in the seasonally warm permanent water of sun-exposed lakes, marshes, and ponds. They also over-winter in the water. For this reason, bullfrogs cannot live in wetlands that routinely or periodically dry up. Summer days are spent basking and feeding in warm shallow water or above water on logs and floating vegetation mats and often in direct sunlight. Activity continues at night, when adults will sometimes emerge at the lake's edge.

EXTENDED DISTRIBUTION

Concerns have been raised for many years that burgeoning populations of imported and released bullfrogs must be damaging to native faunas, and particularly native frog populations (Hammerson 1982; Hayes and Jennings 1986; Stumpel 1992). In 1989 a reexamination of all of the known causes of amphibian declines was begun as part of a coordinated international research effort to determine why numerous frog species were mysteriously disappearing (Halliday and Heyer 1997; Phillips 1994). Consequently, in the spring of 1995, the Invasive Species Specialist Group of the Species Survival Commission (SSC), World Conservation Union (IUCN), sent out a questionnaire world-wide to collate information on bullfrog introductions and thereby establish the true magnitude of the problem. Michael J. Lannoo (1995), Ball State University, produced a summary of the survey results revealing that bullfrog populations are now very widely distributed around the globe, including locations in Europe, Asia, South America, and Africa.

American bullfrogs have been thriving in British Columbia since they were first brought here over 120 years ago (Carl and Guiguet 1972). They are now common in the vicinity of Vancouver, and the Lower Fraser River Valley in the southwest corner of British Columbia (where they were first released), and on southern and central Vancouver Island (in the last 50 years). They are now also established on several of the islands in the Strait of Georgia in southwestern British Columbia and at Osoyoos Lake in the southern Dry Interior. Other records exist from across the southern part of this province but need to be verified.

WHY WERE THEY INTRODUCED?

There are many stories to account for when and how these populations arose. Initially, they seem to have been imported with the vain hope of setting up frog farms. Subsequently, they were transported about largely by children who enjoy catching the very large tadpoles, as well as by garden pond enthusiasts who just wanted to "put something" in their new constructions. The frogs, however, seldom stay where they are put. I also remember, from when I was a child, seeing bullfrog tadpoles being offered in local pet shops. Often imported incidentally with shipments of goldfish, they were then sold for garden pools, but just how many of these tadpoles went on to establish populations will never be known. A few new disciples of the misguided notion of frog farming spring up every year or so, and this also keeps bullfrogs on the move.

WHAT ARE THE ECOLOGICAL RAMIFICATIONS?

Almost 30 years ago, Lawrence Licht, now at York University, Ontario, was studying the ecological relationship between two closely related species, the red-legged frog (*R. aurora*) and spotted frog (*R. pretiosa*). His study sites were located just south of the City of Vancouver, where the habitat was stable, accessible, and both species were easily found in numbers. Spotted frogs and red-legged frogs are both average-sized anurans, native to British Columbia, and both live in the same sort of habitat, though their habits are not identical. In the course of his studies, Licht (1972, 1974) discovered that spotted frogs are more aquatic than red-legged frogs. He also found that American bullfrogs were moving into the area, and he theorized that, because of their aquatic habits, spotted frogs would be more vulnerable than red-legged frogs to being displaced through competition and predation from American bullfrogs.

Though the habitat has changed little over the past 20 years, several recent searches through Licht's original study sites have had no difficulty finding red-legged frogs and an abundance of American bullfrogs, but have failed to turn up any spotted frogs. Admittedly, there is reason to believe that other factors may have been involved in the decline of spotted frogs in western North America, but there is ample reason to suppose that, where present, American bullfrogs will competitively exclude spotted frogs in any event. Complicating our ability to observe and quantify the problem is the fact that adult bullfrogs are difficult to study because they are wary, are powerful jumpers and swimmers, and the vegetation cover in local lakes is often abundant and difficult for herpetologists to manouever in.

At many locations tadpoles and juveniles crowd the shallows to such an extent that the water boils as they scatter upon your approach. Though quick to flee from danger, the tadpoles are nevertheless armed with noxious chemicals in their skins that predators seem to find quite unpalatable (Kats, Petranka, and Sih, 1988; Werner and McPeek, 1994). The transformed juveniles are even more conspicuous because they make a loud and distinctive "MEEP" sound as they propel themselves off like fleeing grasshoppers. Given the phenomenal survivorship of the most vulnerable age classes and the size of the adult frogs and their ravenous appetite, it is reasonable to believe that many native animals that live in and around, or visit, lake and pond edges are falling easy prey to the bullfrogs. For example, the common garter snake (*Thamnophis sirtalis*) and the western garter snake (*T. elegans*) are primarily aquatic foragers and are consequently vulnerable to hungry bullfrogs, as are the hatchlings of painted turtles (*Chrysemys picta*). One area resident who lives just north of the City of Victoria has been shooting bullfrogs as they migrate into his backyard pond. He has taken the trouble to examine their stomach contents and has apparently identified several songbird species. It is probable that many waterfowl are also at risk, particularly the chicks and ducklings. Conceivably, the prey base at these lakes could become so depleted by adult bullfrogs that their diet will be made up largely of juveniles of their own species.

Predation and competition are the most obvious problems that introduced bullfrog populations can create; however, diseases and parasites, though little studied, are also an issue worthy of consideration (Green 1994; Smith 1994). The effect of huge numbers of large imported bullfrogs and tadpoles suddenly taking up residence in a relatively closed system certainly increases the opportunity for pathogenic viruses and bacteria (Gruia-Gray, Petric and Desser 1989; Gruia-Gray and Desser 1992), fungi, and parasites to mutate, multiply, and disperse. The bullfrogs could have some natural resistance to infections that they carry with them, while native species may be hit very hard. In Australia, for example, 14 species of rainforest frogs that have disappeared or declined in seemingly undisturbed habitats may be succumbing, in some cases, to diseases introduced into that country through the tropical fish trade (Laurance, McDonald and Spear 1996).

On March 8, 1986, Frederick W. Schueler, while searching for amphibians in southwestern British Columbia, observed an interesting case of interspecific amplexus (pers. commun.). This is where a male frog of one species has grasped, in a mating posture, a frog (generally female) of a different species. In this case, he observed a male red-legged frog (*Rana aurora*) clasping a juvenile female American bullfrog. In British Columbia, native red-legged frogs mate relatively early in the year, almost invariably before the water is sufficiently warm for American bullfrogs to reproduce. However, it is also the time that juvenile bullfrogs emerge from winter torpor, before the adults do (Willis et al. 1956). Consequently, in some years, particularly when water temperatures are rising rapidly, the reproductive period of red-legged frogs could coincide with the emergence of juvenile American bullfrogs. Schueler has speculated that the effect of interspecific amplexus could actually be related to a local decline in the native frog population. He says, "It has occurred to me that fat compliant (juvenile) frogs of introduced species might divert enough males of the declining species into interspecific amplexus to leave a significant fraction of the egg masses of the declining species unfertilized." While this does not appear to be the case for the red-legged frogs in British Columbia, who are still co-existing with bullfrogs, it could be part of the reason why spotted frog populations have become locally extinct in the past few decades.

BACKGROUND ON THE HUMAN FACTOR

People have many mercenary interests in frogs, and this is the reason that American bullfrogs have been released into so many exotic locations. The commercial trade in amphibians is not popularly recognized as a major conservation issue, but it is nevertheless a serious threat to the survival of many species. There is a large market demand for amphibian specimens for biological research and educational dissection, and many species are also collected for human consumption and the pet trade. Bullfrogs in several of the U.S. states in which they were introduced are now thriving and are managed as game animals with set bag limits. There is even evidence that fish and wildlife agencies may have, in some instances, themselves released the bullfrogs in an attempt to add a new game species to the area (Lannoo 1996).

The commercial trade in frogs' legs for human consumption continues to grow world-wide in spite of the obvious negative consequences to frog populations. France, for example, banned the commercial collecting of its native species in 1979, due to the population-depleting effects of the trade. France now imports all of its frogs' legs, primarily from Asia (Niekisch 1986). From 1973 to 1987, France imported a total of 46,579 tons of frozen frogs' legs. It has been estimated that this represents somewhere between 1 and 2 billion individual frogs (Jorgenson 1985; Le Serrec

1988). In 1983, Bangladesh, India, and Indonesia exported about 11,000 tons of frogs' legs. In addition, losses to wastage were estimated at 10–20% for India and 40–50% for Indonesia (Niekisch 1986).

A 1983 report prepared for the Food and Agriculture Organization of the United Nations discussed the ecological consequences of continued excessive exploitation. The report pointed out that frogs are an effective biological check against destructive agricultural pests, and thus any decrease in frog numbers will translate into an increased need for insecticides. By 1987, however, India had banned the export of frogs' legs entirely (Anonymous 1988).

There are also public health issues concerning the consumption of frog meat. In Taiwan, it was discovered that approximately 14% of bullfrogs reared on farms carried typhoid or the bacterium *Vibrio parahaemolyticus* (Anonymous 1985), and in 1973 Asian frogs' legs were banned from the U.S. for 1 year due to potential *Salmonella* infections (Niekisch 1986).

Accordingly, the concept resurfaces from time to time in British Columbia, and elsewhere, that frogs can and should be farmed for human consumption and for the biological supply trade. The argument is made that animals supplied in this way will relieve collecting pressures on populations of native frogs; however, virtually all of the frogs native to British Columbia are too small, too scarce, or develop too slowly to produce a profitable and sustainable yield. Today there are many expanding populations of large, voracious American bullfrogs and green frogs in southern British Columbia that compete with native species and are the direct result of erstwhile frog farmers. It is noteworthy that after 120 years of trying, no frog farms exist in British Columbia, and it is compelling testimony to the inevitable failure of this form of enterprise; a conclusion that has been extensively documented (Bury and Whelan 1984; Storer 1933). At this latitude, in particular, bullfrogs probably take 7 or 8 years to reach marketable size, which is too long to wait for a return on investment. Other problems include disease and cannibalism under conditions of crowding.

THE LAW

At the Rio Conference in 1992, Canada joined most of the governments of the world in signing the Convention on Biological Diversity. The convention states in Article 8, Item (h) — In-situ Conservation — "Each Contracting Party shall, as far as possible and as appropriate: Prevent the introduction of, control or eradicate those alien species which threaten ecosystems, habitats or species." Unfortunately, except in the case of migratory birds and issues related to international waters, Canada's federal government has very limited authority over the nation's wildlife. Under our British North America Act (1867), jurisdiction for wildlife rests primarily with the provinces, and therefore each province has its own Wildlife Act or equivalent, and is not bound by letters of intent signed by the federal government.

Currently, in British Columbia, laws prohibit the release or transport of bullfrogs, and anyone can kill a bullfrog with impunity — providing that it is done humanely and not within a park. However, our laws have proven to be quite ineffective and unenforceable in controlling the illicit transport and release of exotic species. Unless people can be united in a conviction that this behavior is environmentally irresponsible, and therefore wrong, they will continue to act in the willful and capricious manner that has allowed these situations to develop.

IS ERADICATION A PRACTICAL OPTION?

Over 20 years ago a research proposal was innocently prepared to eradicate 2000 introduced rabbits from Round Island in the Indian Ocean. At stake were 17 endemic plants and animals whose survival was severely threatened by the malignant effects of too many rabbits. Conservationists had few doubts about the ethics of the plan, but animal welfare groups rallied, concerned about cruelty to the rabbits, and held up the project for 12 years while a unique flora and fauna were pushed nearer to the brink (Temple 1990).

Personally, I find killing frogs in large numbers (or any numbers for that matter) to be distasteful, even if they are displacing species that were previously established; and perhaps it is true that the forest-cleared habitats that we have created, and continue to create, seem to favor the sun-loving bullfrogs over our native frogs. However, in spite of all points to the contrary, I am of the firm conviction that eradication is worth attempting. The prodigious success of bullfrogs is clearly at the expense of local species diversity, and anyone who is familiar with the situation in British Columbia and the western U.S. must be impressed by the magnitude of this problem. Unfortunately, the final judgment on whether to eradicate is not entirely in the hands of amphibian experts and informed conservationists. Today, our policies are driven by public opinion, and any proposal that suggests the purposeful killing of animals is bound to raise intense controversy and face organized opposition.

Also, there is the issue of whether public cooperation in locating populations for eradication and in preventing reintroductions is possible. Many people like bullfrogs because they are large and conspicuous, while our native species tend to be small and unobtrusive.

THE APPROPRIATION OF WETLANDS, LAKES AND PONDS

Of great concern to me is the fact that the rapidly diminishing resource of freshwater is being intensively managed for almost the exclusive benefit of three predator groups: fish, waterfowl, and people. In 1990, over $1.5 billion was spent in North America to improve waterways and wetlands for waterfowl (*Victoria Times-Colonist* newspaper 1991). It is likely that at least as much was invested in boosting up the numbers of game fish on this continent.

Low on the food chain, amphibians are a plentiful and nutritious source of protein to predators, such as fish and waterfowl, that live in and around lakes, ponds, marshes, and wetlands. Thus, there is good reason, even for waterfowl and fisheries managers, to work to preserve amphibian diversity, survival, and productivity.

In the past, there were many bodies of water, large and small, that were isolated from connected waterways and consequently did not have predatory fish in them. In many cases, these have been important or vital habitats for amphibian reproduction. Sadly, people have acted obsessively in draining away surface water, removing forest shade, releasing game fish into every water body that will support them, and devoting a disproportionate amount of attention to the management of waterfowl. The upshot of this has been the rapid evolution of people-modified environments that are increasingly hostile and hazardous toward moisture- and shade-loving amphibians. Thrust into this equation comes the American bullfrog, along with a host of other aquatic predators exotic to British Columbia, such as green frogs (*Rana clamitans*), red-eared slider turtles (*Pseudemys scripta*), goldfish (*Carassius auratus*), bass (*Micropterus dolomieui* and *M.salmoides*), sunfish (*Lepomis gibbosus*), black crappies (*Pomoxis nigromaculatus*), and catfish (*Ictalurus melas* and *I. nebulosus*).

CONCLUSION

American bullfrogs were brought to British Columbia because, on the surface, frog-farming seemed like a feasible and profitable idea. Bullfrogs are meaty and productive and the commercial demand for frogs seems to be insatiable. Unfortunate for the hopeful farmers is the fact that, particularly at this latitude, bullfrogs are not adaptable to the crowded conditions necessary for farming. In any case, it would be inefficient to raise a relatively small animal for 8 years before it grows to a marketable size. It creates an enormous backlog of generations that must be contained and fed. However, beyond mere practical considerations, there are ethical questions such as What are the environmental costs of the trade in frogs and frog parts versus the economic benefits? Is the trade in frogs and frogs' legs a vital and sustainable industry or is it simply another frivolous assault on the balance of nature? There is little general knowledge in our society about the ecological importance of amphibians, and ignorance and environmental irresponsibility have certainly exacerbated this "sea of troubles."

There are several positive actions that can be taken. We can be more vigilant about policing the laws that aim to prevent people from importing and releasing any more exotic species. We can determine to hire specialists in amphibian biology who have the knowledge to predict or at least to recognize problems as they arise, to study them, and to devise and articulate solutions. Museums and universities must support basic natural history research and in as comprehensive a range of biological specialities as possible. Amphibian biology should be routinely taught in wildlife management courses. The general public and land developers should be encouraged to create ponds and to preserve the integrity of local ecosystems to the greatest possible extent.

The problems with bullfrogs have been ignored for so long because traditionally amphibian biology and conservation have been treated very lightly by environmental planners. Amphibians were widely perceived as biological lint, scattered about, abundant everywhere, represented by relatively few species, and therefore in no need of protection. It is only in the last decade that an issue has been made of the loss of biological diversity — and diversity is certainly what is at stake when one species displaces many.

Introduced American bullfrogs should be eradicated in British Columbia because their rate of survivorship is, like their appetite, too great. The resulting intense competition and predation are clearly damaging to native faunas — the spotted frog being a conspicuous example. However, bear in mind that the bullfrogs are not vermin; they are the victims of organized frognapping that should never have been permitted in the first place. Eradicating them is an extremely unpleasant prospect, but it must be undertaken if we intend to do more than pay lip service to the proposition that preserving biological diversity is a moral responsibility.

Global warming, forest clearing, water channeling, fish stocking, pollution, and other effects of human overpopulation, plus our naturally chaotic inclinations, are radically changing environmental conditions from what many native species require. Unless we can begin to control human behavior more effectively, we have little hope of solving the manifold environmental problems that people (the primary pests) seem compelled by their nature to create. Accordingly, urban and semi-urban lakes, ponds, marshes, and wetlands will continue to either be drained away or become unmanageable biological waste dumps for exotic plants and animals that are whimsically liberated by their so-called owners.

REFERENCES

Anonymous. 1985. Bullfrogs carry typhoid in Taiwan. *Traffic Bulletin* 7(3/4):60.

Anonymous. 1988. India bans export of frogs' legs. *Traffic Bulletin* 9(1):1.

Bury, R.B. and J.A. Whelan. 1984. *Ecology and management of the bullfrog.* United States Department of the Interior, Fish and Wildlife Service, Resource Publication 155, Washington, DC.

Carl, G.C. and C.J. Guiguet. 1972. *Alien animals in British Columbia.* Handbook no. 14, British Columbia Provincial Museum, Department of Recreation and Conservation, Victoria, BC.

Green, D. E. 1994. Are virus infections contributing to amphibian declines? *Froglog* 9(1994):3.

Gruia-Gray, J. and S.S. Desser. 1992 Cytopathologic observations and epizootiology of frog erythrocytic virus in bullfrogs (*Rana catesbeiana*). *Journal of Wildlife Disease* 28(1):34–41.

Gruia-Gray, J., M. Petric, and S. Desser. 1989. Ultrastructural, biochemical and biophysical properties of an erythrocytic virus of frogs from Ontario, Canada. *Journal of Wildlife Disease* 25(4):487–506.

Halliday, T.R. and W.R. Heyer. 1997. *Are amphibian populations disappearing?: A Task Force Report 1996-1997.* Unpublished report from the Office of the IUCN/SSC Task Force on Declining Amphibian Populations, Species Survival Commission, World Conservation Union, Milton Keynes, U.K.

Hammerson, G.A. 1982. Bullfrog eliminating leopard frogs in Colorado? *Herpetological Review* 13(4):115–116.

Hayes, M.P. and M.R.Jennings. 1986. Decline of ranid species in western North America: are bullfrogs (*Rana catesbeiana*) responsible? *Journal of Herpetology* 20(4):490–509.

Holbrook, J.E. 1842 (reprinted 1976). *Holbrook's North American Herpetology*. Facsimile Reprints in Herpetology, Miscellaneous Publications, Society for the Study of Amphibians and Reptiles.

Jorgenson, A. 1985. Biologists express concern for huge trade in bullfrogs. *Traffic (USA)* 6(2):25–26.

Kats, L., J. Petranka, and A. Sih. 1988. Antipredator defenses and the persistence of amphibian larvae with fishes. *Ecology* 69(6): 1865–1870.

Lannoo, M.J. 1996. *Okoboji Wetlands: a lesson in natural history*. University of Iowa Press, Iowa City, IA.

Lannoo, M.J. 1995. *Summary report of American bullfrog questionnaire*. Unpublished report to the Invasive Species Specialist Group of the Species Survival Commission (SSC), World Conservation Union (IUCN), Switzerland.

Laurance, W.F., McDonald, K.R., and Speare, R. 1996. Epidemic disease and the catastrophic decline of Australian rain forest frogs. *Conservation Biology* 10:406–413.

Le Serrec, G. 1988. France's frog consumption. *Traffic Bulletin* 10(1/2):17.

Licht, L.E. 1972. *Ecology and co-existence of ranid frogs in southwestern British Columbia*. Ph.D. thesis. University of British Columbia, Vancouver, BC.

Licht, L.E. 1974. Survival of embryos, tadpoles, and adults of the frogs *Rana aurora aurora and Rana pretiosa* sympatric in southwestern British Columbia. *Canadian Journal of Zoology*, 52:613–627.

Niekisch, M. 1986. The international trade in frogs' legs. *Traffic Bulletin* 8(1):7–10.

Phillips, K. 1994. *Tracking the vanishing frogs: an ecological mystery*. Penguin Books Canada, Toronto, ON.

Smith, A.W. 1994. A primer on infectious disease. *FROGLOG* 9(1994):2–3.

Storer, T.I. 1933. Frogs and their commercial use. *California Fish and Game* 19(3):203–213

Stumpel, A.H.P. 1992. Successful reproduction of introduced bullfrogs in northwestern Europe: a potential threat to indigenous amphibians. *Biological Conservation* 60:61–62.

Temple, S.A. 1990. The nasty necessity: eradicating exotics. *Conservation Biology* 4(2):113–115.

Victoria Times-Colonist Newspaper. 1991. Monday, 26 August: B16.

Werner E. and M. McPeek. 1994. Direct and indirect effects of predators on two anuran species along an environmental gradient. *Ecology* 75(5): 1398–1382.

Willis, Y.L., D.L.Moyle, and T.S.Baskett. 1956. Emergence, breeding, hibernation, movements, and transformation of the bullfrog, *Rana catesbeiana*, in Missouri. *Copeia* 1956(1):30–41.

Wright, A.H. and A.A. Wright. 1949. *Handbook of the frogs and toads of the United States and Canada*. Comstock Publishing Associates, Cornell University Press, Ithaca, NY.

20 Summary of North American Introductions of Fish through the Aquaculture Vector and Related Human Activities

E.J. Crossman and B.C. Cudmore

The aquaculture vector is apparently responsible, directly or indirectly, for 96 forms of fishes (Table 20.1), and for fishes, ranks second in number to those in the category "Intentional." The reproductive status of each species is provided in the table.

In this category we include a large number of species that have escaped directly from commercial fish farms raising fishes for the aquarium, water garden, biocontrol, bait, and food industries. Some of these forms will appear also in other categories (e.g., aquarium) if they were released into public waters by "customers" of those industries. Many of the publications of Dr. W. Courtenay and his coauthors will provide details on the releases or escapes from the extensive commercial aquarium industry in Florida. One of the better-known cases of the establishment of an introduced species, after escape from commercial establishments, is that of the walking catfish, *Clarias batrachus*.

The concern for chemical control of aquatic vegetation has led to considerable interest in the introduction of herbivorous fishes as a control mechanism. This has led to the introduction of aquacultured grass carp in a number of locations, including Alberta, Canada. Individuals released for biocontrol are supposed to be triploid and incapable of reproduction, but the species is now known to be reproducing at least in the Mississippi and Illinois rivers (Raibley et al. 1995).

Commercial aquaculture firms in Ontario are hoping to be authorized to culture and sell bait species in addition to fishes for human food. A concern in this regard is that the purchaser will take them to a location distant from the area of distribution of the species, release unused bait, and thus introduce nonindigenous species (see discussions of bait fishes by Litvak and Mandrak, Chapter 11). Another concern is that recreational species from aquaculture facilities with long history of domestication will be sold to operators of private fishing ponds and then escape, or be transferred, into public waters containing the same species.

Government fish culture facilities are not normally considered commercial. In the past some have sold, or made available, surplus stock of recreational species. In some cases such fish went to private individuals and anglers' organizations for release in public waters. One of the most unusual origins of the establishment of an introduced species was the escape from, or unauthorized release from, a fish culture station. That incident is associated with the establishment of the pink salmon in Lake Superior and its gradual downstream movement in the Great Lakes. This particular situation did not arise from a commercial aquaculture establishment, was intentional, and did not involve anglers or "citizens." It is included here as the only one of its kind known to us, and exemplifies the difficulty of classifying a broad spectrum of cases under a limited number of vectors.

0-15667-0449-9/99/$0.00+$.50
© 1999 by CRC Press LLC

TABLE 20.1
**List of Fishes Introduced in Ecological North America (Excluding Mexico)
Outside Their Native Range through Commercial Aquacultural Releases or
Escapes and Their Reproductive Status in the Nonindigenous Area(s)**

Scientific Name	Common Name	Status[a]
Acipenseridae — sturgeons		
Acipenser transmontanus	white sturgeon	little to no reproduction
Scaphyrhynchus platorhynchus	shovelnose sturgeon	unknown
Osteoglossidae — bonytongues		
Osteoglossum bicirrhosum	arawana	not established
Anguillidae — freshwater eels		
Anguilla anguilla	European eel	not established
Anguilla australis	shortfin eel	not established
Anguilla rostrata	American eel	established in most areas
Anguilla sp.	unidentified anguillid eels	not established
Cyprinidae — carps and minnows		
Brachydanio rerio	zebra danio	established in some areas
Carassius auratus	goldfish	established
Ctenopharyngodon idella	grass carp	established in many areas
Cyprinella lutrensis	red shiner	established in many areas
Danio malabaricus	Malabar danio	not established
Hypophthalmichthys nobilis	bighead carp	established in some areas
Leusicsus idus	ide	established in some areas
Morulius chrysophekadion	black sharkminnow	not established
Mylopharyngodon piceus	black carp	not established
Puntius conchonius	rosy barb	not established
Puntius gelius	dwarf barb	not established
Puntius schwanenfeldi	tinfoil barb	not established
Puntius tetrazona	tiger barb	not established
Scardinius erythrophthalmus	rudd	established in most areas
Tinca tinca	tench	established in most areas
Catostomidae — suckers		
Catostomus fumeiventris	Owens sucker	established
Ictiobus cyprinellus	bigmouth buffalo	established in some areas
Cobitidae — loaches		
Misgirnus anguillicaudatus	Oriental weatherfish	established
Characidae — characins		
Aphyocharax anisitsi	bloodfin tetra	not established
Colossoma and/or *Piaractus* sp.	unidentified pacus	not established
Gymnocorymbus ternetzi	black tetra	not established
Hemigrammus cf. *ocellifer*	head-and-taillight tetra	not established
Hoplias malabaricus	trahira	not established
Hyphessobrycon cf. *serpae*	serpae tetra	not established
Moenkhausia sanctaefilomenae	redeye tetra	not established
Paracheirodon innesi	neon tetra	not established
Piaractus brachypomus	redbellied pacu	not established
Serrasalmus humeralis	pirambeba	not established

TABLE 20.1 (CONTINUED)
List of Fishes Introduced in Ecological North America (Excluding Mexico)
Outside Their Native Range through Commercial Aquacultural Releases or
Escapes and Their Reproductive Status in the Nonindigenous Area(s)

Scientific Name	Common Name	Status[a]
Ictaluridae — bullhead catfishes		
Ameiurus platycephalus	flathead bullhead	established in most areas
Clariidae — labyrinth catfishes		
Clarias batrachus	walking catfish	established in some areas
Callichthyidae — plated catfishes		
Callichthys callichthys	cascarudo	not established
Corydoras sp.	corydoras	not established
Loricariidae — suckermouth catfishes		
Hypostomus sp.	suckermouth catfishes	established in some areas
Liposarcus disjunctivus	vermiculated sailfin catfish	established
Liposarcus multiradiatus	sailfin catfish	established
Otocinclus sp.		not established
Esocidae — pikes		
Esox masquinongy	muskellunge	established in some areas
Esox reicherti	Amur pike	established in some areas
Salmonidae — trouts		
Oncorhynchus gorbuscha	pink salmon	established
Salmo salar	Atlantic salmon	not established
Salvelinus alpinus	Arctic char	established in some areas
Melanotaenidae — rainbowfishes		
Melanotaenio nigrans	black-banded rainbowfish	not established
Aplocheilidae — rivulines		
Rivulus harti	giant rivulus	established in some areas
Poeciliidae — livebearers		
Belonesox belizanus	pike killifish	established in some areas
Poecilia latipinna X *P. velifera*		not established
Poecilia latipunctata	broadspotted molly	not established
Poecilia mexicana	shortfin molly	established
Poecilia reticulata	guppy	established in most areas
Poecilia sphenops	liberty molly	established in most areas
Poeciliopsis gracilis	porthole livebearer	possibly established in some areas
Xiphophorus helleri	green swordtail	established in some areas
Xiphophorus maculatus	southern platyfish	established in most areas
Xiphophorus variatus	variable platyfish	established
Cyprinodontidae — pupfishes		
Cyprinodon nevadensis	Amargosa pupfish	established in some areas
Gasterosteidae — sticklebacks		
Culaea inconstans	brook stickleback	established in some areas

TABLE 20.1 (CONTINUED)
List of Fishes Introduced in Ecological North America (Excluding Mexico)
Outside Their Native Range through Commercial Aquacultural Releases or
Escapes and Their Reproductive Status in the Nonindigenous Area(s)

Scientific Name	Common Name	Status[a]
Moronidae — temperate basses		
Morone chrysops X M. saxatilis	whiper	not established
Cichlidae — cichlids		
Aequidens pulcher	blue acara	not established
Astronotus ocellatus	oscar	established in some areas
Cichlasoma bimaculatum	black acara	established in some areas
Cichlasoma citrinellum	midas cichlid	established in some areas
Cichlasoma cyanoguttatum	Rio Grande cichlid	established in some areas
Cichlasoma labiatum	red devil	not established
Cichlasoma meeki	firemouth cichlid	established in some areas
Cichlasoma octofasciatum	Jack Dempsey	established in some areas
Cichlasoma urophthalmus	Mayan cichlid	established
Geophagus brasiliensis	pearl eartheater	not established
Geophagus surinamensis	redstriped eartheater	possibly established in some areas
Hemichromis letournauxi	African jewelfish	established in most areas
Oreochromis aureus	blue tilapia	established in some areas
Oreochromis mossambica	Mozambique tilapia	established in many areas
Oreochromis niloticus	Nile tilapia	established in some areas
Pterophyllum scalare	freshwater angelfish	not established
Pterophyllum sp.	freshwater angelfish	not established
Sarotherodon melanotheron	blackchin tilapia	established
Symphysodon discus	red discus	not established
Tilapia mariae	spotted tilapia	established in some areas
Tilapia sparrmannii	banded tilapia	not established
Tilapia urolepis	Wami tilapia	established
Tilapia zilli	redbelly tilapia	established in some areas
Anabantidae — climbing gouramies		
Anabas testudineus	climbing perch	not established
Ctenopoma nigropanosum	twospot ctenopoma	not established
Helostomatidae — kissing gourami		
Helostoma temminck	kissing gourami	not established
Belontiidae — gouramies		
Betta splendens	Siamese fightingfish	some reproduction
Colisa labiosa	thicklip gourami	not established
Colisa lalia	dwarf gourami	not established
Macropodus opercularis	paradisefish	possibly some reproduction
Trichogaster leeri	pearl gourami	not established
Trichogaster trichopterus	threespot gourami	not established
Trichopsis vittata	croaking gourami	established

[a] Status refers to the reproductive nature of "populations" of that fish in areas where it has been introduced, regardless of vector(s) used. Many statements on status were derived from information in the NAS database. Some notations may be open to criticism (e.g., hybrids).

A recent complication faced by the Greater Toronto Area (GTA) in Ontario, Canada, and possibly other cities in Canada and the U.S., arose from the aquaculture industry. Live fishes for human consumption, aimed mainly at the segment of the population of Oriental origin, are being trucked into Canada from fish farms in the southern U.S. for sale alive. There are as many as 14 such dealers in the GTA. One dealer's voluntary declaration listed, by common name only, seven "species" (Table 20.2) imported in this way in 1995. Included were the common names used for four forms of carps. White amur and grass carp, names included in the list, are alternate names for the same species of southeast Asian carp, *Ctenopharyngodon idella*. Bighead carp and black carp are identifiable names of other southeast Asian carps. Possibly neither common name on the list, referrable to *Ctenopharyngodon idella,* actually applied to that species. As a result, we have no way of knowing whether those entries refer to one or several species. With one dealer importing over 1 million lb into the GTA in 1 year, there is considerable concern that an accident or breakdown will result in the liberation of a truckload of these nonindigenous fishes into public waters. Present information suggests that the individuals of the large Asian species brought in this way are triploid and are, therefore, supposedly not reproductive. Individuals of species native to North America arriving in the same way are probably able to reproduce.

TABLE 20.2
List of Fishes and Pounds of Fish Imported Alive by One Company During 1995 into the Greater Toronto Area for Human Consumption

Fish Name	Pounds
white amur [*Ctenopharyngodon idella*?]	48,820
grass carp [*Ctenopharyngodon idella*?]	49,145
bighead carp [*Hyphphthalmichthys nobilis*]	544,739
black carp [*Mylopharyngodon piceus*]	1,038
catfish [?]	56,499
largemouth bass [*Micropterus salmoides*]	5,127
tilapia [?]	307,786
Total Pounds:	**1,013,154**

A second complication is that there can be no guarantee of the final disposition of such live fish; that is not the responsibility of the vendor. One adult bighead carp, *Hypothalmichthys nobilis*, was discovered alive in a fountain pool on a main street in Toronto not far from Lake Ontario. It was dealt with appropriately, but someone concerned for the welfare of animals could easily have transferred that individual to the nearby lake. This species, and other Asian carps arising from aquaculture somewhere, have been found at large in the Great Lakes.

A recent shipment of live largemouth bass, mainly to New York City and Toronto, from the eastern U.S. was questioned. After considerable checking it was determined that they were wild fish poached from the Potomac River.

An actual or potential vector that may have some part in this category involves fishes released alive as "good deeds." An ancient practice, by Chinese people, endorsed and encouraged by the Buddhists, is regarded as a highly approvable act of compassion. They regarded it as a way to accumulate merits so they would be judged favorably in their afterlife. Nowadays, people believe that doing good deeds, such as releasing captive animals, will lengthen their life span. Such interest would be particularly true of releasing alive species of turtles thought to be long-lived. This could be the source of nonindigenous turtles appearing in the wild in British Columbia. The turtles could also have been released by Western peoples when the "pets" became too large for an aquarium, or when children lost interest. Apparently, there is a similar tradition in the Czech Republic that involves buying a live "Christmas carp" (common carp) and releasing it into a nearby river on the morning of December 24. These practices could lead people arriving in North America from those areas to release into public waters nonindigenous aquacultured fishes sold alive.

UNAUTHORIZED ANGLER OR CITIZEN TRANSPLANTS

As stated above, it is often difficult to decide how to organize into categories (vectors) certain types of introductions. We have listed 12 species in the category "Angler or Citizen Transplants" (Table 20.3). They are included here as a subsection, since they probably arose from "Aquaculture," and they were the result of "Human Interests." The reproductive status of those species introduced in this way is provided in the table as well. The list includes an unusual array of native and aquarium

TABLE 20.3
List of Fishes Introduced in Ecological North America (Excluding Mexico) Outside their Native Range by Angler or Other Citizen Transplants, and their Reproductive Status in the Nonindigenous Area(s)

Scientifc Name	Common Name	Status[a]
Anguillidae — freshwater eels		
Anguilla rostrata	American eel	established in most areas
Cyprinidae — carps and minnows		
Orthodon microlepidotus	Sacramento blackfish	established in many areas
Platygobio gracilis	flathead chub	not established
Tinca tinca	tench	established in most areas
Catostomidae — suckers		
Catostomus tahoensis	Tahoe sucker	established in some areas
Ictiobus cyprinellus	bigmouth buffalo	established in some areas
Salmonidae — trouts		
Prosopium coulteri	pygmy whitefish	unknown
Gasterosteidae — sticklebacks		
Gasterosteus aculeatus	threespine stickleback	established in most areas
Centrarchidae — sunfishes		
Micropterus dolomieu	smallmouth bass	established in most areas
Cichlidae — cichlids		
Astronotus ocellatus	oscar	established in some areas
Oreochromis aureus	blue tilapia	established in some areas
Oreochromis mossambica	Mozambique tilapia	established in many areas

[a] Status refers to the reproductive nature of "populations" of that fish in areas where it has been introduced, regardless of vector(s) used. Many statements on status were derived from information in the NAS database. Some notations may be open to criticism (e.g., hybrids).

fishes, so there may be some overlap with the "Aquarium" vector. The list and percentage are small, but must be looked upon as a minimum number, especially with the definition of introduced used here for fishes. In the past there was no great concern for the movement, or transfer, of native fishes either to distant places or to waters within their native distribution that did not contain them. The walleye is probably the most favored recreational species in the Kawartha Lakes on the Trent-Severn Waterway in Ontario, Canada. The species was originally absent from much of the system and the cisco, *Coregonus artedi*, was abundant. One story of the origin of the walleye in the Kawartha Lakes is that many years ago anglers catching walleye downriver, where they did occur, transferred them up over an impassable dam giving them access to the lakes above. The cisco is now gone from the lakes and the native muskellunge, *Esox masquinongy*, may be suffering from the competition from an added piscivorous species and the absence of an appropriate forage species.

One aspect of this vector is involved in what is best called rumor. The unique blue pike, *Stizostedion vitreum glaucum*, of lakes Erie and Ontario is now either extinct or extirpated. Specimens of a gray morph of the walleye are regularly submitted as putative blue pike but no specimen identifiable as blue pike, has appeared (see Billington 1994). There are persistent rumors that populations of that form still exist in Lake Erie, and elsewhere in "Canada" (Ontario and Quebec) as a result of individuals from Lake Erie which were taken north by U.S. anglers and released in lakes there. Another suggests that the blue pike was moved from Lake Erie in the 1950s to inland lakes in New York, Pennsylvania, and Ohio. It seems unlikely, however, that such populations would have escaped attention in the past.

REFERENCES

Billington, N. *Analysis of Mitochondiral DNA of Walleye Specimens Suspected of Being Blue (Walleye) Pike*. Technical Report Prepared for U.S. Fish and Wildlife Service. Amherst, New York by Cooperative Fishery Research Laboratory and Illinois Aquaculture Research Demonstration Center, pp. 1–7, 1994.

Raibley, P.T., D. Blodgett, and R. Sparks. Evidence of Grass Carp (*Ctenopharyngodon idella*) Reproduction in the Illinois and Upper Mississippi Rivers. *Journal of Freshwater Ecology*. 10, pp. 65–74, 1995.

21 Introduction of Molluscs through the Import for Live Food

Gerald L. Mackie

INTRODUCTION

Of the 22 species of molluscs that have been introduced, five appear to have been brought in for their value as food. Four of these are gastropods. The only molluscan bivalve that has been introduced for its value as food is *Corbicula fluminea*. It is discussed extensively in Chapter 22 by McMahon. As one might expect of species imported for their food value, all are relatively large. Most of the species have had either little or no impact or some socioeconomic benefit.

Each species introduced through this vector was examined for the following ten attributes, when known:

1. species characteristics
2. mode of life and habitat
3. most common dispersal mechanisms
4. morphological, behavioral, and physiological adaptations to their habitats
5. predators
6. parasites
7. reproductive potential and life cycle
8. food and feeding habits
9. ecological and/or socioeconomic impact potential
10. a summary of control measures used if the species is a biofouler.

Thorough description of the above attributes is given by the author in Chapter 15 of this book.

DETAILED REVIEW OF MOLLUSCS INTRODUCED TO NORTH AMERICA AS LIVE FOOD

CLASS: GASTROPODA

Subclass: Prosobranchia

Shells small to large; dextral; operculum present with either concentric, paucispiral, or multispiral growth lines; respiration by gills; mostly dioecious, except for Valvatidae.

Family: Pilidae (Ampulariidae)
Mostly amphibious snails with large globose or subglobose shells. Mantle cavity divided into two compartments, the left containing a gill for extracting dissolved oxygen from water and the right serving as a lung for air breathing.

0-15667-0449-9/99/$0.00+$.50
© 1999 by CRC Press LLC

Pomacea bridgesi (Spiketop Applesnail)

Introduced into Florida from Brazil, apparently in the early 1960s (Clench 1966). Little is known about its ecology, but it is probably similar to other pomaceans, such as *P. urceus* (Burky 1974). Pomaceans are well known for their amphibious nature. The mantle cavity is divided into two chambers, the left one containing a ctenidium and the right one being modified as a gas-filled lung (Burky 1974). During the dry season, the snail burrows into the mud and respires by extending a long siphon from its left side to admit air into the pulmonary chamber. The ctenidium is used when immersed in water.

FIGURE 21.1   *Pomacea bridgesi*, or spiketop applesnail.

Species characteristics: Shell less than 50 mm high, whorls strongly shouldered, body whorl rather narrow with a broadly oval aperture. The operculum is large, broadly oval, and with concentric lines of growth.

Mode of life and habitat: Ampullariids are so large that most vegetation will not support them. Usually they are found living on the sediment feeding on epipsammic algae and detritus. Like all pomaceans, *P. bridgesi* is amphibious and has a lung for respiring while in the atmosphere and a ctenidium for respiring while submerged. The species prefers slow-moving, warm water and is common in the eulittoral zones of lakes, ponds, and rivers, many of which dry up for part of the year.

Dispersal mechanisms: Probably relies on humans for dispersal of adults, but birds and mammals may disperse juvenile and younger stages.

Adaptations: Its possession of both a gill and a lung enables the species to survive long periods of drought and is well adapted to life in temporary aquatic habitats. Life in temporary aquatic habitats is rare in prosobranchs but common in pulmonates which have other mechanisms to survive the dry periods.

Predators: Humans are the main predator of adult snails, but birds apparently feed on the juveniles and smaller specimens.

Parasites: Many gastropods are intermediate hosts of trematode worms, and it is possible that *P. bridgesi* is among these. Although there has been no reports of parasites that cause disease in humans, the species may have the same potential as *P. canaliculata* (see below) to cause leptospirosis and meningoencephalitis in humans.

Reproductive potential and life cycle: The species is dioecious and like all pomaceans, it oviposits its shelled eggs on land. The newly hatched snails seek water when born or, like the adults, burrow into the mud for several months until water reappears.

Food and feeding habits: The species apparently feeds mostly on algae and detritus in the sediments, but they probably feed on leaf tissue of macrophytes as well. Sand and grit are eaten as well and used as triturating agents to grind the food in the gizzard-like part of the stomach. See *P. canaliculata* for other details.

Impact potential: Little is known about the ecological impact of the *Pomacea bridgesi* in North America, but any impacts would probably include competition with other amphibious snails living in warm waters.

Control: None, since there is no evidence to suggest that the species requires immediate control.

Pomacea canaliculata (canaliculate applesnail)

Introduced into Hawaii in the late 1980s, the species was introduced either by the aquarium trade or as a food; it is currently being sold in Hawaii as escargot (Wayne Kobayashi, Department of Agriculture, State of Hawaii, pers. commun.). Apparently the species has been reported in Florida but appears not to have caused the same concerns as in Hawaii (Terry Bills, National Biological Service, National Fisheries Research Center, La Crosse, WI, pers. commun.).

Species characteristics: Shell subglobose in shape, very large, base of aperture being canaliculate.

Mode of life and habitat: The species has the same amphibious adaptations as other pomaceans and lives in tropical climates in rivers, streams and ponds, many of which are temporary aquatic habitats. The snail can burrow into the mud and survive for several months without water.

Dispersal mechanisms: The species appears to rely mainly on humans as its dispersal agent. It is possible that young snails are dispersed by waterfowl and mammals, but adults are probably too large to be dispersed by any animal except humans.

Adaptations: It can be transported great distances out of water because it has physiological adaptations to survive long periods of drought. Like all pomaceans, the species is amphibious and has a lung for respiring while in the atmosphere and a ctenidium for respiring while submerged. It has well developed anal glands for eliminating calcium and iron salts and purines that probably accumulate during estivation and intermittent feeding (Andrews 1965).

Predators: Mainly humans, who eat them as escargots.

Parasites: Like *P. bridgesi*, *P. canaliculata* is not commonly reported as an intermediate host of digean trematodes, or other parasites for that matter. However, the species has the potential to cause a bacterial disease called leptospirosis in humans. Some snails carry *Leptospira* bacteria that are thin, helical cells with bent or hooked ends. Infection usually results in fever in humans. The species is also a potential carrier of the rat lungworm, *Angiostrongylus cantonensis*, a nematode parasite that causes meningoencephalitis in humans (Wayne Kobayashi, pers. commun. and brochure, "Apple Snail (*Pomacea canaliculata*)," Department of Agriculture, State of Hawaii).

Reproductive potential and life cycle: The species is dioecious, females laying bright, pink egg clusters on plants and other objects. Mating occurs and fertilization is internal. Very little else is known about its reproductive potential and life cycle. Andrews (1964) described the functional anatomy and histology of the reproductive system of *P. canaliculata*, but little is known about the reproductive biology of ampullariids in general. The only significant studies have been for *P. urceus* by Burky (1974) and Lum-Kong and Kenny (1989). In that species, the reproductive cycle is annual; spawning begins at the end of the rainy season, eggs hatch and develop during the dry season while the adult aestivates, the total development period being 22–30 days, and the fecundity ranges from 21 to 200 eggs/female.

Food and feeding habits: The gut of most pilids is specialized for a macrophagous diet, usually of aquatic angiosperms. *Pomacea canaliculata* feeds on large pieces of leaves of *Pistia* (water lettuce) and *Vallisneria* (eel grass), using sand and grit as triturating agents to grind the plant material against the gastric shield in its gizzard (Andrews 1965). The crop is essentially a storage organ in pilids. The leaf is grasped between the sensory palps on each side of the mouth, then perforated by its radular teeth and large pieces are torn off by action of the jaws (Andrews 1965). Digestion is extracellular, occurring in the stomach where plant cellulose is reduced to a broth by amylase supplied by salivary glands (Andrews 1965). In Hawaiian marshes the species feeds voraciously on rice plants and taro (Wayne Kobayashi, pers. commun. and brochure, "Apple Snail (*Pomacea canaliculata*)," Department of Agriculture, State of Hawaii).

Impact potential: In addition to its potential to carry the bacterial and nematode diseases described above, the species is a voracious feeder on rice plants and taro leaves and threatens to destroy a $2 million taro industry in Hawaii (Wayne Kobayashi, pers. commun. and brochure, "Apple Snail (*Pomacea canaliculata*)," Department of Agriculture, State of Hawaii).

Control: Molluscicides appear to be the most promising control agents.

Family: Viviparidae
Operculum with concentric lines of growth. Sexes separate, some parthenogenetic. Males with modified right tentacle, which serves as a copulatory organ. Ovoviparous, giving birth to young crawling snails.

Cipangopaludina chinensis malleatus (Chinese mystery snail)
The species is more widely introduced than Cipangopaludina japonicus, and occurs in western (California, Arizona, Colorado), eastern (Florida north to Ontario, Quebec, and Nova Scotia, Canada), and most central states, while the latter has been recorded only from Massachusetts, Michigan, and Oklahoma (Burch 1989). Both species are large and were introduced by the Chinese for their food value.

Species characteristics: Shell large, up to about 70 mm high. About 7 whorls, the body whorl inflated, nuclear whorl small, flattened to truncated. Periostracum greenish to brownish, smooth and glossy on upper whorls, lower whorls less glossy and with distinct malleations. Growth rest lines very prominent.

Mode of life and habitat: The species appears to prefer shallow waters with muddy or sandy bottoms containing some organic material. Large populations are often found in beach areas with considerable leaf litter. They occur in large rivers or in streams of order 3 or larger but in pools or eddy areas. Clarke (1980) also reports the species in ponds and marshes.

FIGURE 21.2 *Cipangopaludina chinensis malleatus*, or Chinese mystery snail.

Dispersal mechanisms: Waterfowl and aquatic mammals (otters, muskrats) probably disperse the species among adjacent water bodies. Because of its size, the species is a favorite aquarium animal and may be dispersed accidentally by aquarium enthusiasts.

Adaptations: The species appears to have no unique physiological adaptations.

Predators: Besides humans, who eat them as escargot, waterfowl and aquatic mammals also probably feed on the species.

Parasites: Olson (1974) and Cheng (1973) state that *C. c. malleatus* (as *Viviparus malleatus*) is an intermediate and final host of the aspidogaster, *Aspidogaster conchiola*. The snails become infected by eating embryonated eggs, and the entire life cycle of the parasite occurs in the snail. The most common definitive host of *A. conchiola* is unionid clams, which filter from the water embryonated eggs that have been passed in the faeces by the snail. Turtles may serve as a temporary host (Olsen 1974).

Reproductive potential and life cycle: Like all viviparids, sexes are separate and the species is ovoviparous, the eggs hatching within the uterus and retained for several months. Parthenogenesis is common in many viviparids and may also be present to some degree in *Cipangopaludina chinensis malleatus*.

Food and feeding habits: Herbivore and detritivore, grazing on algae, bacteria, fungi, and other material attached to sand, rocks, and leaf litter. Feeding is probably similar to species of *Viviparus* in which a ctenidial ciliary feeding mechanism is used. Food particles are collected on elongated ctenidial filaments and the floor of the mantle cavity and are then passed to a ciliated food groove on the floor of the mantle cavity, which direct the food particles to the mouth (Cook 1949).

Impact potential: Probably none. In fact, the species has some positive socioeconomic value, still being sold as food in Chinese markets in San Francisco (Hanna 1966). The species is also cultured by Orientals, and more recently by North American entrepreneurs for escargot.

Control: None, since there is no evidence to suggest that the species requires immediate control, especially since its reproductive potential is very low.

Cipangopaludina japonicus (Japanese mystery snail)

The species was probably introduced by the Chinese about 1892 when they were first found in a Chinese vegetable store in San Francisco (Hanna 1966). Both *Cipangopaludina* species occur sporadically but are abundant when they do occur. Nothing is known about the ecological impact of either species, but competition with other common viviparids like *Campeloma* and *Viviparus* species is possible. Like *C. c. malleatus*, *C. japonicus* has a positive socioeconomic value, being sold as food in Chinese markets in San Francisco (Hanna 1966). The biological and ecological attributes of both species summarized below are based mostly on descriptions in Jokinen (1983) and Clarke (1980).

FIGURE 21.3   *Cipangopaludina japonicus*, or Japanese mystery snail.

Species characteristics: Shell not as inflated and with a more acute spire than *C. c. malleatus*. Most whorls have a low carina or acute shoulder, but malleations are absent.

Mode of life and habitat: Like *C. c. malleatus*, the species appears to prefer shallow waters with muddy or sandy bottoms containing some organic material like leaf litter. They occur in large rivers in pools and eddy areas.

Dispersal mechanisms: Like *C. c. malleatus*, the species probably relies on birds and mammals (otters, muskrats) to disperse the species among adjacent water bodies. The species is also a favorite aquarium animal and may be dispersed accidentally by aquarium enthusiasts.

Adaptations: *Cipangopaludina japonicus* also appears to have no unique physiological adaptations. Apparently, the species cannot tolerate low oxygen levels, because winterkill has been reported for the species in Lake Erie by Wolfert and Hiltunen (1968).
Predators: As with *C. c. malleatus*, humans who eat the species as escargot are as much a predator as waterfowl and aquatic mammals.

Parasites: The species is probably intermediate host to aspidogastrids and to sporocysts, rediae and cercariae of some digean trematodes.

Reproductive potential and life cycle: Sexes are separate and the species is ovoviparous, the eggs hatching within the uterus and retained for several months. Parthenogenesis may also be present in *Cipangopaludina japonicus*.

Food and feeding habits: Like *C. c. malleatus*, the species is a herbivore and detritivore, grazing on algae, bacteria, fungi, and other material attached to sand, rocks, and leaf litter. See notes for *C. c. malleatus* regarding feeding mechanisms.

Impact potential: As with *C. c. malleatus*, there is probably little or no ecological impact potential. In fact, the species has some positive socioeconomic value, being cultured by Orientals, and more recently by North American entrepreneurs for escargot.

Control: None, since there is no evidence to suggest that the species requires immediate control, especially since its reproductive potential is very low.

CLASS: BIVALVIA

Family: Corbiculidae

Corbicula fluminea was also introduced for its food value, but the species is discussed by McMahon in Chapter 21 of this Section.

SUMMARY OF IMPACT POTENTIAL OF SPECIES INTRODUCED FOR THEIR FOOD VALUE

Since *Corbicula fluminea* is discussed in detail by McMahon, only the remaining four species are discussed below.

SPECIES WITH LITTLE OR NO ECOLOGICAL IMPACT

Most species introduced for their food value will have little or no obvious impacts on the ecology of aquatic ecosystems. Although most freshwater molluscs are intermediate hosts of numerous species of trematodes, whose definitive hosts are mostly fish or waterfowl and occasionally humans, there are few studies that have assessed the added impact of exotic molluscs with their trematodes on the definitive hosts. It is not anticipated that the exotic species of molluscs will add significantly to the infestation of definitive hosts that are already infected by trematodes of native species of molluscs. All exotic species introduced for their food value will have to compete with their native counterparts for food and space, but very few are known to have replaced native species. Rather, they seem to have added to the diversity of molluscan fauna. Included in this group are

> *Cipangopaludina chinensis malleatus*
> *Cipangopaludina japonicus*

In summary, two of the four species introduced for their food value appear to be of little or no potential ecological risk.

SPECIES OF POTENTIAL SOCIOECONOMIC BENEFIT

Species known to have some beneficial uses (given in parentheses) include

> *Cipangopaludina chinensis malleatus* (as food, especially for immigrants)
> *Cipangopaludina japonicus* (as food, especially for immigrants)

Pomacea bridgesi (as food, especially for immigrants, and escargot)
Pomacea canaliculata (as food, especially for immigrants, and escargot)

In summary, all four of the exotic species have been demonstrated to have some beneficial use as escargot for humans.

SPECIES OF POTENTIAL ECOLOGICAL AND/OR SOCIOECONOMIC HARM

The only potentially harmful socioeconomic impacts from molluscs introduced for their food value are (i) Vectors of Human Parasites, and (ii) Agricultural Impacts. The ecological impacts of molluscs introduced for their food value are poorly understood and, as stated above, appear so far to be minimal.

Socioeconomic Impacts

(i) Vectors of Human Parasites:

Species known to act as intermediate hosts of parasites harmful to man include (parasites given in parentheses)

  Pomacea canaliculata (rat lungworm, Angiostrongylus cantonensis, causes a form of meningitis in man)

(ii) Agricultural Impacts

Pomacea canaliculata is known to cause crop damage and should be eliminated in watersheds used for irrigating agricultural crops of taro or rice.

SUMMARY

Table 21.1 summarizes the country of origin, date of entry, first documented location in North America and their reported impact(s), when known, for molluscs introduced for their food value.

TABLE 21.1
Summary of Species Introduced for their Food Value and their Origin, Date of Entry, First Documented Location in North America, and Reported Impact(s), When Known

Species	Origin	Date	First Location	Impact(s)
Pomacea bridgesi	Brasil	early 1960s	Florida	None reported to date
Pomacea canaliculata	South America	late 1980s	Hawaii	Carrier of bacterial and nematode diseases; potential to destroy rice and taro crops
Cipangopaludina chinensis malleatus	China	1890s	California	None reported to date
Cipangopaludina japonicus	China	1892	San Fransisco	None reported to date

---

REFERENCES

Andrews, E.B. 1964. The functional anatomy and histology of the reproductive system of some pilid gastropod molluscs. *Proc. Malacol. Soc. London* 36: 121–140.

Andrews, E.B. 1965. The functional anatomy of the gut of the prosobranch gastropod *Pomacea canaliculata* (D'Orb) and of some other pilids. *Proc. Zool. Soc. London* 145: 19–36.

Burch, J.B. 1989. North American freshwater snails. *Malacological Publ.*, Hamburg, MI.

Burky, A.J. 1974. Growth and biomass production of an amphibious snail, *Pomacea urceus* (Muller), from the Venezuelan Savannah. *Proc. Malacol. Soc. London* 41: 127–144.

Clarke, A.H. 1980. *The Freshwater Molluscs of Canada.* National Museums of Canada, Ottawa, Ontario.

Clench, W.J. 1966. *Pomacea bridgesi* (Reeve) in Florida. *Nautilus* 79: 105.

Cheng, T.C. 1973. *General Parasitology.* Academic Press, New York.

Cook, P.M. 1949. A ciliary feeding mechanism in *Viviparus viviparus* (L.). *Proc. Malacol. Soc. London.* 27: 265–271.

Hanna, G.D. 1966. Introduced molluscs of western North America. *Occ. Papers Calif. Acad. Sci.* no. 48.

Jokinen, E.H. 1983. The freshwater snails of Connecticut. *St. Geol. Nat. Hist. Surv. Connecticut*, Bull. No. 109.

Lum-Kong, A. and J.S. Kenny. 1989. The reproductive biology of the ampulariid snail *Pomacea urceus* (Muller). *J. Moll. Stud.* 55: 53–65.

Olsen, O.W. 1974. *Animal Parasites. Their Life Cycles and Ecology.* 3rd ed., University Park Press.

Wolfert, D.R. and J.K. Hiltunen. 1968. Distribution and abundance of the Japanese snail, *Viviparus japonicus*, and associated macrobenthos in Sandusky Bay. *Ohio J. Sci.* 68: 32–40.

22 Invasive Characteristics of the Freshwater Bivalve *Corbicula fluminea*

Robert F. McMahon

INTRODUCTION

The freshwater bivalve *Corbicula fluminea* (Müller) (Figure 22.1a,b), was introduced to North America from southeast Asia near the beginning of the 20th century. Its native range includes southeastern Asia west to Turkey, Japan, Indonesia, northern and eastern Australia, and Africa except for the Sahara (Figure 22.2) if, as proposed by Morton (1979, 1986) and supported by allozyme studies (Kijviriya et al., 1991), all freshwater members of *Corbicula* that incubate larvae in the gill are *C. fluminea* (for synonyms, see Morton, 1979, 1986). *C. fluminea* spread rapidly through U.S. drainage systems (McMahon, 1982, 1983a; Counts, 1986), now occurring in 36 continental states, Hawaii, and northern and central Mexico (Taylor, 1981; McMahon, 1982; Hillis and Mayden, 1985; Nelson and McNabb, 1994; Torres-Orozco and Revueltas-Valle, 1996) (Figure 22.2), where it often accounts for over 99% of macrobenthos dry weight biomass (McMahon, 1983a, 1991). It is an economically costly macrofouler of industrial, potable water treatment, agricultural and power station raw water systems (Johnson et al., 1986). Thus, it is considered the most important nonindigenous aquatic animal in North America (McMahon, 1983a, 1991).

    C. fluminea has also been introduced to Europe and South America (Figure 22.2). Unintentionally introduced to Europe in 1980 (Kinzelbach, 1991), it was first recorded in the Dordogne and Tagus rivers of western France and Portugal, respectively (Mouthon, 1981). It has more recently invaded Germany (bij de Vaate, 1991) and the Netherlands (bij de Vaate and Greijdanus-Klass, 1990) and continues to expand in Portugal (Nagel, 1989) and France (Gruet, 1992). It may have been introduced near Trapani, Sicily, before 1940, although no Sicilian populations are presently extant (Mienis, 1991). *C. fluminea* was introduced to South America in the Rio de la Plata at Punta Atalaya, Argentina (Ituarte, 1981, 1985; Darrigran, 1991), and into rivers in the adjacent State of Rio Grand do Sul, Brazil (Mansur and Garces, 1988) in the late 1970s. It has been in the Caripe and San Juan rivers, State of Monagas, Venezuela, since the early to mid-1980s (Martínez, 1987). Thus, *C. fluminea* now appears to have a relatively broad geographic distribution in South America. Continued intercontinental introductions are likely to eventually lead to *C. fluminea* having a transhemispheric, world-wide distribution.

    The rapid North America spread of *C. fluminea* involved human vectors (McMahon, 1982, 1983a, 1991; Counts, 1986), including dispersal as a tourist curiosity, by fish stocking, as juveniles in bilge water of trailered water craft, with or as fish bait, in dredged river sand and gravel, as juveniles byssally attached to boat/barge hulls, and by aquarium hobbyists (McMahon, 1982, 1983a; Counts, 1986). Although the Asian clam's North American dispersal involved anthropomorphic vectors, it also has extensive natural dispersal capacities (see section on Dispersal). Hillis and Mayden (1985) found *C. fluminea* in the Rio Carrizal drainage of northeastern Mexico, which was not impounded, navigable, or fished, had no human settlements, and was accessible by a single

FIGURE 22.1   The nonindigenous Asian clam, *Corbicula fluminea.* (2) Two specimens in natural substratum (gastropod on shell is *Physella* sp.) (largest specimen has a shell length of ≈ 20 mm. (B) Details of shell sulcations, foot and siphons (specimen has a shell length of 15 mm long).

road crossing. Lack of human activity suggested that *C. fluminea* invaded this drainage by a natural dispersal mechanism. More recent expansions of *C. fluminea* in Mexico also have occurred where anthropomorphic vectors were unlikely (Torres-Orozco and Revueltas-Valle, 1996). Similar *C. fluminea* invasions have occurred in Texas drainages isolated from human activity (McMahon, unpublished observations). Thus, investigators (McMahon, 1982; Hillis and Mayden, 1985; Torres-Orozco and Revueltas-Valle, 1996) have concluded that *C. fluminea* has extensive dispersal adaptations reflected not only by colonization of drainages isolated from human activity, but also by fossil-aminostrategraphic evidence that a *Corbicula* species reinvaded English fresh waters during the last three to four interglacial periods (Miller et al., 1979).

     This chapter describes the mode of introduction of the common light morph of *C. fluminea* to North America (for information on *Corbicula* morphs see McMahon, 1991) and reviews its physiology, ecology, life history traits, and dispersal adaptations with special reference to its success as an invasive species and as a macrofouler of raw water systems. Its North American ecological and economic impacts are also discussed.

MODE OF INTRODUCTION TO NORTH AMERICA

Electrophoretic data indicates that North American *C. fluminea* display little or no genetic variation, all tested alleles being highly monomorphic (Smith et al., 1979; McLoed, 1986). Interpopulation genetic variation and allele heterozygosity are greater among populations from Japan, China,

FIGURE 22.2 Present world-wide distribution of *Corbicula fluminea* and distribution in North America (inset box). Original endemic distribution of this species is indicated by darkly shaded areas in Asia, Africa, and Australia, while its distributions in continents it has invaded since 1900 (i.e, North America, South America, and Europe) are indicated by lightly shaded areas.

Thailand, Taiwan, and the Philippines (Smith et al., 1979; Kijviriya et al., 1991; Tsoi et al., 1991; Wu, 1994). North American populations are most genetically related to those from China and the Philippines, suggesting that *C. fluminea* was introduced to North America from southeast Asia. This is likely, as Japanese *Corbicula* are a separate species with a planktonic veliger larva not found in *C. fluminea* (Fusiwara, 1975, 1977; Ikematsu and Kamakura, 1976; Ikematsu and Yamane, 1977). Nearly complete lack of enzyme heterozygosity among North America *C. fluminea* populations suggest that they were derived from a single introduction, in the vicinity of northeastern Washington State, where they were first found (Counts, 1981, 1986). Indeed, because *C. fluminea* is hermaphroditic, all North American populations may have descended from a single, self-fertilizing individual (McMahon, 1983a, 1991) with subsequent genetic drift and founder effects fixing single alleles at many loci (Smith et al., 1979; Hillis and Patton, 1982; McLeod, 1986). Allozyme variation studies suggest that self-fertilization is the common reproductive mode, reducing genetic variation among even native Asian populations (Kijviriya et al., 1991).

In China and the Philippines, *Corbicula* is wild-harvested or cultured for human consumption (Villadolid and Del Rosario, 1930; Miller and McClure, 1931; Morton, 1973; Chen, 1976). In high relative humidities, *C. fluminea* survives emersion over 40 days (McMahon, 1979b; Byrne et al., 1988), allowing its transportation on transoceanic vessels from Asia to western North America by immigrants as a food source. Once introduced, the Asian clam's well-developed capacities for reproduction, growth and dispersal allowed it to rapidly spread through North American inland waters (McMahon, 1982, 1983a).

The Asian clam's ability to rapidly develop dense populations would soon have drawn attention to it. Thus, because it was first reported on Victoria Island, British Columbia, in 1924 (Counts, 1981), it was unlikely to have been introduced in the late 19th century by Chinese immigrants as suggested by Fox (1969). It was more likely introduced to North America in the 1920s (Counts, 1981) unintentionally by Asian immigrants using them as food (Britton and Morton, 1979) or intentionally to create harvestable North America populations (McMahon, 1983a). As a freshwater species (McMahon, 1983a), it could not have been introduced with Japanese marine oyster spat, *Crassostrea gigas* (Thunberg) (Filice, 1958), or as part of a marine fouling community on trans-oceanic vessel hulls (Ingram, 1948), nor could it have entered North America in guts of migratory water fowl (Thompson and Sparks, 1977; Dreier, 1977). Continuing anthropomorphic transcontinental introductions into Europe, Hawaii, and South America assure that *C. fluminea* will continue to expand its range in both hemispheres.

PHYSIOLOGY

Species' distributions are determined by their resistance to the physical extremes of their habitats. The physiological limits of the light morph of *C. fluminea* (McMahon, 1991) are reviewed below and compared to those of native North American freshwater bivalve species.

Hypoxia

C. fluminea is intolerant of even moderate hypoxia, being the most anoxia-intolerant of all tested bivalves, excepting the zebra mussel, *Dreissena polymorpha* (Pallas) (Matthews and McMahon, 1994). It is much less hypoxia-tolerant than most freshwater unionids and sphaeriids (Dietz, 1974; McMahon, 1991). Its O_2 uptake rates decline disproportionately with decline in P_{O_2} (partial pressure of O_2), catabolism apparently becoming anaerobic at $P_{O_2} < 112$ Torr (i.e., < 70% of full air O_2 saturation) (McMahon, 1979a). Thus, *C. fluminea* is generally restricted to well-aerated, lotic habitats or shallow, oxygenated, epilimnetic lentic waters (Fast, 1971; White and White, 1977; White, 1979; Aldridge and McMahon, 1978; McMahon, 1979a, 1991). Poor hypoxia tolerance excludes it from waters with high natural or sewage organic loads (Weber, 1973; Horne and McIntosh, 1979) and reducing sands or muds (Fast, 1971; Aldridge and McMahon, 1978; Eng, 1979; Lenat and Weiss, 1973; Rinne, 1974). Its restriction to oxygenated hyperlimnetic waters was demonstrated when artificial aeration allowed establishment of *C. fluminea* in hypolimnetic waters from which hypoxic conditions previously excluded it (Fast, 1971).

Salinity Tolerance and Osmoregulation

C. fluminea is restricted to freshwater (McMahon, 1983a). It can occassionally occur in upper portions of estuaries, but it is generally not found at salinities > 5‰ (Filice, 1958; Heinsohn, 1958; Copeland et al., 1974; Diaz, 1974). Thus, *C. fluminea* did not colonize a newly constructed reservoir until salinity fell below 2‰ (Morton, 1977). Most freshwater animals inhabit salinities up to 3–8% (Remane and Schlieper, 1971). As the salinity limit of *C. fluminea* falls in this range, it is considered a freshwater species. However, *C. fluminea* is generally more salinity-tolerant than other freshwater bivalves. It can tolerate 10–14‰ (Morton and Tong, 1985), above which capacity to regulate hemolymph osmolarity is lost (Gainey and Greenberg, 1977). In contrast, unionids cannot regulate hemolymph osmolarity or volume above 3‰ (Hiscock, 1953). *C. fluminea,* when hyperosmotically stressed, increases hemolymph free amino acid concentration to remain isosmotic to the medium, thereby, regulating fluid volume (Gainey and Greenberg, 1977; Gainey, 1978a, 1978b; Matsushima et al., 1982). Such hyperosmotic volume regulation is unusual among freshwater bivalves and reflects its recent invasion of freshwater from an estuarine ancestor.

The mechanism of hyperosmotic regulation in *C. fluminea* is different from unionids and sphaeriids. Hemolymph osmotic and Na^+, Ca^{2+}, K^+, and Cl^- concentrations are elevated and HCO_3^- concentrations reduced compared to those of sphaeriids and unionids (Dietz, 1977, 1979). Na^+ influx and efflux rates are higher than those of unionids. A higher proportion of Na^+ uptake results from exchange diffusion than in unionids, and the ion transport system has a higher affinity for Na^+. Na^+ uptake is more dependent on degree of salt depletion than in unionids (McCorkle and Dietz, 1980). Salt depletion stimulates active Na^+ uptake in *C. fluminea,* but not unionids (Dietz, 1985).

Osmoregulatory capacities of *C. fluminea* are more like those of estuarine bivalves than of freshwater unionids or sphaeriids, including capacity for hyperosmotic volume regulation; elevated hemolymph osmotic and ionic concentrations; increased ion flux rates; and retention of Na^+/NH_4^+ exchange diffusion as an auxiliary mode of Na^+ uptake (McCorkle and Dietz, 1980). The fossil history of the Corbiculidae extends from the middle to late Jurassic. Early species were estuarine with freshwater *Corbicula* evolving only in geologically recent times (Keen and Casey, 1969). In contrast, unionids and sphaeriids have longer freshwater fossil histories, extending from the Cretaceous (Keen and Dance, 1969) and Triassic (Haas, 1969), respectively. With a short freshwater fossil history, *C. fluminea* has retained many physiological attributes of its estuarine ancestors not found in unionids and sphaeriids.

Elevated hemolymph osmotic concentrations allow *C. fluminea* to maintain higher metabolic rates (McMahon, 1979a), filtration rates (Mattice, 1979; Foe and Knight, 1986; Lauritsen, 1986), and rates of pedal activity (Kraemer, 1979a) than other freshwater bivalves. Increased metabolic rates give Asian clams a distinct advantage over unionids and sphaeriids in temporally unstable lotic habitats, subject to periodic flooding, as they can more rapidly reburrow when dislodged from the substratum (McMahon, 1983a, 1991).

TEMPERATURE

C. fluminea evolved in semitropical/tropical southeast Asia (Morton, 1979, 1986) where it is rarely exposed to temperature extremes. Thus, a Hong Kong population experienced an annual ambient temperature range of only 13–30°C (Morton, 1977). Its tropical origins are associated with reduced temperature tolerance. While it survives short-term exposures to 0–43°C, the long-term tolerated range is 2–34°C in the laboratory (Mattice and Dye, 1976) and 36°C in the field (McMahon and Williams, 1986a). Both O_2 uptake rates (McMahon, 1979a) and filtration rates (Mattice, 1979) are depressed above 30°C, suggestive of metabolic suppression.

The Asian clam's lower lethal limit of 2°C results in extensive winter population die-offs on the northern edge of its range in North America (McMahon, 1983a). However, recent reports indicate that some northern U.S. populations may tolerate <2°C, suggesting that intolerance of low winter temperatures is not limiting its northern North American distribution (Burky and White, 1989; Janech and Hunter, 1995; Kreiser and Mitton, 1995). Instead, it has been hypothesized that northern waters do not reach temperature thresholds for reproduction (Burky and White, 1989; Janech and Hunter, 1995). However, the Asian clam's threshold reproductive temperature is 15–16°C (McMahon and Williams, 1986b), which falls well below typical summer water temperatures on the northern limit of its North American range (Counts, 1986). Indeed, Lake Erie populations are restricted to thermal discharges during the winter even though ambient temperatures remain above the 15–16°C threshold for reproduction from June–September (Garton and Haag, 1993). Similarly, French and Schloesser (1991) report winter extirpation of *C. fluminea* from downstream portions of a Lake St. Clair thermal effluent that remained below 2°C for over 2 months. In the northern U.S., *C. fluminea* survives overwinter in small streams (Burky and White, 1989; Janech and Hunter, 1995), but not in larger water bodies (French and Schloesser, 1991), perhaps because winter water temperatures in small lotic habitats are unstable, rising above the 2°C lower limit often enough during winter months to prevent mortality. They may also survive over winter in the deeper waters (>10–15 m depth) of iced-over northern lakes (Kreiser and Mitton, 1995) if those waters are oxygenated and inversely stratified so that they remain above 2°C. In any case, the tolerated temperature range of *C. fluminea* is less than that of sympatric North American unionids and sphaeriids (McMahon, 1983a).

C. flumina displays "typical" metabolic temperature compensation, whereby clams laboratory acclimated to low temperatures have higher metabolic rates than when acclimated to higher temperatures (McMahon, 1979a). However, clams that were field-collected at low winter water temperatures had lower metabolic rates than those taken at warmer periods (Williams, 1985; Williams and McMahon, 1985), an unusual "reverse" pattern of metabolic temperature compensation opposite that recorded in the laboratory (McMahon, 1979a). In addition, metabolic rates are suppressed above 33°C in field-collected clams and above 30°C in laboratory-acclimated clams (Williams, 1985). Such data indicate that temperature responses in laboratory-acclimated individuals may not be representative of those occurring under field conditions. Thus, under field conditions, this species may have greater upper and lower thermal limits (Burky and White, 1989) and a capacity to sustain metabolic activity at higher temperatures than laboratory-acclimated individuals in addition to displaying reverse rather than typical metabolic temperature acclimation.

C. fluminea experiences major seasonal changes in tissue condition, overwintering clams having twice the tissue biomass and elevated nonprotein energy stores relative to summer-conditioned individuals (Williams and McMahon, 1989). Thus, reverse acclimation in natural populations may result from winter accumulation of nonrespiring organic energy stores causing computation of lower weight-specific O_2 uptake rates than in summer-conditioned individuals with reduced tissue organic energy stores (i.e., a greater concentration of respiring tissues). Such seasonal variation in nonrespiring tissue mass could mask the typical metabolic temperature compensation characteristic of laboratory-acclimated individuals with similar energy stores, accounting for some of the physiological differences between laboratory and field-conditioned individuals.

EMERGENCE TOLERANCE

Because freshwater bivalves are relatively immobile, they can be emersed for long periods by receding water levels. Many unionid and sphaeriid bivalves have remarkable emersion tolerance, extending for months or even years (Dance, 1958; Hiscock, 1953; Collins, 1967; White and White, 1977; White, 1979; Burky, 1983; Byrne and McMahon, 1994). In contrast, *C. fluminea* is relatively emersion intolerant, surviving means of 26.8 days and 13.9 days in humid and dry air at 20°C, respectively, survival being reduced to 8.3 and 6.7 days at 30°C (McMahon, 1979b). Survival times increase with increasing humidity and decreasing temperature (Byrne et al., 1988). Emersed unionids and *C. fluminea* periodically gape the shell valves and expose mantle tissues to maintain aerobic gas exchange (Byrne and McMahon, 1994). Emersed sphaeriids exchange gas across closed valves through specialized mantle pyramidal cells extending through shell punctae (Collins, 1967). Consumption rates of O_2 are elevated during mantle edge exposure in *C. fluminea* and associated with bursts of metabolic heat production, suggesting payment of an O_2 debt accumulated by anaerobiosis during valve closure periods (McMahon and Williams, 1984; Byrne et al., 1990). Frequency and duration of mantle edge exposure in *C. fluminea* (Byrne et al., 1988) and unionids (Byrne and McMahon, 1994) is reduced with increased temperature, decreased relative humidity, and increasing duration of emersion duration, suggesting that both groups adaptively respond to increasing desiccation pressure by closing valves for longer periods, which increases dependence on anaerobic catabolism, but reduces evaporative water loss rates (Byrne et al., 1990; Byrne and McMahon,

1994). Periodic mantle edge exposure reduces evaporative water loss rates, and thus, increases emersion survival times in freshwater bivalves relative to continuously gaping intertidal/estuarine bivalves (Byrne and McMahon, 1994). Emersed unionids keep their valves closed between bouts of mantle edge exposure longer than *C. fluminea* and are generally more emersion tolerant, allowing them to survive in habitats with greater seasonal water level variation than can Asian clams (Byrne and McMahon, 1994).

CONCLUSIONS, PHYSIOLOGICAL ADAPTATIONS

Physiological resistance/compensation adaptations appear less developed in *C. fluminea* than in other freshwater bivalves. It is less tolerant of hypoxia, temperature extremes, and emersion (McMahon, 1991). Lack of extensive resistance adaptations may reflect the relatively short fossil history of *C. fluminea* in freshwater compared to unionids and sphaeriids. Instead, it remains physiologically allied to its estuarine ancestors reflected by retention of relatively high capacities for osmotic and volume regulation (McMahon, 1983a, 1991). Therefore, success of *C. fluminea* as a freshwater invasive species does not appear to be based on resistance or capacity adaptations. Thus, its ecological and life history adaptations are the likely basis for its success as an invasive species. These are reviewed in the section on Ecology below.

ECOLOGY

Habitat Preference

Unlike most unionids and spheroids with rather narrow microhabitats (Burky, 1983; McMahon, 1991), *C. fluminea* is euryoecious. It is successful in habitats ranging from small, spring-fed streams to major rivers (Morton, 1979, 1986; McMahon, 1982, 1983a, 1991) and ponds (Sinclair and Isom, 1963) to large lakes (Aldridge and McMahon, 1978; Abbott, 1979; Scott-Walsilk et al., 1986). It colonizes agricultural and drainage canals, underground water piping and industrial, potable water and power station raw water systems. It inhabits a variety of substrata, including bare rock, boulders, cobbles (Figure 22.3), gravel, sand, mud, and fine silt, but is most successful in sand and gravel (McMahon, 1983a; Neck, 1986). It thrives in oligotrophic to eutrophic conditions (McMahon, 1983a) and tolerates mildly acidic and low calcium waters inhospitable to other bivalves (Kat, 1982).

Life Cycle and Population Dynamics

Within its wide habitat range, *C. fluminea* displays extensive interpopulation life cycle variation. Typically, it has spring and fall reproductive periods. Maximum adult life span is 3 years. Maturity occurs in 3–12 months at a shell length (SL) of 6–10 mm. Maximum adult size varies from <20 to >50 mm SL and is affected by habitat productivity, ambient temperature, substratum type, calcium concentration, and other biotic and abiotic parameters (McMahon, 1983a, 1991). While such life cycle features are typical of *C. fluminea* in Asia and North America (McMahon, 1983a), they are not fixed. Populations with high individual growth rates can reach a maximum SL of 60–70 mm (Britton and Morton, 1979). In slower growing populations, life span may be 4 years (Eng, 1979), while in others it may be only 1–2 years, with most individuals being annuals (Heinsohn, 1958; Aldridge and McMahon, 1978). Clams in Lake Conroe, Texas, under oligotrophic conditions were semelparous annuals, never reaching an SL >25 mm, while, elsewhere in the same drainage, populations were iteroparous, surviving for 3 years and reaching an SL >35 mm (McMahon, unpubl. observ.). Even in populations with maximum life spans of 3–4 years, high mortality rates cause most individuals to be semelparous annuals. In a Texas lotic population, mortality was 96%, 59%, and 93% during the first, second and third years of life, respectively (Williams and McMahon, 1986). As the temperature threshold for reproduction is >15–16°C, northern U.S. populations in

FIGURE 22.3 A dense population of living adult *Corbicula fluminea* (> 4000 adult clams m^{-2}) in cobble-sand substratum in the Clear Fork of the Trinity River in Benbrook, Tarrant County, Texas. Note that clams survive equally well whether buried in or lying on top of sediments.

colder waters may have a single, annual, mid-summer reproductive period, rather than the spring and fall periods typical of southern and coastal U.S. populations (McMahon, 1983a; Scott-Walsilk et al., 1986).

REPRODUCTION

In North America, *C. fluminea* is a simultaneous hermaphrodite that both cross- and self-fertilizes (Kraemer, 1986). Mature eggs are found in the gonads throughout the year, but sperm are produced only during reproductive periods (Eng, 1979; Kraemer and Galloway, 1986). Fertilized eggs are brooded in the interlammellar spaces of the inner demibranch (gill) through the trochophore, veliger, and pediveliger larval stages that are planktonic in estuarine species of *Corbicula* (Morton, 1982, 1986). Pediveligers are released through the exhalent siphon at approximately 200 μm SL with a fully formed foot and ctenidia (gills). They settle and rapidly metamorphose into an umbonally shelled juveniles (King et al., 1986; Kraemer and Galloway, 1986). The pediveliger shell hinge is straight (i.e., shell is D-shaped), similar to that of the early veliger stage that precedes the pediveliger in most bivalves, but it lacks the veliger's ciliated velum (Kraemer and Galloway, 1986). Incubation time is short, development to pediveliger release requiring <5–6 days (King et al., 1986). Lack of a ciliated velum prevents active swimming (Kraemer and Galloway, 1986); thus, the pediveliger can settle immediately, attaching to the substratum with a single byssal thread retained in juveniles <5 mm SL (Kraemer, 1979b). Although unable to swim due to lack of a ciliated velum, Asian clam pediveligers and juveniles <2 mm SL can be suspended and dispersed long distances in turbulent, flowing water (McMahon and Williams, 1986a; Williams and McMahon, 1986).

Spermiogenesis is initiated above 10°C (Kraemer and Galloway, 1986) and embryo incubation and pediveliger release, above 13–19°C (Kraemer and Galloway, 1986; McMahon and Williams, 1986b). Shell growth and reproduction are suppressed above 30°C (Aldridge and McMahon, 1978), causing mid-summer reproductive suppression in Asian and southern North American populations (McMahon, 1983a), while in cooler northern U.S. waters there is a single abbreviated summer reproductive period.

Fecundity exponentially increases with size (Aldridge and McMahon, 1978) and is estimated to be 97–570 pediveligers adult^{-1} day^{-1} (Heinsohn, 1958). An average adult in a Texas lotic population released 387 pediveligers day^{-1} during a 94-day spring reproductive period and 320 pediveligers day^{-1} during a 101-day fall reproductive period, yielding an annual fecundity of 68,678 pediveligers adult^{-1} yr^{-1} (Aldridge and McMahon, 1978). The abbreviated 5–6 day larval incubation period may allow a single clam to produce many broods during a reproductive season, increasing fecundity. High fecundities and growth rates allow rapid development of dense populations soon after invasion of a new habitat (Figure 22.3, McMahon, 1991).

Growth Rates

C. fluminea has high shell growth rates compared to unionid or sphaeriid bivalves. It attains an SL of 16–30 mm in the first year of life, an 80–150-fold increase in SL relative to the 0.2-mm-long pediveliger (McMahon, 1983a), and representing an 18,000–144,000-fold increase over pediveliger dry tissue mass of 0.0028 mg (Aldridge and McMahon, 1978). Shell growth rate declines linearly with size and age and increases exponentially with temperature. McMahon and Williams (1986b) modeled the shell growth rates of a natural C. fluminea population in relation to both size and temperature, yielding the following equation:

$$\text{Log}_{10} \text{ mm SL day}^{-1} = -2.621 - 0.034(\text{mm SL}) + 0.065(°C).$$

Reported field shell growth rates for specimens <10 mm SL range from 2.0–6.5 mm SL 30 days^{-1} (Dreier and Tranquilli, 1981; Mattice and Wright, 1986; McMahon and Williams, 1986b). Shell growth rates and maximum SL vary within populations and are temperature-dependent, the largest individuals developing in warmer waters, particularly if they occur in the first year of life (McMahon and Williams, 1986b). Since C. fluminea populations in North America and Asia are genetically similar (Smith et al., 1979; McLeod, 1986; Kijviriya et al., 1991; Tsoi et al., 1991; Wu, 1994), such interpopulation variations in shell growth rate, life span, maximum size, and population dynamics must represent ecophenotypic (i.e., nongenetic, environmentally induced) plasticity. Smith et al. (1979) found no electrophoretic variation in 18 enzyme systems among five North America populations, all tested loci being monomorphic. Similarly, examination of 21 enzyme loci showed almost no genetic variation among 14 North America populations. Only 0–19% of loci were polymorphic in any one population, and only 4 of 14 populations had heterozygous individuals, with heterozygosity being quite low at 0.13–0.37% (McLeod, 1986). While genetic variation is higher in Asian C. fluminea populations, it is still low compared to other freshwater molluscs, perhaps due to a tendency to self-fertilize (Kijviriya et al., 1991; Tsoi et al., 1991; Wu, 1994). Thus, interpopulation genetic variation appears too limited to account for interpopulation variation in the population dynamics of C. fluminea. Indeed, when clams from two isolated populations were transferred into a third Corbicula habitat, their growth rates and reproductive patterns became modified to resemble those of the recipient population, suggesting that differences between them were of ecophenotypic rather than genetic origin (McMahon and Britton, unpubl. data).

POPULATION BIOENERGETICS

Population bioenergetic partitioning has been investigated in a Texas lentic *C. fluminea* population (Aldridge and McMahon, 1978) that had high shell growth rates, two annual reproductive periods, an attenuated 1.5–2-year life span, and a terminal SL of 45 mm. Organic carbon allocation patterns were estimated over the life spans of the spring and fall generations and annually for the entire population (Table 22.1). Allocation patterns of assimilated organic carbon to respiration, growth, and reproduction in the spring and fall generations were similar to each other and the entire population (Table 22.1). The population had relatively high assimilation rates (Aldridge and McMahon, 1978) corresponding to the Asian clam's comparatively high rates of filtration (Mattice, 1979; Foe and Knight, 1986; Lauritsen, 1986). Maintenance demands were low, with only 29% of carbon assimilation devoted to respiration, allowing allocation of a large proportion of assimilated carbon (71%) to growth and reproduction (i.e., nonrespired assimilation). The percentage of total assimilation devoted to growth and reproduction is known as the "net production efficiency," which at 71% was elevated compared to other bivalves. The exception is the fast-growing zebra mussel whose assimilation efficiencies range from 44.6–69.8% (Waltz, 1978).

Energy allocation has been studied in sphaeriids. Mean net production efficiency for 13 sphaeriid species was 39.5% (s.d. = ±19.1, range = 12.5–79%) (Hornbach, 1985), considerably lower than in *C. fluminea*. The 71% field-based net production efficiency of *C. fluminea* (Aldridge and McMahon, 1978) has been confirmed by a laboratory study in which 47.4–57.7% of ingested algae ^{14}C was assimilated, and net production efficiency was 59.4–78.2% (Lauritsen, 1986). Assimilation efficiency in small individuals of *C. fluminea* (SL = 8–11 mm) at 16–24°C ranged from 36–51%, and net production efficiency from 58–89% (Foe and Knight, 1986).

The lentic Texas *C. fluminea* population devoted a large proportion of nonrespired assimilation (85%) to tissue growth, or 60% of total assimilation (Table 22.1). Allocation of a high proportion of nonrespired assimilation to tissue growth sustained rapid individual growth rates in this population, giving it one of the highest annual production rates recorded among freshwater bivalves (10.4 g C m^{-2} y^{-1} at an average density of 32.1 clams m^{-2}) (Aldridge and McMahon, 1978). Based on this result, Asian clam populations with higher densities of 3,000–14,000 clams m^{-2} (Heinsohn, 1958; O'Kane, 1976; McMahon and Williams, 1986b) could have annual production rates as high as 1.0–4.5 kg C m^{-2} yr^{-1}, falling among the highest values recorded for any mollusc species. Among freshwater bivalves, only *D. polymorpha* has comparable densities and annual production rates (0.05–14.7 g C m^{-2} yr^{-1}, computed from data in Mackie et al., 1989).

The proportion of assimilated energy allocated by this Texan *C. fluminea* population to reproduction (11% of total assimilation or 15% of nonrespired assimilation) (Table 22.1) falls within that of other freshwater bivalves (mean = 14% of nonrespired assimilation, range = 5–24%) (Burky, 1983). However, the Asian clam's elevated assimilation rates (Mattice, 1979; Foe and Knight, 1986; Lauristen, 1986) may allow the actual amount of energy it allocates to reproduction to be equivalent to or greater than other species. The low tissue biomass of the pediveliger (0.0028 mg dry weight or 0.0014 mg C) (Aldridge and McMahon, 1978) allows high fecundity even though the proportion of nonrespired assimilation allocated to reproduction is relatively low. Elevated allocation of nonrespired assimilation to growth allows the pediveliger to grow rapidly, maturing at an SL of 6–10 mm within 2 months (Aldridge and McMahon, 1978; McMahon, 1983a; McMahon and Williams, 1986a, 1986b; Williams and McMahon, 1986).

Hermaphroditism allows all adults to produce pediveligers, resulting in higher effective fecundities than in dioecious freshwater unionids in which females release 7500–25,000 glochidia adult^{-1} yr^{-1} (from the data of Negus, 1966). The unionid glochidium larval stage is obligately parasitic on fish so that the low proportions of glochidia successfully encysting in a fish host and metamorphosing into juveniles greatly reduce effective fecundities. Like *C. fluminea*, sphaeriids are monoecious and ovoviviparus, brooding embryos in the inner demibranch until release as juvenile clams. Unlike Asian clams, they have relatively low fecundities, producing only 1–136 large juveniles per

TABLE 22.1
Bioenergetic Analysis of Organic Carbon Allocation Rates of Assimilated Organic Carbon (Also Expressed as Percent of Assimilated Carbon Or Percent of Nonrespired Assimilated Carbon) in a Lentic Corbicula Fluminea Population in Lake Arlington, Texas

Generation	Assimilation (A) mg C m^{-2} yr^{-1}	Respiration mg C m^{-2} yr^{-1}	%A	Nonrespired Assimilation (NRA) mg C m^{-2} yr^{-1}	%A	Growth mg C m^{-2} yr^{-1}	%A	%NRA	Reproduction mg C m^{-2} yr^{-1}	%A	%NRA
Spring generation	18.2	6.1	34%	12.1	66%	10.3	57%	85%	1.8	15%	15%
Fall generation	13.4	3.0	23%	10.4	78%	8.7	65%	84%	1.7	13%	16%
Entire population (annual)	40.2	11.6	29%	28.6	71%	24.2	60%	85%	4.4	11%	15%

Source: Aldridge, D.W. and R.F. McMahon, Growth, Fecundity, and Bioenergetics in a natural population of Freshwater Clam, *Corbicula manilensis* Philippi, from North Central Texas. *Journal of Molluscan Studies*, 44, pp. 49–70, 1978. With permission.

reproductive cycle (Burky, 1983). Larval incubation periods in sphaeriids are much longer than the 5–6 days in *C. fluminea* (King et al., 1986), being 1–9 months in *Musculium partumeium* (Say) before juvenile release at SL ≈ 1.4 mm. High growth rates allow juvenile *C. fluminea* to attain an SL of 5–20 mm and reproduce within the 1–9 month brooding period of sphaeriids. Even zebra mussel larval development requires 14–21 days in the plankton before pediveliger settlement at 170–250 μm SL (Nichols, 1996). Thus, while *C. fluminea* devotes a low to average proportion of nonrespired assimilation to reproduction, its high assimilation rates, small eggs, highly attenuated development times, rapid juvenile growth, and early maturity give it much greater reproductive potential than other freshwater bivalves.

Energetic allocation patterns can vary widely among *C. fluminea* populations. Aldridge and McMahon (1978) analyzed organic carbon allocation in a Texas population of *C. fluminea* (Table 22.1) in 1975. When it was reinvestigated in 1982 (Williams and McMahon, 1986), shell growth rates had slowed, and life span increased to 3 years rather than the 1.5–2 years recorded 7 years earlier (Aldridge and McMahon, 1978). Growth rate reduction and life span extension appeared to be a result of a 4.6-fold reduction in phytoplankton densities between sampling periods (Williams and McMahon, 1986). In 1987, the organic carbon allocation budget of this population was again reexamined (Long, 1989). Reduction in algal density appeared to induce further changes in allocation patterns with maintenance accounting for 81% of assimilation, leaving only 17% for growth and 2% for reproduction. Corresponding values recorded in 1975 (Aldridge and McMahon, 1978) under elevated algal densities were 29%, 60%, and 11%, respectively, suggesting that reduction in food availability reduced assimilation rates to the point that maintenance demands dominated the energy budget, reducing allocations to growth and reproduction by 2.8- and 5.5-fold, respectively, with that allocated to reproduction being reduced by a greater proportion than to growth. Similarly, the organic carbon assimilation rate of adult clams enclosed in a power station's heated effluent was reduced by 99.7% compared to individuals enclosed in a cooler portion of the receiving water body. The percentage of nonrespired organic carbon devoted to growth was 89.8% vs. 92.6% and to reproduction 10.2% vs. 7.4% for individuals enclosed in heated and unheated habitats, respectively (Long, 1989), again suggesting that clams in the heated discharge responded to a reduction in assimilation by disproportionately reducing the proportion of nonrespired assimilation devoted to reproduction. Because *C. fluminea* is iteroparous, reproducing twice a year over a 3–4 year life span, such reduction in energy allocation to reproductive effort during periods of low food availability or environmental stress may be adaptive as it increases probability of survival to a future reproductive period when more favorable conditions may allow greater energy allocation to reproduction. This type of life history tactic is called "bet hedging" (Stearns, 1976, 1977, 1980).

ADAPTATION TO UNSTABLE HABITATS

McMahon (1983a, 1991) has suggested that *C. fluminea* is adapted for life in temporally unstable habitats subject to unpredictable, perturbation-induced faunal reductions. Thus, it has successfully colonized North American drainages subject to anthropomorphic interference such as channelization, dredging, impoundment, sand and gravel mining, recreational and commercial ship traffic, and pollution (McMahon, 1983a). These anthropomorphic influences are generally detrimental to all freshwater bivalves including *C. fluminea* (Sickel, 1986). Indeed, the Asian clam may be more susceptible to environmental disturbance/stress than most native North America bivalve species (see section on "Physiology"). Thus, *C. fluminea* population die-offs have been associated with ambient temperature extremes (Dreier and Tranquilli, 1981; McMahon, 1983a; McMahon and Williams, 1986a), hypoxia (Horne and McIntosh, 1979); rapid temperature fluctuation (McMahon and Williams, 1986b), and post-reproductive depletion of organic energy reserves (Williams, 1985; Williams and McMahon, 1985; McMahon and Williams, 1986b). This latter factor may account for summer die-offs attributed to elevated temperatures, disease, pollution, predators, or density-induced stress (for mortality factors in *C. fluminea* see Sickel, 1986).

If *C. fluminea* is less tolerant of environmental stress than unionid and sphaeriid bivalves, why is it so successful in habitats where anthropomorphic and/or natural perturbations extirpate native North American unionid and sphaeriid species (van der Schalie, 1973; Gardner et al., 1976; Kraemer, 1979a)? The answer is that Asian clam life history adaptations allow it to recover from catastrophic density reductions more rapidly than unionids and sphaeriids. Life history traits of *C. fluminea* reviewed above include (1) short life spans (3–4 years); (2) small size at maturity (SL = 6–10 mm); (3) rapid growth to reproductively competent sizes within 1–3 months and to 35–45 mm SL within 2–4 years; (4) simultaneous hermaphroditism with capacity for self-fertilization; (5) twice-annual reproduction; (6) high fecundity; and (7) extensive ecophenotypic plasticity. In contrast, a closely related southeast Asian estuarine corbiculid, *Corbicula fluminalis* (Müller), has (1) a longer, 10-year life span; (2) maturity at a larger SL (16–18 mm); (3) slower growth, maturing after 1 year of life; (4) larger maximum SL (>60 mm); (5) separate sexes with external fertilization; (6) a single annual reproductive period; and (7) an extended planktonic larval life prior to pediveliger settlement (Morton, 1982, 1986). Thus, relative to the estuarine *C. fluminalis*, *C. fluminea*'s elevated growth rates, attenuated development, and early maturation allow it to rapidly re-establish populations after catastrophic density reductions.

C. fluminea's capacity for rapid population recovery is highly adaptive in its favored, unstable habitats and leads to rapid establishment of thriving populations after introduction to a new water body. Such life-history traits are generally characteristic of invasive species (for reviews, see Stearns, 1976, 1977, 1980; Charlesworth, 1980). They result in invasive species' populations having higher proportions of immature than mature individuals, an almost universal characteristic of *C. fluminea* (McMahon, 1983a). Being a self-fertilizing hermaphrodite allows Asian clams to reproduce in isolation from other individuals, leading to establishment of new populations from introduction of single individuals. High fecundity and passive hydrological pediveliger/juvenile dispersal allow rapid recolonization of habitats from which populations have been catastrophically extirpated. Juveniles grow rapidly and mature early, allowing reestablishment of populations with normal age/size structures within 2–4 years and its twice yearly "bet hedging" (Stearns, 1980) reproduction assures survival of unpredictable environmental perturbations that would otherwise result in loss of an entire year's reproductive effort.

There are examples of rapid *C. fluminea* population recoveries. A New River, Virginia, population recovered to 1000 clams m^{-2} within 5 months of nearly complete winter extirpation (Rodgers et al., 1979). Similarly, an irrigation canal population recovered to 10,000–20,000 clams m^{-2} within 2 years of mechanical removal (Prokopovich and Hebert, 1965; Eng, 1979). A power station discharge canal population recovered in winter to 418–441 clams m^{-2} after being reduced to 6–8 clams m^{-2} by high summer temperatures (Dreier and Tranquilli, 1981). A Texas lotic population recovered to 12,000 clams m^{-2} by mid-winter subsequent to a catastrophic reduction in density from 2,655 clams m^{-2} to 305 clams m^{-2} induced by a short-term spring decline in water temperature (McMahon and Williams, 1986b). In contrast, the delayed maturity, extended life spans, high levels of iteroparity, low effective fecundities, and large, well-developed juveniles of sphaeriids and unionids are life history traits that optimize production and survival of offspring to maturity in stable habitats where populations remain near carrying capacity, and in which episodic disturbance such as drying or low winter temperatures are highly predictable (Stearns, 1976, 1977, 1980; Charlesworth, 1980). The ability of *C. fluminea* to rapidly recover from unpredictable density reductions allows it to form dense populations in habitats where periodic disturbance reduces interspecific competition and increases resource availability. Its capacity to rapidly reestablish populations after extirpation may account for the extreme temporal density fluctuations that have been reported for this species (McMahon and Williams 1986a, 1986b).

In contrast to *C. fluminea*, species of freshwater bivalves adapted to stable habitats are characterized by populations with greater proportions of mature than immature individuals (McMahon, 1983a, 1991). Unionids are generally dioecious, iteroparous, obligate cross-fertilizers, with life spans >10–15 years, requiring 6–12 years to mature and have a single annual reproductive period per year (Negus, 1966; Trdan, 1981; for a review see McMahon, 1991), life history traits that make them much less capable of recovering from catastrophic density reductions (McMahon, 1991). Indeed, 20–50 years of habitat stability may be required for unionid recoveries. Their long population recovery times make anthropomorphic activities such as pollution or channelization highly detrimental to unionid populations. Sphaeriids also appear adapted to stable habitats. Like *C. fluminea*, they are monecious, with similar life spans and maturation times (Burky, 1983). However, their very low fecundities and long larval incubation periods may preclude most species from rapid recovery after catastrophic density reduction.

Thus, as native unionid and sphaeriid populations decline in anthropomorphically perturbed North American inland waterways, those of *C. fluminea* have thrived and expanded. It has been claimed that *C. fluminea* out-competes North America unionids (Clarke, 1986; Sickel,1986). Sympatric filter-feeding bivalves compete to a degree for food and space. However, in North America, anthropomorphic activities simultaneously make freshwater habitats unsuitable for native unionid and sphaeriid species adapted to stable habitats, and susceptible to invasion by species such as *C. fluminea* and *D. polymorpha*, whose life history traits allow rapid recovery from disturbance (McMahon, 1983a, 1991). It is almost certain that anthropomorphic disturbance is primarily responsible for the 60+ unionid species presently on the U.S. Register of Endangered Species (McMahon, 1991) rather than detrimental interspecific competition with *C. fluminea.*

Ecophenotypic Plasticity

As described in earlier, North American *C. fluminea* populations display ecophenotypic intrapopulation and interpopulation variation in life history traits. Most of this plasticity is environmentally induced, particularly as this species has little genetic variation throughout its North American range (McLeod, 1986). Reduction in life span, growth rate, maximum size, and number of spawning periods in a Texas lotic *C. fluminea* population were associated with a reduction in average annual water temperature (McMahon and Williams, 1986b). Maximum size was reduced after a catastrophic mortality of large adult clams in the tail waters of Kentucky Dam on the Tennessee River (Sickel, 1986). Williams and McMahon (1986) observed a reduction in growth rate, and increase in life span in a Texas lentic population associated with a 50% decline in phytoplankton productivity. That individuals of *C. fluminea* transferred from one habitat to another rapidly acquire the life history traits of the receiving population is strong evidence for extensive ecophenotypic plasticity in its population dynamics (McMahon and Britton, unpubl. data).

Such nongenetic, ecophenotypic variation is well documented in freshwater molluscs (for reviews see Russell-Hunter, 1964, 1978; Aldridge, 1983; Burky, 1983; McMahon, 1983b; Russell-Hunter and Buckley, 1983) and allows invasive species such as the Asian clam to colonize, survive, and reproduce in highly variable, unstable habitats (Russell-Hunter, 1964, 1978; McMahon, 1983a, 1991). Thus, freshwater basommatophoran pulmonates, which are the most invasive freshwater gastropods, correspondingly display the highest levels of ecophenotypic plasticity (Russell-Hunter, 1978; McMahon, 1983b). Such plasticity allows *C. fluminea* to invade and survive in disturbed habitats from which less plastic "climax" species like unionids and sphaeriids are excluded.

Conclusions, Ecology

While circumstantial evidence suggests that intraspecific competition with *C. fluminea* has led to declines in North American unionid populations (Fuller and Richardson, 1977; Clarke, 1986; Sickel, 1976, 1986), there is an equally large or larger body of evidence indicating that Asian clams are not successful in the stable, unperturbed habitats favored by unionids and sphaeriids (Isom, 1974; Fuller and Imlay, 1976; Klippel and Parmalee, 1979; Kraemer, 1979a; Taylor, 1980; Taylor, 1981; Sickel, 1986). Indeed, co-occurrence of sympatric, highly dense *C. fluminea* populations and declining unionid and sphaeriid populations are likely to be evidence of anthropomorphic disturbance.

The well-developed physiogical resistance adaptations and high level of species diversity of North American sphaeriids and unionids (Burch, 1975a, 1975b; McMahon, 1991) suggest that these groups have long evolutionary histories in which isolated species have become highly adapted to specific freshwater habitats and microniches. Such specialized species are sensitive to environmental perturbation and recover slowly from it. Conversely, *C. fluminea*, adapted for life in temporally disturbed habitats, has poorly developed resistance adaptations (see section on "Physiology") that are of little selective value in their unstable habitats, where they are subject to catastrophic density reductions regardless of capacity to tolerate environmental stress. Instead, *C. fluminea* has evolved a suite of life history traits and capacity for ecophenotypic plasticity that allow rapid recolonization and population recoveries after disturbance-induced density reductions or extirpation (McMahon, 1991).

DISPERSAL

PEDIVELIGER AND JUVENILE DISPERSAL

Invasive species such as *C. fluminea* characteristically have well-developed capacities for dispersal. Dispersal adaptations allow rapid reinvasion of habitats after disturbance-induced extirpation (Pianka, 1978; McNaughton and Wolf, 1979). The dispersal capacities of pediveliger and juvenile Asian clams are extraordinary. The pediveliger's small size (SL $\approx$ 200 μm) and mass (total dry weight $\approx$ 0.1 mg) (Aldridge and McMahon, 1978) allow it to remain suspended in even minimally turbulent water, although it is incapable of active swimming. Thus, pediveligers can be transported long distances on water currents prior to settlement. Juveniles are also hydrologically transported, the single, long juvenile byssal thread (Kraemer, 1979b) acting as drag line. In addition, juvenile byssal attachment to floating objects (i.e., debris, macrophytes, algal mats, or boat/barge hulls) facilitates their dispersal (Counts, 1986). Thus, channelization of waterways for flood control or navigation increases flow velocities and turbulence improving conditions for *C. fluminea* colonization and dispersal while simultaneously rendering them unsuitable for native bivalves (McMahon, 1982, 1991).

In a power station intake canal, densities of suspended Asian clam pediveligers and juveniles ranged between 77–9154 m^{-3} in a population of 168 clams m^{-2}. Maximum size of suspended juveniles was 2.0 mm SL (Williams and McMahon, 1986). Densities of suspended pediveligers/juveniles peaked during reproductive periods, but were greatest during winter months at ambient water temperatures <10°C, below which juvenile burrowing and/or capacity for byssal attachment may be inhibited (Williams and McMahon, 1986). Hypoxia appears to stimulate entry of juvenile *C. fluminea* into the water column (McMahon and Williams, 1986), suggesting that water quality may affect dispersal rates.

Passive downstream pediveliger/juvenile dispersal allows *C. fluminea* to rapidly recolonize portions of lotic habitats after extirpation. A *C. fluminea* population was completely eliminated from a thermal discharge canal during two successive summers when ambient temperatures rose above the chronic upper lethal limit of 36°C. In both years, the canal was rapidly recolonized by pediveliger and juvenile clams transported through the power station's raw water systems into the discharge canal. Fall recolonization rates were 319 clams m^{-2} day^{-1} and 522 clams m^{-2} day^{-1} in the first and second year, respectively. With a benthic surface area of (6.75×10^4 m^2), recolonization rates for the entire canal were estimated to be 2.15×10^7 clams over 24 days in the first year and 3.52×10^7 clams over 40 days in the second year (McMahon and Williams, 1986a).

The large numbers of pediveligers/juveniles suspended in the water column suggest that passive hydrological transport of pediveligers/juveniles is the main dispersal mode of Asian clams, allowing rapid downstream colonization of lotic habits and reestablishment of disturbance decimated populations by dispersal from upstream populations. Thus, downstream invasion rates of North American lotic habitats by *C. fluminea* have generally been more rapid than upstream rates (McMahon, 1982).

While upstream dispersal of *C. fluminea* clearly has involved anthropomorphic activities in North America, it has also been suggested that juvenile clams (SL <5 mm) may be transported upstream in the guts of fish (Ingram, 1959; Sinclair and Isom, 1963; Grantham, 1967; Rinne, 1974; Britton and Murphy, 1977; Dreier, 1977). They may also be transported upstream and between isolated drainage systems when their byssal threads become entangled in the feet of wading birds and water fowl. Juvenile Asian clams also attach to filamentous algae, which can also become entangled in the feet, limbs, and feathers of wading birds and water fowl. While the desiccation resistance of juvenile *C. fluminea* is low, they are likely to survive short flights by birds upstream or between closely adjacent upper reaches of isolated drainage systems. As *C. fluminea* is a self-fertilizing hermaphrodite, introduction of only a single individual could initiate a new population. Juvenile byssal attachment to macrophytic vegetation could allow dispersal between drainages or upstream on infested vegetation used for nest-building by aquatic birds (McMahon, 1982). In contrast, studies indicate that juvenile transport in duck digestive tracks is highly unlikely (Dreier, 1977; Thompson and Sparks, 1977). Thus, while much of the dispersal of *C. fluminea* in North America may have involved anthropomorphic vectors (Counts, 1986), as an invasive species, *C. fluminea* has extensive natural dispersal capacities. These natural dispersal capacities have allowed it to invade drainages in Mexico isolated from anthropomorphic activity (Hillis and Mayden, 1985; Torres-Orozco and Revueltas-Valle, 1996).

ADULT DISPERSAL

Adult *C. fluminea* are also capable of dispersal. They produce a long, mucus dragline from the exhalent siphon which can suspend them in turbulent water (Figure 22.4), allowing downstream dispersal (Prezant and Chalermwat, 1984). However, a zooplankton net suspended 1 m above the bottom monthly over a 1.5 year period in the rapid, turbulent flow of a power station intake canal drawing water from a source infested with *C. fluminea* trapped no specimen larger than 2.0 mm SL. Rather, larger clams were taken in a clam trap anchored to the substratum. This observation suggested that adults are more often hydrologically transported across the substratum surface than in the water column (Williams and McMahon, 1986). A mucus dragline would facilitate hydrological transport across the substratum as well as suspension in the water column.

Simultaneous monitoring of adult and pediveliger/juvenile hydrological transport in a power station intake canal by zooplankton net, clam trap anchored to the substratum, and recording of numbers of dispersing adult clams captured on the plant's intake traveling screens, indicated that numbers of dispersing adults and subadults were minuscule compared with pediveligers and juveniles (Williams and McMahon, 1986). The distances that adult clams could be hydrologically transported are also likely to be much less than those of pediveligers and juveniles suspended in water currents. Thus, adult dispersal behavior is unlikely to be an adaptation for long-distance dispersal in *C. fluminea*.

Adult dispersal is distinctly seasonal. Dispersing adults and subadults in a power station intake canal were collected in low numbers most of the year, but peaked just prior to onset of the spring and fall reproductive periods (Williams and McMahon, 1986). Dispersing adults had lower tissue biomass, higher tissue organic nitrogen to carbon ratios (Williams and McMahon, 1989) and lower molar O_2 uptake rate to ammonia N_2 excretion rate ratios than individuals remaining in the substratum (Williams, 1985; Williams and McMahon, 1985), indicating that dispersing adults had reduced energy stores, reflecting poor nutritional condition. The initiation of spring reproduction in *C. fluminea* is associated with massive transferral of tissue organic energy stores to gamete production marked by 50% reduction in dry tissue biomass (Williams and McMahon, 1989) and shifting from a carbohydrate-dominated metabolism to one dominated by proteins (Williams, 1985; Williams and McMahon, 1985). Individuals in poor nutritional condition prior to reproduction may not be able to sustain the energetic demands of gametogenesis and larval incubation (Williams,

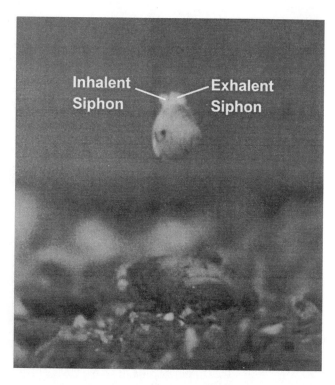

FIGURE 22.4   Small individual of *Corbicula fluminea* (shell length ≈ 10 mm) suspended above the substratum in a turbulent water column from a mucus dragline secreted from the opening of the exhalent siphon. Note that the posteriorly extending siphons are directed upward as the clam is being suspended below a mucus dragline extending from the exhalant siphon, whose transparency makes it invisible in this photograph.

1985; Williams and McMahon, 1985, 1986, 1989). Thus, leaving the substratum to be carried hydrologically into microhabitats more favorable to energy acquisition would be adaptive, particularly as *C. fluminea* supplements filter feeding with pedal feeding on sediment organic detritus (Way et al., 1990; Reid et al., 1992). Such detritus would tend to accumulate in the low-flow areas into which clams would be transported. Hydrological transport could also function to allow dispersal of nutritionally deficient adults away from areas of high clam density, reducing interspecific competition for limited phytoplankton and sediment organic detritus food resources.

CONCLUSIONS, DISPERSAL

The extensive capacities of *C. fluminea* for natural and anthropomorphic dispersal, and its rapid growth, early maturity and high fecundities make it one of the world's most successful aquatic invaders (see Introduction to this chapter). Its invasive nature is enhanced by adaptations for life in habitats subjected to unpredictable natural and/or anthropomorphic disturbance from which native bivalves are often eliminated (McMahon, 1983a, 1991). The main dispersal stage is the pediveliger/small juvenile (SL <2 mm) passively transported downstream on water currents or upstream and across drainages on feet or feathers of aquatic birds. Juvenile byssal attachment to macrophytic vegetation, floating debris, and boat/barge hulls facilitates long-distance dispersal. Natural cross-drainage dispersal is well documented for *Corbicula* in the fossil record (Miller et al., 1979) and by its introduction into isolated Mexican drainages not subject to human activity (Hillis and Mayden, 1985; Torres-Orozco and Revueltas-Valle, 1996).

In contrast, capacity of adult *C. fluminea* to leave the substratum and be hydrologically transported on mucus dragline in the water column (Figure 22.4) or across the sediment surface is unlikely to be a mode of long-distance dispersal. Instead, it may allow clams nutritionally unprepared for gametogenesis and larval incubation to disperse short distances into microhabitats with greater food availability and/or lower levels of intraspecific competition where energy reserves can be accumulated before reproduction.

When the natural dispersal capacities of *C. fluminea* are supplemented by anthropomorphic vectors and disturbance of aquatic habitats, colonization rates can be extraordinary, as evidenced by its rapid spread through North American inland waters over the last 70 years (McMahon, 1982; Counts, 1986) and its more recent invasions of South America and Europe (see Introduction to this chapter). The spread of *C. fluminea* through eastern U.S. drainages after introduction to the lower Ohio River in 1957 (Sinclair and Isom, 1961, McMahon, 1982) was more rapid than that of zebra mussels in the same drainages (Ram and McMahon, 1996), suggesting that it may be the world's most invasive freshwater bivalve species.

IMPACTS

ECOLOGICAL IMPACTS

Ecological impacts of *C. fluminea* in North America have been less studied than those of the zebra mussel, *D. polymorpha* (MacIsaac, 1996), perhaps because the zebra mussel was first found in the Great Lakes in 1989, at a time and place where the public was more aware of the dangers of nonindigenous species introductions. The Asian clam's high filtering rates and densities (Figure 22.3) make it a major consumer of phytoplankton. Phytoplankton densities and chlorophylla concentrations declined by 20–75% in Potomac River reaches with dense *C. fluminea* populations, which filtered the equivalent of the entire water column every 3–4 days (Cohen et al., 1984). Similar seston reductions over isolated beds of *C. fluminea* occurred in the Savannah River, South Carolina (Leff, 1990). A population of 350 clams m^{-2} in the Chowan River, North Carolina, filtered the equivalent of a 5.25 m deep water column every 1–6 days (Lauritsen, 1986), while a lotic clam population averaging 3705 clams m^{-2} filtered the equivalent of the water column every 304 m of stream reach or once every 16–17 min in a current of 18.5 m min^{-1} (McMahon and Williams, 1986b). *C. fluminea* filtering could limit seston available to other suspension-feeding species, affecting structure of planktonic communities. Thus, laboratory studies indicated that *C. fluminea* filtering increased copepod dominance and reduced rotifer densities (Beaver et al., 1991) as reported for *D. polymorpha* (MacIsaac, 1996). *C. fluminea* may also compete with native unionid and sphaeriid clams for limited phytoplankton. However, where unionids co-exist with dense *C. fluminea* populations, no negative impacts appear to have occurred (Beaver et al., 1991; Miller and Payne, 1994). *C. fluminea* efficiently filters bacteria as well as algae (Way et al., 1989; 1990; Silverman et al., 1995), and thus could negatively impact bacterial seston feeders. Consumption of the majority of bacterial and phytoplankton production by *C. fluminea* and resultant accumulation of long-lived, predator-resistant adult clams could also divert primary productivity from higher trophic levels, leading to reduction of game and commercial fish stocks while favoring less desirable fish species that feed on bivalves (Robinson and Wellborn, 1988). In addition, Asian clam sediment detrital pedal feeding could negatively impact burrowing detritivores. Water clarification by clam filtering favors growth of rooted macrophytes, shifting primary production from planktonic to benthic communities.

C. fluminea also affects habitat nutrient dynamics. Its catabolism of seston returns nitrogen to the water column, affecting nitrogen cycling rates and cycles of phytoplankton productivity (Lauritsen and Mozely, 1989; Beaver et al., 1991). Dense *C. fluminea* populations and accumulations of empty shells provide hard substratum in soft sediments, leading to their invasion by hard substratum plant and animal species. Accumulations of clam shells obstruct water flow, impounding lotic habitats (Prokopovich and Hebert, 1965; Prokopovich, 1969).

ECONOMIC IMPACTS

The Asian clam's greatest economic impact in North America has been macrofouling of raw water systems (McMahon, 1983a), particularly fossil-fueled or nuclear power stations where annual U.S. costs for control/repair/replacement were estimated in 1986 at $1 billion (Isom, 1986). Because it does not byssally attach or cement to hard substratum, Asian clam macrofouling is different from that of oysters or mussels (Neitzel et al., 1984). Carried into raw water systems on intake flows, nonswimming pediveligers (200–250 μm) and juveniles (<5 mm) (McMahon, 1983a; Page et al., 1986; Williams and McMahon, 1986) pass traveling screens and tertiary strainers to settle in low-flow areas (<0.3 m/sec or 1 ft/sec, Neitzel et al., 1984; Page et al., 1986). However, settlement in clam-infested systems has been reported at velocities up to 1.2–1.5 m/sec (4–5 ft/sec) (Eng, 1979; Page et al., 1986), suggesting that accumulations of adult shells locally reduce flow velocities below the threshold for settlement.

Initially, settlement rates are greatest on the floors of intake structures where low flows allow sand and silt to accumulate in which clams burrow (Smithson, 1986). In intake pipes and embayments where flow is continually below the 1.2–1.5 m/sec threshold for settlement, adult populations reach >20,000 clams m^{-2} (unpublished observations of the author). Settlement occurs where sediment and corrosion products accumulate (i.e., intake structures, dead end piping, partially closed valves, pipe tees and bends, and nonoperational, redundant, or emergency systems with reduced water flows). In raw water systems, juvenile clams byssally attach to hard substrata through early growth (up to 5 mm SL) from which they release to take up adult life in sediments (Eng, 1979; McMahon, 1983a). After settlement, juveniles grow rapidly to shell lengths (>2 cm or 0.75 inch in 6–12 months) capable of fouling small-diameter components (Aldridge and McMahon, 1978; McMahon, 1983a; Williams and McMahon, 1986b).

Unlike the planktonic larvae of mussels and oysters, Asian clams in a raw water system produce large numbers of pediveligers capable of immediate settlement within that system. Thus, the 4.7 × 9.1 m floor of a low-flow, intake embayment harbored 20,000 adult clams m^{-2} which released ≈ 3.5 × 10^{10} settlement competent pediveligers annually into a power station service water system (unpublished data of the author). Juveniles also invade raw water systems transported on intake currents (Williams and McMahon, 1986). If small enough to pass traveling screens, translocating juveniles settle and grow in intake embayments or, if there are no downstream strainers, can be carried deeper into fouling-sensitive, low-flow components.

Because Asian clams settle only in low-flow areas and adults do not byssally attach, they cannot restrict the diameter of large lines as can oysters or mussels (McMahon, 1983a; Neitzel et al., 1984; Page et al., 1986). Instead, adult shells lodge in and foul small diameter components such as narrow gage lines, heat exchanger/condenser tubing, fire protection lines, and course strainers (diameter <4 cm or 1.6 in.) (Figure 22.5). Adult Asian clams pedally locomote to the substratum surface or into areas of higher flow, where they can be hydrologically transported into small diameter components. Empty clam shells can be similarly transported (McMahon 1983a; Page et al., 1986). If a hydrologically transported shell has a height equivalent to a component's internal diameter, it becomes lodged where that component's diameter is constricted (Figure 22.5), or where pipe diameter is abruptly reduced (i.e., at bends, valves, and joints) (McMahon, 1977; McMahon, 1983a). In contrast, if shell diameter is less than the component, it passes through it to the discharge (McMahon, 1977). Smaller debris (including small clam shells) become trapped upstream of lodged shells, eventually completely occluding the component (Figures 22.6a,b).

Asian clam fouling is particularly severe in peak-load power stations and redundant or emergency systems (i.e., fire protection systems) operated intermittently. If low flow is maintained when these systems are nonoperational by design or incomplete valve sealing, pediveligers and juveniles

FIGURE 22.5 Adult shells of *Corbicula fluminea* lodged in the cold-water box side openings of heat exchanger tubing. Note that shells have diameters just large enough to allow them to become wedged within the tubes after entering them.

settle in them and rapidly grow to fouling sizes. When the system becomes operational, adult clams are transported by increased flows into small diameter components leading to almost immediate fouling (McMahon, 1977, 1983a; Neitzel et al., 1984). Thus, in peak-load generating units, massive clam populations develop in low-flow embayments and intake lines during nonoperational periods and are entrained in huge numbers on traveling screens and/or into small diameter components when the system becomes operational.

 C. fluminea macrofouling also greatly increases sedimentation rates. Their filter-feeding removes suspended silt and organic detritus from intake water which would normally be transported through the system. Filtered silt and detritus is consolidated into dense, mucus-bound feces or "pseudofeces" (pseudofeces contain filtered material rejected before ingestion; Morton, 1983) that sediment at high rates. Sedimentation is also facilitated by low-flow areas between adjacent shells where feces, pseudofeces, and silt accumulate. Clam-facilitated sedimentation rates can be prodigious. In the Delta Mendota Canal, California, sediment accumulation induced by Asian clams required dewatering and removal every 2 years (Prokopovich and Herbert, 1965). An Asian clam with a wet weight of 1 g (shell length $\approx$ 10 mm) was estimated to filter 0.5 l/day, producing 5.4 g of sediment per year such that canal populations with densities of 1000–5000 clams m^{-2} produced 5.4–27.0 kg sediment m^{-2} yr^{-1} (12–60 lb m^{-2} yr^{-1}) or 16–73 g sediment m^{-2} day^{-1} (0.6–2.6 oz m^{-2} day^{-1}) (Prokopovich, 1969). Clam-facilitated sedimentation occurred in this canal even though its current velocities were originally designed to prevent sedimentation (>1 m/sec or 3.3 ft/sec) (Prokopovich, 1969). Clam feces and pseudofeces have high organic contents. Sediments organi-

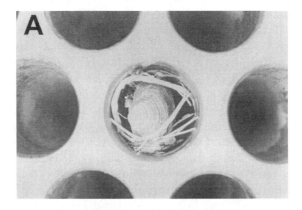

FIGURE 22.6 Examples of upstream accumulation of debris above *Corbicula fluminea* shells wedged in heat exchanger tubing. (a). Macrophytic plant debris accumulated upstream of a lodged adult shell (b) Small *C. fluminea* shells accumulated upstream of a lodged adult shell. Note that as debris continues to accumulate on the upstream side of the lodged shells, the heat exchanger tubes will eventually become completely occluded.

cally enriched with marine mussel feces and pseudofeces displayed dramatic increases in bacterial activity (Genz et al., 1990) which could accelerate microbiologically influenced metallic corrosion (MIC) in raw water systems (McMahon and Lutey, 1996).

Common macrofouling control technologies for Asian clams include oxidizing molluscicides (i.e., chlorine, bromine, potassium permanganate, and ozone), a variety of proprietary nonoxidizing molluscicides including quaternary and polyquaternary ammonium surfactants, and nonchemical controls including, strainers, shell traps, thermal treatment, line pigs, and oxygen deprivation, among others (reviewed in Gross et al., 1979; Mattice, 1979; McMahon, 1983a; Cherry et al., 1986; Mussalli et al., 1986; Potter and Liden, 1986; Neitzel et al., 1986; Johnson and Neitzel, 1987; Neitzel and Johnson, 1988). Such technologies control Asian clam macrofouling in North America industrial, potable water, agricultural and power-generating facilities, but have increased the costs of their operations (Isom, 1986).

CONCLUSIONS, IMPACTS

Because the zebra mussel has not developed dense populations throughout the Mississippi Drainage and its tributaries, *C. fluminea* remains the most widely spread and economically damaging North American nonindigenous aquatic animal. Water-using facilities must allocate substantial funds for

Asian clam fouling control/repair/replacement (Isom, 1986). The North American distribution of
C. fluminea has not substantially changed since the late 1970s (Counts, 1986; McMahon, 1991;
Figure 22.2), allowing enough time for facilities experiencing its macrofouling to have developed
routine control procedures. Thus, this species' macrofouling is publicly less visible than that of the
more recently introduced zebra mussel (Ram and McMahon, 1986), for which several North
American fouling incidents have been widely publicized (Claudi and Mackie, 1994; O'Neill, 1996).
However, *C. fluminea* macrofouling problems are being reported where its range is expanding in
western Europe and are likely to occur as its range expands in South America (Figure 22.2)

In contrast to its macrofouling, there are few studies of the ecological impacts of *C. fluminea*
on native North American biota. Most studies center on post-invasion descriptions without comparable pre-invasion biotic and abiotic data. Available data are mixed, some suggesting that Asian
clams negatively impact native species and communities while others reporting minimal impacts.
One well-supported conclusion is that it is most successful where anthropomorphic and/or natural
disturbance have reduced native benthic fauna and is relatively unsuccessful in undisturbed aquatic
habitats (McMahon, 1983a, 1991). The impacts of *C. fluminea* on North American freshwater
drainages clearly warrant further experimental study before hard conclusions can be drawn.

ACKNOWLEDGMENTS

I wish to express my appreciation to Dr. David W. Aldridge, Dr. Rodger A. Byrne, Carol J. Williams,
John D. Cleland, Dr. Robert G.B. Reid, Dr. Joseph C. Britton, Dr. Thomas H. Dietz, and David W.
Long for their collaborative research on some aspects of the biology of *Corbicula fluminea* discussed
in this chapter. Their shared enthusiasm for this interesting and economically important nonindigenous species has been the inspiration for this chapter.

REFERENCES

Abbott, T.M., Asiatic Clam (*Corbicula fluminea*) Vertical Distributions in Dale Hollow Reservoir, Tennessee,
    in *First International Corbicula Symposium*, J.C. Britton, Ed. Texas Christian University Research Foun-
    dation, Fort Worth, TX, pp. 111–118, 1979.
Aldridge, D.W., Physiological Ecology of Freshwater Prosobranchs, in *The Mollusca*, vol. 6, *Ecology*, W.D.
    Russell-Hunter, Ed. Academic Press, Orlando, FL, pp. 330–358, 1983.
Aldridge, D.W. and R.F. McMahon. Growth, Fecundity, and Bioenergetics in a Natural Population of the
    Freshwater Clam, *Corbicula manilensis* Philippi, from North Central Texas. *Journal of Molluscan Studies*.
    44, pp. 49–70, 1978.
Beaver, J.R., T.L. Crisman, and R.J. Brock. Grazing Effects of an Exotic Bivalve (*Corbicula fluminea*) on
    Hypereutrophic Lake Water. *Lake and Reservoir Management*. 7, pp. 45–51, 1991.
Britton, J.C. and B. Morton, *Corbicula* in North America: the Evidence Reviewed and Evaluated, in *First
    International Corbicula Symposium*, J.C. Britton, Ed. Texas Christian University Research Foundation,
    Fort Worth, TX, pp. 250–287, 1979.
Britton, J.C. and C.E. Murphy. New Records and Ecological Notes on *Corbicula manilensis* in Texas. *The
    Nautilus*. 91, pp. 20–23, 1977.
Burch, J.B., *Freshwater Sphaeriacean Clams (Mollusca: Pelecypoda) of North America*. Malacological Pub-
    lications, Hamburg, MI, 1975a, p. 96.
Burch, J.B., *Freshwater Unionacean Clams (Mollusca: Pelecypoda) of North America*. Malacological Publi-
    cations, Hamburg, MI, 1975b, p. 204.
Burky, A.J., Physiological Ecology of Freshwater Bivalves, in *The Mollusca*, vol. 6, *Ecology*, W.D. Russell-
    Hunter, Ed. Academic Press, Orlando, FL, pp. 281–327, 1983.
Burky, A.J. and P.A. White. Seasonal Low and High Temperature Tolerance of *Corbicula fluminea* from a
    Northern Habitat. *Bulletin of the North American Benthological Society*. 6, p. 59, 1989.
Byrne, R.A. and R.F. McMahon. Behavioral and Physiological Responses to Emersion in Freshwater Bivalves.
    American Zoologist. 34, pp. 194–204, 1994.

Byrne, R.A., R.F. McMahon, and T.H. Dietz. 1988. Temperature and Humidity Effects on Aerial Exposure Tolerance in the Freshwater Bivalve, *Corbicula fluminea. Biological Bulletin* (Woods Hole, MA). 175, pp. 253–260, 1988.

Byrne, R.A., R.F. McMahon, and T.H. Dietz. Behavioral and Metabolic Responses to Emersion and Subsequent Reimmersion in the Freshwater Bivalve, *Corbicula fluminea. Biological Bulletin* (Woods Hole, MA). 178, pp. 251–259, 1990.

Charlesworth, B., 1980. *Evolution of Age Structured Populations.* Cambridge University Press, Cambridge, England, 1980, p. 162.

Chen, T.P., *Aquaculture Practices in Taiwan.* Page Brothers, Norwich, England, 1976, p. 161.

Cherry, D.S., R.L. Roy, R.A. Lechleitner, P.A. Dunhart, G.T. Peters, and J. Cairns, Jr. *Corbicula* Fouling and Control Measures at the Celco Plant, West Virginia. *American Malacological Bulletin.* Special ed. no. 2, pp. 69–81, 1986.

Clarke, A.H., Competitive Exclusion of *Canthyria* (Unionidae) by *Corbicula fluminea. Malacological Data Net.* 1, pp. 3–10, 1986.

Claudi, R. and G.L. Mackie, *Practical Manual for Zebra Mussel Monitoring and Control.* Lewis Publishers, Boca Raton, FL, 1994, p. 227.

Cohen, R.R.H., P.V. Dresler, E.P. Phillips, and R.L. Cory. The Effect of the Asiatic Clam, *Corbicula fluminea,* on Phytoplankton of the Potomac River, Maryland. *Limnology and Oceanography.* 29, pp. 170–180, 1984.

Collins, W.T., "Oxygen-uptake, shell morphology and desiccation of the fingernail clam, *Shaerium occidentale* Prime," Ph.D dissertation, University of Minnesota, 1967.

Copeland, B.J., K.R. Tenore, and D.B. Horton, Oligohaline Regime, in *Coastal Ecological Systems of the United States,* vol 2, H.T. Odum, B.J. Copeland, and E.A. McMahan, Eds. The Conservation Foundation, in co-operation with the National Oceanic and Atmospheric Administration, Office of Coastal Environment, Washington, DC, pp. 315–357, 1974.

Counts, C.L. III., *Corbicula fluminea* (Bivalvia: Sphaeriacea) in British Columbia. *The Nautilus.* 95, pp. 12–13, 1981.

Counts, C.L., III., The Zoogeography and History of the Invasion of the United States by *Corbicula fluminea* (Bivalvia: Corbiculidae). *American Malacological Bulletin.* Special ed. no. 2, pp. 7–39, 1986.

Dance, S.P., Drought Resistance in an African Freshwater Bivalve. *Journal of Conchology.* 24, pp. 281–282, 1958.

Darrigran, G.A., Competencia entre Dos Especies de Pelecipodos Invasores, *Corbicula fluminea* (Müller, 1774) and *C. largillierti* (Philippi, 1844), en el Litoral Argentino del Estuario del Rio De La Plata. *Biología Acuática.* 15, pp. 214–215, 1991.

Diaz, R.J., Asiatic Clam, *Corbicula fluminea* (Philippi) in the Tidal James River, Virginia. *Chesapeake Science.* 15, pp. 118–120, 1974.

Dietz, T.H., Body Fluid Composition and Aerial Oxygen Consumption in the Freshwater Mussel, *Ligumia subrostrata* (Say): Effects of Dehydration and Anoxic Stress. *Biological Bulletin* (Woods Hole, MA). 147, pp. 560–572, 1974.

Dietz, T.H., Solute and Water Movement in Freshwater Bivalve Mollusks, in *Water Relations in Membrane Transport in Plants and Animals,* A.M. Jungreis, Ed. Academic Press, Orlando, FL, pp. 111–119, 1977.

Dietz, T.H., Uptake of Sodium and Chloride by Freshwater Mussels. *Canadian Journal of Zoology.* 57, pp. 156–160, 1979.

Dietz, T.H., Ionic Regulation in Freshwater Mussels: a Brief Review. *American Malacological Bulletin.* 3, pp. 233–242, 1985.

Dreier, H., 1977. Study of *Corbicula,* in Lake Sangchris, in Section 7 of *The Annual Report for Fiscal Year 1976, Lake Sangchris Project,* Illinois Natural History Survey, Urbana, IL, 1977.

Dreier, H. and J.A. Tranquilli. Reproduction, Growth, Distribution and Abundance of *Corbicula* in an Illinois Cooling Lake. *Bulletin of the Illinois Natural History Survey.* 32, pp. 378–393, 1981.

Eng, L.L., Population Dynamics of the Asiatic Clam, *Corbicula fluminea* (Müller) in the Concrete-lined Delta-mendota Canal of Central California, in *First International Corbicula Symposium,* J.C. Britton, Ed. Texas Christian University Research Foundation, Fort Worth, TX, pp. 39–68, 1979.

Fast, A.W., The Invasion and Distribution of the Asiatic Clam (*Corbicula manilensis*) in a Southern California Reservoir. *Bulletin of the Southern California Academy of Science.* 70, pp. 91–98, 1971.

Filice, F.P., Invertebrates from the Estuarine Portion of San Francisco Bay and Some Factors Influencing Their Distribution. *Wasmann Journal of Biology.* 16, pp. 159–211, 1958.

Foe, C. and A. Knight. A Thermal Energy Budget for Juvenile *Corbicula fluminea. American Malacological Bulletin.* Special ed. no. 2, pp. 143–150, 1986.

Fox, R.O., The *Corbicula* Story: a Progress Report. *Second Annual Meeting of the Western Society of Malacologists.* 1969, pp. 1–11, 1969.

French, J.R.P., III, and D.W. Schloesser. Growth and Overwinter Survival of the Asiatic Clam, *Corbicula fluminea*, in the St. Clair River, Michigan. *Hydrobiologia.* 219, pp. 165–170, 1991.

Fuller, S.L.H. and M.J. Imlay. Spacial Competition Between *Corbicula manilensis* (Philippi), the Chinese Clam (Corbiculidae), and Fresh-water Mussels (Unionidae) in the Waccamaw River Basin of the Carolinas (Mollusca: Bivalvia). *ASB (Association of Southeastern Biologists) Bulletin.* 23, pp. 60, 1976.

Fuller, S.L.H. and J.W. Richardson. Amensalistic Competition between *Corbicula manilensis* (Philippi), the Asiatic Clam (Corbiculidae), and Freshwater Mussels (Unionidae) in the Savannah River of Georgia and South Carolina (Mollusca: Bivalvia). *Association of Southeastern Biologists (ASB) Bulletin.* 24, p. 52, 1977.

Fuziwara, T., On the Reproduction of *Corbicula leana* Prime. *Venus.* 34, pp. 54–56, 1975.

Fuziwara, T., On the Growth of the Young Shell of *Corbicula fluminea* Prime. *Venus.* 36, pp. 19–24, 1977.

Gainey, L.F., Jr., The Response of the Corbiculidae (Mollusca: Bivalvia) to Osmotic Stress: the Organismal Response. *Physiological Zoology.* 51, pp. 68–78, 1978a.

Gainey, L.F., Jr., The Response of the Corbiculidae (Mollusca: Bivalvia) to Osmotic Stress: the Cellular Response. *Physiological Zoology.* 51: 79–91, 1978b.

Gainey, L.F., Jr. and M.J. Greenberg. Physiological Basis of the Species Abundance-Salinity Relationship in Molluscs: a Speculation. *Marine Biology* (Berlin). 40, 41–49, 1977.

Gardner, J.A., Jr., W.R. Woodall, Jr., A.A. Staats, Jr. and J.F. Napoli. The Invasion of the Asiatic Clam (*Corbicula manilensis* Philippi) in the Altamaha River, Georgia. *The Nautilus.* 90, pp. 117–125, 1976.

Garton, D.W. and W.R. Haag, Seasonal Reproductive Cycles and Settlement Patterns of *Dreissena polymorpha* in Western Lake Erie, in *Zebra Mussels: Biology, Impacts, and Control*, T.F. Nalepa and D.W. Schloesser, Eds. Lewis Publishers, Boca Raton, FL, pp. 111–128, 1993.

Genz, C., M-N. Hermin, D. Baudinet, and R. Daumas. *In Situ* Biochemical and Bacterial Variation of Sediments Enriched with Mussel Biodeposits. *Hydrobiologia.* 207, pp. 153–160, 1990.

Grantham, B.J., The Asiatic clam in Mississippi, in *Proceedings of the Mississippi Water Resources Conference.* Mississippi Water Resources Institute, Mississippi State University, Jackson, MS, pp. 81–85, 1967.

Gross, L.B., J.M. Jackson, H.B. Flora, B.G. Isom, C. Gooch, S.A. Murray, C.G. Burton, and W.S. Bain, Control Studies on *Corbicula* for Steam-electric Generating Plants, in *First International Corbicula Symposium*, J.C. Britton, Ed. Texas Christian University Research Foundation, Fort Worth, TX, pp. 139–151, 1979.

Gruet, Y. Un Nouveau Mollusgue Bivalve pour Notre Région: *Corbicula* sp. (Heterodonta: Sphaeriacea). *Bulletin de la Société de Sciences Naturelles de l'Quest de la France.* 14, pp. 37–43, 1992.

Hass, F., Superfamily Unionacea Fleming, 1828, in *Treatise on Invertebrate Paleontology, Part N, Mollusca 6*, R.C. Moore, Ed. Geological Society of America, Boulder, CO, pp. 411–467, 1969.

Heinsohn, G.E., "Life history and ecology of the freshwater clam, *Corbicula fluminea*," M.S. thesis, University of California, Santa Barbara, p. 64, 1958.

Hillis, D.M. and J.C. Patton. Morphological and Electrophoretic Evidence for Two Species of *Corbicula* (Bivalvia: Corbiculidae) in North Central America. *American Midland Naturalist.* 108, pp. 74–80, 1982.

Hillis, D.M. and R.L. Mayden. Spread of the Asiatic Clam, *Corbicula* (Bivalvia: Corbiculacea), into the New World Tropics. *Southwestern Naturalist.* 30, pp. 454–456, 1985.

Hiscock, I.D., Osmoregulation in the Australian Freshwater Mussels (Lamellibranchiata). I. Water and Chloride Exchange in *Hybridella australis. Australian Journal of Marine and Freshwater Research.* 1, pp. 317–329, 1953.

Horne, F.R. and S. McIntosh. Factors Influencing the Distribution of Mussels in the Blanco River of Central Texas. *The Nautilus.* 94, pp. 119–133, 1979.

Ikematsu, W. and M. Kamakura. Ecological Studies of *Corbicula leana* Prime. I. On the Reproductive Season and Growth. *Bulletin of the Faculty of Agriculture Miyazaki University.* 22, pp. 185–195, 1976.

Ikematsu, W. and S. Yamane. Ecological Studies of *Corbicula leana* Prime. III. On Spawning Throughout the Year and Self-fertilization in the Gonad. *Bulletin of the Japanese Society of Scientific Fisheries.* 43, pp. 1139–1146, 1977.

Ingram, W.M., The Larger Freshwater Clams of California, Oregon and Washington. *Journal of Entomology and Zoology.* 40, pp. 72–93, 1948.

Ingram, W.M., Asiatic Clams as Potential Pests in California Water Supplies. *Journal of the American Water Works Association.* 51, pp. 363–370, 1959.

Isom, B.G., Mussels of the Green River, Kentucky. *Transactions of the Kentucky Academy of Science.* 35, pp. 55–57, 1974.

Isom, B.G., Historical Review of Asiatic Clam (*Corbicula*) Invasion and Biofouling of Waters and Industries in the Americas. *American Malacological Bulletin.* Special ed. no. 2, pp. 1–5, 1986.

Ituarte, C.F. Primera Noticia Acerca de la Introdúccion´ de Pelecípodos Asiáticos en el Área Rioplatense (Moll. Corbiculidae). *Neotropica.* 27, pp. 79–83, 1981.

Ituarte, C.F., Growth Dynamics in a Natural Population of *Corbicula fluminea* (Bivalvia Sphaeriacea) at Punta Atalaya, Rio Del La Plata, Argentina. *Studies Neotropical Fauna and Environment.* 20, pp. 217–225, 1985.

Janech, M.G. and R.D. Hunter. *Corbicula fluminea* in a Michigan River: Implications for Low Temperature Tolerance. *Malacological Review.* 28, pp. 119–124, 1995.

Johnson, K.I. and D.A. Neitzel. *Improving the Reliability of Open-cycle Water Systems: Application of Biofouling Surveillance and Control to Sediment and Corrosion Fouling at Nuclear Power Plants.* NUREG/CR-4626, PNL-5876, U.S. Nuclear Regulatory Commission, Washington, DC, p. 37, 1987.

Johnson, K.I., C.H. Henager, T.L. Page, and P.F. Hayes. Engineering Factors Influencing *Corbicula* Fouling in Nuclear Service Water Systems. *American Malacological Bulletin.* Special ed. no. 2, pp. 47–52, 1986.

Kat, P.W., Shell Dissolution as a Significant Cause of Mortality for *Corbicula fluminea* (Bivalvia: Corbiculidae) Inhabiting Acidic Waters. *Malacological Review.* 15, pp. 129–134, 1982.

Keen, M. and R. Casey, Family Corbiculidae Gray, 1847, in *Treatise on Invertebrate Paleontology, Part N, Mollusca 6,* R.C. Moore, Ed. Geological Society of America, Boulder, CO, pp. 664–669, 1969.

Keen, M. and P. Dance, Family Pisidiidae Gray, 1857, in *Treatise on Invertebrate Paleontology, Part N, Mollusca 6,* R.C. Moore, Ed. Geological Society of America, Boulder, CO, pp. 669–670, 1969.

Kijviriya, V., E.S. Upatham, V. Viyanant, and D.S. Woodruff. Genetic Studies of Asiatic Clams, *Corbicula*, in Thailand: Allozymes of 21 Nominal Species Are Identical. *American Malacological Bulletin.* 4, pp. 97–106, 1991.

King, C.A., C.J. Langdon, and C.L. Counts, III. Spawning and Early Development of *Corbicula fluminea* (Bivalvia: Corbiculacea) in Laboratory Culture. *American Malacological Bulletin.* 4, pp. 81–88, 1986.

Kinzelbach, R., Die Körbchenmuscheln *Corbicula fluminalis, Corbicula fluminea* und *Corbicula fluviatilis* in Europe (Bivalvia: Corbiculidae). *Mainzer Naturwwissenschaftliches Archiv.* 29, pp. 215–228, 1991.

Klippel, W.E. and P.W. Parmalee. The Naiad Fauna of Lake Springfield, Illinois: an Assessment after Two Decades. *The Nautilus.* 94, pp. 189–197, 1979.

Kraemer, L.R., *Corbicula* (Bivalvia: Sphaeriacea) vs. Indigenous Mussels (Bivalvia: Unionacea) in U.S. Rivers: a Hard Case for Interspecific Competition? *American Zoologist.* 19, pp. 1085–1096, 1979a.

Kraemer, L.R., Juvenile *Corbicula*: Their Distribution in the Arkansas River Benthos, in *First International Corbicula Symposium*, J.C. Britton, Ed. Texas Christian University Research Foundation, Fort Worth, TX, pp. 89–97, 1979b.

Kraemer, L.R. and M.L. Galloway. Larval Development of *Corbicula fluminea* (Müller) (Bivalvia: Corbiculacea): an Appraisal of its Heterochrony. *American Malacological. Bulletin.* 4, pp. 61–79, 1986.

Kraemer, L.R., M.L. Galloway, and R. Kraemer. Biological Basis of Behavior in *Corbicula fluminea*. II. Functional Morphology of Reproduction and Development and Review of Evidence for Self-fertilization. *American Malacological Bulletin.* Special ed. no. 2, pp. 193–201, 1986.

Kreiser, B.R. and J.B. Mitton. The Evolution of Cold Tolerance in *Corbicula fluminea* (Bivalvia: Corbiculidae). *The Nautilus.* 109, pp. 111–112, 1995.

Lauritsen, D.D., Assimilation of Radiolabeled Algae by *Corbicula*. *American Malacological Bulletin.* Special ed. no. 2, pp. 219–222, 1986.

Lauritsen, D.D. and S.C. Mozley. Nutrient Excretion in the Asiatic Clam, *Corbicula fluminea*. *Journal of the North American Benthological Society.* 8, pp. 134–139, 1989.

Leff, L.G., J.L. Burch, and J.V. McAuthur. Spatial Distribution, Seston Removal, and Potential Competitive Interactions of the Bivalves *Corbicula fluminea* and *Elliptio complanata*, in a Coastal Plain Stream. *Freshwater Biology.* 24, pp. 409–416, 1990.

Lenat, D.R. and C.M. Weiss. *Distribution of Macroinvertebrates in Lake Wylie North Carolina – South Carolina.* ESE Publication No. 331, Department of Environmental Science and Engineering, School of Public Health, University of North Carolina, Chapel Hill, NC, p. 75, 1973.

Long, D.P., "Seasonal variation and the influence of thermal effluents on the bioenergetics of the introduced Asian clam, *Corbicula fluminea* (Müller)," M.S. thesis, University of Texas at Arlington, p. 212, 1989.

MacIsaac H.J., Potential Abiotic and Biotic Impacts of Zebra Mussels on the Inland Waters of North America. *American Zoologist*. 36, pp. 287–299, 1996.

Mackie, G.L., W.N. Gibbons, B.W. Muncaster, and I.M. Gray. *The Zebra Mussel, Dreissena polymorpha: a Synthesis of European Experiences and Preview for North America*. O-7729-5647-2, Water Resources Branch, Ontario Ministry of the Environment, Ontario, Canada, 1989, p. 122.

Mansur, M.C.D. and L.M.M. Garces. Ocorrência e Densidadae de *Corbicula fluminea* (Mueller, 1774) e *Neocorbicula limosa* (Maton, 1811) na Estacão Ecológica do Taim e Áreas Adjacentes, Rio Grande do Sul, Brazil (Mollusca, Bivalvia, Corbiculidae). *Iheringia, Série Zoologia*. 68, pp. 99–115, 1988.

Martinez, R.E., *Corbicula manilensis*, Mollusco Introducido en Venezuela. *Acta Científica Venezolana*. 38, pp. 384–385, 1987.

Matsushima, O., F. Sakka, and Y. Kado. Free Amino Acid Involved in Intracellular Osmoregulation in the Clam, *Corbicula*. *Journal of Science of Hiroshima University*, Series B, Division I. 30, pp. 213–219, 1982.

Matthews, M.A. and R.F. McMahon, The Survival of Zebra Mussels (*Dreissena polymorpha*) and Asian Clams (*Corbicula fluminea*) Under Extreme Hypoxia, in *Proceedings: Fourth International Zebra Mussel Conference '94*. Wisconsin Sea Grant, Madison, WI, pp. 231–249, 1994.

Mattice, J.S., Interactions of *Corbicula* sp. with Power Plants, in *First International Corbicula Symposium*, J.C. Britton, Ed. Texas Christian University Research Foundation, Fort Worth, TX, pp. 119–138, 1979.

Mattice, J.S. and L.L. Dye, Thermal tolerance of Asiatic clam, in *Thermal Ecology II*, G.W. Esch and R.W. McFarlane, Eds. CONF-750425, U.S. Energy Research and Development, National Technological Information Service, Department of Commerce, Springfield, VA, pp. 130–135, 1976.

Mattice, J.S. and L.L. Wright. Aspects of Growth of *Corbicula fluminea*. *American Malacological Bulletin*. Special ed. no. 2, pp. 167–178, 1986.

McCorkle, S. and T.H. Dietz. Sodium Transport in the Freshwater Clam *Corbicula fluminea*. *Biological Bulletin* (Woods Hole, MA). 159, pp. 325–336, 1980.

McLeod, M.J., Electrophoretic Variation in North American *Corbicula*. *American Malacological Bulletin*. Special ed. no. 2, pp. 125–132, 1986.

McMahon, R.F., Shell Size-Frequency Distributions of *Corbicula manilensis* Philippi from a Clam-Fouled Steam Condenser. *The Nautilus*. 91, pp. 54–59, 1977.

McMahon, R.F., Response to Temperature and Hypoxia in the Oxygen Consumption of the Introduced Freshwater Clam *Corbicula fluminea* (Müller). *Comparative Biochemistry and Physiology*. 63A, pp. 383–388, 1979a.

McMahon, R.F., Tolerance of Aerial Exposure in The Asiatic Freshwater Clam, *Corbicula fluminea* (Müller), in *First International Corbicula Symposium*, J.C. Britton, Ed. Texas Christian University Research Foundation, Fort Worth, TX, pp. 227–241, 1979b.

McMahon, R.F., The Occurrence and Spread of the Introduced Asian Freshwater Bivalve, *Corbicula fluminea* (Müller) in North America: 1924-1982. *The Nautilus*. 96, pp. 134–141, 1982.

McMahon, R.F., Ecology of the Invasive Pest Bivalve, *Corbicula*, in *The Mollusca*, vol. 6, *Ecology*, W.D. Russell-Hunter, Ed. Academic Press, Orlando, FL, pp. 505–561, 1983a.

McMahon, R.F., Physiological Ecology of Freshwater Pulmonates, in *The Mollusca*, vol. 6, *Ecology*, W.D. Russell-Hunter, Ed. Academic Press, Orlando, FL, pp. 359–430, 1983b.

McMahon, R.F., Mollusca: Bivalvia, in *Ecology and Classification of North American Freshwater Invertebrates*, J.H. Thorp and A.P. Covich, Eds. Academic Press, Orlando, FL, pp. 315–399, 1991.

McMahon, R.F. and R.W. Lutey, Review of the Effects of Invertebrate Macrofouling on Microbiologically Influenced Corrosion in Raw Water Systems, in *Official Proceedings of the International Water Conference*, N.C. Millhouse, P. O'Boyle and J. Schubert, Eds. Paper IWC-96-70, Engineers Society of Western Pennsylvania, Pittsburgh, PA, pp. 650–658, 1996.

McMahon, R.F. and C.J. Williams. A Unique Respiratory Adaptation to Emersion in the Introduced Asian Freshwater Clam, *Corbicula fluminea* (Müller) (Lamellibranchia: Corbiculacea). *Physiological Zoology*. 57, pp. 274–279, 1994.

McMahon, R.F. and C.J. Williams. Growth, Life Cycle, Upper Thermal Limit and Downstream Colonization Rates in a Natural Population of *Corbicula fluminea* (Müller) Receiving Thermal Effluents. *American Malacological Bulletin*. Special ed. no. 2, pp. 231–239, 1986a.

McMahon, R.F. and C.J. Williams. A Reassessment of Growth Rate, Life Span, Life Cycles and Population Dynamics in a Natural Population and Field Caged Individuals of *Corbicula fluminea* (Müller) (Bivalvia: Corbiculacea). *American Malacological Bulletin*. Special Ed., no. 2, pp.151–166, 1986b.

McNaughton, S.J. and L.L. Wolf, *General Ecology*. Holt, Rhinehart and Winston, New York, p. 702, 1979.

Mienis, H.K. Some Remarks Concerning Asiatic Clams Invading Europe with a Note on Sample of *Corbicula fluminea* (Müller, 1774) from Trapani, Sicily. *Notiziario S.I.M.* 9, pp.137–139, 1991.

Miller, A.C. and B.S. Payne. Co-occurrence of Native Freshwater Mussels (Unionidae) and the Non-indigenous *Corbicula fluminea* at Two Stable Shoals in the Ohio River, U.S.A. *Malacological Review.* 27, pp. 87–97, 1994.

Miller, G.H., J.T. Hollin, and J.T. Andrews. Aminostratigraphy of UK Pleistocene Deposits. *Nature* (Lond.). 281, pp. 539–543, 1979.

Miller, R.C. and F.A. McClure. The Fresh-water Clam Industry of the Pearl River. *Lingnan Science Journal.* 10, pp. 307–322, 1931.

Morton, B., Analysis of a Sample of *Corbicula manilensis* Philippi from the Pearl River, China. *Malacological Review.* 6, pp. 35–37, 1973.

Morton, B., The Population Dynamics of *Corbicula fluminea* (Bivalvia: Corbiculacea) in Plover Cove Reservoir, Hong Kong. *Journal of Zoology.* 181, pp. 21–42, 1977.

Morton, B., *Corbicula* in Asia, in *First International Corbicula Symposium*, J.C. Britton, Ed. Texas Christian University Research Foundation, Fort Worth, TX, pp. 15–38, 1979.

Morton, B., Some Aspects of the Population Structure and Sexual Strategy of *Corbicula* cf. *fluminalis* (Bivalvia: Corbiculacea) from the Pearl River, Peoples Republic of China. *Journal of Molluscan Studies.* 48, pp. 1–23, 1982.

Morton, B., Feeding and Digestion in Bivalvia in *The Mollusca*, vol. 5, *Physiology,* part 2, A.S.M. Saleuddin and K.M. Wilbur, Eds. Academic Press, New York, pp. 65–147, 1983.

Morton, B., *Corbicula* in Asia — an Updated Synthesis. *American Malacological Bulletin*. Special ed. no. 2, pp. 113–124, 1986.

Morton, B. and K.Y. Tong. The Salinity Tolerance of *Corbicula fluminea* (Bivalvia: Corbiculoidea) from Hong Kong. *Malacological Review.* 18, pp. 91–95, 1985.

Mouthon, J., Sur la Présence en France et au Portugal de *Corbicula* (Bivalvia: Corbiculidae) Originaire d'Asia. *Basteria.* 45, pp. 109–116, 1981.

Mussalli, Y.G., I.A. Dias-Tous, and J.B. Sickle. Asiatic Clam Control by Mechanical Straining and Organotin Toxicants. *American Malacological Bulletin*. Special ed. no. 2, pp. 83–88, 1986.

Nagel, K-O. Ein Weiterer Fundort von *Corbicula fluminalis* (Müller 1774) (Mollusca: Bivalvia) in Portugal. *Mitteilungen der Deutschen Malakozoologischen Gesellschaft.* 44/45, p. 17, 1989.

Neck, R.W., *Corbicula* in Public Recreation Waters of Texas: Habitat Spectrum and Clam-Human Interactions. *American Malacological Bulletin*. Special ed. no. 2, pp. 179-184, 1986.

Negus, C.L., A Quantitative Study of Growth and Production of Unionid Mussels in the River Thames at Reading. *Journal of Animal Ecology.* 35, pp. 513–532, 1966.

Neitzel, D.A. and K.I. Johnson. *Technical Findings Document for Generic Issue 51: Improving the Reliability of Open-cycle Service Water Systems.* NUREG/CR-5210, PNL-6623, U.S. Nuclear Regulatory Commission, Washington, DC, p. 99, 1988.

Neitzel, D.A., K.I. Johnson, T.L. Page, J.S. Young, and P.M. Daling. *Bivalve Fouling of Nuclear Plant Service Water Systems*, Vol. 1, *Correlation of Bivalve Biological Characteristics and Raw Water System Design.* NUREG/CR-4070, PNL-5300, U.S. Nuclear Regulatory Commission, Washington, DC, p. 119, 1984.

Neitzel, D.A., K.I. Johnson, and P.M. Daling. *Improving the Reliability of Open-cycle Water Systems: an Evaluation of Biofouling Surveillance and Control Techniques for Use at Nuclear Power Plants.* NUREG/CR-4626, PNL-5876, U.S. Nuclear Regulatory Commission, Washington, DC, p. 37, 1986.

Nelson, S.M. and C. McNabb. New Record of Asiatic Clam in Colorado. *Journal of Freshwater Ecology.* 9, pp. 79, 1994.

Nichols, S.J., Variations in the Reproductive Cycle of *Dreissena polymorpha* in Europe, Russia and North America. *American Zoologist.* 36, pp. 311–325, 1996.

O'Kane, K.D., "A population study of the exotic bivalve *Corbicula manilensis* (Philippi, 1841) in selected Texas reservoirs," M.S. thesis, Texas Christian University, p. 134, 1976.

O'Neill, C.R., Jr., *The Zebra Mussel: Impacts and Control.* Cornell Cooperative Extension Information Bulletin 238, New York Sea Grant, Cornell University, State University of New York, Ithaca, NY, p. 62, 1996.

Page, T.L., D.A. Neitzel, M.A. Simmons, and P.F. Hayes. Biofouling of Power Plant Service Systems by *Corbicula. American Malacological Bulletin.* Special ed. no. 2, pp. 41–45, 1986.

Pianka, E.R. *Evolutionary Ecology,* 2nd ed. Harper and Row, New York, p. 397, 1978.

Potter, J.M. and L.H. Liden. *Corbicula* Control at the Potomac River Steam Electric Station Alexandria, Virginia. *American Malacological Bulletin.* Special ed. no. 2, pp. 53–58, 1986.

Prezant R.S. and Chalermwat, K. Flotation of the Bivalve *Corbicula fluminea* as a Means of Dispersal. *Science* 225, pp. 1491–1493, 1984.

Prokopovich, N.P., Deposition of Clastic Sediments by Clams. *Sedimentary Petrology.* 39, pp. 891–901, 1969.

Prokopovich, N.P. and D.J. Hebert. Sedimentation in the Delta-Mendota Canal. *Journal of the American Water Works Association.* 57, pp. 375–382, 1965.

Ram, J.L. and R.F. McMahon. Introduction: the Biology, Ecology, and Physiology of Zebra Mussels. *American Zoologist.* 36, pp. 239-243, 1996.

Remane, A. and C. Schlieper. *Biology of Brackish Water.* Wiley Interscience, New York, p. 372, 1971.

Ried, R.B.G., R.F. McMahon, D.O. Foighil, and R. Finnigan. Anterior Inhalant Currents and Pedal Feeding in Bivalves. *The Veliger.* 35, pp. 93–104, 1992.

Rinne, J.N. The Introduced Asiatic Clam, *Corbicula,* in Central Arizona Reservoirs. *The Nautilus.* 88, pp. 56–81, 1974.

Robinson, J.V. and G.A. Wellborn. Ecological Resistance to the Invasion of a Freshwater Clam, *Corbicula fluminea*: Fish Predation Effects. *Oecologia.* 77, pp. 445–452, 1988.

Rodgers, J.H., D.S. Cherry, K.L. Dickson, and J. Cains, Jr., Elemental accumulation of *Corbicula fluminea* in the New River at Glen Lyn, Virginia, in *First International Corbicula Symposium,* J.C. Britton, Ed. University Research Foundation, Fort Worth, TX, pp. 99–110, 1979.

Russell-Hunter, W., Physiological Aspects of Ecology in Nonmarine Molluscs, in *Physiology of the Mollusca,* vol. 1, K.M. Wilbur and C.M. Yonge, Eds. Academic Press, Orlando, FL, pp 83–128, 1964.

Russell-Hunter, W.D., Ecology of Freshwater Pulmonates, in *Pulmonates,* vol. 2A, *Systematics, Evolution and Ecology,* V. Fretter and J. Peake, Eds. Academic Press, Orlando, FL, pp. 335–383, 1978.

Russell-Hunter, W.D. and D.E. Buckley, Actuarial Bioenergetics of Nonmarine Molluscan Productivity, in *The Mollusca,* vol. 6, *Ecology,* W.D. Russell-Hunter, Ed. Academic Press, Orlando, FL, pp. 463–503, 1983.

Scott-Wasilk, J., J.S. Lietzow, G.G. Downing, and K.L. (C.) Nash. The Asiatic Clam in Lake Erie. *American Malacological Bulletin.* Special ed. no. 2, p. 185, 1986.

Sickel, J.B., "An ecological study of the Asiatic clam, *Corbicula manilensis* (Philippi, 1841), in the Altamaha River, Georgia, with emphasis on population dynamics, productivity and control methods," Ph.D dissertation, Emory University, 1976.

Sickel, J.B., *Corbicula* Population Mortalities: Factors Influencing Population Control. *American Malacological Bulletin.* Special ed. no. 2, pp. 89–94, 1986.

Sinclair, R.M. and B.G. Isom, *A Preliminary Report on the Introduced Asiatic Clam Corbicula in Tennessee.* Tennessee Stream Pollution Control Board, Tennessee Department of Public Health, p. 31, 1961.

Sinclair, R.M. and B.G. Isom, *Further Studies on the Introduced Asiatic Clam* (Corbicula*) in Tennessee.* Tennessee Stream Pollution Control Board, Tennessee Department of Public Health, p. 76, 1963.

Silverman, H., E.C. Achberger, J.W. Lynn, and T.H. Dietz. Filtration and Utilization of Laboratory-cultured Bacteria by *Dreissena polymorpha, Corbicula fluminea,* and *Carunculina texasensis. Biological Bulletin* (Woods Hole, MA). 189, pp. 308–319, 1995.

Smith, M.H., J.C. Britton, P. Burke, R.K. Chesser, M.W. Smith, and J. Hagen, Genetic Variability in *Corbicula,* an Invading Species, in *First International Corbicula Symposium,* J.C. Britton, Ed. Texas Christian University Research Foundation, Fort Worth, TX, pp. 243–248, 1979.

Smithson, J.A., Development of a *Corbicula* Control Treatment at the Baldwin Power Station. *American Malacological Bulletin.* Special ed. No. 2, pp. 63–67, 1986.

Stearns, S.C. Life-history Tactics: a Review of the Ideas. *Quarterly Review of Biology.* 51, pp. 3–47, 1976.

Stearns, S.C. The Evolution of Life History Traits: a Critique of the Theory and Review of the Data. *Annual Reviews of Ecology and Systematics.* 8, pp. 145–171, 1977.

Stearns, S.C. A New View of Life-history Evolution. *Oikos.* 35, pp. 266–281, 1980.

Talyor, D.W., Freshwater Mollusks of California: a Distributional Checklist. *California Fish and Game.* 67, pp. 140–163, 1981.

Talyor, R.W., Freshwater Bivalves of Tygart Creek, Northeastern Kentucky. *The Nautilus.* 94, pp. 89–91, 1980.

Thompson, C.M. and R.E. Sparks. Improbability of Dispersal of Adult Asiatic Clams, *Corbicula manilensis*, via the Intestinal Tract of Migratory Water Fowl. *American Midland Naturalist*. 98, pp. 219–223, 1977.

Torres-Orozco, R. and E. Revueltas-Valle. New Southernmost Record of the Asiatic Clam, *Corbicula fluminea* (Bivalvia: Corbiculidae), in Mexico. *Southwestern Naturalist*. 41, pp. 60–98, 1996.

Trdan, R.J., Reproductive Biology of *Lampsilis radiata siliquoidea* (Pelecypoda: Unionidae). *American Midland Naturalist*. 106, pp. 243–248, 1981.

Tsoi, S.C.M., S-C. Lee, W-L. Wu, and B. Morton. Genetic Variation in *Corbicula fluminea* (Bivalvia: Corbiculoidea) from Hong Kong. *Malacological Review*. 24, pp. 25–34, 1991.

van der Schalie, H., The Mollusks of the Duck River Drainage in Central Tennessee. *Sterkiana*. 52, pp. 45–55, 1973.

bij de Vaate, A. Colonization of the German Part of the Rhine River by the Asiatic Clam, *Corbicula fluminea* Müller, 1774 (Pelecypoda, Corbiculidae). *Bulletin Zoölogisch Museum Universiteit van Amsterdam*. 13, pp. 13–16, 1991.

bij de Vaate, A. and M. Greijdanus-Klaas. The Asiatic Clam, *Corbicula fluminea* (Müller, 1774) (Pelecypoda, Corbiculidae), a New Immigrant in the Netherlands. *Bulletin Zoölogisch Museum Universiteit van Amsterdam*. 12, pp. 173–177, 1990.

Villadolid, D.V. and F.G. Del Rosario. Some Studies on the Biology of Tulla (*Corbicula manilensis* Philippi) a Common Food Clam of Laguna De Bay and Its Tributaries. *Philippine Agriculturalist*. 19, pp. 355–382, 1930.

Waltz, N., The Energy Balance of the Freshwater Mussel *Dreissena polymorpha* Pallas in Laboratory Experiments and in Lake Constance. III. Growth under Standard Conditions. *Archiv fur Hydrobiologie*. Suppl. 55, pp. 121-141, 1978.

Way, C.M., D.J. Hornbach, T. Deneka, and R.A. Whitehead. A Description of the Ultrastructure of the Gills of Freshwater Bivalves, Including a New Structure, the Frontal Cirrus. *Canadian Journal of Zoology*. 67, pp. 357-362, 1989.

Way, C.M., D.J. Hornbach, C.A. Miller-Way, B.S. Payne, and A.C. Miller. Dynamics of Filter-feeding in *Corbicula fluminea* (Bivalvia: Corbiculidae). *Canadian Journal of Zoology*. 68, pp. 115-120, 1990.

Weber, C.I., *Biological Field and Laboratory Methods for Measuring the Quality of Surface Waters and Effluents*. EPA-670/4-73-001, National Environment Research Center, Office of Research and Development, U.S. Environmental Protection Agency, Cincinnati, OH, p. 38, 1973.

White, D.S., The Effect of Lake-level Fluctuation on *Corbicula* and Other Pelecypods in Lake Texoma, Texas and Oklahoma, in *First International Corbicula Symposium*, J.C. Britton, Ed. University Research Foundation, Fort Worth, TX, pp. 82-88, 1979.

White, D.S. and S.J. White. The Effect of Reservoir Fluctuations on Populations *Corbicula manilensis* (Pelecypoda: Corbiculidae). *Proceedings of the Oklahoma Academy of Science*. 57, pp. 106-109, 1977.

Williams, C.J. The population biology and physiological ecology of *Corbicula fluminea* (Müller) in relation to downstream dispersal and clam impingement of power station raw water systems, M.S. thesis, University of Texas at Arlington, p. 152, 1985.

Williams, C.J. and R.F. McMahon. Seasonal Variation in Oxygen Consumption Rates, Nitrogen Excretion Rates and Tissue Organic Carbon:Nitrogen Ratios in the Introduced Freshwater Bivalve, *Corbicula fluminea* (Müller) (Lamellibranchia: Corbiculidae). *American Malacological Bulletin*. 3, pp. 267–268, 1985.

Williams, C.J. and R.F. McMahon. Power Station Entrainment of *Corbicula fluminea* (Müller) in Relation to Population Dynamics, Reproductive Cycle and Biotic and Abiotic Variables. *American Malacological Bulletin*. Special ed. no. 2, pp. 99–111, 1986.

Williams, C.J. and R.F. McMahon. Annual Variation in Tissue Condition of the Asian Freshwater Bivalve, *Corbicula fluminea*, in Terms of Dry Weight, Ash Weight, Carbon and Nitrogen Biomass and Its Relationship to Downstream Dispersal. *Canadian Journal of Zoology*. 67, pp. 82–90, 1989.

Wu, W.L., T.C. Chen, and K.Y. Liao. Population Genetics on *Corbicula fluminea* of Taiwan — Isozyme Study. *Bulletin of Malacology, Republic of China*. 8, pp. 29–53, 1994.

Section VI

Canals and Diversions

23 The Role of Canals in the Spread of Nonindigenous Species in North America

Edward L. Mills, Jana R. Chrisman, and Kristen T. Holeck

SECTION INTRODUCTION

Glaciers blanketed the North American landscape 10,000 years ago and, as they receded, natural waterways were created allowing freshwater organisms to move freely within individual subbasins such as the Great Lakes, Mississippi, and Atlantic drainages (Bailey and Smith 1981). When European explorers and settlers arrived in the late 16th century in what was to be known as the New World (Hadfield 1986), an era of exploration and colonization began. Canal construction was initiated soon after colonization by the first settlers. North American colonists utilized canals to improve navigation for commerce and immigration into unsettled areas, to provide more direct transportation routes, and to provide protection and escape routes during war-time. Among the earliest canals built were those from Plymouth Bay to Green Harbor in Massachusetts for use by Jesuit missionaries to carry stone and timber (Hadfield 1986). By the late 1700s, a vast network of canals was constructed in northeastern North America, thereby dissolving natural barriers to the dispersal of freshwater organisms into the Great Lakes (Mills et al. 1993). In 1779, the first locked canal was built to bypass rapids near Lake St. Francis on the St. Lawrence River (Hadfield 1986). In Canada, the Northwest Trade Company built the Sault St. Marie Canal along the St. Mary's River between lakes Huron and Superior (Legget 1976; Hadfield 1986) and, by the 1800s, one could travel throughout all of Canada on continuous waterways via canals and preexisting water routes. In the American Colonies, most canals were built to provide direct routes for expansion westward into unsettled areas, as canal construction was a less expensive alternative to roads (Hadfield 1986). Smith (1985) indicated that in the late 18th century the future of New York State and the development of the Midwest depended on canals as efficient and economical means to transport goods and people inland. At the peak of canal activity, New York State had 10 major canals and many of these figured prominently in the movement of organisms, particularly fish (Smith 1985) (Figure 23.1).

The construction of canals provided a pathway for the invasion of hundreds and perhaps thousands of organisms into new ecosystems. Unfortunately, we will never know the precise number of species that invaded new areas as a result of canals, since most invasions likely occurred before biologists and limnologists began making biological surveys and documenting the spread of exotic species. Canals are considered a primary vector of introduction for some organisms, such as the sea lamprey (*Petromyzon marinus*) and alewife (*Alosa pseudoharengus*). These are thought to have migrated through the Erie Canal from their native habitat in the Atlantic drainage into the Great Lakes (Dymond 1922; Smith 1972; Emery 1985; Hocutt and Wiley 1986) and both organisms have had significant impacts on native fish species. Most exotic species, however, have used canals as a secondary vector to expand their range. The invasion of North America by the zebra mussel

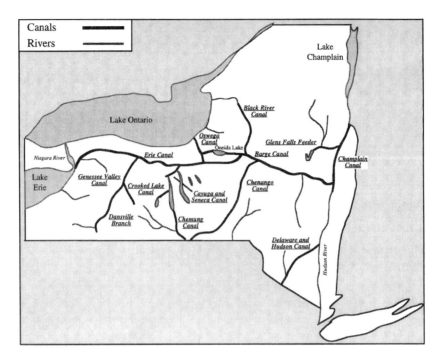

FIGURE 23.1   Canals of New York State (after Smith 1985).

(*Dreissena polymorpha*) is an excellent example. Zebra mussels were introduced into North America's Great Lakes in the ballast water of ships (primary vector) and utilized canals (secondary vector) and other waterways to spread rapidly throughout North America. Canals have also been an important secondary vector linked to the range expansion of plants. Purple loosestrife (*Lythrum salicaria*) was introduced to the Atlantic Coast from Europe in the solid ballast of ships in the early 1800s and likely spread into the Great Lakes basin as a result of water transport through canals and soil disturbance associated with the construction of canals (Thompson et al. 1987).

Evidence for the introduction of nonindigenous species associated with canals is often circumstantial. Because canals are built across drainage divides and are often associated with glacial outlets, it is not always certain that the presence of an organism can be attributed to its migration through a canal. However, if records indicate the presence of an organism near a canal terminus, it is likely that the canal is responsible for its presence there (Smith 1985). The primary objective of this chapter is to examine the role of canals as a pathway for the introduction and spread of exotic species in North America. More specifically, we develop a historical chronology of major canals in North America (Table 23.1) and focus on canal introductions of aquatic organisms that have had significant impacts. Our approach was to focus on canals in six different faunal regions of North America (Mayden 1992) (Figure 23.2). These faunal regions include the Great Lakes, Northern Appalachian, Mississippi, Southeastern, Colorado, and Great Basin–Baja–Klamath–Sacramento.

CANALS OF THE GREAT LAKES BASIN

Canals are important waterways on the Laurentian Great Lakes landscape. They link the Great Lakes to the Atlantic Ocean (St. Lawrence Seaway and Erie-Barge Canal), they link each of the Great Lakes to each other (Welland Canal and Sault St. Marie Canal), and they link the Great Lakes to inland waters (Rideau and Ottawa River Canals and the Trent-Severn Waterway) (Figure 23.3).

TABLE 23.1
Dates of Opening and Terminals for North American Canal

Canal	Opening Date	Terminals	
		From	To
St. Lawrence River		Atlantic Ocean	Lake Ontario
Sault St. Marie	1798	Lake Huron	Lake Superior
Enlargement	1855	Lake Huron	Lake Superior
Champlain	early 1820s	Hudson River	Lake Champlain
Erie			
First Opening	1819	Hudson River	Lake Ontario
Second Opening	1825	Hudson River	Lake Erie
Welland			
First Opening	1829	Lake Ontario	Welland River
Second Opening	1833	Lake Ontario	Port Colbourne, Lake Erie
Third Improvements	1881	Lake Ontario	Port Colbourne, Lake Erie
Fourth Opening	1921	Port Colbourne, Lake Erie	Lake Ontario
Ottawa River Canals	1832	Montreal, St. Lawrence River	Ottawa, Ottawa River
Rideau Canal	1832	Ottawa, Ottawa River	Kingston, Lake Ontario
Ohio and Erie	1832	Portsmouth, Ohio River	Cleveland, Lake Erie
Miami and Erie	1845	Cincinnati, Ohio River	Toledo, Lake Erie
Wabash and Erie		Evansville, Ohio River	Miami and Erie Canal
Chicago	1848	Illinois River	Lake Michigan
Illinois and Michigan	1871	Lake Michigan	Illinois River
Chicago Sanitary and Ship Channel	1900	Lake Michigan	Illinois River
Enlargement	1933	Lake Michigan	mouth of Illinois River at the Mississippi River
Trent-Severn Waterway	1920		
Trent Portion		Kawartha Lakes	Lake Ontario
Severn Portion		Western Haliburton	Georgian Bay, Lake Huron
Tennessee-Tombigbee Waterway	1985	Tennessee River	Tombigbee River
Southeastern Florida canals	varies	southern and eastern Lake Okeechobee	metropolitan areas of southeastern Florida
Channelization of the Kissimmee River	1971	Lake Okeechobee	Lake Istokpoga and Lake Kissimmee
Tamiami Canal		eastern coast of South Florida at Miami	western coast of south Florida near Naples
St. Lucie Canal		eastern coast of Florida at Fort Pierce	Lake Okeechobee
Central Arizona Project	1990s	Colorado River at Lake Havasu	Salt, Verde, and Gila Rivers of central Arizona
California Aqueduct		Sacramento basin of northern California	San Joaquin, Tulare Lake, and Los Angeles basins
Central Valley Project			
Delta-Mendota Canal	1951	Sacramento basin	San Joaquin basin
Friant-Kern Canal		San Joaquin River	Tulare Lake basin
All-American Canal	1940	Colorado River at Yuma	Imperial Valley
Los Angeles Aqueduct	1913	Owens Valley	Los Angeles
extension	1940	Mono basin	Los Angeles
Colorado River Aqueduct	proposed 1928	Colorado River at Parker Dam	Los Angeles, Orange, and San Bernadino Counties

FIGURE 23.2   Faunal regions of North America (after Mayden 1992).

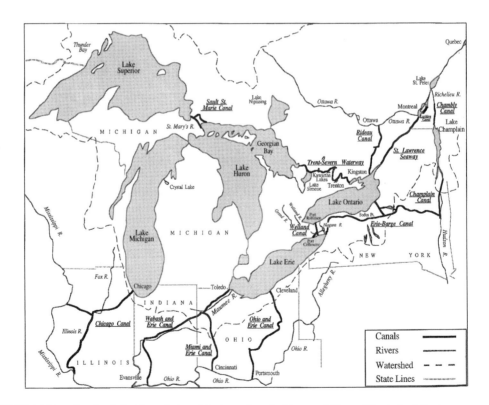

FIGURE 23.3   Canals of the Great Lakes, Northern Appalachian, and Upper Mississippi Basins.

St. Lawrence Seaway

The natural waterway formed by the St. Lawrence River, the Great Lakes, and their connecting channels stretches 3,700 km from the Atlantic Ocean to the heartland of North America. The St. Lawrence River, a critical link in this waterway, began receiving considerable attention as early as 1680 as an economical transport route for the movement of goods into and out of the interior of North America. In the early 16th century, French explorer Jacques Cartier was turned back by the Lachine Rapids in the St. Lawrence River. It was soon recognized by others that such natural obstacles would impede navigation (St. Lawrence Seaway Authority 1985). The first major canal built to bypass the Lachine Rapids was the Lachine Canal, which was completed in 1825. Over time, additional canals were constructed, and the waterway was deepened and widened to accommodate navigational changes — from fur trader canoes and small sailing vessels to schooners, steamers, and finally to large ocean-going vessels (St. Lawrence Seaway Authority 1985).

By the end of the 19th century, rapid industrial growth and population growth in North America's interior combined with the desire to harness the turbulent waters of the river for electric power prompted interest in the construction of a deeper waterway on the St. Lawrence River. This deep waterway became known as the St. Lawrence Seaway. In total, seven locks (St. Lambert, Cort Ste. Catherine, Lower and Upper Beauharnois, Bertrand H. Snell, Dwight D. Eisenhower, and Iroquois) were built on the Montreal–Lake Ontario section of the Seaway (Figure 23.4) (St. Lawrence Seaway

FIGURE 23.4   The St. Lawrence Seaway at Ste. Catherine d'Alexandrie, Ontario, Canada (photo courtesy of the St. Lawrence Seaway Authority, Ottawa, Ontario, Canada).

Authority 1985). Construction of the St. Lawrence Seaway and hydroelectric facility was complete by April of 1959, permitting transoceanic ships to ply the largest freshwater resources in North America (St. Lawrence Valley Souvenir Company 1978). The Great Lakes were now vulnerable to invasion by organisms from throughout the world! Almost one-third of the exotic species in the Great Lakes have been introduced in the last 30 years. This surge corresponds with the opening of the St. Lawrence Seaway in 1959 (Mills et al. 1993). Many of these species are associated with ship ballast water as the primary vector of introduction and have made effective use of connecting waterways and canals as a secondary vector to expand their range.

WELLAND CANAL

Before construction of the Welland Canal, aquatic organisms and humans did not have easy access between the upper and lower Great Lakes because of a major escarpment and barrier, Niagara Falls. Once this natural barrier was bypassed and final improvements were made, the Welland Canal provided the gateway for transportation and commerce into North America's heartland. The Welland Canal was first proposed in 1799, and construction began in 1824 (Whitford 1906). In 1829 the canal was opened for commerce (complete with 26 locks) (McCombie 1968; Ashworth 1986) (Figure 23.5) and provided a transportation route to the Welland River and then to the Niagara River. In 1830, a more direct route was constructed from Port Robinson, Lake Ontario to Port Colbourne, Lake Erie (McCombie 1968). At this time, the Grand River provided water to the Welland Canal at its high point, with water flowing downward to Lake Erie to the south and downward to Lake Ontario to the north (Whitford 1906; Ashworth 1986). Final improvements to the Welland Canal were completed in 1921 when the Grand River was cut off from the waterway, thereby allowing Lake Erie water to flow freely through the Canal from Lake Erie into Lake Ontario (Ashworth 1986). Completion of the Welland Canal was a significant event in North American history, for it opened the interior of the U.S. and Canada to global commerce and transportation, and greatly increased the risk of exotic species invasion into the upper Great Lakes.

FIGURE 23.5 A lock on the Welland Canal south of Port Weller, Ontario, Canada (photo courtesy of the St. Lawrence Seaway Authority, Ottawa, Ontario, Canada).

Sault St. Marie (Soo) Canal

In 1798, the Northwest Fur Company built the first lock on the St. Mary's River at Soo Falls near Sault St. Marie (McCombie 1968; Hadfield 1986) permitting easy travel by small craft between lakes Huron and Superior (Ashworth 1986). By 1837, a canal large enough for transit of steamships was proposed, and the need became more apparent once copper and iron were discovered in the mining districts near Lake Superior (Ashworth 1986). Construction of this larger canal on the U.S. side was completed in 1855 (Legget 1976) with six additional locks built between 1855 and 1919. In 1895, the largest lock in North America for that time was built on the Canadian portion of the St. Mary's River. The Sault St. Marie Canal is currently the most heavily used canal in the world (Legget 1976; Ashworth 1986) and has provided exotic species with a pathway of invasion into Lake Superior.

Rideau Canal and the Ottawa River Canals

After the War of 1812, fear of an American ambush on military supplies being transported up the St. Lawrence River prompted the British, responsible for Canada's defense, to seek an alternative route. A route further away from the American colonies and safer than the St. Lawrence River was needed between Montreal and Kingston (Hadfield 1986). This route was constructed along the Ottawa and Rideau Rivers. The Ottawa River, 1126 km long, is a main tributary of the St. Lawrence River. Its elevation drops 335 m, with the last 122 m being rapids west of the city of Ottawa where the Rideau River joins the Ottawa River (Legget 1976). Canalization of the Ottawa River included a series of three small canals, collectively known as the Ottawa River Canals, to circumvent the Long Sault and adjacent rapids (Legget 1976; Hadfield 1986). The Rideau Canal, from Ottawa to Kingston on Lake Ontario, was built concurrently with the Ottawa River Canals and was finished in 1832. The canal system from Montreal to Kingston was completely navigable by 1834 (Legget 1976; Hadfield 1986). The Rideau Canal is used extensively by recreational boaters, some of which use other connecting canals and reach the Rideau from ports as far away as Florida.

Trent-Severn Waterway

The water route along the Trent and Severn Rivers in Ontario, Canada has been used for transportation, communication, and trade since as early as 9000 BC (Daw 1985). This "Ottawa Waterway," used by Samuel de Champlain in 1615, provided a water route with a few short portages from the Bay of Quinte, Lake Ontario to Georgian Bay, Lake Huron (Legget 1976). The first lock of the Trent-Severn Waterway was built in 1833 in the Kawartha Lakes Region. In 1914, plans were made for the construction of a lock at Big Chute, between Lake Simcoe and Georgian Bay; instead a lift was constructed to prevent sea lamprey invasion into Lake Simcoe (Legget 1976). Water in the Trent-Severn Waterway flows in two directions: the Trent portion flows southward from eastern Haliburton and the Kawartha lakes into Lake Ontario and the Severn portion flows from western Haliburton through Lake Simcoe to Georgian Bay (Daw 1985). The entire system was completed in 1920 (Ireland and Schepanek 1993). Though the Trent-Severn Waterway has a steep gradient, making migration of organisms through it more difficult; plants and organisms are easily dispersed by intensive recreational boating.

NONINDIGENOUS SPECIES OF THE GREAT LAKES BASIN

At least 140 exotic species have entered the Great Lakes, with canals being implicated as a primary or secondary vector of their introduction. The most notable primary invaders into the Great Lakes proper are the sea lamprey (*Petromyzon marinus*), the alewife (*Alosa pseudoharengus*), rainbow smelt (*Osmerus mordax*), and white perch (*Morone americanus*). Most exotics used canals as a secondary vector for invasion into the Great Lakes, including purple loosestrife (*Lythrum salicaria*) and dreissenid mussels (*Dreissena polymorpha* and *D. bugensis*).

SEA LAMPREY

The sea lamprey (Figure 23.6) was first discovered in the Great Lakes in the 1830s, when it was reported in Lake Ontario and was thought to have migrated from the Atlantic drainage either through the Hudson River and Erie Canal or the St. Lawrence River (Aron and Smith 1971; Emery 1985). Morman et al. (1980) suggest that the sea lamprey attached to boats transiting the Erie and St. Lawrence canal systems. The first report of sea lamprey in Lake Erie was in 1921 (Dymond 1922), where it probably entered via the Welland Canal (Applegate and Moffet 1955; Trautman 1957; Lawrie 1970). The delayed invasion may have been a result of changes in drainage patterns after

FIGURE 23.6   Sea lamprey (*Petromyzon marinus*) attached to a Great Lakes lake trout (*Salvelinus namaycush*) (photo courtesy of the USGS Biological Resources Division, Oswego, NY).

modifications in the Welland Canal. Before improvements were made to the Welland Canal, migrating lamprey were likely diverted into the Grand River because of their instinct to swim upstream during spawning. After the Grand River was cut off from the Welland Canal in 1921, making Lake Erie the water source, lamprey appeared in Lake Erie and subsequently spread to all of the Great Lakes (Ashworth 1986). Once sea lamprey infested Lake Erie, they could expand their range into all the remaining Great Lakes (Applegate and Moffet 1955; Aron and Smith 1971). Spawning populations of sea lamprey were first observed in Lake Huron tributaries in 1937 (Emery 1985), where they may have originated from the Trent-Severn Waterway or Lake Erie. Sea lamprey reached Lake Superior, the western-most Great Lake, by 1946 (Lawrie and Rahrer 1973; Emery 1985). Invasion of the Great Lakes by sea lamprey is associated with the decline of several fish species, including lake trout, and has caused millions of dollars in damage to Great Lakes sport and commercial fisheries (Applegate and Moffet 1955; Fetteroff 1980). Millions of dollars have been and will continue to be spent on lamprey control (Baldwin 1968; Lawrie 1970; Emery 1985).

ALEWIFE

The alewife (Figure 23.7) is native to the lakes and rivers along the Atlantic Coast (Scott and Crossman 1973). Most evidence suggests that alewife entered Lake Ontario through the Erie Canal, even though they were not recorded until 54 years after completion of the canal (Miller 1957; Smith 1970; Aron and Smith 1971; Lawrie and Rahrer 1973). Scott and Crossman (1973) suggested that alewife may have invaded Lake Ontario through the St. Lawrence River. Alternatively, Smith (1970) suggested alewife are native to Lake Ontario, but populations were suppressed by Atlantic salmon and lake trout until their decline in the late 1800s. Like sea lamprey, alewife expanded into Lake Erie through the Welland Canal (Dymond 1932; Miller 1957; Trautman 1957) after alterations changed water flow (Ashworth 1986). Alewife were first recorded from Lake Erie in 1931 (Dymond 1932). Once alewife became established in Lake Erie, it expanded its range to other Great Lakes (Miller 1957). They were first recorded from Lake Huron in 1933 and from Lake Michigan in 1949 (Miller 1957; Lawrie and Rahrer 1973). Alewife likely spread from lakes Michigan and Huron through the Sault St. Marie Canal into Lake Superior, where they were first reported around 1953 (Smith 1970; Lawrie and Rahrer 1973). Sea lamprey caused a severe decline in alewife predators (Aron and Smith 1971), causing alewife to become extremely abundant. Subsequently, massive die-offs occurred, resulting in polluted shores and clogged municipal and industrial intakes. The alewife also preyed on native fishes, suppressing coregonids, yellow perch, and emerald shiners, and competed with rainbow smelt for zooplankton. Today, alewife are prey for lake trout and introduced salmon (Emery 1985) and support a multimillion dollar sport fishery.

FIGURE 23.7   Alewife (*Alosa pseudoharengus*) collected in midwater trawls from Lake Ontario in the 1980s (photo courtesy of the USGS Biological Resources Division, Oswego, NY).

Rainbow Smelt

In the early 1900s, rainbow smelt (Figure 23.8) were unsuccessfully stocked into the St. Mary's River between lakes Huron and Superior (Creaser 1926; Van Oosten 1937). Rainbow smelt stocked into Crystal Lake, Michigan in 1912 are probably responsible for the established populations in Lake Michigan (Creaser 1926; Scott and Crossman 1973; Emery 1985). The first record of rainbow smelt in Lake Huron occurred in 1925 (Van Oosten 1937; Berst and Spangler 1973; Emery 1985). The Soo Canal provided the pathway into Lake Superior (Van Oosten 1937), where they became abundant (Lawrie and Rahrer 1973). Rainbow smelt were first seen in Lake Erie in 1936 (Van Oosten 1937). Though the origin of rainbow smelt in Lake Ontario is not known, Christie (1973) suggested that rainbow smelt entered Lake Ontario through the Welland Canal, while Scott and Crossman (1973) suggested the Erie Canal. Rainbow smelt were first recorded in Lake Ontario at Sodus Point in 1929 (Christie 1973). Some suggest that this species was introduced to Lake Ontario (Dymond 1944), while others believe that it was native to this system (Hubbs and Lagler 1958). Rainbow smelt can now be found throughout the Great Lakes, and even though they have become a valuable sport and commercial fish, they compete with and prey on native fishes (Christie 1973; Emery 1985). Rainbow smelt may have had the greatest impact on native fishes of Lake Erie and caused the extinction of blue pike (Christie 1973; Regier and Hartman 1973; Emery 1985).

FIGURE 23.8   Rainbow smelt (*Osmerus mordax*) from Lake Ontario (photo courtesy of the USGS Biological Resources Division, Oswego, NY).

White Perch

The white perch, native to the Atlantic coastal plain, was first observed in the Lake Ontario watershed in 1950 (Dence 1952) and in Lake Erie in 1953 (Larsen 1954; Busch et al. 1977). White perch probably expanded its range into Lake Ontario through the Erie-Barge Canal system from the Hudson River (Scott and Christie 1963; Scott and Crossman 1973). Evidence suggests that the Erie Canal is the most likely route for the white perch invasion into Lake Erie (Scott and Christie 1963), even though the Welland is the most likely route of invasion for sea lamprey and alewife (Busch et al. 1977). The impact of white perch on native fishes has been difficult to assess (Christie 1973; Emery 1985; Hurley 1986) but the potential for competition for food with other fish species has been strongly suggested (Christie 1972; Scott and Crossman 1973; Elrod et al. 1981; Prout et al. 1990).

PURPLE LOOSESTRIFE

Thompson et al. (1987) reviewed the introduction and spread of purple loosestrife (Figure 23.9) into North America and Canada. Purple loosestrife is thought to have entered the Atlantic Coast of North America in the early 1800s and spread into the Great Lakes basin through soil disturbance associated with railroad and canal construction (e.g., Erie Canal). The rapid spread of this wetland plant throughout North America occurred soon after its initial invasion in the Great Lakes (Thompson et al. 1987). Monospecific stands of this plant now occur throughout North America and, ecologically, they outcompete native plants such as cattails, resulting in the loss of prime habitat for waterfowl and other marsh animals (Rawinski and Malecki 1984).

FIGURE 23.8 Purple loosestrife (*Lythrum salicaria*) from the Oneida Lake, New York watershed (photo courtesy of the Cornell Biological Field Station, Bridgeport, NY).

ZEBRA MUSSEL AND QUAGGA MUSSEL

Zebra mussel (*Dreissena polymorpha*) and quagga mussel (*D. bugensis*) (Figure 23.10) are two of the most important exotic species to use canals as a secondary vector for invasion. *Dreissena polymorpha* became established in Lake St. Clair by 1986 (Hebert et al. 1989), and by 1990 had spread throughout the all of the Great Lakes and its major canals, and into the Mississippi drainage (Hebert et al. 1989; Griffiths et al. 1991; Martell 1995). The quagga mussel, first sighted at Port Colbourne on Lake Erie in 1989, is primarily restricted to the lower Great Lakes (Mills et al. 1993), although one sighting of *D. bugensis* has been confirmed outside the Great Lakes basin in the Mississippi drainage near St. Louis, Missouri (O'Neill 1995). Recent evidence suggests that the quagga mussel may be displacing the zebra mussel, as the quagga mussel is the most abundant dreissenid in areas of lakes Erie and Ontario that once dominated by zebra mussels (Mills et al. unpublished). Both dreissenids are filter feeders and are having significant impacts on ecosystem processes (Leach 1993; Nichols and Hopkins 1993; Madenjian 1995). These impacts are also examined in detail by Nalepa and colleagues in Chapter 16 of this volume.

FIGURE 23.10   Dreissenid mussels collected in bottom trawls from Lake Ontario (photo courtesy of the USGS Biological Resources Division, Oswego, NY).

NONINDIGENOUS SPECIES AND THE INLAND WATERS OF THE GREAT LAKES BASIN

NONINDIGENOUS SPECIES OF THE RIDEAU AND OTTAWA RIVER CANALS

Canals, such as the Rideau and Ottawa River canals, have provided native and nonnative organisms access to inland waters from the Great Lakes. Alewife was found in Lake Opinicon in the early 1900s and reached the Upper Rideau by the 1960s (J. Casselman, Ontario Ministry of Natural Resources, Picton, Ontario, pers. commun.). Alewife may have entered the Ottawa lakes through the Rideau or were introduced as bait by anglers. Carp were found in autumn of 1994 in the lower reaches of the Rideau Canal near Kingston (G. Smith, Kawartha Lakes Fisheries Assessment Unit, Lindsay, Ontario, pers. commun.). The Rideau Canal has been very susceptible to invasion by plants. European frog-bit (*Hydrocharis morsus-ranae*) is a free-floating aquatic plant imported from Switzerland to Ottawa in 1932 (Minshall 1940; Roberts et al. 1981; Lumsden and McLachlin 1988). By 1944, frog-bit had spread into the Rideau Canal (Catling and Porebski 1995) and by 1958, it was well established in the St. Lawrence River near Montreal (Dore 1954, 1968). Eurasian water-milfoil (Figure 23.11) gained access inland from the Great Lakes and is currently present in the Rideau Canal (N. McClean, Ontario Ministry of Natural Resources, pers. commun.). The exotic invertebrate with the greatest impact on the canal system is the zebra mussel. This species was first sighted on the hull of a boat dry-docked near Ottawa in December 1990 (Smith 1992; Martel 1995). In 1993, the first larval and adult zebra mussels were found in the Rideau lakes (W. G. Smith, Kawartha Lakes Fisheries Assessment Unit, Lindsay, Ontario, pers. commun.).

NONINDIGENOUS SPECIES OF THE TRENT-SEVERN WATERWAY

The Trent-Severn Waterway (TSW) has also provided an important connection for aquatic organisms to gain access to inland waters of Canada that were isolated until the waterway was constructed. An excellent example is the invasion of organisms from the Great Lakes into the Kawartha Lakes Region via the TSW (L. Deacon, Ontario Ministry of Natural Resources, Lindsay, Ontario, pers. commun.). The Kawartha lakes are part of the TSW linking Georgian Bay, Lake Huron with the

FIGURE 23.11 Eurasian watermilfoil (*Myriophyllum spicatum*) in Fish Lake, Dane County, Wisconsin (photo courtesy of Wisconsin Department of Natural Resources, Madison, WI).

Bay of Quinte, Lake Ontario. Consequently, inland waters such as the Kawarthas are vulnerable to invasions of fish, plants, and invertebrates that use interconnecting canals to expand their range. The common carp (*Cyprinus carpio*) gained access to the Kawartha lakes through the Trent-Severn (probably from Georgian Bay) in the late 1800s. Three sport fish have invaded lakes along the waterway, including bluegill (*Lepomis macrochirus*), black crappie (*Pomoxis nigromaculatus*), and northern pike (*Esox lucius*). Bluegill was native to the Great Lakes but expanded its range in the mid-1960s into the Kawarthas, where it is now the most abundant fish in angler harvest. Both black crappie and northern pike are relative newcomers. Black crappie, first seen in the mid-1980s, likely invaded the Kawarthas from the Trent River to the east and Lake Simcoe from the west. Northern pike has been present in the TSW since at least 1980, when one was captured in Stony Lake (L. Deacon, Ontario Ministry of Natural Resources, Lindsay, Ontario, pers. commun.). Northern pike are now abundant and have a negative impact on a former keystone predator, native muskellunge (*Esox masquinongy*). Both black crappie and northern pike compete for forage with popular sport fish such as the walleye (*Stizostedion vitreum vitreum*) and muskellunge (J. Casselman, Ontario Ministry of Natural Resources, Picton, Ontario, pers. commun.).

Eurasian watermilfoil (*Myriophyllum spicatum*) invaded the TSW and its lakes in the mid-1960s (Aiken 1981; J. Norris, Trent-Severn Waterway, Peterborough, Ontario, pers. commun.).

Dense areas of watermilfoil growth have created problems for recreational users and compete with native plants (J. Norris, Trent-Severn Waterway, Peterborough, Ontario, pers. commun.; S. Painter, Canada Centre for Inland Waters, Burlington, Ontario, pers. commun.).

The two most recent invaders are invertebrates, the zebra mussel and spiny waterflea (*Bythotrephes cederstroemi*). The first sighting of zebra mussel was in Balsam, Big Bald, and Rice lakes in 1991; since then it has spread rapidly throughout the Trent-Severn system (L. Deacon, Ontario Ministry of Natural Resources, Lindsay, Ontario, pers. commun.). The spiny water flea, first recorded from the Great Lakes Region in 1984 (Burr et al. 1986), was observed in Pigeon Lake in 1996, where it likely invaded via the TSW (L. Deacon, Ontario Ministry of Natural Resources, Lindsay, Ontario, pers. commun.).

CANALS OF THE EASTERN UNITED STATES

ERIE-BARGE CANAL

Construction of the Erie Canal began in 1817. The Erie Canal opened to Lake Ontario in 1819, and by 1825 it formed a transportation route for commerce, travel, and migration between the Hudson River and Lake Erie (Aron and Smith 1971; Hadfield 1986). The canal stretched from Albany to Buffalo, a total of 563 km. Celebrations marking the completion of the Erie Canal in 1825 illustrated the potential impact the canal would have on the Great Lakes. Upon arrival of the first boat to officially navigate the Erie Canal from Buffalo, New York, the Governor of New York and other celebrants poured a keg of Lake Erie water into the Atlantic Ocean (Mills 1910). The canal crossed major streams, the largest of which were the Genessee River at Rochester and the Mohawk River (Figure 23.12). In 1836, the Erie Canal was shortened by 21.7 km and routed through, rather than south of, Oneida Lake (Smith 1985). Today, the canal is heavily used by boaters transiting the Great Lakes to the Atlantic Ocean and is now referred to as the Erie-Barge Canal (Ashworth 1986). The success of the Erie Canal resulted in the development of other canals along the eastern coast of U.S. to meet growing demands associated with industrial and agricultural development.

FIGURE 23.12   The Waterford Flight on the Erie-Barge Canal near Troy, NY (photo courtesy of the USGS Biological Resources Division, Oswego, NY).

"Feeder" Canals of the Erie-Barge Canal

New York has a rich history of canals, and at one time all the major drainages in the state were linked by canals (Figure 23.1). In the early to mid-1800s, numerous canals were built to provide water to the Erie Canal, including the Cayuga-Seneca Canal, the Chemung Canal, the Genessee Valley Canal, the Black River Canal, the Chemung Canal, and the Chenango Canal. These "feeder canals" figured prominently in the distribution of organisms inland (Smith 1985) by providing new habitat for plants and fish immigrants (Thompson et al. 1987). The Oswego Canal connected the Erie-Barge Canal with Lake Ontario at Oswego, New York and provided the shortest route for vessels to transit from Albany to the upper Great Lakes via Lake Ontario and the Welland Canal (Smith 1985). Except for the Oswego Canal and the Cayuga-Seneca Canal, all other canals are no longer in use today.

Champlain and Chambly Canals

The Champlain and Chambly Canals were built along an Indian trail, turned trade route, from the Hudson River northward to Lake Champlain, along the Canadian border and up the Richelieu River (Hadfield 1986) (Figure 23.3). The French Canadians began improvements on the Richelieu River near the Chambly Rapids in the 1750s. By 1818 the Chambly Canal opened between the north end of Lake Champlain and the Richelieu River. The Champlain Canal opened from the Hudson River to Lake Champlain in the early 1820s (Legget 1976; Smith 1985; Hadfield 1986). Both canals linked Canada with the Atlantic seaboard, thereby allowing Atlantic Ocean species to travel through the Hudson River, the Champlain Canal, Lake Champlain and into the St. Lawrence River (Hocutt and Wiley 1986). Opening of the Champlain Canal and the Erie-Barge Canal with its many feeder canals led to a great increase in shipping activity in New York by the mid-19th century, with the Hudson River becoming an important gateway into North America (Bruce 1907; Adams 1981).

Canals of the Atlantic Coast

In New England, three canals built in early 1800s were intended primarily for navigation from the Atlantic Ocean inland (Hadfield 1986). These included the Cumberland and Oxford Canal in Maine, the New Haven and Northampton Canal in Connecticut, and the Blackstone Canal which connected tidewater at Providence, Rhode Island inland to Worcester, Massachusetts. Further south, numerous small canals were built to transport commodities, such as coal, to major cities and markets. For example, the Dismal Swamp Canal, the Chesapeake and Delaware Canal, and the Delaware and Raritan Canal formed the Atlantic Waterway, thereby allowing inland navigation between North Carolina and New York (Hadfield 1986).

NONINDIGENOUS SPECIES AND CANALS OF THE EASTERN UNITED STATES

Nonindigenous Species and the Erie-Barge Canal

The breaching of the Appalachian Divide by both the Erie and Champlain Canals caused the exchange of a large number of fishes and molluscs. Of the exotic introductions into the Hudson River Drainage, 19% occurred as a result of canals (Mills et al. 1996), primarily the Erie Canal. Improvements of the New York State Erie-Barge Canal in 1918 played a major role in the movement of fish species (Smith 1985). Historically, the numbers of exotics using the Erie Canal as a primary vector to invade the Great Lakes have not equaled the numbers contributed to the Hudson River; more species have invaded the Hudson River from the Interior Basin than have invaded the Great Lakes from the Atlantic Slope (Mills et al. 1996). This unequal exchange has likely occurred because the species diversity of freshwater biota of the Atlantic Slope was much lower than that of the Interior Basin (Ortmann 1913; Hocutt and Wiley 1986). A comprehensive list of fish and aquatic plant and invertebrate species that have moved into the Hudson River system as a result of canals is found in Mills et al. (1996). The first section of the Erie Canal, connecting the Mohawk and Seneca Rivers, opened a pathway of species invasion into the Finger Lakes and many other New York State rivers and lakes. Sea lamprey have been present in these and other Finger Lakes since the 1870s (Aron and Smith 1971; Lark 1973). It is likely that the sea lamprey migrated upstream into the "feeder" canals to spawn (Ashworth 1986). Alewife entered Cayuga and Seneca lakes (Aron and Smith 1971), where they have been present since at least 1868 (Christie 1973). Rainbow smelt and gizzard shad (*Dorosoma cepedianum*) likely invaded the Finger Lakes through the Erie-Barge Canal (Scott and Crossman 1973). The zebra mussel has spread throughout New York State via the canal system and, to date, quagga mussel has also gained access to Cayuga and Seneca lakes in the Finger Lakes through inland canals including the Erie-Barge, Seneca, and Oswego Canals (R. Tuttle, New York State Gas & Electric, Binghamton, New York, pers. commun.).

NONINDIGENOUS SPECIES AND THE CHAMPLAIN CANAL

The Champlain Canal is the second major New York canal that breached the divide between the Great Lakes and Appalachian drainages. George (1985) and Reider (1979) attribute the presence of silver lamprey (*Icthyomyzon unicuspis*) in the Hudson River to its passing through the Champlain Canal, since Greeley (1930) found it to be common in Lake Champlain where it is native. Bowfin (*Amia calva*) was not observed in the Hudson River until 1988 (Mills et al. 1996); this fish species may have moved through the Champlain Canal (Greeley 1930) or was introduced by people (Smith 1985). The zebra mussel is found throughout Lake Champlain (Serrouya et al. 1995), the Hudson River estuary (Strayer et al. 1996; Walton 1996), and the Champlain Canal.

CANALS OF THE MISSISSIPPI RIVER DRAINAGE

CHICAGO CANAL SYSTEM

The Chicago Canal was designed to link the Great Lakes and Mississippi drainages; for purposes of this chapter we consider the Chicago Canal in the Mississippi drainage. The Chicago Canal was first suggested by the French explorer, Joliet, in 1673 to provide a continuous water route from Lake Michigan to Florida. The Illinois and Michigan Canal Company was chartered to begin construction in 1835 (Changnon and Harper 1994) and by 1848, the Chicago Canal connecting Chicago on Lake Michigan to LaSalle on the Illinois River was opened (Hadfield 1986) (Figure 23.3). The city of Chicago, having grown significantly after construction of the canal, needed a source to remove sewage effluent. In 1866, construction of a new channel through Mud Lake and the Des Plaines River was initiated, and when completed in 1871, the flow of the Chicago River was reversed from flowing into Lake Michigan to flowing into the Illinois River. A new canal was needed to isolate sewage effluent from drinking water and was built parallel to the Illinois and Michigan Canal, with construction complete in 1900. This canal became known as the Drainage Canal, Chicago Sanitary and Ship Channel, or the Main Channel, and diverted sewage away from Lake Michigan, Chicago's water supply (Changnon and Harper 1994). In 1915, the Illinois Waterway Commission was formed to build an enlarged waterway from Chicago to the Mississippi River. The Chicago Sanitary and Ship Canal and the Chicago River became part of this enlarged waterway, from Lake Michigan to the mouth of the Illinois River on the Mississippi, which was completed by 1933 (Hadfield 1986). The Illinois (Chicago) Canal is no longer in use; however, the Chicago Sanitary and Ship Canal is fully operational today and represents a major link between the Great Lakes and the Mississippi drainages (Ashworth 1986). With completion of this link, vessels could reach the Mississippi River from the Atlantic Ocean via the Erie Canal, Great Lakes, and the Chicago Sanitary and Ship Canal.

OHIO CANAL SYSTEM

On February 4, 1825, the Ohio legislature passed "an act to provide for the internal improvement of the state of Ohio by navigable canals" to mark the beginning of canal construction in the state. The Ohio and Erie Canal and the Miami and Erie Canal were the first to be authorized by this act (Hadfield 1986). The Ohio and Erie Canal, connecting Portsmouth on the Ohio River to Cleveland on Lake Erie, was completed in 1832 (Regier and Hartman 1973) (Figure 23.3). In 1845, Ohio's Miami and Erie Canal was completed extending from the Ohio River at Cincinnati to Toledo on Lake Erie (Hadfield 1986). The Wabash and Erie Canal, connecting the Ohio River at Evansville, Indiana to the Miami and Erie Canal was abandoned by 1905 (Hadfield 1986). The Ohio and Erie Canal is no longer in use today (Ashworth 1986).

NONINDIGENOUS SPECIES AND THE MISSISSIPPI RIVER DRAINAGE CANALS

NONINDIGENOUS SPECIES AND THE CHICAGO CANAL

Construction of the Chicago Canal was a major activity that linked the Mississippi and the Great Lakes drainages. Organisms could now move freely between these two large drainages. Some fish species such as gizzard shad and blue herring (*Pomolobus chrysochloris*) entered Lake Michigan from the Mississippi drainage via the Chicago Canal (Bean 1903; Good and Gill 1903; Dymond 1922; Miller 1957), while rainbow smelt invaded the Mississippi drainage from Lake Michigan and the Chicago Canal (Hocutt and Wiley 1986). The round goby (*Neogobius emlanostomus*), first discovered in the St. Clair River in 1990 (Crossman et al. 1992), was found in abundance near Chicago in 1994 (Ghedotti et al. 1995). Though its presence there may be due to ballast water introduction, it will likely spread through the Chicago Canal and into the Mississippi drainage. Another recent invader to the Mississippi basin in the 1990s is the zebra mussel (*Dreissena polymorpha*), which is present throughout the drainage as far south as New Orleans (Louisiana Sea Grant Communications Office 1995).

NONINDIGENOUS SPECIES AND THE OHIO CANAL SYSTEM

Some fish species may have used the Ohio Canal system to reach the Great Lakes from the Mississippi drainage. Gizzard shad, first recorded in Lake Erie in 1848, and the blue herring may have migrated from the Mississippi River basin to Lake Erie through the Ohio Canals (Bean 1903; Goode and Gill 1903; Dymond 1922; Greeley 1929; Miller 1957). According to Goode and Gill (1903), paddlefish (*Polyodon spathula*), a Mississippi Valley native, was rare in Lake Erie at that time. Greeley (1929) suggested that paddlefish entered Lake Erie through the Wabash and Erie Canal.

CANALS OF THE SOUTHEASTERN UNITED STATES

TENNESSEE-TOMBIGBEE WATERWAY

The completion of the Tennessee-Tombigbee Waterway (TTW) in 1985 (Stine 1993) provided a shortcut linking the Tennessee and Ohio Rivers with the Gulf of Mexico (Figure 23.13). In doing so, it provided the first link between waterways of the Mississippi and Southeastern drainages. The 377 km-long TTW consists of ten locks and five dams, and involved extensive alteration of large sections of the Tombigbee River through widening, deepening, and straightening. It was believed that the use of the waterway, as opposed to the Mississippi River route, would significantly decrease the transportation costs of commodities between the Gulf of Mexico and the Tennessee, Ohio, and upper Mississippi River valleys. Unlike most other basins in the U.S. at that time, the Mobile Basin, which houses the Tombigbee River, was unique in that, prior to 1985, it was free of canals connecting

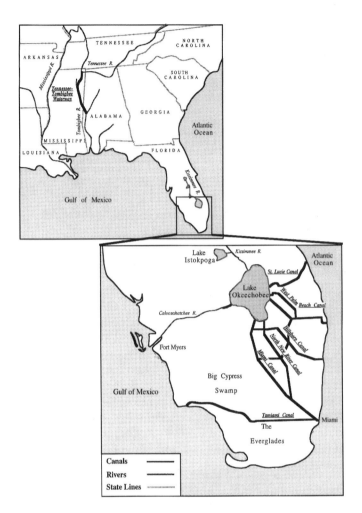

FIGURE 23.13   Canals of the southeastern United States.

it to other major drainage basins. The construction of the TTW opened the door for the interbasin transfer of nonindigenous species between the Mobile and Tennessee River basins.

CANALS OF FLORIDA

Predominant geographic features of Florida are its natural and man-made waterways, into which many nonindigenous species have been deliberately and accidentally introduced (U.S. Congress 1993). Unlike the canals of the Northeast, which were built primarily for navigational purposes, canals in Florida, those of the southeastern portion of the state in particular, were built to supply water to meet the demands of the increasing urban population, for flood control, and for agriculture. Currently, a network of canals (Figure 23.13) stretches from Lake Okeechobee to southern Dade County, just east of the Everglades National Park. The West Palm Beach, Hillsboro, North New River, and Miami canals are some of the larger canals that are a part of this network. Other major Florida canals include the Tamiami Canal, which passes through southern portions of the Big Cypress Swamp and Everglades; the St. Lucie Canal, which connects Lake Okeechobee to the Atlantic coast; the channelization of a portion of the Caloosahatchee River from Lake Okeechobee to the Gulf coast; and the channelization of the Kissimmee River from northern Lake Okeechobee, northward to Lake Istokpoga and Lake Kissimmee. The Cross-Florida Barge Canal was initiated in 1964 to provide a barge waterway between the St. Johns River and the Gulf of Mexico, but in 1971 after 40 km of the channel was completed, further construction was halted because of the potential for environmental damage (U.S. Congress 1986).

NONINDIGENOUS SPECIES OF THE SOUTHEASTERN UNITED STATES

NONINDIGENOUS SPECIES AND THE TENNESSEE-TOMBIGBEE WATERWAY

Among many ecological concerns for the Tombigbee River System were the possibilities that Eurasian watermilfoil would be introduced, and that native fish species would be lost due to competition with species invading from the Tennessee River (Stine 1993). McCann et al. (1996) expressed concern that the zebra mussel, which is well established in the Tennessee River, would be introduced into parts of Alabama and Georgia through the TTW. Two fish species known to have moved from the Mobile basin to the Tennessee River drainage through the TTW are the blacktail shiner (*Cyprinella venusta*) and weed shiner (*Notropis texanus*) (Etnier and Starnes 1993). Also, Billington and Strange (1995) expressed concern about the possible invasion of northern walleye, strains since genetic testing has confirmed the presence of a unique southern strain of walleye in the Tombigbee River. Although movement of species to this point has been documented only from the Tombigbee to the Tennessee, it is likely that interbasin transfers of species between both drainages will occur.

NONINDIGENOUS SPECIES AND THE CANALS OF FLORIDA

Canals and channelization of rivers provide suitable habitat where nonindigenous species can gain a foothold, leading to eventual establishment. For example, the channelization of the Kissimmee River, which flows from Lake Kissimmee to Lake Okeechobee, was completed in 1971 and provided flood control for central and southern Florida. One effect of the channelization was the alteration of the hydrological regime of the river, which allowed the invasion of nuisance exotic plants such as *Hydrilla* (Koebel 1995). The establishment of nonnative species in canals in Florida is aided by a number of factors, including the subtropical climate, prolonged growing season, releases from a growing aquarium trade industry, growth of the aquaculture industry, and urbanization (U.S. Congress 1993). The network of canals in Florida has provided a secondary vector for species to extend their range. Canals also provide suitable habitat for many exotic species. For example, canals in urbanized areas may have poor water quality, and predator populations are low, thereby favoring exotics (Loftus and Kushlan 1987). When Kushlan (1986) monitored changes in species composition in Little River Canal, Dade County, Florida between 1964 and 1982, he found that the number of exotic species increased (1 to 5) while the number of native species declined (23 to 16). In 1964, the fish fauna in the canal was dominated by centrarchids and poeciliids, all native fishes. By 1982, the number of exotics had increased and the fish community was dominated by the nonnative spotted tilapia (*Tilapia mariae*). Exotic species dominate canals in developed areas, such as those near the southeastern coast of Florida, but native species dominate canals in natural areas such as those that cross the Everglades, where exotics are present but do not outnumber native species.

Courtenay et al. (1974) give a detailed account of exotic fishes of Florida, many of which have dispersed through the massive network of canals in this state. Loftus and Kushlan (1987) indicate that canal excavation, in conjunction with levee construction and water management practices, have been the most consequential factors affecting the distribution of southern Florida freshwater fishes. They also suggest that many nonnative species associated with canals, especially in southern Florida, were originally introduced through the aquarium industry. Possible impacts of two fish species, blue tilapia (*Oreochromis aureus*) and the walking catfish (*Clarias batrachus*), are described by Hale et al. (1995) and Courtenay and Miley (1975).

Blue Tilapia

Blue tilapia (Figure 23.14) were deliberately released into Florida in 1961 when the Florida Game and Fresh Water Fish Commission imported this fish as a food and sport fish (Crittenden 1965). By 1984 blue tilapia became (and presently is) the most widely distributed exotic fish species in Florida (Courtenay et al. 1984; Hale et al. 1995). Courtenay et al. (1974) indicate that blue tilapia were also introduced in many areas of the southern U.S. for vegetation control purposes. Blue tilapia are known to have a negative impact on native species through competition for food. For example, Hale et al. (1995) indicated that age-0 tilapia had similar food habits to age-0 centrarchids, and adult blue tilapia almost completely displaced largemouth bass (*Micropterus salmoides*) during spawning. Since efforts to control this nonindigenous species in Florida have failed and management options are few, this fish is commercially harvested as a low-cost food for consumers.

FIGURE 23.14   A blue tilapia (*Oreochromis aureus*) collected from a pool on the east side of the Colorado River, north of the Morelos dam (photo courtesy of Dr. Walter R. Courtenay, Jr., Florida Atlantic University).

Walking Catfish

The walking catfish (*Clarias batrachus*) (Figure 23.15) was an aquarium escapee in the mid-1960s that used canals to expand its range throughout Florida (Courtenay et al. 1974; Courtenay and Miley 1975). The walking catfish is an air-breathing fish that is able to crawl on land (thereby bypassing flooding and salinity control structures). This fish is highly fecund and preys on native fishes, especially during the dry season when low water conditions concentrate fish (Courtenay and Miley 1975). By 1974, sightings were made as far north as 26 km northwest of Lake Okeechobee, and as far south as the northern Everglades. A fish-kill made up of 90% walking catfish occurred in Big Cypress Swamp in early 1976, and an overland migration of walking catfish was observed in Fort Myers in 1977 (Courtenay 1978). The Tamiami Canal likely contributed to the westward expansion of the walking catfish, while its northward expansion has been attributed to a series of interconnected agricultural and drainage canals (Courtenay 1979). Specimens of walking catfish have also been collected in the All America Canal of southern California (Minckley 1973), although there is no evidence for its establishment there (Courtenay et al. 1984).

FIGURE 23.15   An amelanic walking catfish (*Clarias batrachus*) (photo courtesy of Dr. Walter R. Courtenay, Jr., Florida Atlantic University).

Invasive Plants

Of all the nonindigenous aquatic species introduced into Florida, exotic plants have perhaps proved to be the most troublesome. Hydrilla (*Hydrilla verticillata*) (Figure 23.16), waterhyacinth (*Eichornia crassipes*) (Figure 23.17), and waterlettuce (*Pistia stratiotes*) are the most notable invaders, all of which grow in and spread through canals (McCann et al. 1996). Hydrilla, an aquarium plant similar to the Canadian waterweed (*Elodea canadensis*), was first introduced to Florida waters into canals near Tampa around 1950, and later into Miami Canals and the Crystal River (Haller 1978; U.S. Congress 1993). The presence of this plant became threatening by the early 1960s, when infestation of flood canals in southeastern Florida became evident (Schmitz et al. 1993). By the mid-1980s, hydrilla had spread to many southeastern states and as far west as California (Steward et al. 1984). Hydrilla commonly spreads when plant fragments are transported to new waters on

FIGURE 23.16   *Hydrilla* infestation at Sheldon Reservoir near Houston, TX (photo courtesy of Mike Grodowitz, U.S. Army Engineers Waterways Experiment Station, Vicksburg, MS).

FIGURE 23.17   Waterhyacinth (*Eichornia crassipes*) infestation in a canal of Lake Theriot, LA (photo courtesy of Alfred Cofrancesco, U.S. Army Engineers Waterways Experiment Station, Vicksburg, MS).

boat trailers. Once established, boat traffic through canals and waterways enhances its dispersal (Schardt and Schmitz 1991). Hydrilla outcompetes native plants because of its ability to adapt to low light levels (Bowes et al. 1977) and its ability to produce a surface canopy that shades out submersed plants. In Florida, a statewide reduction of eel grass in 1983 was attributed to competition with hydrilla (van Dijk 1985). Dense stands of this plant can also impact zooplankton and fish populations (Richard et al. 1985). Hydrilla and other nuisance aquatic plant species continue to expand their range (41% of Florida's public waters have this plant), and control costs are extremely high (Schmitz et al. 1993).

Two other exotic plant species that have invaded the waterways of Florida are waterlettuce and waterhyacinth. The weedy characteristics and prolific growth of waterlettuce have enabled it to become a problem in Florida (Schmitz et al. 1993). This nuisance plant has used waterways and canals to expand its range, and waterlettuce mats can impact native plant species such as bulrush (*Scirpus maritimus*) because of shading. However, waterlettuce provides a habitat for many macroinvertebrates of which the amphipod, *Hyallela azteca*, is the most abundant (Dray et al. 1988). Waterhyacinth is an impressive floating exotic species that produces mats that often degrade water quality and can lead to dramatic changes in plant and animal communities (Schmitz et al. 1993). Dense stands can impact critical wildlife habitat areas (Griffen 1989) and impact beds of submersed vegetation (Tabita and Woods 1962; Chestnut and Barman 1974). Problems associated with waterhyacinth in Florida's waterways were noted as early as 1894 in the St. Johns River (Raynes 1964).

OTHER NONINDIGENOUS SPECIES

Exotic amphibians have also been known to expand their ranges through the use of canals. The marine toad (*Bufo marinus*) was released by a former importer at the Miami International Airport prior to 1955 (King and Krakauer 1966). In 1958, a canal was constructed linking the introduction site to the extensive Florida canal system, and the toad began appearing in regions further from the airport (Krakauer 1968). The marine toad poses a threat to the native southern toad (*Bufo terrestris*), and has replaced the native in many residential areas (McCann et al. 1996).

CANALS AND DIVERSIONS OF THE WESTERN UNITED STATES

At the time the western U.S. was settled, railroads were well under construction, so the idea of using canals for navigational purposes was never well developed. However, the need to irrigate arid lands for agricultural purposes initiated the concept of diversion of water from major western rivers and its redistribution among adjacent drainage basins through the construction of canals. Two major drainage systems in the western U.S. are the Colorado River Basin and the Great Basin–Baja–Klamath–Sacramento system (Mayden 1992) (Figure 23.2). Of particular interest are areas where basins share common boundaries with each other; major water diversion projects often traverse these boundaries providing routes for freshwater organisms to spread. For example, the Colorado River Basin shares borders with six other major drainages (Figure 23.2).

CALIFORNIA WATER PROJECTS

One of the earliest interbasin water transfer projects to be completed in California was the construction of the Los Angeles Aqueduct from the Owen's Valley to Los Angeles in 1913 (Howe and Easter 1971) (Figure 23.18). However, major diversions of water had been occurring even before

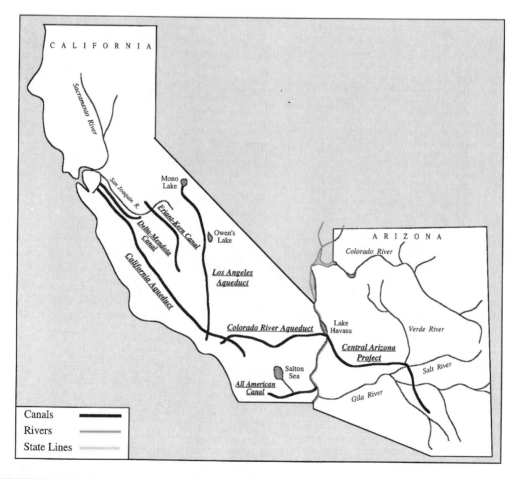

FIGURE 23.18   Canals and water diversions in the western United States.

this date. For example, in 1901 a primitive irrigation canal was dug from the Colorado River at Yuma, Arizona westward to California's Imperial Valley (Pearce 1992). During improvement of the canal (known as the Alamo Channel) in 1905, heavy flooding caused water from the Colorado River to flow into the Salton Sink, creating the Salton Sea (Courtenay and Robins 1989; Pearce 1992). Other major diversions of the lower Colorado River include the 389-km-long Colorado River Aqueduct, which runs from the Colorado River at Parker Dam on Lake Havasu to metropolitan areas of Los Angeles, and the All American Canal, which runs from the Colorado River at Imperial Dam to the Imperial Valley of California (Howe and Easter 1971).

The Central Valley Project (CVP) (Figure 23.19 and 23.20), which is contained largely in the Sacramento-San Joaquin drainage system of central California, consists of a number of canals whose general purpose is the diversion and redistribution of water from the Sacramento drainages in northern California to the San Joaquin and Tulare Lake drainages in the south. Major CVP canals are the Friant Kern, which transfers water from the San Joaquin River southward to the Tulare Lake subbasin, and the Delta-Mendota, which diverts water southward from the Sacramento sub-basin; the California Aqueduct, a state project that shares part of its course with the CVP, diverts water from the Sacramento subbasin south through the San Joaquin subbasin along the west side of the San Joaquin Valley and into the Tulare Lake subbasin (El-Ashry and Gibbons 1986). Water is then pumped over the Tahachapi Mountains and continues its course through the aqueduct to the Los Angeles subbasin of the south coastal drainage system. There were more than 120,000 km of main canals in the west by 1930 (Pearce 1992). This and subsequent constructions have undoubtedly contributed to the interbasin transfer of nonindigenous species.

FIGURE 23.19 Madera Canal, Central Valley Project, California; Friant Dam and Millerton Lake on the San Joaquin River are seen in the background (photo courtesy of J.C. Dahilig, U.S. Bureau of Reclamation, Sacramento, CA).

FIGURE 23.20   Tehama-Colusa Canal, Central Valley Project, California (photo courtesy of J.C. Dahilig, U.S. Bureau of Reclamation, Sacramento, CA).

ARIZONA AND COLORADO WATER PROJECTS

A major project involving the diversion of water from the Colorado River for use in the state of Arizona is the Central Arizona Project (CAP) (Figure 23.21). The purpose of the CAP was to supply water to central Arizona, originally for agriculture, but eventually to meet the demands of the increasing urban population (Johnson 1977). The project was designed to deliver 2.714×10^9 m³ per year of Colorado River to central and southern Arizona (Grabowski et al. 1984). Construction of the CAP was authorized in September 1968, environmental impact statements were filed by the Bureau of Reclamation in 1972, and construction began in May 1973 (Johnson 1977). The project diverted water from the lower Colorado River at the southern end of Lake Havasu into the south-central interior of Arizona (Figure 23.18). Water from Lake Havasu would be mixed with that of the Salt and Gila Rivers. Grabowski et al. (1984) recognized the potential for the introduction of tilapia from the lower Colorado into the Salt and Gila Rivers, and Courtenay and Robins (1989) warned of the potential for the introduction of rainbow smelt if Utah were to carry out a stocking program in a major reservoir upstream in the Colorado River. One project involving the interbasin transfer of water from the Colorado River Basin to the Mississippi River Basin is the Colorado-Big Thompson Project; it was opened in 1957 with the original purpose of supplementing surface and groundwater supplies in eastern Colorado (Howe and Easter 1971). The project transfers water from the upper Colorado River to four tributaries of the South Platte River in the Mississippi River Basin (Howe and Easter 1971).

NONINDIGENOUS SPECIES OF THE WESTERN UNITED STATES

The spread of nonindigenous species in the western U.S. has been accelerated due to species movement through irrigation canals and diversions of major river systems for municipal, industrial, and agricultural water supply. The problem has been exacerbated by habitat disturbance associated

FIGURE 23.21   The main canal of the Central Arizona Project, Sonoran Desert, south of Phoenix near the Picacho Mountains (photo courtesy of Tom Burke, U.S. Bureau of Reclamation, Boulder City, NV).

with the construction of large water projects. For example, Minckley (1991) attributed the extirpation of native fish species of the Colorado River downstream from the Glen Canyon Dam to stabilization of the system caused by damming, diversion, and other human regulation. The effects of damming include destruction of spawning areas, blocking of migration runs, and the creation of new habitats (e.g., reservoirs) which favor the survival of exotics. Moyle and Williams (1990) determined that the decline of California's native fish fauna could be attributed primarily to habitat disruption resulting from the construction of large water projects coupled with the introductions of fish species better able to cope in altered habitats. Minckley (1991) stated that the demise of the fauna of the lower Colorado River started just after 1900, as the river was progressively harnessed for water supply, flood control, and power production. Sigler and Sigler (1987) cite nonindigenous species as the "second most important decimating factor" of desert fishes in the Great Basin, given that many of the native fish are endemic, having evolved without competition, and are therefore not equipped to handle such pressure from exotic, predaceous fish. Moyle (1976) stated that "interbasin transfers of native fishes can have potentially as much impact on resident fish populations as can the introduction of exotics."

Nonindigenous species introductions in the western U.S. are characterized by interbasin transfers of native species, and spread of exotic species introduced by other means, both through canals and diversions constructed for water projects. Most western states have had their share of both types of introductions.

NONINDIGENOUS FISH

Moyle (1976) lists the introduced fishes of California as well as the reasons for their introduction and locations. The Mississippi silversides (*Menidia audens* Hay) has spread rapidly since its introduction into Clear Lake in 1967, possibly moving downstream from Clear Lake via irrigation canals (Moyle 1974). Canal and diversions have not been implicated as the primary vector of introduction in any case except in that of the Owen's sucker (*Catostomus fumeiventris*). An example of an interbasin transfer of a native species, the Owen's sucker traveled from Owen's Lake in the

Death Valley drainage system to June Lake in the Mono Lake basin and to the Los Angeles Basin via the Los Angeles Aqueduct (Hubbs et al. 1943). The Los Angeles Aqueduct, one of the first water transfer projects in California, completed in 1913, transferred water from the Owen's Valley on the eastern slopes of the Sierra Nevada to Los Angeles. It was extended in 1940 to the Mono Basin in the Sierras. Hubbs et al. (1943) reported extensive hybridization between the Owen's sucker and the Santa Ana sucker in the Santa Clara River Basin.

Two tilapia species, Mozambique tilapia (*Tilapia mossambica*) and Zill's cichlid (*T. zilli*), that were introduced into irrigation ditches in southern California for control of aquatic weeds (Moyle 1976) have the potential to spread through these canals and into connecting drainage systems where temperatures will allow them to do so. Mozambique tilapia were common in the lower Colorado River before the 1970s (Minckley 1973) and may have entered the Salton Sea via the Alamo and All American canals (Courtenay and Robins 1989). Since the arrival and establishment of the Mozambique tilapia and other species in the Salton Sea, desert pupfish populations have declined; it is possible that this decline is due at least partially to competition with juvenile Mozambique tilapia (Courtenay and Robins 1989).

NONINDIGENOUS MOLLUSC — *CORBICULA*

Exotic species other than fish have also made their way through canals and diversions in the western U.S., although documentation is poorer than for fish. One exception is the history of the Asiatic clam (*Corbicula fluminea*) (Figure 23.22) invasion into U.S. waters (McMahon 1982; Counts 1986). It was first observed in North America in British Columbia in 1924 and later spread down the West Coast to California (McMahon 1982; Counts 1986). Severe biofouling problems with Asiatic clams were first reported in the U.S. in 1956 (Isom 1986). *Corbicula fluminea* was first discovered in California in the Sacramento River in 1945 (Hanna 1966) and later infested many of the canals surrounding the San Francisco Bay estuary and the Central Valley. Infestation of the clam was reported in the water supply of the Metropolitan Water District of Southern California in 1956, and Dundee and Dundee (1958) extended the known range of *Corbicula* outside coastal areas of the west and central California to an irrigation canal in Phoenix, Arizona.

FIGURE 23.22 *Corbicula* infestation in a canal of the Central Valley Project (photo courtesy of Doug Ball, U.S. Bureau of Reclamation, Sacramento, CA).

Clam infestations were reported in 1961 in irrigation canals, and numerous reports of clams biofouling irrigation canals in the west appeared in subsequent years (Isom 1986). The most significant early problem with Asiatic clams was in the Delta-Mendota Canal, a part of the California Central Valley project (Prokopovich and Hebert 1965). After founder populations became established in California and Washington state, Asiatic clams utilized canals and aqueducts as one means to spread to other locations in the U.S. including the southwest and the midwest (Counts 1986). *Corbicula* is a major fouling organism of industrial and municipal intake structures and the cost of control has been enormous. *Corbicula* is examined in detail by McMahon in Chapter 22 of this volume.

SUMMARY

Canals and diversions have become important features on the North American landscape for navigation, recreation, commerce, water supply, flood control, and agriculture. The infestation and range expansion of nonnative organisms into new waters has been a negative by-product of canal and waterway development in North America. Canals have provided the pathway through which hundreds and perhaps thousands of organisms have invaded new waters. Some of these organisms such as *Corbicula*, sea lamprey, and hydrilla have had significant ecological and economic impacts. Canals and diversions are less often primary pathways of the introduction of nonindigenous species, as they are secondary vectors for the spread of organisms introduced by other methods. The best documentation of the spread of nonindigenous species through canals is for plants, fish, and molluscs, while the association of canals with invertebrates and phytoplankton is less documented.

Presently, there is heightened awareness of the importance of canals and diversions in the spread of exotic species throughout North America. For example, the Canadian government has expressed its concern about the Garrison Diversion Unit, a proposed water irrigation project which would use reservoir water from behind the Garrison Dam on the Missouri River to irrigate areas in the Hudson Bay drainage (Wright and Franzin, Chapter 24 of this volume). The threat of nonindigenous species transfers between the two basins has led to the creation of the Interbasin Water Transfer Studies Program (IWTSP) in 1985. The IWTSP and the governments of North Dakota and Canada continue researching possible economic and ecological impacts. To date, their concerns over the spread of nonindigenous species into the Hudson Bay drainage have prevented construction of the Garrison Diversion Unit. Creation of programs like IWTSP demonstrates that concerns of invading nonindigenous and exotic species is growing. With the continued threat of human-mediated transfers of organisms through such mechanisms as baitbucket transfers and ballast water introductions, canals and diversions will continue to play a significant role in the movement of aquatic organisms throughout North America.

REFERENCES

Adams, A.G. *The Hudson: A Guidebook to the River.* State University of New York Press, Albany, NY, 1981, p. 424.

Aiken, S.G. A Conspectus of Myriophyllum (Haloragidaceae) in North America. *Brittonia.* 33, pp. 57–69, 1981.

Applegate, V.C. and J.W. Moffet. The Sea Lamprey. *Scientific American.* 192, pp. 36–41, 1955.

Aron, W.I. and S.H. Smith. Ship Canals and Aquatic Ecosystems. *Science.* 174, pp. 13–20, 1971.

Ashworth, W. *The Late, Great Lakes: An Environmental History.* Alfred A. Knopf, New York, 1986, p. 274.

Bailey, R.M. and G.R. Smith. Origin and Geography of Fish Fauna of the Laurentian Great Lakes Basin. *Canadian Journal of Fisheries and Aquatic Sciences.* 38, pp. 1539–1561, 1981.

Baldwin, N.S. Sea Lamprey in the Great Lakes. *Limnos.* 1, pp. 20–27, 1968.

Bean, T.H. *The Food and Game Fishes of New York.* J.B. Lyon, Albany, NY, 1903, p. 460.

Berst, A.H. and G.R. Spangler. Lake Huron: The Ecology of the Fish Community and Man's Effects on It, in *Great Lakes Fishery Commission. Technical Report No. 21.*, Great Lakes Fishery Commission, Ann Arbor, MI, 1973.

Billington, N. and R.M. Strange. Mitochondrial DNA Analysis Confirms the Existence of a Genetically Divergent Walleye Population in Northeastern Mississippi. *Transactions of the American Fisheries Society.* 124, pp. 770–776, 1995.

Bowes, G., T.K. Van, L.A. Garrard, and W.T. Haller. Adaptation to Low Light Levels by Hydrilla. *Journal of Aquatic Plant Management.* 15, pp. 32–35, 1977.

Bruce, W. *The Hudson. Three Centuries of History, Romance, and Invention.* Bryant Union County, NY, 1907, p. 224.

Burr, M.T., D.M. Klarer, and K.A. Krieger. First Records of a European Cladoceran, *Bythotrephes cederstroemi*, in Lakes Erie and Huron. *Journal of Great Lakes Research.* 12, pp. 144–146, 1986.

Busch, W.N., D.H. Davies, and S.J. Nepszy. Establishment of White Perch, *Morone americana*, in Lake Erie. *Journal of the Fisheries Research Board of Canada.* 34, pp. 1039–1041, 1977.

Catling, P.M. and Z.S. Porebski. The Spread and Current Distribution of the European Frog-bit, *Hydrocharis morsus-ranae* L., in North America. *Canadian Field Naturalist.* 109, pp. 236–241, 1995.

Changnon, S.A. and M.E. Harper. History of the Chicago Diversion. In *The Lake Michigan Diversion at Chicago and Urban Drought: Past, Present, and Future Regional Impacts and Responses to Global Climate Change, Final Report*, Changnon, S.A., Ed., Prepared for the Great Lakes Environmental Research Laboratory, Ann Arbor, MI, 1994.

Chestnut, T.L. and E.H. Barman, Jr. Aquatic Vascular Plants of Lake Apopka, Florida. *Florida Scientist.* 37, pp. 60–64, 1974.

Christie, W.J. Lake Ontario: Effects of Exploitation, Introduction, and Eutrophication on the Salmonid Community. *Journal of Fisheries Research Board of Canada.* 29, pp. 913–929, 1972.

Christie, W.J., A Review of the Changes in Fish Species Composition of Lake Ontario. In *Great Lakes Fishery Commission. Technical Report No. 23*, Great Lakes Fishery Commission, Ann Arbor, MI, 1973.

Counts, C.L., III. The Zoogeography and History of the Invasion of the United States by *Corbicula fluminea* (Bivalvia: Corbiculidae). *American Malacological Bulletin.* Special Edition No. 2, pp. 7–39, 1986.

Courtenay, W.R., Jr. and C.R. Robins. Fish Introductions: Good Management, Mismanagement, or No Management? *Reviews in Aquatic Sciences.* 1, pp. 159–172, 1989.

Courtenay, W.R., Jr., H.F. Sahlman, W.M. Miley II, and D.J. Herrema. Exotic Fishes in Fresh and Brackish Waters of Florida. *Biological Conservation.* 6, pp. 292–302, 1974.

Courtenay, W.R., Jr., D.A. Hensley, J.N. Taylor, and J.A. McCann. Distribution of Exotic Fishes in the Continental United States. In *Distribution, Biology, and Management of Exotic Fishes,* Courtenay, W.R., Jr. and J.R. Stauffer, Jr., Eds., The Johns Hopkins University Press, Baltimore, MD, 1984.

Courtenay, W.R., Jr. and W.M. Miley II. Range Expansion and Environmental Impress of the Introduced Walking Catfish in the United States. *Environmental Conservation.* 2, pp. 145–148, 1975.

Courtenay, W.R., Jr. Additional Range Expansion in Florida of the Introduced Walking Catfish. *Environmental Conservation.* 5, pp. 273–275, 1978.

Courtenay, W.R., Jr. Continued Range Expansion in Florida of the Walking Catfish. *Environmental Conservation.* 6, p. 20, 1979.

Creaser, C.W. The Establishment of the Atlantic Smelt in the Upper Waters of the Great Lakes. *Michigan Academy of Science, Arts and Letters.* 5, pp. 405–424, 1926.

Crittenden, E., Status of *Tilapia nilotica* (Linnaeus) in Florida. In *Proceedings of the Annual Conference, Southeastern Association of Game and Fish Commissioners.* Southeastern Association of Game and Fish Commissioners, Frankfort, KY, pp. 257–262, 1965.

Crossman, E.J., E. Holm, R. Cholmondeley and K. Tuininga. First Record for Canada of the Rudd, *Scardinius erythrophthalmus*, and Notes on the Introduced Round Goby, *Neogobius melanostomus. Canadian Field-Naturalist.* 106, pp. 206–209, 1992.

Daw, J.S. *A Resource Guide to the Trent-Severn Waterway.* Prepared for the Tent-Severn Waterway Interpretive Program, Parks Canada, Peterborough, Ontario, Canada, 1985.

Dence, W.A. Establishment of White Perch, *Morone americana*, in Central New York. *Copeia.* 1952, pp. 200–201, 1952.

Dore, W.G. Frog-bit (*Hydrocharis morsus-ranae* L.) in the Ottawa River. *Canadian Field-Naturalist.* 68, pp. 180–181, 1954.

Dore, W.G. Progress of the European Frog-bit in Canada. *Canadian Field-Naturalist.* 82, pp. 76–87, 1968.

Dray, F.A., Jr., C.R. Thompson, D.H. Habeck, et al. A survey of the fauna associated with *Pistia stratiotes* L. (waterlettuce) in the United States. USACE Technical Report A-88-6, Vicksburg, MS, 1988.

Dundee, D.S. and H.A. Dundee. Extension of Known Ranges of Four Mollusca. *Nautilus.* 72, pp. 51–53, 1958.

Dymond, J.R. A Provisional List of the Fishes in Lake Erie. *University of Toronto Studies, Biological Serial 20, Publications of the Ontario Fisheries Laboratory.* 4, pp. 57–73, 1922.

Dymond, J.R. Records of the Alewife and Steelhead (Rainbow) Trout from Lake Erie. *Copeia.* 1932, pp. 32–33, 1932.

Dymond, J.R. Spread of Smelt (*Osmerus mordax*) in the Canadian Waters of the Great Lakes. *Canadian Field-Naturalist.* 58, pp. 12–14, 1944.

El-Ashry, M.T. and D.C. Gibbons, *Troubled Waters: New Policies for Managing Water in the American West.* World Resources Institute, Holmes, PA, 1986, p. 89.

Elrod, J.H., Busch, W.D.N., Griswold, B.L., Schneider, C.P., and Wolfert, D.R. Food of White Perch, Rock bass, and Yellow Perch in Eastern Lake Ontario. *New York Fish and Game Journal.* 28, pp. 191–201, 1981.

Emery, L., Review of Fish Species Introduced into the Great Lakes, 1819-1974. *Great Lakes Fishery Commission. Technical Report No. 45,* Great Lakes Fishery Commission, Ann Arbor, MI, 1985.

Etnier, D.A. and W.C. Starnes. *Fishes of Tennessee.* The University of Tennessee Press, Knoxville, TN, 1993, p. 681.

Fetteroff, C.M. Jr. Why a Great Lakes Fishery Commission and Why a Sea Lamprey International Symposium. *Canadian Journal of Fisheries and Aquatic Sciences.* 37, pp. 1588–1593, 1980.

George, C.J. Occurrence of the Silver Lamprey in the Stillwater Sector of the Hudson River. *New York Fish and Game Journal.* 32, p. 95, 1985.

Ghedotti, M.J., J.C. Smihula, and G.R. Smith. Zebra Mussel Predation by Round Goby in the Laboratory. *Journal of Great Lakes Research.* 21, pp. 665–669, 1995.

Goode, G.B. and T. Gill. *American Fishes: A Popular Treatise Upon the Game and Food Fish of North America with Especial Reference to Habits and Methods of Capture.* L.C. Page, Boston, MA, 1903, p. 562.

Grabowski, S.J., S.D. Hiebert, and D.M. Lieberman, *Potential for the Introduction of Three Species of Nonnative Fishes into Central Arizona via the Central Arizona Project — a Literature Review and Analysis.* Standard Technical Report Number REC-ERC-84-7. Environmental Science Section, United States Bureau of Reclamation, Denver, CO, 1984.

Greeley, J.R. Fishes of the Erie-Niagara Watershed. In *A Biological Survey of the Erie-Niagara System,* Supplement to Eighteenth Annual Report, State of New York Conservation Department, J.B. Lyon, Albany, NY, 1929.

Greeley, J.R. Fishes of the Lake Champlain Watershed. In *A Biological Survey of the Champlain Watershed,* Supplemental to Nineteenth Annual Report, New York State Conservation Department, J.B. Lyon, Albany, NY, 1930.

Griffen, J. The Everglades Kite. *Aquatics.* 11, pp. 17–19, 1989.

Griffiths, R.W., W.P. Kovalak, and D.W. Schloesser. The Zebra Mussel, *Dreissena polymorpha* (Pallas 1771), in North America: Impact on Raw Water Users. In *Proceedings: EPRI Service Water System Reliability Improvement Seminar.* Electric Power Research Institute, Palo Alto, CA, pp. 11–27, 1991.

Hadfield, C. *World Canals: Inland Navigation Past and Present.* David and Charles Publishers, Devon, Great Britain, 1986, p. 432.

Hale, M.M., J.E. Crumpton, and R.J. Schuler, Jr. From Sportfishing Bust to Commercial Fishing Boon: A History of the Blue Tilapia in Florida. *American Fisheries Society Symposium.* 15, pp. 425–430, 1995.

Haller, W.T. *Hydrilla: A New and Rapidly Spreading Aquatic Weed Problem.* Circular S-245, University of Florida, Gainesville, FL, 1978.

Hanna, D.G. Introduced Mollusks of Western North America. *California Academy of Sciences Occasional Papers.* 48. pp. 1–108, 1966.

Hebert, P.D.N., B.W. Muncaster, and G.L. Mackie. Ecological and Genetic Studies on *Dreissena polymorpha* (Pallas): A New Mollusk in the Great Lakes. *Canadian Journal of Fisheries and Aquatic Sciences.* 46, pp. 1587–1591, 1989.

Hocutt, C.H. and E.O. Wiley, *The Zoogeography of North America Freshwater Fishes.* John Wiley and Sons, Canada, 1986, p. 866.

Howe, C.W. and K.W. Easter, *Interbasin Transfers of Water.* The Johns Hopkins Press, Baltimore, MD, 1971, p. 196.

Hubbs, C.L. and K.F. Lagler. *Fishes of the Great Lakes Region*. Cranbrook Institute of Science, Bloomfield Hills, MI, 1958, p. 213.

Hubbs, C.L. and R.E. Johnson. Hybridization in Nature Between Species of Catostomid Fishes. *Contributions of the Laboratory of Vertebrate Biology, University of Michigan*. 22, pp.1–76, 1943.

Hurley, D.A. Effects of Nutrient Reduction on the Diets of Four Fish Species in the Bay of Quinte, Lake Ontario. *Canadian of the Journal of Fisheries and Aquatic Sciences Special Publication*. 86, pp. 237–246, 1986.

Ireland, R.R. and M.J. Schepanek. The Spread of the Moss *Hyophila involuta* in Ontario. *The Bryologist*. 96, pp. 132–137, 1993.

Isom, B.G. Historical Review of Asiatic Clam (*Corbicula*) Invasion and Biofouling of Waters and Industries in the Americas. *American Malacological Bulletin. Special Edition*. No. 2, pp 1–5, 1986.

Johnson, R. *The Central Arizona Project: 1918-1968*. The University of Arizona Press, Tucson, AZ, 1977, p. 242.

King, W. and T. Krakauer. The Exotic Herpetofauna of Southern Florida. *Quarterly Journal of the Florida Academy of Science*. 29, pp. 144–154, 1966.

Koebel, J.W., Jr. An Historical Perspective on the Kissimmee River Restoration Project. *Restoration Ecology*. 3, pp. 149–159, 1995.

Krakauer, T. The Ecology of the Neotropical Toad, *Bufo marinus*, in South Florida. *Herpetologica*. 24, pp. 214–221, 1968.

Kushlan, J.A. Exotic Fishes of the Everglades: A Reconsideration of Proven Impact. *Environmental Conservation*. 13, pp. 67–69, 1986.

Lark, J.G. An Early Record of the Sea Lamprey (*Petromyzon marinus*) from Lake Ontario. *Journal of Fisheries Research Board of Canada*. 30, pp. 131–133, 1973.

Larsen, A. First Record of the White Perch (*Morone americana*) in Lake Erie. *Copeia*. 1954, p. 154, 1954.

Lawrie, A.H. The Sea Lamprey in the Great Lakes. *Transactions of the American Fishery Society*. 1970, pp. 766–775, 1970.

Lawrie, A.H. and J.F. Rahrer. Lake Superior: A Case History of the Lake and Its Fisheries. *Great Lakes Fishery Commission. Technical Report No. 19*. Great Lakes Fishery Commission, Ann Arbor, MI, 1973.

Leach, J.H. Impacts of the Zebra Mussel, *Dreissena polymorpha*, on Water Quality and Spawning Reefs in Western Lake Erie, in *Zebra Mussels: Biology, Impacts and Control*, Nalepa, T.F. and D.W. Schloesser, Eds., Lewis Publishers, Boca Raton, FL, 1993.

Legget, R.F. *Canals of Canada*. Douglas, David, & Charles Ltd., North Vancouver, Canada, 1976, p. 261.

Loftus, W.F. and J.A. Kushlan. Freshwater Fishes of Southern Florida. *Bulletin of the Florida State Museum, Biological Sciences*. 31, pp.147–344, 1987.

Louisiana Sea Grant Communications Office. (1995). *Invasion of Zebra Mussels*. Prepared by the Louisiana Sea Grant College Program. Located on the Internet (http://www.ansc.pur-due.edu/sgnis/update/zebra.html).

Lumsden, H.G. and D.J. McLachlin. European Frog-bit, *Hydrocharis morsus-ranae* L., in Lake Ontario Marshes. *The Canadian Field-Naturalist*. 102, pp. 261–263, 1988.

Madenjian, C.P. Removal of Algae by Zebra Mussel (*Dreissena polymorpha*) Population in Western Lake Erie: A Bioenergetics Approach. *Canadian Journal of Fisheries and Aquatic Sciences*. 52, pp. 381–390, 1995.

Martel, A. Demography and Growth of the Exotic Zebra Mussel (*Dreissena polymorpha*) in the Rideau River (Ontario). *Canadian Journal of Zoology*. 73, pp. 2244–2250, 1995.

Mayden, R.L. *Systematics, Historical Ecology, and North American Freshwater Fishes*. Stanford University Press, Stanford, CA, 1992, p. 969.

McCann, J.A., L.N. Arkin, and J.D. Williams. *Nonindigenous Aquatic and Selected Terrestrial Spaces of Florida: Status, Pathway and Time of Introduction, Present Distribution, and Significant Ecological and Economic Effects*. Prepared by University of Florida, Center for Aquatic Plants, Located on the Internet, March 1996.

McCombie, A.M. Changes in the Physical and Chemical Environment of the Laurentian Great Lakes. In *A Symposium on Introductions of Exotic Species*, Ontario Department of Lands and Forests, Research Branch, Research Report No. 82. p. 111, 1968.

McMahon, R.F. The Occurrence and Spread of the Introduced Asiatic Freshwater Clam, *Corbicula fluminea* (Müller) in North America: 1924–1982. *Nautilus*. 96, pp. 134–141, 1982.

Miller, R.R. Origin and Dispersal of the Alewife, *Alosa pseudoharengus,* and the Gizzard Shad, *Dorosoma cepedianum,* in the Great Lakes. *Transactions of the American Fishery Society.* 86(1956), pp. 97–111, 1957.

Mills, J.C. *Our Inland Seas: Their Inland Shipping and Commerce for Three Centuries.* A.C. McClurg, Chicago, IL, 1910.

Mills, E.L., J.H. Leach, J.T. Carlton, and C.L. Secor. Exotic Species in the Great Lakes: A History of Biotic Crises and Anthropogenic Introductions. *Journal of Great Lakes Research.* 19, pp. 1–54, 1993.

Mills, E.L., M.D. Scheuerell, J.T. Carlton, and D. Strayer. Exotic Species in the Hudson River Basin: A History of Invasions and Introductions. *Estuaries.* 19, pp.814–823, 1996.

Minckley, W.L. *Fishes of Arizona.* Sims Printing, Phoenix, AZ, 1973, pp. 293.

Minshall, W.H. Frog-bit (*Hydrocharis morsus-ranae*-L.) at Ottawa. *Canadian Field- Naturalist.* 54, pp. 44–45, 1940.

Morman, R.H., D.W. Cuddy, and P.C. Rugen. Factors Influencing the Distribution of Sea Lamprey (*Petromyzon marinus*) in the Great Lakes. *Canadian Journal of Fisheries and Aquatic Sciences.* 37, pp. 1811–1826, 1980.

Moyle, P.B. Mississippi Silversides and Logperch in the Sacramento-San Joaquin River System. *California Fish and Game.* 60, pp. 144–149, 1974.

Moyle, P.B. *Inland Fishes of California.* University of California Press, Berkeley, CA, 1976, p. 405.

Moyle, P.B. and J.E. Williams. Biodiversity Loss in the Temperate Zone: Decline of the Native Fish Fauna of California. *Conservation Biology.* 4, pp. 275–284, 1990.

Nichols, K.H. and G.J. Hopkins. Recent Changes in Lake Erie (North Shore) Phytoplankton: Cumulative Impacts of Phosphorus Loading Reductions and Zebra Mussel Introduction. *Journal of Great Lakes Research.* 19, pp. 637–647, 1993.

O'Neill, C. *Dreissena.* 6(3):6. Zebra Mussel Clearing House. SUNY Brockport, Brockport, NY, 1995.

Ortmann, A.E. The Alleghany Divide, and Its Influence Upon the Freshwater Fauna. *Proceedings of the American Philosophical Society.* 52, pp. 287–390, 1913.

Pearce, F. *The Dammed.* The Bodley Head, London, 1992, p.376.

Prokopovich, N. and D.J. Hebert. Sedimentation in the Delta-Mendota Canal. *Journal of the American Water Works Association.* 57, pp. 375–382, 1965.

Prout, M.W., E.L. Mills, and J.L. Forney. Diet, Growth, and Potential Competitive Interactions Between Age-0 White Perch and Yellow Perch in Oneida Lake, New York. *Transactions of the American Fishery Society.* 119, pp. 966–975, 1990.

Rawinski, T.L. and R.A. Malecki. Ecological Relationships among Purple Loosestrife, Cattails, and Wildlife at the Montezuma Wildlife Refuge. *New York Fish and Game Journal.* 31, pp. 81–87, 1984.

Raynes, J.J. Aquatic Plant Control. *Hyacinth Control Journal.* 3, pp. 2–4, 1964.

Regier, H.A. and W.L. Hartman. Lake Erie's Fish Community: 150 Years of Cultural Stresses. *Science.* 180, pp. 1248–1255, 1973.

Reider, R.H. Occurrence of the Silver Lamprey in the Hudson River. *New York Fish and Game Journal.* 26, pp. 93, 1979.

Richard, D.I., J.W. Small and J.A. Osborne. Response of Zooplankton to the Reduction and Elimination of Submerged Vegetation by Grass Carp and Herbicide in Four Florida Lakes. *Hydrobiologia.* 123, pp. 97–108, 1985.

Roberts, M.L., R.L. Stuckey, and R.S. Mitchell. *Hydrocharis morsus-ranae* (Hydrocharitaceae): New in the United States. *Rhodora.* 83, pp. 147–148, 1981.

St. Lawrence Seaway Authority. *The Montreal-Lake Ontario Section of the Seaway.* The St. Lawrence Seaway Authority. Ottawa, Canada, 1985, p. 12.

St. Lawrence Valley Souvenir Company, *The Billion Dollar Story: The International St. Lawrence Seaway and Power Development.* St. Lawrence Valley Souvenir Company, Massena, NY, 1978, vol. 10, p. 76.

Schardt, J.D. and D.C. Schmitz. *1990 Florida Aquatic Plant Survey, the Exotic Aquatic Plants. Technical Report 91-CGA* . Florida Department of Natural Resources, Tallahassee, FL, 1991, p. 89.

Schmitz, D.C., J.D. Schardt, A.J. Leslie, F.A. Dray, Jr., J.A. Osborne, and B.V. Nelson. The Ecological Impact and Management History of Three Invasive Alien Aquatic Plant Species in Florida. In *Biological Pollution: The Control and Impact of Invasive Aquatic Species,* McKnight, B.N., Ed., Indiana Academy of Science, Indianapolis, IN, 1993.

Scott, W.B. and W.J. Christie. The Invasion of the Lower Great Lakes by the White Perch, *Roccus americanus* (Gmelin). *Journal of Fisheries Research Board of Canada.* 20, pp. 1189–1195, 1963.

Scott, W.B. and E.J. Crossman. *Freshwater Fishes of Canada.* Fisheries Research Board of Canada. Bulletin 184. Ottawa, 1973. p. 966.

Serrouya R., A. Ricciardi, and F.G. Whoriskey. Predation on Zebra Mussel (*Dreissena polymorpha*) by Captive-Reared Map Turtles (*Graptemys geographica*). *Canadian Journal of Zoology.* 73, pp. 2238–2243, 1995.

Sigler, W.F. and J.W. Sigler. *Fishes of the Great Basin: A Natural History.* University of Nevada Press, Reno, NV, 1987, p. 425.

Smith, S.H. Species Interactions of the Alewife in the Great Lakes. *Transactions of the American Fishery Society.* 70, pp. 754–765, 1970.

Smith, S.H. Factors of Ecological Succession in Oligotrophic Fish Communities of the Laurentian Great Lakes. *Journal of Fisheries Research Board of Canada.* 29, pp. 717–730, 1972.

Smith, C.L. *The Inland Fishes of New York State.* New York State Department of Environmental Conservation, Albany, NY, 1985, p. 522.

Smith, W.G. 1991 Zebra Mussel Invasion Monitoring of Big Rideau Lake, Upper Rideau Lake, and Opinicon Lake. In *Rideau Lakes FAU: Update '92-1.* Rideau Lakes Fisheries Assessment Unit, Sharbot Lake, Ontario, 1992.

Steward, K.K., T.K. Van, V. Carter, and A.H. Pieterse. Hydrilla Invades Washington, D.C. and the Potomac. *American Journal of Botany.* 71, pp. 162–163, 1984.

Stine, J.K. *Mixing the Waters: Environment, Politics, and the Building of the Tennessee-Tombigbee Waterway.* The University of Akron Press, Akron, OH, 1993, p. 336.

Strayer, D.L., J. Powell, P. Ambrose, L.C. Smith, M.L. Pace, and D.T. Fischer. Arrival, Spread, and Early Dynamics of a Zebra Mussel (*Dreissena polymorpha*) Population in the Hudson River Estuary. *Canadian Journal of Fisheries and Aquatic Science.* 53, pp. 1143–1149, 1996.

Tabita, A. and J.W. Woods. History of Waterhyacinth Control in Florida. *Hyacinth Control Journal.* 1, pp. 19–23, 1962.

Thompson, D.Q., R.L. Stuckey, and E.B. Thompson. *Spread, Impact, and Control of Purple Loosestrife (Lythrum salicaria) in North American Wetlands.* U.S. Department of the Interior, United States Fish and Wildlife Service, Washington, DC, 1987, p. 55.

Trautman, M.B. *The Fishes of Ohio with Illustrated Keys.* Waverly Press, Baltimore, MD, 1957, pp. 683.

United States Congress. Cross-Florida Barge Canal, *Hearing before the Subcommittee on Water Resources of the Committee on Public Works and Transportation,* House of Representatives, Ninety-Ninth Congress, First Session, June 10, 1985, Palatka, FL. U.S. Government Printing Office, Washington, DC, 1986.

United States Congress, Office of Technology Assessment. *Harmful Nonindigenous Species in the United States.* OTA-F-565, Washington, DC: U.S. Government Printing Office, September 1993.

van Dijk, G. *Vallisneria* and Its Interactions with other Species. *Aquatics.* 7, pp. 6–10, 1985.

Van Oosten, J.V. The Dispersal of Smelt, *Osmerus mordax* (Mitchill), in the Great Lakes Region. *Transactions of the American Fisheries Society.* Sixty-sixth Annual Meeting, pp. 160–171, 1937.

Walton, W.C. Occurrence of Zebra Mussel (*Dreissena polymorpha*) in the Oligohaline Hudson River, New York. *Estuaries.* 19, pp. 612–618, 1996.

Whitford, N.E., History of the Canal System of the State of New York Together with Brief Histories of the Canals of the United States and Canada. In *Supplement to the Annual Report of the State Engineer and Surveyor of the State of New York*, vol. 2. Brandow Printing, Albany, NY, 1906.

24 The Garrison Diversion and the Interbasin Biota Transfer Issue: Case Study

Dennis G. Wright and William G. Franzin

INTRODUCTION

The effects of the introduced species in the Great Lakes and the rapidity with which these species can spread through an ecosystem have been well documented (Crossman 1991). In this chapter we present an overview of another potential biota transfer situation — the Garrison Diversion Unit — and what has been and is being done to counter this threat.

HISTORY OF THE GARRISON DIVERSION UNIT

The Garrison Diversion Unit is a multipurpose water resource project designed to divert Missouri River water into central and eastern North Dakota to provide water for irrigation and for municipal and industrial uses. The Missouri River has long been viewed in the State of North Dakota as a source of irrigation water. In the 1940s, a proposal was initiated known as the Pick-Sloan Plan whereby a series of dams was to be constructed on the Missouri River to control downstream flooding and provide stable water levels for navigation. One of these dams, the Garrison Dam at Riverdale, North Dakota, was begun in 1946 and completed in 1956. A side benefit of this flood control dam was to be the provision of water to irrigate up to 1 million acres of land to compensate in part for the inundation of rich river bottom lands by the creation of the reservoir. However, funding for the project failed to receive congressional approvals for a number of years in the late 1950s and early 1960s (Kelly and Leitch 1990).

In 1965, the U.S. Congress authorized a reduced plan that would use water stored behind the Garrison Dam to irrigate 101,200 ha (250,000 acres) of land, 87% of which lay not in the Missouri River drainage basin but in the Hudson Bay drainage basin, and 82% of which would have discharged irrigation return flows and operational wastes into streams ultimately draining into Canada (Figure 24.1) (U.S. Bureau of Reclamation 1974). Under this plan, water was to be pumped from Garrison Reservoir (Lake Sakakawea) to Snake Creek Reservoir (Lake Audubon). The McClusky Canal was to carry water 117 km (73 miles), crossing the continental divide between the Missouri and Hudson Bay drainages, to the Lonetree Reservoir situated at the headwaters of the Sheyenne River, a tributary of the Red River of the North. From there, water was to be distributed north through the Velva Canal to irrigate lands in the loop of the Souris River, east through the New Rockford Canal to irrigate lands in the Warick-McVille and New Rockford areas, and south through the James River Feeder Canal to irrigate lands along the James River.

As early as 1970, the government of Canada formally expressed its concerns to the U.S. government about the potential impacts of the project in Canada. Both Canada and the Province of Manitoba were concerned that leachates from irrigated soils would degrade the water quality of shared watersheds; that return flows would increase flooding; and that the possible introduction of

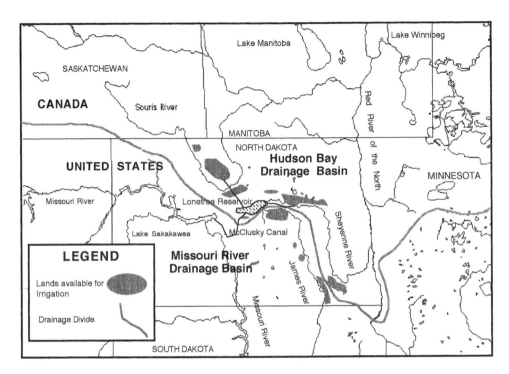

FIGURE 24.1 The Garrison Diversion Unit—major features of the original (1965) plan.

nonnative biota such as fish and fish eggs, parasites, and diseases could affect existing aquatic ecosystems and the subsistence, commercial, and recreational fisheries that were dependent upon them. As more and more information became available regarding those impacts, Canada's concerns increased, until on October 23, 1973, the Canadian Embassy sent a Diplomatic Note to the U.S. State Department, stating, in part, that

> The Government of Canada has concluded that based on studies conducted in both countries the [Garrison Diversion Unit] proposal would run counter to the obligations assumed by the United States under Article IV of the Boundary Waters Treaty of 1909. . . . Accordingly, the Government of Canada requests urgently that the Government of the United States establish a moratorium on all further construction of the Garrison Diversion Unit until such time as the United States and Canada can reach an understanding that Canadian rights and interests have been fully protected in accordance with the Boundary Waters Treaty.

On October 22, 1975, the government of Canada and the government of the U.S. referred the issue of the transboundary impacts of the Garrison Diversion Unit to the International Joint Commission (IJC)

> ...to examine into and report upon the transboundary implications of the proposed completion and operation of the Garrison Diversion Unit in the State of North Dakota; and to make recommendations as to such measures, including modifications, alterations or adjustments to the Garrison Diversion Unit, as might be taken to assist governments in ensuring that the provisions of Article IV of the Boundary Waters Treaty are honored.

The IJC established the International Garrison Diversion Study Board and charged the Board to

…advise the Commission as to measures including but not limited to modifications, alternatives or adjustments to the Garrison Diversion Unit which could be taken to avoid or relieve adverse impacts, if any, of water uses in Canada… .

Why this concern? Following the retreat of the last great continental ice sheets, the Hudson Bay drainage basin was recolonized for the most part by fish species from the Missouri–Mississippi refugia (Crossman 1976). Although having a common heritage, the two drainage basins have developed distinct ichthyofaunas. A number of fish species that are present in the Missouri drainage are not present in the Hudson Bay drainage (Table 24.1). Because Garrison would provide a direct link between the two basins, these species were considered to be potential problem species. Life history information on these species was reviewed in detail. The conclusion was that nine fish species present in the Missouri drainage, if introduced to the Hudson Bay basin, might cause adverse impacts on indigenous fish resources (International Garrison Diversion Study Board 1976). These species were pallid sturgeon (*Scaphirhynchus albus*), shovelnose sturgeon (*S. platorynchus*), paddlefish (*Polyodon spathula*), shortnose gar (*Lepisosteus platostomus*), gizzard shad (*Dorosoma cepedianum*), rainbow smelt (*Osmerus mordax*), Utah chub (*Gila atraria*), smallmouth buffalo (*Ictiobus bubalus*), and river carpsucker (*Carpiodes carpio*). Carp (*Cyprinus carpio*) was added to the list, since diversion of Garrison Reservoir waters would have enhanced conditions for the introduction of this species into the Souris River in North Dakota and Saskatchewan, where it does not occur.

TABLE 24.1
Summary of Survival Potential and Ecosystem Impacts of Interbasin Fish Introductions Based on Life History Information

Species	Potential for Introduction	Survival Potential in the Hudson Bay Basin	Potential for Biological Impact	Potential for Adverse Impact on Manitoba	Availabilty of Information for Evaluation
Pallid sturgeon	low	high	medium	low	low
Shovelnose sturgeon	low	high	medium	low	low
Paddlefish	medium	high	medium	low	medium
Shortnose gar	medium	high	medium	medium	low
Gizzard shad	high	medium	high	medium	high
Rainbow smelt	high	high	high	high	high
River carpsucker	medium	high	medium	low	low
Smallmouth buffalo	medium	high	medium	low	low
Utah chub	medium	high	high	high	medium

After this qualitative assessment of the potential impact of each of the nine species, quantitative assessment of their introductions on commercially important species in the two major lakes, Lake Winnipeg (north and south basins) and Lake Manitoba, was undertaken.

The four commercially important species considered were lake whitefish (*Coregonus clupeaformis*), lake herring (now known as *C. artedi*), walleye (*Stizostedion vitreum*), and sauger (*S. canadense*). The predicted impact was estimated as a percentage reduction in population size. The estimated range of low, most likely and maximum reductions were as follows:

	Percentage Reduction in Population Size (Lowest - Most Likely - Maximum)		
	Lake Whitefish	Lake Herring	Walleye and Sauger
Lake Winnipeg— North Basin	25 - 50 - 75	50 - 75 - 99	25 - 50 - 75
Lake Winnipeg— South Basin	0 - 5 - 10	0 - 5 - 10	25 - 50 - 75
Lake Manitoba	10 - 30 - 50	50 - 75 - 99	50 - 75 - 99

The Study Board report concluded that construction and operation of the project would have adverse effects on Manitoba, causing injury to health and property as a result of adverse impacts on the water quality and biological resources of the province. Specifically, the IJC concluded that

"…The possibility of a transfer of exotics, that is the transfer of foreign fish species, fish diseases and fish parasites indigenous to the Missouri River Drainage Basin into the Hudson Bay Drainage Basin has been a major concern of the Biology Committee, the Board and the Commission itself (International Joint Commission, 1977).

In fact, overriding everything else since release of that report has been the necessity that such introductions be prevented. Modifications to the project design that will eliminate direct connections between the Missouri and Hudson Bay Drainage basins would reduce some of the adverse impacts. The IJC recommended that those portions of the project affecting water flowing into Canada not be built until the issue had been resolved.

In 1984, the findings and conclusions of the IJC were reexamined. The distributions of the nine species were reevaluated to determine if there had been any range extensions and if the conclusions reached were still valid. Some range extensions in the species were observed, but the conclusions drawn in the IJC report were still valid. The introduction of nonnative biota was still predicted to have serious negative impacts on Canadian waters.

In order to meet the contemporary water needs of North Dakota, the Garrison Diversion Project was totally redesigned under the Garrison Reformulation Act of 1986 (Figure 24.2) (Public Law 99-294) (Garrison Diversion Unit Commission 1984). Among these changes were

- Reduction of the irrigation development from 250,000 acres to 130,940 acres, and elimination of discharge of irrigation return flows to streams ultimately draining to Canada;
- Expansion of the municipal, rural, and industrial water (MRI) supply component;
- Authorization of the Sykeston Canal to be constructed as a functional replacement for the Lonetree Reservoir to link the McClusky Canal and the New Rockford Canal.

The U.S. Bureau of Reclamation developed another plan to incorporate the recommendations and conditions outlined in the Reformulation Act. Canada and the U.S. undertook a risk analysis of the Reformulated Plan and reported in 1990 that three of the ten options were *potentially* acceptable from the standpoint of risk to Canadian waters (Garrison Diversion Unit Joint Technical Committee 1990). While the risk of direct introduction of Missouri River biota would be virtually eliminated, a risk of introduction by means of "bait bucket transfer" still remained. At the present time, there has been no transfer of water between the Missouri and Hudson Bay basins. Funding for the project has not been approved and it is uncertain when or if this portion of the Garrison Diversion will be completed.

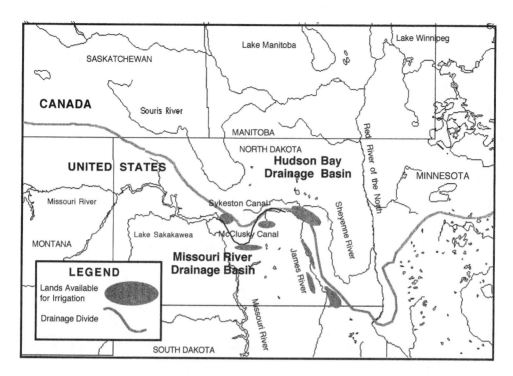

FIGURE 24.2   The Garrison Diversion Unit—major features of the reformulated 91986) plan.

However, the MRI supply aspects of the Garrison Diversion are proceeding. The Northwest Area Water Supply (NAWS) project would supply MRI water to ten counties north and west of Lake Sakakawea (Engineering-Biology Task Group 1994). One of its components, the East System, would transfer raw water by pipeline from Lake Sakakawea to the City of Minot for treatment and distribution. Most of this system is located in the Hudson Bay basin, and the Minot water treatment plant is located on the banks of the Souris River, which flows into Canada. A system failure at the treatment facility or at some point along the pipeline could transfer raw water, potentially containing nonnative fish, fish eggs, fish disease pathogens, and fish parasites, from the Missouri River basin to the Hudson Bay basin. To alleviate concerns about fish diseases, the project proponent agreed to treat the raw water in the Missouri River basin to minimize the biota transfer risk to Canada. Screens will be utilized to eliminate or reduce the possibility of transfer of fish, and a study has been undertaken to develop a method to disinfect the water to drinking water standards before it crosses the drainage divide to minimize the risk of larval fish, fish egg, fish disease, and parasite transfer. Because the raw water must be disinfected to reduce the build-up of slimes in the pipeline, reduction of the threat of biota transfer can be realized with little or no additional cost or modifications to the project design. Preliminary results of the study are encouraging, and addition of a low level of chloramine has been shown to provide a 99.9% removal of viruses within 9 minutes of contact time. Ozonation is also being examined as a treatment option.

The change in the focus of the Garrison Diversion project from irrigation to mainly a MRI water supply project has been welcomed by Canada and Manitoba because of the lessened probability of biota transfer. In recent reviews of the Garrison Diversion Unit project, Clarkson (1992) and Krenz (1994) interpreted the current status of negotiations between the parties as two major remaining issues:

- That North Dakota has a right to increased quantities and quality of MRI water in the Hudson Bay basin, and
- That benefits to North Dakota from its use of Missouri River water should not accrue at the expense of Canada and Manitoba.

INTERBASIN BIOTA TRANSFER ISSUES

What then has been determined about the potential risks of fish transfers from Missouri River waters to the Hudson Bay drainage basin? What are the mechanisms by which fish transfers might occur? Table 24.2 summarizes the mechanisms of species movements across drainage basin boundaries. The following examples pertinent to Garrison illustrate some of these mechanisms.

TABLE 24.2
Mechanisms of Transfer of Fish Species Across Drainage Basin Boundaries

1. Intentional introduction by official agency
2. Unauthorized introduction by individuals
3. Release of live bait
4. Escape from culture
5. Invasion with the aid of humans, e.g., following an intentional or unauthorized introduction
6. Accidental introduction by transportation: ballast waters, live wells, vehicle accident, pipeline hydrostatic testing
7. Natural invasion; continuing postglacial dispersion
8. Natural expansion of range; climatically moderated dispersion

Source: After Crossman 1991, and Litvak and Mandrak 1993.

NATURAL DISPERSION/NATURAL EXPANSION OF RANGE

There are a number of studies that bear on these issues. First, it has not been very long (<10,000 years) in the geologic sense since the two basins were connected through postglacial drainages. Indeed there is continuing infrequent connection between the Mississippi River and Red River headwaters at Lake Traverse, most recently in 1997 during high spring runoff (Figure 24.3). There are unconfirmed reports that similar connections occur between the Red and Missouri rivers in the region of Wild Rice and James rivers and between Mississippi River and Winnipeg/Rainy River headwaters. The result of these connections is that the native biota of these systems are imperfectly separated. This is evident because several species of fish occur in headwater tributaries of the Red River in Minnesota that do not yet occur in Manitoba (Koel and Peterka 1994) and apparently are limited in their distributions by habitat requirements. Occasionally, new species are found in Manitoba and attributed to movements out of these small reservoirs of species which are, in fact, in headwaters of the Hudson Bay drainage but not in Manitoba (e.g., *Cyprinella spiloptera*, Franzin, unpublished data). Additional discoveries of species new to Manitoba have occurred that are better attributed to postglacial distributions followed by severe truncation of ranges, leaving remnant populations in habitat-specific areas throughout the Hudson Bay basin (e.g., *Notropis texanus*; Stewart 1988). These species have been discovered by increased collecting effort in recent decades. However there are a few species that have arrived in Manitoba with the aid of humans. The mechanisms by which this has occurred are of concern in the Garrison Diversion scheme.

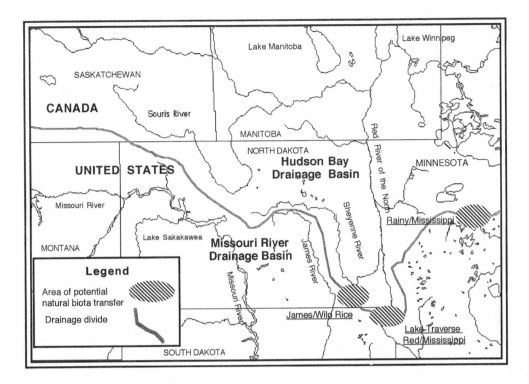

FIGURE 24.3   Areas of potential natural interbasin biota transfer.

INTENTIONAL INTRODUCTIONS

Rainbow Smelt

Rainbow smelt (*Osmerus mordax*) is a species exotic to inland North America west of Appalachia. It was introduced to Lake Sakakawea from the Great Lakes (an introduction in themselves) in the spring of 1971. Rainbow smelt was introduced into Lake Sakakawea to provide a forage fish for Pacific salmon (*Oncorrynchus* spp.) that had been stocked into the newly developed deep cold-water reservoir behind the Garrison Dam. Rainbow smelt rapidly colonized most of the mainstem Missouri river reservoirs downstream of Lake Sakakawea and by 1979 was collected as far south as Louisiana. The species has been in the Hudson Bay basin for sometime also (since perhaps 1962; Campbell et al. 1991) but apparently has spread recently by multiple intentional and unintentional introductions followed by rapid downstream colonization in the English and Rainy river watersheds (Franzin et al. 1994). Rainbow smelt appeared in Lake Winnipeg in 1990 (Campbell et al. 1991), and now are widespread in the lake. The species is moving down the Nelson River (Derksen, pers. commun.) toward Hudson Bay, where it is expected that it would resume an anadromous lifestyle.

We do not know by which mechanism rainbow smelt arrived in Lake Winnipeg, but we suspect bait-bucket transfer as the species appeared in Lake Winnipeg without having been observed in the connecting waters between Lake Winnipeg and English and Rainy River watersheds. Rainbow smelt has had deleterious effects on other fish species, especially in smaller lakes (Evans and Loftus 1987; Evans and Waring 1987; Loftus and Hulsman 1986; Wain 1993) and has been implicated in a reduction in the total value of Lake Erie fisheries (Remnant 1991).

White Bass (*Morone chrysops*)

The State of North Dakota introduced white bass (*Morone chrysops*) to Lake Ashtabula, a reservoir on the Sheyenne River within the Hudson Bay drainage basin, in the 1960s to enhance the recreational fishery. Within several years this species moved downstream into the Red River and into Lake Winnipeg, where it is now present in substantial numbers. Coincident with the arrival of the white bass there has been a decline in the landed weight of walleye and sauger from the commercial fishery. However, several other factors such as habitat degradation, overfishing, and climatic changes preclude establishing a direct correlation between the arrival of white bass and the decline in the walleye/sauger fishery. However, unlike rainbow smelt, white bass are native to the Mississippi River, from which most of the Hudson Bay fauna are derived. Because most of the Hudson Bay basin ichthyofauna has co-evolved with white bass in glacial times, its impact may be less severe than for a true exotic. White bass has recently completed the full colonization of Lake Winnipeg following years of presence in small numbers. The recent increase in numbers probably is due in part to the warm spell of the late 1980s favoring reproduction and successful overwintering of its young-of-the-year. Whether white bass affects the reproductive success of other species in Lake Winnipeg is unclear, because it is a predator of the open water pelagic zone, a niche that previously was unoccupied.

Other Species

Black crappie (*Pomoxis nigromaculatus*), another Mississippi species introduced to the Red River, also has increased its range in Lake Winnipeg after the warm years. Several other species, including stonecat (*Noturus flavus*), spotfin shiner (*Cyprinella spilotera*), and golden redhorse (*Moxostoma erythrurum*), seem to have spread northward into Manitoba from upstream in the Red River in response to a general warming trend. If the stonecat is an example of the threat of native species colonization, then studies by McCulloch and Stewart (1994) suggest that the further dispersal of *native* Mississippi (or Missouri) faunas into the Hudson Bay drainage may not have impacts as adverse as would a true exotic.

Until recently, deliberate introductions still were a significant threat. For example, North Dakota, in the 1980s attempted to introduce a European percid, zander (*Stizostedion leucioperca*). Canada and Manitoba mounted strong opposition to this project, as there were very serious potential ecological implications to Canadian waters and throughout the Missouri–Mississippi basin. As a result of these efforts and those of some North Dakotans, the Governor of North Dakota ordered a halt to the project. North Dakota Department of Game and Fish subsequently initiated an interagency committee with representatives of the fisheries agencies of Manitoba, Minnesota, Montana, North Dakota, South Dakota, and Saskatchewan to examine any future fish introductions in the region. The committee agreed that before any new species may be introduced by any jurisdiction, a risk analysis must be undertaken, and the proposed introduction must have the consent of all potentially affected neighboring jurisdictions.

UNAUTHORIZED INTRODUCTION/ESCAPE FROM CULTURE/INVASION WITH THE AID OF HUMANS

Fish

Several exotics have appeared in the Mississippi–Missouri drainage by means of several different invasion mechanisms. Utah chub was introduced from the Bonneville Basin in Utah to the upper Missouri, most likely as an unauthorized introduction by an individual. Grass carp (*Ctenopharyngodon idella*) and bighead carp (*Hypophthalmichthys nobilis*) have escaped from culture in the lower Missouri and Mississippi rivers.

Parasites and Diseases

Fish pathology is an infant science, and previously unknown disease organisms still are being discovered. Pathogen testing, for example, is done only for known bacteria, viruses, and parasites, and little information is available on the interbasin transfer of pathogens and parasites. A recent estimate is that less than 2% of the fish diseases are known and even for these the knowledge is incomplete (Kinne 1984). All organisms carry a suite of other organisms including viruses, rickettsia, bacteria, yeasts, fungi, protozoans, and larger parasites. Some are harmful only to the species carrying them, while others within this suite can cause the host little or no harm regardless of circumstance, but may be highly pathogenic for other species with which the host may come in contact (Stewart 1991). Failure to find known disease organisms in a stock of fish from outside the drainage basin is of little or no assurance that an unsuspected disease problem will not show up later. The introduction of a disease pathogen or parasite could be catastrophic to wild or cultured fish.

RELEASE OF LIVE BAIT

Among the many issues that have arisen during the development of the Garrison Diversion project, the one that keeps emerging is bait bucket transfer. Large-scale water diversions can create enhanced recreational fishing opportunities. For example, the original 1965 plan and a recent plan put forth by North Dakota proposed the creation of a large reservoir known as Lonetree or Mid-Dakota in the headwaters area of the Sheyenne River and filled with Missouri River water (Figure 24.1). There the Missouri and Hudson Bay drainages would have been separated by the width of the reservoir embankment. The creation of a reservoir inevitably leads to the development of a recreational fishery and the potential for bait bucket transfer. Despite all that engineering can do to eliminate or reduce the risk of interbasin biota transfer, the actions of sometimes well-intentioned but ill-advised anglers is a major vector for the introduction of nonnative species. In addition to baitfish introductions themselves, there is also the potential for the incidental transfer of other nonnative species such as zebra mussels (*Dreissena polymorpha*) and spiny water fleas (*Bythotrephes cederstoemi*) or pathogens in bait buckets and live wells. The effects of these incidental introductions may be as significant as the nonnative fish themselves.

Ludwig et al. (1994) surveyed bait dealers and anglers in the North Dakota/Minnesota area near the boundaries of the Hudson and Mississippi drainages. They found that baitfish wholesalers and retailers were buying and selling baitfish across drainage basin boundaries. Also, anglers were moving across boundaries with both captured and purchased baitfish. Neither Minnesota nor North Dakota has regulations regarding transbasin fish movements, but there are regulations on the movement of baitfish among states. Some baitfish caught in the Mississippi Basin were transferred to the Hudson Bay basin especially in fishing areas near the basin junctions. In addition, the advent of boats with livewells capable of holding fish for several days, even while being transported on a trailer, and the significant increase in the popularity of fishing tournaments have increased the probability that baitfish and other biota would be introduced to new ecosystems where they may survive, reproduce, and possibly have deleterious effects. Recent increases in the number of inland Wisconsin lakes adjacent to Lake Michigan that have been invaded by zebra mussel (Wiland 1995) illustrate the potential of this problem.

What can be done about it? There is only one real solution and that is to ban the use of live bait. However, this seems unlikely, given the economic value of the live bait industry (perhaps $30 million per year in Ontario alone). Education of anglers and boaters to sterilize live wells upon leaving lakes and not to transport bait may be the only feasible approach, but the effectiveness of this approach is doubtful. Unless firm action is taken by fishery management agencies and resource users themselves, the fish fauna of North America could become more homogeneous, in which competition from exotics, climate, and habitat features would control fish distributions.

ADDRESSING THE ISSUES

The major problem, from a Canadian perspective, that precludes the use of Missouri River water in the Hudson Bay basin has been identified as the transfer of potentially injurious aquatic biota across the basin boundaries (International Joint Commision 1977).

How has North Dakota sought to alleviate these concerns? The Garrison Diversion Unit Commission, recognizing the problem as partly a lack of information about the biota of the two drainage basins, recommended a one-time baseline survey of fish species and pathogens in waters of the Hudson Bay drainage to be administered by an international organization not affiliated with the governments involved in the debate. The result was the creation by the Governor of North Dakota of the Interbasin Water Transfer Studies Program (IWTSP) in 1985 with the mandate to study the problem of possible adverse effects of transfers of fish species, biota, and pathogens from the Missouri River to the Hudson Bay drainage. The task of managing the IWTSP program was given to the Water Resources Research Institute at North Dakota State University in Fargo. Two committees were set up: a Technical Advisory Team (TAT) comprising scientists from U.S. and Canadian universities, U.S. state and federal government officials, and private-sector members to determine what research was necessary and to review the merits of research proposals received; and an Oversight Committee comprising a broad cross-section of involved government agencies provided program review functions.

The IWTSP has focused on four areas: (1) how to keep Missouri River biota out of the Hudson Bay drainage; (2) if Missouri biota were introduced to the Hudson Bay drainage, would they survive and reproduce there; (3) how could or would aquatic biota from the Missouri get into the Hudson Bay drainage, and (4) what economic/ecological effects would biota transfer have in the Hudson Bay drainage.

Approximately U.S. $500,000 was spent in the five years between 1989 and 1994 on research covering the broad spectrum of these biota transfer issues from methods of water filtration and sterilization to remove pathogens (Area 1), parasites of sturgeon and paddlefish, and rainbow smelt distribution and biology (Area 2), bait bucket transfer issues (Area 3), to historic records of commercial fishing and a synthesis of biological knowledge on Lake Winnipeg (Area 4). Scientists from Universities in North Dakota and Manitoba participated in the research, training graduate students and accumulating new knowledge, much of which has been published in the primary literature. Givers and Leitch (1992) reviewed progress of the research against the 23 objectives of the major proposals. They determined that 10 objectives had been met, four were partially met, five are in progress, and four remained to be addressed. Their analysis suggested that more work remained ahead than lay behind. The goal to have in hand a conceptual model to predict effects of biota transfer has remained elusive, due to lack of sufficient information on biota of the two drainage basins. Work has been completed on both sides of the border to better understand the distributions of fishes, fish parasites, and pathogens, but much more data are required to develop the model. Work on water treatment issues has been promising but treatment at source rather than at destination will be required to reduce even more the risk of potential biota transfer.

The Garrison Diversion has been a very interesting case from the standpoint of interbasin biota transfer. A problem was identified and, under the terms of an international treaty, a solution is being developed jointly and to the satisfaction of both countries.

REFERENCES

Campbell, K.B., A.J. Derksen, R.A. Remnant, and K.W. Stewart. 1991. First specimens of the rainbow smelt, *Osmerus mordax*, from Lake Winnipeg. *Can. Field-Nat.* 105: 568–570.

Clarkson, R.N. 1992. Canadian concerns regarding the Garrison Diversion Unit. North Dakota Water Quality Symp. Proc. March 25-26, 1992. Bismarck, ND. NDSU Extension Service. pp. 168–174.

Crossman, E.J. 1976. *Quetico Fishes*. Roy. Ont . Mus. Life Sci. Misc. Publ.

Crossman, E.J. 1991. Introduced freshwater fishes: a review of the North American perspective with emphasis on Canada. *Can. J. Fish. Aquat. Sci.* 48 (Suppl.1): 46–57.

Engineering-Biology Task Group. 1994. Report of the Engineering-Biology Task Group to the Garrison Diversion Unit Joint Technical Committee on the Northwest Area Water Supply. May 1994.

Evans, D.O. and D.H. Loftus. 1987. Colonization of inland lakes in the Great Lakes region by rainbow smelt, *Osmerus mordax*: their freshwater niche and effects on indigenous fishes. *Can. J. Fish. Aquat. Sci.* 44 (Suppl. 2): 249–266.

Evans, D.O. and P. Waring. 1987. Changes in the multispecies, winter angling fishery of Lake Simcoe, Ontario. 1961-83: invasion by rainbow smelt, *Osmerus mordax*, and the roles of intra- and interspecific interactions. *Can. J. Fish. Aquat. Sci.* 44 (Suppl. 2): 182–197.

Franzin, W.G., B.A. Barton, R.A. Remnant, D.B. Wain, and S.J. Pagel. 1994. Range extension, present and potential distribution, and possible effects of rainbow smelt in Hudson Bay drainage waters of North-western Ontario, Manitoba and Minnesota. *N. Amer. J. Fish. Mgmt.* 14: 65–76.

Garrison Diversion Unit Commission. 1984. Final Report to the Secretary of the Interior, Senate Committee on Energy and Natural Resources, Senate Committee on Appropriations, House Committee on Interior and Insular Affairs, and House Committee on Appropriations. December 20, 1984.

Garrison Diversion Unit Joint Technical Committee. 1990. Garrison Diversion Unit Joint Technical Committee Report to the United States-Canada Consultative Group. November 1990.

Givers, D.R. and J.A. Leitch. 1992. Assessment of progress toward accomplishing interbasin biota transfer study program (IBTSP) objectives. North Dakota Water Quality Symp. Proc. March 25-26, 1992. Bismarck, ND. NDSU Extension Service. 218–227.

International Joint Commission. 1977. Transboundary Implications of the Garrison Diversion Unit.

International Garrison Diversion Study Board. 1976. Report to the International Joint Commission.

Kelly P.E. and J.A. Leitch. 1990. History of Missouri River and biota transfer. North Dakota Water Quality Symp. Proc. March 20-21, 1990 Fargo, ND. NDSU Extension Service. 234–251

Kinne, O. 1984. *Diseases of Marine Animals*. vol. IV, pt. 1, *Introduction*, Pisces. Biologische Anstalt, Helgoland, Hamburg.

Koel, T.M. and J.J. Peterka. 1994. Distribution and dispersal of fishes in the Red River of the North basin: a progress report. North Dakota Water Quality Symp. Proc. March 30-31, 1994 Fargo, ND. NDSU Extension Service. 159–168.

Krenz, G. 1994. Interbasin biota transfer study program: an overview and progress report. North Dakota Water Quality Symp. Proc. March 30-31, 1994 Fargo, ND. NDSU Extension Service. 128–131.

Litvak, M.K. and N.E. Mandrak. 1993. Ecology of freshwater baitfish use in Canada and the United States. *Fisheries* 18(12): 6–13.

Loftus, D.H. and P.F. Hulsman. 1986. Predation on boreal lake whitefish *Coregonus clupeaformis* and lake herring *C. artedii* by adult rainbow smelt *Osmerus mordax*. *Can. J. Fish. Aquat. Sci.* 43: 812–818.

Ludwig, H.R., D.R. Givers, and J.A. Leitch. 1994. Bait bucket movement of fish across basin boundaries. North Dakota Water Quality Symp. Proc. March 30-31, 1994 Fargo, ND. NDSU Extension Service. 150–157.

McCulloch, B.R. and K.W. Stewart. 1992. Habitat and diet comparisons between longnose dace and stonecat in the Little Saskatchewan River: interactions between a native and an invading species. North Dakota Water Quality Symp. Proc. March 25-26, 1992. Bismarck, ND. NDSU Extension Service. 176–193.

Remnant, R.A. 1991. An assessment of the potential impact of the rainbow smelt on the fishery resources of Lake Winnipeg. Master's thesis. University of Manitoba, Winnipeg.

Stewart, J. E. 1991. Introductions as factors in diseases of fish and aquatic invertebrates. *Can. J. Fish. Aquat. Sci.* 48 (Suppl. 1): 110–117.

Stewart, K.W. 1988. First collection of the weed shiner, *Notropis texanus*, in Canada. *Can. Field-Nat.* 102: 657–660.

Taylor, J.N., W.R. Courtnay, Jr., and J.A. McCann. 1984. Known impacts of exotic fish introductions in the continental United States. In W. R. Courtnay, Jr. and J.R. Stouffer, Jr. (Eds.), *Distribution, Biology and Management of Exotic Fishes*. Johns Hopkins University Press, Baltimore, MD.

U.S. Bureau of Reclamation. 1974. Final Environmental Statement, Initial Stage, Garrison Diversion Unit, Pick-Sloan Missouri Basin Program, North Dakota. INT FES 74-3. U.S. Department of the Interior.

Wain, D.B. 1993. The effects of introduced rainbow smelt (*Osmerus mordax*) on the indigenous pelagic fish community of an oligotrophic lake. Master's thesis. University of Manitoba, Winnipeg.

Wiland, L. 1995. Zebra mussel larvae found in two new Wisconsin lakes. Littoral Drift July 1995. Wisconsin Sea Grant Institute.

25 Summary of Fish Introductions Through Canals and Diversions

E.J. Crossman and B.C. Cudmore

Some fishes listed in Table 25.1 as utilizing this vector may have been included by other authors in the category "Multiple," since some have been introduced originally at one place and were able to move to others by means of canals and diversions. In this treatment, 37 forms (Table 25.1) have been judged to have been involved, to varying extents, in this vector. The reproductive status of these forms is also indicated.

Canals occur in many parts of North America, but the largest number providing access into or out of a single system are those associated with the Great Lakes (see detailed review in Chapter 23 of this volume). The well-documented destructive impact of the sea lamprey, *Petromyzon marinus*, in freshwater is based largely on its utilization of canals to reach the upper lakes. The redfin pickerel, *Esox americanus*, exists in parts of Québec, including the St.Lawrence River, as a result of the artificial connection of the Hudson River and Lake Champlain. The alewife, *Alosa pseudoharengus*, rainbow smelt, and the white perch, *Morone americana*, have all reached areas beyond their natural distribution by canals, resulting in both positive and negative impacts. The rapid, downstream spread of the pink salmon, *Oncorhynchus gorbuscha*, from an original introduction in Lake Superior to eastern Lake Ontario must have involved canals.

Meador (1996) listed important diversions of water in the U.S. as follows: the California State Water Project, Garrison Diversion Project, Lake Texoma Water Transfer Project, Santee-Cooper Diversion and Rediversion Projects, and the Tristate Comprehensive Study. All of these have potentials for the introduction of fishes. The Garrison Diversion Project (see detailed review Wright and Franzin, Chapter 24 of this volume) has the greatest potential for international complications. It would connect the Missouri River and Hudson Bay. If it is implemented, at least nine species of fishes in the Missouri River that are absent from Manitoba could invade northward. An example of the reality of this risk is the fact that one of the species listed, the rainbow smelt, has invaded the Nelson River system and in a surprisingly short time has reached the delta of the Nelson at Hudson Bay.

Species can, without additional human intervention, move only downstream in the Ogoki and Longlac Diversions. These diversions connect the Albany River (Hudson Bay System) with lakes Nipigon and Superior. Comparing species recorded for the Albany River in 1964 (Ryder et al. 1964) with those absent in lakes Nipigon and Superior suggests the possibility that threespine stickleback, rock bass *Ambloplites rupestris*, river darter *Percina shumardi*, johnny darter *Etheostoma nigrum*, Iowa darter *E. exile*, blacknose shiner *Notropis heterolepis*, mimic shiner *N. volucellus*, and finescale dace *Phoxinus neogaeus* could reach the Great Lakes by those diversions. Hubbs and Lagler (1958) suggested that the appearance of the fallfish, *Semotilus corporalis*, in Lake Superior seemed to have been the result of those connections.

TABLE 25.1
List of Fishes Introduced in Ecological North America (Excluding Mexico) Outside Their Native Range Through Canals and Diversions and Their Reproductive Status in the Nonindigenous Area(s)

Scientific Name	Common Name	Status[a]
Petromyzontidae — lampreys		
Petromyzon marinus	sea lamprey	established
Anguillidae — freshwater eels		
Anguilla anguilla	European eel	not established
Anguilla rostrata	American eel	established in most areas
Clupeidae — herrings		
Alosa pseudoharengus	alewife	established in most areas
Alosa sapidissima	American shad	established in most areas
Dorosoma cepedianum	gizzard shad	established in most areas
Cyprinidae — carps and minnows		
Cyprinella venusta	blacktail shiner	established in some areas
Cyprinus carpio	common carp	established in most areas
Lythrurus atrapiculus	blacktip shiner	established
Nocomis biguttatus	hornyhead chub	established
Notropis atherinoides	emerald shiner	established
Notropis texanus	weed shiner	established
Semotilus corporalis	fallfish	established in most areas
Ictaluridae — bullhead catfishes		
Ameiurus melas	black bullhead	established in most areas
Ameiurus nebulosus	brown bullhead	established in most areas
Ictalurus punctatus	channel catfish	established in most areas
Noturus flavus	stonecat	not established
Esocidae — pikes		
Esox americanus americanus	redfin pickerel	established in some areas
Esox niger	chain pickerel	established in most areas
Osmeridae — smelts		
Osmerus mordax	rainbow smelt	established
Salmonidae — trouts		
Oncorhynchus gorbuscha	pink salmon	established
Oncorhynchus mykiss	rainbow trout	established in some areas
Oncorhynchus tshawytscha	chinook salmon	not established
Salmo salar	Atlantic salmon	not established
Salvelinus fontinalis	brook trout	established in most areas
Belonidae — needlefishes		
Strongylura marina	Atlantic needlefish	not established
Gasterosteidae — sticklebacks		
Gasterosteus aculeatus	threespine stickleback	established in most areas
Moronidae — temperate basses		
Morone americana	white perch	established in most areas

TABLE 25.1 (CONTINUED)
List of Fishes Introduced in Ecological North America (Excluding Mexico) Outside Their Native Range Through Canals and Diversions and Their Reproductive Status in the Nonindigenous Area(s)

Scientific Name	Common Name	Status[a]
Morone chrysops	white bass	established in most areas
Centrarchidae — sunfishes		
Ambloplites rupestris	rock bass	established
Micropterus salmoides	largemouth bass	established in most areas
Pomoxis annularis	white crappie	established in most areas
Pomoxis nigromaculatus	black crappie	established in some areas
Percidae — perches		
Perca flavescens	yellow perch	established in most areas
Stizostedion vitreum	walleye	established in most areas
Cichlidae — cichlids		
Hemichromis letournauxi	African jewelfish	established in most areas
Gobiidae — gobies		
Neogobius melanostomus	round goby	established

[a] Status refers to the reproductive nature of "populations" of that fish in areas where it has been introduced, regardless of vector(s) used. Many statements on status were derived from information in the NAS database. Some notations may be open to criticism (e.g., hybrids).

There was an unusual case of potential transfer of species by water diversion almost continent wide. It involved a request to carry out a pressure test on a transcontinental pipeline using water from western Canada that would have been released at the eastern end of the pipeline into the Great Lakes system.

REFERENCES

Hubbs, C.L. and K.F. Lagler. Fishes of the Great Lakes Region. Cranbrook Institute of Science: Bulletin 26. Bloomfield Hills, MI. 1978.

Meador, M. Water Transfer Projects and the Role of Fisheries Biologists. *Fisheries.* 21(9), pp. 18–23, 1996.

Ryder, R.A., W.B. Scott, and E.J. Crossman. Fishes of Northern Ontario, North of the Albany River. *Royal Ontario Museum, Life Science Division, Contribution 60.* 1964.

Section VII

Climate Change as a Vector of Range Expansion

26 Climate Change and the Future Distribution of Aquatic Organisms in North America

J.H. Leach

SECTION INTRODUCTION

Human activity has resulted in rapid increases in global atmospheric concentrations of carbon dioxide, methane, nitrous oxide, and chlorofluorocarbons over the last 100 years. Global warming associated with increased levels of these greenhouse gases has been predicted (Kellogg and Schware, 1981) and measured (Gates, 1993). On a global scale, increases in atmospheric temperature of from 1.5 to 5.5°C have been predicted (Bolin et al., 1986). This is in the middle range of most predictions by models, for the next 100 years. If these increases actually occur, the global atmosphere will be warmer than at any time during the past 200,000 years, with temperature increase rates 15 to 40 times faster than previous natural changes (U.S. Congress, 1993).

Thermal structure in aquatic environments will follow closely the trend in atmospheric temperatures, especially at higher latitudes. This warming trend will provide opportunities for nonindigenous species to invade North America from Central and South America and other warmer regions of the world. It will also accelerate the northern shift of species already established in North America. Altered thermal regimes in more northern lakes and rivers will reduce environmental resistance to invasions by southern species, particularly through range expansion. Aquatic communities could also be altered through local extirpation of coolwater and coldwater species. These ecological disturbances could further enhance the success of invasion by nonindigenous species through reduced competition and predation.

As the largest freshwater system in North America, the Laurentian Great Lakes have attracted interest and speculation in the movement of aquatic species, especially fish species, due to climate changes. Mandrak (1989) has identified 19 species of fish from the Mississippi and Atlantic coastal basins as potential invaders of the lower Great Lakes (Michigan, Erie, Ontario) and has predicted that the invading species would greatly alter existing fish communities. He also listed 27 species of fish that could rapidly extend their ranges from the lower Great Lakes to the upper Great Lakes (Huron, Superior) under climate warming. Crossman and Cudmore have reviewed current and potential range extensions of fish species in the Great Lakes basin and other regions of North America in Chapter 27 of this volume.

Future distributions of aquatic invertebrates and plants will also be influenced by climate change. Mills et al. (1993a) found that 22% of nonindigenous invertebrate species colonizing the Great Lakes basin had originated in the Mississippi and Atlantic coastal areas. Several others, such as the Asiatic clam (*Corbicula fluminea*) had originated elsewhere, invaded the western part of North America, and migrated eastward and northward to the Great Lakes basin. Distribution of the Asiatic clam is limited to the southern areas of the Great Lakes basin (primarily in heated effluents) but this range could be extended under climate warming. The zebra mussel (*Dreissena polymorpha*), which was a direct ballast water inoculation from Eurasia to the Great Lakes, has migrated southward to the Gulf of Mexico. Northward distribution of the mussel is expected to be assisted by climatic warming provided that other ecological requirements (i.e., calcium, pH) are favorable (Strayer, 1991; Neary and Leach, 1992).

Only 11% of nonindigenous species of aquatic flora now found in the Great Lakes basin arrived from the Mississippi and Atlantic coastal areas; the majority arrived from Eurasia (Mills et al., 1993a). However, plants respond rapidly to climatic changes that permit them to survive winter temperatures, and many native and foreign species now found in the southern part of the continent are expected to expand their geographic ranges under warmer conditions. Some of these will be ecologically disruptive (as they are now in Florida) in new ecosystems. Incidences of diseases and parasites could also increase under climate change as vector organisms extend their ranges northward.

Predictions of invasions and colonization by nonindigenous species and their possible damage to ecosystem health is difficult (Mills et al., 1993b; Pimm, 1991; Leach, 1995). If climate change leads to ecological disruption as forecast, the task of predicting the probability of successful invasions by nonindigenous species and their potential damage becomes more difficult. Moreover, the migration of organisms into new ecosystems due to climate warming adds to the complexity of defining and managing the invasions through policy development and legislation (Leach et al., in press).

REFERENCES

Bolin, B., J. Jager, and B.R. Doos. The greenhouse effect, climatic change, and ecosystems. In *The Greenhouse Effect, Climatic Change, and Ecosystems*. B. Bolin et al. (Eds.). John Wiley and Sons, Toronto, 1986.

Gates, D.M. *Climate Change and Its Biological Consequences*. Sinauer Associates, Sunderland, MA.

Kellogg, W.W. and R. Schware. *Climate Change and Society, Consequences of Increasing Atmospheric Carbon Dioxide*. Westview Press, Boulder, CO.

Leach, J.H. Nonindigenous species in the Great Lakes: were colonization and damage to ecosystem health predictable? *J. Aquatic Ecosystem Health* 4:117–128. 1995.

Leach, J.H., E.L. Mills, and M.A. Dochoda. Nonindigenous species in the Great Lakes: Ecosystem impacts, binational policies, and management. In *Great Lakes Fisheries Policy and Management — a Binational Perspective*. Taylor, W. (Ed.), Michigan State University Press, Lansing, MI (in press).

Mandrak, N.E. Potential invasion of the Great Lakes by fish species associated with climatic warming. *J. Great Lakes Res.* 15:306–316. 1989.

Mills, E.L., J.H. Leach, J.T. Carlton, and C.L. Secor. Exotic species in the Great Lakes: a history of biotic crises and anthropogenic introductions. *J. Great Lakes Res.* 19:1–54. 1993a.

Mills, E.L., J.H. Leach, C.L Secor, and J.T. Carlton. What's next? The prediction and management of exotic species in the Great Lakes (report of the 1991 workshop). Great Lakes Fishery Commission, Ann Arbor, MI. 1993b.

Neary, B.P. and J.H. Leach. Mapping the potential spread of the zebra mussel (*Dreissena polymorpha*) in Ontario. *Can. J. Fish. Aquat. Sci.* 49:406–415. 1992.

Pimm, S.L. *The Balance of Nature?* The University of Chicago Press, Chicago. 1991.

Strayer, D.L. Projected distribution of the zebra mussel, *Dreissena ploymorpha*, in North America. *Can. J. Fish. Aquat. Sci.* 48:1389–1395. 1991.

U.S. Congress. *Harmful Nonindigenous Species in the United States*. Office of Technology Assessment F-565, U.S. Government Printing Office, Washington, D.C., p. 391. 1993.

27 Invasive Habits of Fishes, Global Warming, and Resulting Range Extensions

E.J. Crossman and B.C. Cudmore

As part of the attempt to standardize the terminology for fishes, associated with the new and broader definition given by the American Fisheries Society of "Introduced," Hocutt (1985) suggested a new term: "Invasive." This term was to include species that expand their range either as a consequence of a natural phenomenon or as a result of human action, deliberate or inadvertent. The phenomena were to include natural dispersal, canals, water diversions, etc. Unfortunately, that definition will be in conflict with the much broader one adopted by the 1996 National Invasive Species Act of the U.S. That new act replaces and expands the Nonindigenous Aquatic Nuisance Prevention and Control Act of 1990. In addition to the conflict in the word "Invasive," the new act seems, by definition, to place very little concern for invasions by means other than ballast water, whereas the previous act seemed to provide opportunity for the control of a much broader group of nuisance species. For this discussion, the term *invasive* will be limited to "natural" range extensions, those without direct intervention by humans.

In North America most of the larger river systems that provide pathways for dispersal of aquatic organisms trend, generally, north–south. The long-term warming of the environment, possibly speeded by greenhouse effect, etc., would seem to be increasing the rate of dispersal northward. All of the so-called indigenous freshwater fishes of the Great Lakes and of Canada are the result of postglacial invasion or reinvasion, principally following the Wisconsinan Glaciation. In a recent presentation, Dr. G.R. Smith of the University of Michigan indicated that the post-Wisconsinan invasion may have been only the most recent of as many as 24 such movements of species into the area now occupied by the Great Lakes and Canada. More recent, unaided invasion of Canadian waters of species indigenous in the U.S., or introduced there, could be said to be part of the natural process that started with the postglacial reinvasions. The rate of invasion has, however, increased. Prior to the 1950s either the rate was slower, or the extent to which such "new" fishes were detected and reported was very much lower. Commercial fishermen are quick to spot something unknown to them previously. After 1950 the rate of reporting such unknown fishes increased significantly (Table 27.1).

Several of the species may have a variety of impacts on the Canadian fauna and habitats. As a result, invasive species are of particular concern, at least to Canada and possibly Alaska. In a similar way, the fauna of the U.S. could be affected by the northward movement of Mexican species. For that reason, we feel it appropriate to use the term "Invasive" in this somewhat unique way.

We have judged that 62 species have gained their present distribution, at least in part, by an invasive process (Table 27.2). That rather large number, when compared to numbers associated with other vectors, includes fishes that appear in one or more of the other vectors also. The list includes introduced species that reached another location, or several others, by this process. One

TABLE 27.1
Fishes Native to the United States that Have Invaded the Canadian Waters of the Great Lakes

Scientific Name	Common Name	Lake Arrived	Date Arrived
Catostomidae — suckers			
Erimyzon sucetta	lake chubsucker	Erie	1949
Ictiobus cyprinellus	bigmouth buffalo	Erie	1957
Ictiobus niger	black buffalo	Erie	1978
Minytrema melanops	spotted sucker	Erie	1962
Ictaluridae — bullhead catfishes			
Noturus stigmosus	northern madtom	St. Clair	1963
Pylodictus olivaris	flathead catfish	Erie	1978
Centrarchidae — sunfishes			
Chaenobyrttus gulosus	warmouth	Erie	1966
Lepomis humilis	orangespotted sunfish	tribs. to Erie	1980

example is the rudd, *Scardinius erythrophthalmus*. That European species had been introduced into an inland lake in New York State (Smith 1985) but seemed not to spread from there. More recently it was imported into western New York State, from fish farms in the southern states, for bait. From there it escaped into the St. Lawrence River, and only after several years dispersed into Ontario waters, were it has become established (Crossman et al. 1992). Another example is the warmouth, *Chenobryttus gulosus*, which was known to occurr on the southern side of the Great Lakes but was unknown in Canada. A specimen kept, as a rock bass, in an aquarium in a Provincial Park on Lake Erie was the first indication that an established population had developed on the Ontario side of the lake (Crossman et al. 1996). As yet, the species has been reported in Ontario only in a limited area of western Lake Erie.

As we anticipate an increasing number of invasions in Canada, possibly influenced by global warming, we should look for species that are likely to invade in any collections made in Canada. A list of the likely invaders (Mandrak 1989; Mandrak and Crossman 1992) made on the basis of proximity and environmental requirements of species not presently known in Ontario includes 41 species (Table 27.3).

These fishes might "move north" in response to increasing habitat temperatures (see also Meisner et al. 1987; Regier and Meisner 1990). Meisner (1990a,b) provided evidence for loss of habitat and reduction of the southern limit of distribution of the brook trout as a result of climatic warming.

A similar interpretation was provided by Rahel et al. (1996), in which they calculated the extent to which habitat for coldwater species would be lost in Wyoming as a result of temperature increases of 1°C, 3°C, and 5°C. The areas lost to coldwater species would be open to invasion by a nonindigenous community of coolwater species. The introduced sea lamprey was used (Holmes 1990) as an indicator of ecological response to climate change.

The presence in British Columbia of tench, *Tinca tinca*, yellow perch, smallmouth bass and largemouth bass, has always been attributed to unaided dispersal north after introductions in Washington State (Carl et al. 1967). Alberta and Saskatchewan reported that there was no record of species having moved north from Montana and North Dakota. One of the most interesting examples of an invasive species is the rainbow smelt in Manitoba. After being introduced in North Dakota that species moved north in the Red River, reaching Lake Winnipeg in 1990. It was quickly captured in locations farther and farther downstream in the Nelson River, reaching the delta of the

TABLE 27.2
Fishes Introduced in Ecological North America (Excluding Mexico)
Outside Their Native Range Through Invasiveness or Range Extensions
and Their Reproductive Status in the Nonindigenous Area(s)

Scientific Name	Common Name	Status[a]
Petromyzontidae — lampreys		
Petromyzon marinus	sea lamprey	established
Clupeidae — herrings		
Alosa aestivalis	blueback herring	established
Alosa pseudoharengus	alewife	established
Dorosoma cepedianum	gizzard shad	established in most areas
Dorosoma petenense	threadfin shad	established in some areas
Cyprinidae — carps and minnows		
Campostoma anomalum	central stoneroller	established in some areas
Ctenopharyngodon idella	grass carp	established in many areas
Cyprinella galactura	whitetail shiner	established
Cyprinella spiloptera	spotfin shiner	established in some areas
Lythrurus ardens	rosefin shiner	established
Nocomis leptocephalus	bluehead chub	established
Notemigonus crysoleucas	golden shiner	established in some areas
Notropis buchanani	ghost shiner	established in some areas
Notropis heterodon	blackchin shiner	established in some areas
Notropis photogenis	silver shiner	established in some areas
Notropis rubricroceus	saffron shiner	established
Notropis texanus	weed shiner	established
Phoxinus oreas	mountain redbelly dace	established in some areas
Scardinius erythrophthalmus	rudd	established in most areas
Tinca tinca	tench	established in most areas
Catostomidae — suckers		
Erimyzon sucetta	lake chubsucker	established
Ictiobus cyprinellus	bigmouth buffalo	established in some areas
Minytrema melanops	spotted sucker	established in some areas
Moxostoma erythrurum	golden redhorse	established
Ictaluridae — bullhead catfishes		
Noturus flavus	stonecat	not established
Noturus gyrinus	tadpole madtom	established in most areas
Pylodictis olivaris	flathead catfish	established in most areas
Esocidae — pikes		
Esox americanus americanus	redfin pickerel	established in some areas
Esox masquinongy	muskellunge	established in some areas
Esox niger	chain pickerel	established in most areas
Osmeridae — smelts		
Osmerus mordax	rainbow smelt	established
Salmonidae — trouts		
Oncorhynchus mykiss	palomino trout	not established
Oncorhynchus gorbuscha	pink salmon	established
Salmo trutta	brown trout	some natural reproduction

TABLE 27.2 (CONTINUED)
Fishes Introduced in Ecological North America (Excluding Mexico) Outside Their Native Range Through Invasiveness or Range Extensions and Their Reproductive Status in the Nonindigenous Area(s)

Scientific Name	Common Name	Status[a]
Salmo salar	Atlantic salmon	not established
Fundulidae — topminnows and killifishes		
Fundulus notatus	blackstripe topminnow	estabished in some areas
Fundulus zebrinus	plains killifish	established in most areas
Poeciliidae — livebearers		
Gambusia affinis	western mosquitofish	established in some areas
Cottidae — sculpins		
Cottus asper	prickly sculpin	established
Moronidae — temperate basses		
Morone americana	white perch	established in most areas
Morone chrysops	white bass	established in most areas
Centrarchidae — sunfishes		
Ambloplites rupestris	rock bass	established
Chaenobryttus gulosus	warmouth	established in most areas
Lepomis auritus	redbreast sunfish	established
Lepomis humilis	orangespotted sunfish	established
Lepomis microlophus	redear sunfish	established in most areas
Micropterus dolomieu	smallmouth bass	established in most areas
Micropterus salmoides	largemouth bass	established in most areas
Pomoxis nigromaculatus	black crappie	established in some areas
Percidae — perches		
Etheostoma edwini	brown darter	established in some areas
Etheostoma zonale	banded darter	established
Gymnocephalus cernuus	ruffe	established
Perca flavescens	yellow perch	established in most areas
Percina macrolepida	bigscale logperch	established
Stizostedion canadense	sauger	established in most areas
Stizostedion vitreum	walleye	established in most areas
Cichlidae — cichlids		
Oreochromis aureus	blue tilapia	established in some areas
Oreochromis mossambica	Mozambique tilapia	established in many areas
Oreochromis niloticus	Nile tilapia	established in some areas
Gobiidae — gobies		
Acanthogobius flavimanus	yellowfin goby	established
Neogobius melanostomus	round goby	established
Proterorhinus marmoratus	tubenose goby	established

[a] Status refers to the reproductive nature of "populations" of that fish in areas where it has been introduced, regardless of vector(s) used. Many statements on status were derived from information in the NAS database. Some notations may be open to criticism (e.g., hybrids).

TABLE 27.3
Potential Fish Invaders into the Great Lakes and Canada with Global Warming

Scientific Name	Common Names
Acipenseridae — sturgeons	
Scaphyrhynchus platorynchus	shovelnose sturgeon
Lepisosteidae — gars	
Lepisosteus oculatus	spotted gar
Lepisosteus platostomus	shortnose gar
Hiodontidae — mooneyes	
Hiodon alosoides	goldeye
Cyprinidae — carps and minnows	
*Exoglossum laurae	tonguetied minnow
*Notropis amblops	bigeye chub
*Notropis buccatus	silverjaw minnow
*Notropis texanus	weed shiner
*Phenacobius mirabilis	suckermouth minnow
*Phoxinus erythrogaster	southern redbelly dace
*Semotilus corporalis	fallfish
Cyprinella venusta	blacktail shiner
Cyprinella whipplei	steelcolor shiner
Hybognathus placitus	plains minnow
Notropis blennius	river shiner
Notropis chalybraeus	ironcolor shiner
Notropis dorsalis	bigmouth shiner
Notropis nubilus	Ozark minnow
Catostomidae — suckers	
Carpiodes carpio	river carpsucker
Cycleptus elongatus	blue sucker
Moxostoma carinatum	river redhorse
Esocidae — pikes	
*Esox americanus americanus	redfin pickerel
Esox americanus vermiculatus	grass pickerel
Esox niger	chain pickerel
Fundulidae — topminnows and killifishes	
Fundulus chrysotus	golden topminnow
Fundulus olivaceus	blackspotted topminnow
Fundulus sciadicus	plains topminnow
Moronidae — temperate basses	
*Morone americana	white perch
Centrarchidae — sunfishes	
*Enneacanthus gloriosus	bluespotted sunfish
*Lepomis megalotis	longear sunfish
*Lepomis microlophus	redear sunfish
Acantharcus pomotis	mud sunfish
Centrarchus macropterus	flier
Elassoma zonatum	banded pygmy sunfish

TABLE 27.3 (CONTINUED)
Potential Fish Invaders into the Great Lakes and Canada with Global Warming

Scientific Name	Common Names
Enneacanthus chaetodon	blackbanded sunfish
Enneacanthus obesus	banded sunfish
Lepomis symmetricus	bantam sunfish
Percidae — perches	
Etheostoma spectabile	orangethroat darter
Etheostoma variatum	variegate darter
Etheostoma zonale	banded darter
Gobiidae — gobies	
Proterorhinus marmoratus	tubenose goby

Note: Fishes marked with an (*) are based on proximity alone, others on proximity and ecological requirements.

Source: Taken from Mandrak (1989) and Mandrak and Crossman (1992).

river at Hudson Bay in 1996 (Campbell et al. 1991; Remnant et al. 1997). Franzin et al. (1994) described possible environmental effects of the arrival of the rainbow smelt in that watershed.

A survey of texts on the fishes of Arizona, New Mexico, Texas, and California revealed that while there are species that have extended their range westward after being introduced in an adjacent state, there seemed little evidence of invasions north from Mexico. Several species in the Yaqui River Basin in Arizona (e.g., *Gila purpurea, Catostomus bernadini*) that occur also in Mexico may be examples of species that invaded some time ago (see especially Dill and Cordone 1997 and Minckley 1973).

REFERENCES

Campbell, K.B., A.J. Derksen, R.A. Remnant, and K.W. Stewart. First Specimens of the Rainbow Smelt, *Osmerus mordax*, from Lake Winnipeg, Manitoba. *Canadian Field-Naturalist.* 105, pp. 568–570, 1991.

Carl, G.C., W.A. Clemens, and C.C. Lindsey. *The Fresh-water Fishes of British Columbia.* Handbook No. 5, 4th ed., British Columbia Provincial Museum, Victoria British Museum, 1967.

Crossman, E.J., E. Holm, R. Cholmondeley, and K. Tuininga. First Record for Canada of the Rudd, *Scardinius erythrophthalmus*, and notes on the Introduced Round Goby, *Neogobius melanostomus. Canadian Field-Naturalist.* 106, pp. 206–209, 1992.

Crossman, E.J., and J. Houston, and R.R. Campbell. Status of the Warmouth *Chaenobryttus gulosus*, in Canada. *Canadian Field-Naturalist.* 110, pp. 495–500, 1996.

Dill, W.A., and A.J. Cordone. History and Status of Introduced Fishes in California, 1871-1996, *California Dept. Fish and Game, Fish Bulletin* 178, 1997.

Franzin, W.G., B.A. Barton, R.A. Remnant, D.R. Wain, and S.J. Pagel. Range Extension, present and potential distribution, and possible effects of rainbow smelt in Hudson Bay drainage waters of Northwestern Ontario, Manitoba and Minnesota. *North American Journal of Fisheries Management* 14, pp. 65–76, 1994.

Hocutt, C.H. Standardization of Terminology — Politics or Biology. *Fisheries* 10(3), p. 49, 1985.

Holmes, J.A. Sea lamprey as an Early Indicator of Ecological Response to Climate Change in the Great Lakes. *Transactions of the American Fisheries Society.* 119, pp. 292–300, 1990.

Mandrak, N.E. Potential Invasion of the Great Lakes by Fish Species Associated with Climatic Warming. *Journal of Great Lakes Research.* 15, pp. 306–316, 1989.

Mandrak, N.E., and E.J. Crossman. *A Checklist of Ontario Freshwater Fishes Annotated with Distribution Maps.* Royal Ontario Museum, Toronto, ON, 1992.

Meisner, J.D. Effects of Climate Warming on the Southern Margins of the Native Range of the Brook Trout, *Salvelinus fontinalis. Canadian Journal of Fisheries and Aquatic Sciences.* 47, pp. 1065–1070, 1990a.

Meisner, J.D. Potential Loss of Thermal Habitat for Brook Trout, Due to Climate Warming, in Two Southern Ontario Streams. *Transactions of the American Fisheries Society.* 119, pp. 282–291, 1990b.

Meisner, J.D., J.L. Goodier, H.A. Regier, B.J. Shuter, and J.W. Christie. An assessment of the effects of climate warming on Great Lakes Basin fishes. *J. Great Lakes Res.* 13:340–352. 1987.

Minckley, W.L. *Fishes of Arizona.* Arizona Game and Fish Department, Phoenix, AZ, 1973.

Rahel, F.J., C.J. Keleher, and J.L. Anderson. Potential Habitat Loss and Population Fragmentation in the North Platte River Drainage of the Rocky Mountains: Response to Climate Warming. *Limnology and Oceanography.* 41, pp. 1116–1123, 1996.

Regier, H.A. and J.D. Meisner. Anticipated effects of climate change on freshwater fishes and their habitat. *Fisheries* 15(6):10–15. 1990.

Remnant, R.A., P.G. Faveline, and R.L. Bretecher. Range extension of the Rainbow Smelt, *Osmerus mordax,* in the Hudson Bay drainage of Manitoba. *Canadian Field-Naturalist.* 111, pp. 660–662, 1997.

Smith, C.L. The inland fishes of New York State. New York State Department of Environmental Conservation, Albany, NY. 1985.

Section VIII

Why Some Introductions Succeed While Others Fail

28 Modeling the Invasion Process

Michael M. Fuller and James A. Drake

SECTION INTRODUCTION

Predictive ability has always been the benchmark against which the utility of a science is measured. Ecology is no exception (Mayr 1988, Grubb 1989, Peters 1991, Drake et al. 1998). For those who manage biological systems, the ability to forecast system vulnerability and reaction to species introductions is an increasing priority. If resource management is to move beyond triage and restoration, ecologists must develop the tools needed to understand and predict the reaction of complex ecological systems to alien species. Easier said than done; the very nature of biological reality has defied prediction at all but the simplest levels. Ecological systems are the most complex of all systems found in nature and yield only superficially to traditional analytical approaches. What does one do with a system where the *mean* is meaningless but the *variance* meaningful, pieces self-organize producing emergent phenomena, and there are far too many parts to model explicitly?

Strategies for prediction generally fall into one of two categories. They may be based on observation, in which case forecasts of the future depend on the existence of a recurrent theme. For example, the accuracy of estimates of an organism's ability to invade a particular system is enhanced by knowledge of the organism's *biology* and the system's *structure*. This approach has the potential to provide very accurate predictions when species biology and system characteristics (e.g., community structure, hydrology, and percolation regime) are well known. Its benefit, however, is also its greatest drawback — it is site-specific. Lacking detailed knowledge of the species involved, the accuracy of forecasts drops precipitously. Thus the effectiveness of the empirical approach is proportional to how much is known about the species and system in question. Even with detailed information, however, it has proven difficult to predict which species will invade, even under the highly controlled conditions of the laboratory (Drake 1991).

Model-based prediction rests on theoretical constructs that may be quantitative, qualitative, or speculative (e.g., see Shigesada and Kawasaki 1997). The complexity of ecological systems provides a serious if not insurmountable challenge to analytical solution. Hence numerical simulation has been employed to analyze possible scenarios involving different levels of trophic complexity. And patterns of interaction and connectance are occasionally integrated with ecosystem processes. Here, the potential outcome of a species introduction to a given system is deduced from a set of assumptions — axioms — concerning how species interact with each other and their environment (Drake et al. 1998). One then turns the crank on the computer, and with any luck, some efficacy is realized.

The goal of the model-based approach is the same as the empirical approach: to examine the current situation as well as predict general patterns for a broad range of systems and species sets. But what models offer in generality, they lack in precision. Model-based prediction does require some knowledge of species' biology and interactions, but because the probable outcomes it seeks to identify are general, this strategy needs less detailed species information. This is not to say that modeling is unsuited to predicting outcome in a particular system. But the level of precision depends on the level of knowledge of the system. Models designed to examine specific ecological systems (e.g., Florida Everglades) can perform well, but they are rarely generalizable. More general models address more general aspects of structure and organization such that they should not be used to predict the outcome of a specific case. Modeling's greatest asset is its ability to explore a range of possible scenarios and alternative communities that differ in interaction signs, interaction strengths, and number of trophic connections. This enlarges the analysis base to include potential management practices that can then be evaluated in the context of the model system.

That both strategies can be put to good use in the analysis of invasion dynamics in real systems is illustrated in the following two chapters. Brown et al. (Chapter 29) use an empirical approach to present evidence that extremes of flow dynamics are important to invasion success in fluctuating environments. Where large fluctuations in flow occur seasonally (California) or unpredictably (Hawaii), native faunas tend to be species poor and restricted to species well adapted to extreme variation in flow that results from seasonal cycles or stochastic spates. Few species it seems, including exotics, are able to establish self-sustaining populations under these conditions. In such instances, the physical hostility of extreme flow regimes appears to be the dominant barrier to successful establishment of nonindigenous species — a condition Baltz and Moyle (1993) termed "environmental resistance." As a consequence, resistance to invasion appears to be greater in unaltered streams where large fluctuations in flow occur naturally.

In contrast, habitat alterations that regulate flow regimes and mitigate the effects of storm events foster conditions suitable for a broader category of potential invaders. Once the physical environment has been tamed by human intervention, the importance of biotic interactions as a filter to community structure increase. Unfortunately, this may tip the scales in favor of exotics, particularly piscivorous species, because species indigenous to depauparate faunas tend to have narrower tolerance ranges and are thus less competitive and more vulnerable to predation. Brown et al. (Chapter 29) pin their hope for predictability in an expanding base of ecological studies of individual species. By cataloging the life-history traits, foraging modes, and range of environment tolerance for difference species, and comparing these traits to the community profile and hydrological conditions of specific systems, it may be possible to distinguish potential matches of exotics to receiving system, and thus the probability of successful establishment.

In sharp contrast to the above approach, Li et al. (Chapter 30) apply digraph models and loop analysis to the problem of forecasting a community's reaction to an invading species. The application of loop analysis to ecological communities was pioneered by Levins (1975). It is a qualitative approach that involves analysis of community stability and equilibria using linear algebra. Li et al. first demonstrate these techniques using general hypothetical situations that involve predator–prey and self-regulation relations. They then apply them to two well-documented invasion events: the invasion of Flathead Lake by the mysid shrimp (*Mystis relicta*) and the introduction of Nile perch (*Lates niloticus*) into East African Rift Lakes. Predictions of future complexity (richness and connectance) and trophic structure of the as-yet-unresolved community reactions of these systems are provided by their models. A feature of their models is that the stability of systems decreases as species richness increases. Most simulated introductions are deleterious to the stability of the status quo. Because they seek only qualitative predictions of community reaction, Li et al.'s approach does not require specific knowledge of transfer coefficients (interaction strength), only the sign of such interactions. This enhances the utility of their approach to complex systems for which detailed information is lacking.

Are the examples presented in these chapters generalized across systems? The impact of the pattern and frequency of high flows on invasion success has been previously shown for streams with extreme dynamics (Strange et al. 1992, Moyle and Light 1996). In such cases, regulation of flows to more consistent levels has been shown to increase invasion success. Variance is meaningful and the mean of questionable value. But most of the case studies for which extreme flow dynamics have been shown to be important (California, Hawaii, and New Mexico) encompass but a subset of the range of flow conditions encountered. While such results may underscore the importance of environmental resistance, it should be noted that the examples are from systems of species-poor recipient communities. Less variable lotic systems, such as characterize many streams and rivers east of the North American continental divide, tend to have richer fish faunas. It is difficult to weigh the relative importance of biotic factors against abiotic features when they are so intimately intertwined.

We believe, as Li et al. have shown (Chapter 30), that models that can function without detailed quantitative information may be among the most useful and cost-effective tools for prediction. This does not mean that detailed biological information is immaterial; quite the opposite is true. Basic population-level information, if not individual-based information, interactions among species, couplings to nutrient pathways, and the operation of emergent properties all provide essential information which can be exploited by modelers. The reduction of assumption by solid biological data greatly enhances model performances.

To conclude, we offer a few observations about the nature of biological invasions and ecological systems that we believe merit consideration for future modeling and empirical programs. Much of the quandary over invasion biology, we believe, is a direct result of an ineffectual paradigm coupled with rampant misuse of concepts and terminology. For example, it should come as no surprise that attempts to examine community resistance and vulnerability to invasion have produced little generality when the community is conceived as a set of fishes, zooplankton, or birds on one hand and a detailed food web on the other hand.

Guilds, taxonomically defined units, or parts of a system of specific interest *are not* communities. Attempts to generalize invasion into such units of the community will necessarily fail.

1. Ecological systems are spatially and temporally extended. Models and experiments that average over space and time, if considering space at all, miss dynamics and processes that operate patchily across the ecological landscape. A lake is a metacommunity with vast differences in organization and levels of coupling between metacomponents. The reductionist program has seriously impeded progress. For examples of models that address system patchiness, see Levin (1994) and Durrett and Levin (1994).

2. Ecological systems are nonequilibrium structures. Yes, if one looks out the window the forest is much the same today as it was yesterday, but human-imposed frames of reference have led us astray (Samuels and Drake 1997).

3. Chance and timing play a huge role in the production of ecological structure. This is not to say that ecological systems are a hodge-podge; quite the contrary. While chance plays a powerful role, mechanisms readily kick-in when species arrive. But fluctuations in source populations and factors important to dispersal success, such as seasonal climate (amount and timing of rainfall, temperature levels, degree of isolation, etc.), strongly influence who arrives and in what order and abundance. As such, the mean is meaningless but the variance meaningful.

4. The regional phylogenetic make-up in which a community is embedded influences its resistance to foreign invaders by placing constraints on the traits of its members, and the functional roles they play (Brooks and McLennan 1991, 1993). Diversity is more than the number of species and their relative abundance. The functional abilities and life histories that characterize the historical set of potential members are coevolved. Thus, a community's internal structure is assembled from its phylogenetic stock. As species from different regions are introduced to local communities, differences in phylogenetic constraints to niche and function between invader and existing species set can influence invasion success and the impact on the rest of the community.

Invasion forecasting — who will be successful and what impact they will have — is complicated by the complexity and historical nature of ecological communities. Modeling promises to illuminate species and community traits that are important to invasion success, but it is unlikely that models will provide the precision coveted by resource managers. Detailed knowledge of species biology and the biotic and abiotic conditions of a particular system offer hope for managers — but only in those rare situations where such knowledge is available. Used in combination, perhaps some degree of generality and precision can be reached. We emphasize the need to acknowledge the stochastic and historical nature of ecological systems when making forecasts of invasions. Regardless of the approach used, the accuracy of predictions can only increase when such factors are taken into account.

REFERENCES

Baltz, D.M. and P.B. Moyle. Invasion resistance to introduced species by a native assemblage of California stream fishes. *Ecological Applications* 3:246–255. 1993.

Brooks, D.R. and D.A. McLennan. *Phylogeny, Ecology and Behavior.* University of Chicago Press. 1991.

Brooks, D.R. and D.A. McLennan. Historical ecology: examining phylogenetic components of community evolution. In R.E. Ricklefs and D. Schluter (Eds.), *Species Diversity in Ecological Communities.* University of Chicago Press. 1993.

Drake, J.A., C. Zimmerman, T. Puruker, and C. Rojo. On the nature of the assembly trajectory. In, E. Weiher and P. Keddy (Eds.), *The Assembly of Ecological Communities.* Cambridge University Press. In press.

Drake, J.A. Community assembly mechanics and the structure of an experimental species ensemble. *American Naturalist* 137:1–26. 1991.

Durrett, R. and S. Levin. Stochastic spatial models: a user's guide to ecological applications. *Phil. Trans. R. Soc. Lond.* B 343:329–350. 1994.

Grubb, P.J. Toward a more exact ecology: a personal view of the issues. In P.J. Grubb and J.B. Whittaker (Eds.), *Toward A More Exact Ecology.* Blackwell Scientific, Oxford. 1989.

Levins, R. Evolution in communities near equilibrium. In M. L. Cody and J. M. Diamond (Eds.), *Ecology and Evolution of Communities.* The Belknap Press of Harvard University. 1975.

Levin, S. Patchiness in marine and terrestrial systems: from individuals to populations. *Phil. Trans. R. Soc. Lond.* B 343:99–103. 1994.

Mayr, E. *Toward A New Philosophy of Biology: Observations of an Evolutionist.* Harvard University Press, Cambridge, MA. 1988.

Moyle, P.B. and T. Light. Fish invasions in California: do abiotic factors determine success? *Ecology* 77:1666–1670. 1996.

Peters, R.H. *A Critique for Ecology.* Cambridge University Press, Cambridge, UK. 1991.

Samuels, C. and J.A. Drake. Divergent perspectives on community convergence. *Trends in Ecology and Evolution* 12:427–432. 1997.

Shigesada, Nanako and Kohkichi Kawasaki. *Biological Invasions: Theory and Practice.* Oxford University Press, Oxford. 1997.

Strange, E.M., P.B. Moyle and T.C. Foin. Interactions between stochastic and deterministic processes in stream fish community assembly. *Environmental Biology of Fishes* 36:1–15. 1992.

29 Success and Failure of Nonindigenous Aquatic Species in Stream Systems: Case Studies from California and Hawaii

Larry R. Brown, Anne M. Brasher,
Bret C. Harvey, and Melody Matthews

INTRODUCTION

Aquatic ecosystems around the world are being altered at an alarming rate (Dudgeon, 1992; Moyle and Leidy, 1992; Allan and Flecker, 1993). The reasons for these changes are varied and interactive, and in any particular ecosystem may include habitat alteration, overharvest of species, and the introduction of nonindigenous species. Nonindigenous species can have many effects on native species including competition, predation, hybridization, and the introduction of new parasites and diseases. Unfortunately, ecologists and resource managers are often unable to predict the outcome of the addition of species to aquatic ecosystems (Li and Moyle, 1981; Moyle et al., 1986; Lodge 1993 a,b). In many cases the success of nonindigenous species is associated with past or ongoing human-induced habitat alterations, making it difficult to assess the relative effects of the two factors on the receiving ecosystem.

One way to measure the success and effects of species introductions in particular is to study the success of nonindigenous species in systems that have not been greatly altered by human activities. Unfortunately, few stream systems have escaped human alteration. The purpose of this paper is to examine the success and effects of nonindigenous aquatic species in stream ecosystems with relatively unaltered flow regimes where other human-induced factors have not played a major role in the success or failure of the invasion. We present two case studies based on our own research. The first involves the introduction of a number of exotic fishes into the Eel River system of northern California. The second examines the relative success of nonindigenous aquatic species in altered and unaltered streams of the Hawaiian Islands. We also discuss these two examples in the context of other studies and attempt to derive some general conclusions about the characteristics of successful and unsuccessful invaders in relatively unaltered stream ecosystems.

The broad hydrologic features of streams appear to be very important in determining species composition of fish communities (Poff and Ward, 1989; Poff and Allan, 1995). Clearly, many kinds of human alterations of stream ecosystems are possible but dams and diversions and the resulting changes in hydrologic regime affect many aspects of stream systems (Ligon et al., 1995; Bain et al., 1988; Cushman, 1985). Understanding the reasons for the success or failure of introduced species in stream systems will become increasingly important for the preservation and management of the few remaining relatively pristine aquatic ecosystems (e.g., Moyle and Yoshiyama, 1994).

The linear, interconnected nature of stream ecosystems, the mobility of many aquatic species, and the possibility of accidental or illegal introductions make likely the invasion by nonindigenous species of even the most closely managed stream preserves (Moyle and Sato, 1991).

FISH INVASIONS INTO THE EEL RIVER SYSTEM

Our studies of the fishes of the Eel River system in northern California were prompted by the unauthorized introduction, in about 1979, of Sacramento squawfish (*Ptychocheilus grandis*). The major concern of resource management agencies was the possible effects of this predatory cyprinid on the anadromous fisheries resources of the drainage, particularly chinook salmon (*Oncorhynchus tshawytscha*), coho salmon (*O. kisutch*), and steelhead rainbow trout (*O. mykiss*). A literature survey revealed that the Eel River had undergone some significant environmental perturbations, including the introduction of a number of nonindigenous species of fish. In this section we synthesize what has been learned to date concerning the success of various introduced species in the Eel River, the effects of the successful introductions on the aquatic ecosystem, and how the present mixture of native and introduced fishes interact to form fish assemblages.

The Eel River system on the north coast of California is the third-largest drainage in the state (Figure 29.1). Most precipitation falls in winter and spring. Because there is little snowpack, precipitation runs off quickly, resulting in high flows following storms. The only major water development in the system is Pillsbury Reservoir, near the headwaters of the mainstem Eel River. Water stored in the reservoir is diverted out of the drainage at a small dam 19 km downstream. Because the reservoir is small and the flow manipulations affect only a small proportion of the drainage, the natural hydrologic regime remains basically intact. The major hydrologic effect on the mainstem Eel River immediately downstream of the diversion dam is some reduction of peak winter and spring flows. The summer low flow regime is probably very similar to natural conditions. The drainage has not escaped other human influences. A combination of human activities, such as logging and grazing, that accelerated erosion processes, combined with major storms in the 1950s and 1960s, led to major inputs of gravel and fine sediment to channels throughout the entire drainage. At present, the stream channels are recovering from these perturbations through both natural processes and stream restoration activities. These habitat changes do not appear to have had a major influence on the success or failure of invading species.

The native fish fauna of the Eel River drainage is dominated by anadromous salmonid species (*Oncorhynchus* spp.) and anadromous Pacific lamprey (*Lampetra tridentata*) and includes only one obligate freshwater species, the Sacramento sucker (*Catostomus occidentalis*). The populations of all of the anadromous species have apparently declined, though only anecdotal evidence of decline exists for many species (Brown and Moyle, 1997). By 1996, 16 species introductions had been documented in the drainage, 10 into Pillsbury Reservoir in particular (Table 29.1). Of the introduced species, 12 are still present in the system, though the status of fathead minnow (*Pimephales promelas*) is unknown because its presence has been established on the basis of just a few collections. Brown and Moyle (1997) also classified the status of white catfish (*Ameiurus catus*) as unknown, but recent collections of several year classes indicate that it has established a reproducing population in the lower mainstem Eel River (B. Harvey, unpubl. data). The success of the remaining introductions has varied and seems to be linked to the similarity of the species' native habitat to that of the Eel River drainage.

Two diadromous species have been introduced. American shad (*Alosa sapidissima*), an anadromous species native to the east coast of North America, invaded the system sometime after its introduction to the Sacramento–San Joaquin River drainage in 1872. The species was well established by 1939 (Shapovalov, 1939). The ayu (*Plecoglossus altivelis*), an amphidromous species imported from Japan, was introduced into the river annually from 1961–1965 (a total of 3.25 million eggs and fry) but never became established (Dill and Cordone, 1997).

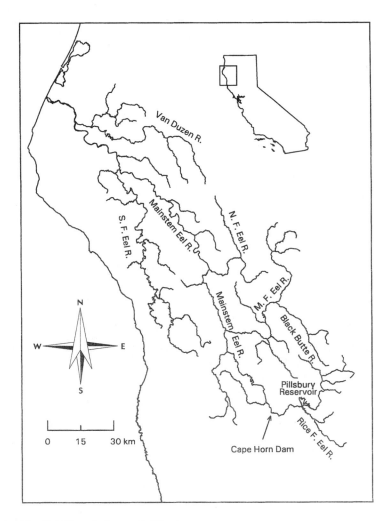

FIGURE 29.1 The Eel River drainage, California.

The success of American shad and failure of ayu are most likely related to differences in their life histories. The Eel River offers American shad appropriate habitat, and the species is temporally separated from the native anadromous salmonids. Shad spawn in the spring and the young over-summer in warm, deep pools, primarily in the mainstem Eel River. The juveniles migrate to the ocean in the fall. Though little is known about the behavior of West Coast shad populations once they enter the ocean, they do make extensive coastal migrations. The natural colonization by shad of the Eel River indicates that ocean conditions in the area are appropriate for this species. In contrast to shad, the anadromous salmonids spawn primarily in the fall and winter and the juveniles either migrate to the ocean or utilize cool headwater and tributary streams. The low spatial overlap between juvenile shad and oversummering salmonids, primarily steelhead rainbow trout and coho salmon, appears explainable by the preference of shad for warmer water. In addition, there are no pelagic freshwater zooplanktivores native to the Eel River system to compete with the young shad for food. Shad apparently invaded a system with few physical or biological barriers to their success.

In contrast, there are several aspects of ayu life history that made a successful introduction unlikely. The ayu is a semelparous, amphidromous species with an annual life cycle (McDowall, 1988; Saruwatari, 1995). In their native streams, the adults move downstream from August to November to spawn adhesive, demersal eggs. Hatching takes place from October to December, and

TABLE 29.1
Invading Species of the Eel River Drainage, California

Species	Group[a]	Success[b]	Source[c]	Year[d]
American shad, *Alosa sapidissima*	D	Y	CA[e]	>1872
Ayu, *Plecoglossus altivelis*	D	N	J	1961
Brown trout, *Salmo trutta*	R	N	EUR	1960s
Brook trout, *Salvelinus fontinalis*	R	N	E	1960s
Kokanee, *Oncorhynchus nerka kennlyi*	R	N	BC	1960s
Threadfin shad, *Dorosoma petenense*	R	Y	E	ca. 1980
Golden shiner, *Notemigonus crysoleucas*	R	Y	E	<1961
Bluegill, *Lepomis macrochirus*	R	Y	E	<1939
Largemouth bass, *Micropterus salmoides*	R	Y	E	1986
Green sunfish, *Lepomis cyanellus*	WS	Y	E	<1939
Brown bullhead, *Ameiurus nebulosus*	WS	Y	E	<1939
California roach, *Hesperoleucas symmetricus*	CS	Y	CA	ca. 1970
Sacramento squawfish, *Ptychocheilus grandis*	CS	Y	CA	ca. 1979
Speckled dace, *Rhinichthys osculus*	CS	Y	CA	<1986
Fathead minnow, *Pimephales promelas*	?	?	E	?
White catfish, *Ameiurus catus*	?	Y	E	<1995

[a] D = diadromous, R = species restricted to Pillsbury Reservoir, WS = warmwater stream species, CS = California stream species, ? = unknown.
[b] Y = yes, N = no, ? = uncertain.
[c] CA = California, J = Japan, EUR = Europe, E = eastern U.S., BC = British Columbia
[d] > = after indicated date, < = before indicated date, ? = unknown.
[e] American shad invaded the Eel River from the Sacramento–San Joaquin River drainage sometime after shad were introduced there. The species was introduced from the east coast of the U.S. to the Sacramento–San Joaquin River system in 1872.

the larvae rear in the ocean from December to February. The juveniles, which are herbivorous, then move into the upper and mid-reaches of streams from January to April, where they stay until maturity. In Japan, the lower reaches of rivers provide the appropriate spawning conditions which include fine gravel (2–40 mm in diameter, optimum <5 mm), depths ranging from 20 to 200 cm, water velocities ranging from 20 to 120 cm/sec, and water temperatures between 10 and 20°C (K. Uchida pers. commun.). The 10–20°C temperature range spans the thermal tolerance limits of the eggs and larvae. Increased stream flow from rain is an important cue for adults to begin the downstream spawning migration, but some studies suggest that large autumn flows cause decreased numbers of juveniles moving upstream in the spring due to some unknown mechanism (K. Uchida, pers. commun.). It is unknown if the ocean currents at the mouth of the Eel River are favorable for retention of larvae and their eventual upstream migration. It is also unknown if algae in the Eel River have the nutritional characteristics appropriate for growth and survival of the juveniles. Broadcast spawning of demersal eggs may not be a viable life history characteristic in the Eel River given the high probability of large fall and winter storms causing scour during the spawning and incubation periods. An annual life cycle may also be disadvantageous, making it difficult for the population to become established or recover from a series of spawning failures, especially if the founding population is small. A major storm in the winter of 1964 likely contributed to the failure of the introduction by destroying eggs and transporting larvae far out to sea. The latter event may have been particularly important, because the ayu stock most commonly used for supplementation in Japan, and the most likely source of eggs for export, comes from land-locked Lake Biwa rather than amphidromous stocks (K. Uchida, pers. commun.). Studies have shown that the land-locked stocks of ayu cannot survive ocean conditions. The 1964 flood would likely have transported the vast majority of surviving larvae into ocean conditions they could not survive. Water temperatures in the Eel River typically fall below 10°C by the end of November just after the onset of rains that would presumably trigger downstream migration and spawning of adults. This would greatly restrict the time period for successful reproduction. Finally, one study in Japan has shown that introduced ayu from the Lake Biwa stock contributed little to reproduction in the Shinano River for unknown reasons (Pastene et al., 1991), suggesting that the probability of a successful introduction was low if the Lake Biwa stock was the source of Eel River eggs.

Three salmonid species, brown trout (*Salmo trutta*), brook trout (*Salvelinus fontinalis*) and kokanee (*Oncorhynchus nerka kennlyi*) were introduced into Pillsbury Reservoir, but none established permanent populations. All three species are widely established in coldwater reservoirs and streams in California and elsewhere (Moyle, 1976). The failure of these species was most likely due to unfavorable habitat conditions in Pillsbury Reservoir, which is often too warm and anoxic for salmonids during the summer (Fisk and Pelgren, 1955). Brown trout and brook trout might have been successful if introduced into the cool headwater streams in the drainage that presently support resident rainbow trout. However, both species spawn in the fall, and winter storms during the egg incubation period have been associated with poor spawning success of brown trout in Sierra Nevada streams (Strange et al., 1992). It seems likely that in the Eel River drainage, where most winter precipitation falls as rain, the probability of such storm effects would be high. The kokanee requires lake or reservoir habitat, so there was no chance of it becoming established elsewhere in the drainage.

The nine nonmigratory introduced species in the drainage form three groups (Table 29.1): (1) a reservoir group comprised of four species from the eastern United States that are largely confined to Pillsbury Reservoir; (2) a warm-water stream group comprised of two species from the eastern U.S., which are present in Pillsbury Reservoir and have small river populations; and (3) a California streams group comprised of three species native to other California stream systems.

The reservoir group includes threadfin shad (*Dorosoma petenense*), golden shiner (*Notemigonus crysoleucas*), bluegill (*Lepomis macrochirus*), and largemouth bass (*Micropterus salmoides*). In their native range, these species are most successful in ponds, lakes, and low velocity habitats of large rivers. All of the species have been present in the system for at least 10 years, but have not established permanent downstream populations (Brown and Moyle, 1997; B. Harvey, unpubl. data). All of the species have been captured or observed in small numbers below Pillsbury Reservoir (Brown and Moyle, 1997; Day, 1968; B. Harvey, unpubl. data) indicating that invasions of the river occur yearly with downstream transport of juvenile or adult fish. The inability of these species to establish reproducing riverine populations probably results from poor survival during the extremely turbid, high discharge conditions that characterize runoff from winter storms in the Eel River drainage. Also, there is little floodplain habitat to provide low velocity refugia. Lack of suitable habitat (e.g., beds of aquatic vegetation) for the larvae or juveniles to avoid predation may also be important.

The warm-water stream group includes green sunfish (*Lepomis cyanellus*) and brown bullhead (*Ameiurus nebulosus*). In their native ranges, these species commonly inhabit small streams and ponds and are commonly associated with disturbed habitat throughout North America. Both have been present in the drainage since the 1930s and are present in Pillsbury Reservoir. These two species have small, scattered populations in downstream areas but have never been abundant enough to be considered more than a minor component of the fish fauna (Brown and Moyle, 1997). It is unclear if the downstream populations are self-maintaining or if they are supported by repeated invasions from Pillsbury Reservoir or other lentic sources (e.g., farm ponds). Both species have successfully invaded small warmwater streams, ponds, lakes, and reservoirs, in other areas of California. However, neither of them has been notably successful in colder, larger streams (Moyle, 1976). As with the reservoir species group, poor adaptation to physical conditions seems to be the limiting factor.

The California streams group includes California roach (*Hesperoleucas symmetricus*), Sacramento squawfish, and speckled dace (*Rhinichthys osculus*). All three species are native to nearby drainages including the Sacramento–San Joaquin River drainage to the east and various coastal drainages to the north (speckled dace) or south (all species). California roach were introduced into the Eel River system around 1970. In the following 10–15 years this species colonized most of the available habitat suitable for the species, though the process of invasion was poorly documented. By 1990, California roach was the most widely distributed and most abundant species in the drainage (Brown and Moyle, 1997). By 1992, about 12 years after their introduction, Sacramento squawfish had invaded over 500 km of the mainstem Eel River and its four major tributaries (Brown and Moyle, 1997). Brown and Moyle (1997) predicted a continued expansion of the squawfish's range, but recent surveys indicate that the upstream limits of squawfish distribution have not changed appreciably since 1992 (B. Harvey, unpubl. data).

Speckled dace were first noted in the Van Duzen River drainage (Figure 29.1), the northernmost tributary to the Eel River, during the summer of 1987. The lack of previous records of this species in the drainage indicated that it was a recent introduction. Brown and Moyle (in press) predicted that the speckled dace would likely become abundant based on its wide temperature tolerance and ability to utilize shallow riffles as habitat. However, as of 1996, individuals have only occasionally been found downstream of the original range described by Brown and Moyle (1997), and even in that reach the species has not become abundant compared to California roach or juvenile Sacramento squawfish in the Eel River or speckled dace in other California streams (B. Harvey, unpubl. data).

The success of the California roach and Sacramento squawfish is easily attributed to their adaptation to the physical regime. Both species are present and abundant in coastal drainages to the south of the Eel River with hydrologic and temperature regimes similar to those of the Eel River. Deposition of gravel in stream channels and the resulting increase in shallow, warm habitat in the larger streams may have contributed to the magnitude of the success of roach and squawfish, but it is likely that the invasions would have been successful even in the absence of such habitat alteration. Given that both species are found and are often abundant in cooler, forested streams in the Sierra Nevada foothills (Moyle, 1976; Brown and Moyle, 1993), it seems probable that the invasions would have been equally successful even in an unaltered Eel River system. In fact, both species reach their greatest local abundances in the Eel River drainage in areas of complex habitat most similar to the unaltered system.

The failure of speckled dace to become abundant or to expand its range is puzzling because the available habitat appears to be suitable for the species. Smith (1982) documented both upstream and downstream dispersal by the species in the Pajaro River system, suggesting no behavioral barrier to range expansion. Recent observations (B. Harvey, unpubl. data) suggest that the presence of unembedded cobbles may be an especially important habitat feature for fishes in the Van Duzen River. Speckled dace, roach, and juvenile squawfish all extensively utilize interstices between cobbles in the upper Van Duzen River, presumably as a refuge from predation. Such habitat is present but not especially abundant within the present range of the speckled dace and could conceivably limit the rate of invasion.

A second possible factor limiting downstream dispersal of speckled dace is the presence of prickly sculpin (*Cottus asper*) and coastrange sculpin (*C. aleuticus*). Baltz et al. (1982) documented that riffle sculpin (*C. gulosus*) were able to exclude speckled dace from cover structures in artificial stream experiments. Field data indicated that riffle sculpin and speckled dace partitioned the habitat with speckled dace limited to shallow edge habitat. The partitioning broke down in downstream areas where temperatures became stressful for sculpins. In these areas dace inhabited all habitats. The present downstream limit of speckled dace approximately coincides with the upstream limit of prickly and coastrange sculpin (Brown et al., 1995). The Van Duzen River never becomes too warm for sculpins in the downstream reaches, and the river supports some of the highest densities of sculpins in the Eel River system (Brown et al., 1995). The two sculpins may be impeding the invasion of speckled dace by excluding them from preferred habitat and perhaps by direct predation as well. If preferred habitat is rare, as indicated earlier, the interaction between habitat limitation and high sculpin abundance may be especially effective in preventing invasion of downstream areas. In any case, the speckled dace situation suggests that predictions about the success of an invasion based on general knowledge of a species' requirements and environmental conditions may often be incorrect.

California roach and Sacramento squawfish, the two highly successful invaders from nearby drainages, have both affected the aquatic ecosystem of the Eel River drainage. Predation by California roach can have an effect on periphyton in some habitats through both grazing and alterations in the benthic insect community (Power, 1990). The invasion of Sacramento squawfish will likely change the dynamics of this relationship. The presence of Sacramento squawfish, which prey on California roach and other small fishes, causes shifts in habitat and microhabitat use by other species (Brown and Moyle, 1991). In particular, California roach and other species reduce their use of deeper habitats, particularly pools, where the effect of California roach on the benthos is the strongest (Power, 1992). These shifts have the added effect of reducing spatial overlap among the different species and size classes of species. Before the invasion of squawfish, the deep pools were used by all species resulting in high overlap.

Besides affecting spatial partitioning among species, the squawfish invasion may have consequences for the geographic distribution of some species, particularly the native threespine stickleback (*Gasterosteus aculeatus*). Studies conducted in experimental streams showed that rainbow trout and California roach exhibit strong behavioral avoidance of deep habitats containing Sacramento squawfish large enough to consume them (Brown and Brasher, 1995). However, experiments with threespine stickleback suggested that the presence of even a minimal amount of cover was sufficient for sticklebacks to remain in deep habitats with squawfish (L. Brown and A. Brasher, unpubl. data), even though remaining there increased the probability of predation by squawfish. The absence of complex cover in large areas of the Eel River system combined with the observation by Smith (1982) that threespine stickleback and Sacramento squawfish do not coexist in pools with sparse cover, prompted Brown and Moyle (1997) to predict that threespine stickleback would disappear from large areas of the drainage. Subsequent observations have indicated that threespine stickleback have become rare in parts of the system containing adult squawfish over the entire year, particularly the lower mainstem Eel River and lower South Fork Eel River. However, threespine stickleback remain abundant in the Van Duzen River. Adult squawfish do not remain abundant in the Van Duzen River during the summer low flow season, although abundant young-of-year squawfish and early summer censuses of adults indicate seasonal use of the stream for spawning (B. Harvey, unpubl. data).

Even species that respond to squawfish by altering their use of habitat may face significant risk of predation from the introduced piscivore. Despite behavior to minimize contact with predatory squawfish, both rainbow trout and California roach are common food items for squawfish (B. Harvey, unpubl. data). Radio-tagging studies have shown that adult squawfish inhabit deep water habitats during the day and move into shallow refuge habitats of these prey at night. Presumably, the diel pattern in use of habitat by adult squawfish is motivated by both the availability of prey in shallow habitats and relatively high risk of predation in such habitats during the day. The adult squawfish only forage in the refuge habitats of rainbow trout and California roach at night when their vulnerability to avian and mammalian predators is low.

The successful invasions by California roach and Sacramento squawfish into the riverine habitats of the drainage have produced a group of species that naturally occurs in other parts of California. These species are presently organizing into fish assemblages that very closely resemble fish assemblages in streams where these species have coexisted for long periods of time (Brown and Moyle, 1997). A biologist unfamiliar with the history of invasions to the system might assume that squawfish and roach were native species in the Eel River drainage and could even construct reasonable explanations for how they dispersed from the Sacramento–San Joaquin river system, where they evolved. In contrast, the nonindigenous species from outside of California are most successful in the most altered habitat in the drainage, Pillsbury Reservoir. These species are better suited for lakes and reservoirs than either the native or introduced California stream species. In the Eel River drainage, suitability of the physical environment appears to be the most important factor in determining the success or failure of nonindigenous species.

INVASIONS OF NONINDIGENOUS SPECIES INTO HAWAIIAN STREAMS

Hawaii is the most isolated island archipelago in the world, located almost 4000 km from the nearest continent. This has resulted in a stream fauna that is depauperate, by continental standards, and highly endemic. Hawaii has only five native stream fishes (four gobies and one eleotrid), and each represents a different genus. In contrast to the Eel River, many Hawaiian streams have been extensively altered by human activities. In this section we review studies that document the success of introduced species in altered streams and altered stream reaches and the failure of most introduced species to invade unaltered habitat where native species remain abundant.

Most streams in Hawaii are small by continental standards. Their drainage basins and flows are functions of the size, elevation, and geological age of the islands, and local rainfall (Timbol and Maciolek, 1978). Tradewinds dominate Hawaiian weather patterns, and rainfall is largely determined by orographic effects; as air masses are lifted over the mountains they cool, creating localized rain, primarily on the windward sides of the islands (Hawaii Stream Assessment, 1991). Streams tend to have steep gradients, rocky channels, heavy vegetative cover and great temporal variability in discharge, fluctuating up to a thousandfold (Timbol and Maciolek, 1978).

Stream channelization has been a common practice in Hawaii in the last few decades. Extremely rapid population growth has been accompanied by increased urban development and the perceived need to control the highly variable discharge in Hawaiian streams to protect developed property and open additional areas for development (Norton et al., 1978). Over 19% of 376 perennial streams have been channelized to some degree (Hawaii Stream Assessment, 1991), and no streams on Oahu, where 80% of the state population resides, can be considered physically pristine. Oahu, the most developed island, has 89% of the lined channel in Hawaii (Timbol and Maciolek, 1978). As of 1978, water was exported from 53% of all perennial streams in Hawaii for agricultural and domestic uses (Timbol and Maciolek, 1978). Channelization, stream realignment, removal of riparian vegetation, and diversion of water all can result in increased temperatures, more exposure to light causing excessive algal growth, and consequently strong diel fluctuations in pH and dissolved oxygen concentration.

The native Hawaiian freshwater fishes are derived from marine forms and have retained a marine larval lifestage as part of an amphidromous life history (McDowall, 1988). Four native macroinvertebrates, two shrimp and two snails, also have this life history. Eggs are laid in the stream, the eggs hatch, and larvae are washed out to the ocean. After spending a larval phase of up to 7 months as marine plankton, post-larvae return to streams (Radtke et al., 1988). The post-larvae do not necessarily return to their home stream but probably enter whatever stream is nearby when the larval period is over. Thus, both the species and number of larvae returning to a particular stream are dependent on recruitment from the pool of larvae present in the ocean at a particular time. Spates (high flow events) may play an important role in determining and maintaining stream community structure by providing a cue for both reproduction and recruitment back to the stream (Kinzie, 1988). Of the native taxa, two of the fish, one of the shrimp, and one of the snails are primarily estuarine, and their ranges do not extend upstream of major waterfalls (Kinzie, 1988). The native stream fauna are well adapted to the flashy nature of Hawaiian streams and the steep topography of the watersheds. All four gobiid fishes have fused pelvic fins with which they cling to the substrate. Both snails are limpet-like and attach firmly to the substrate, and the eleotrid and shrimps are also strongly demersal.

The effect of channelization on stream fauna of oceanic islands such as Hawaii can be especially severe, because the most extensive modifications occur on the lower reaches of streams (Resh et al., 1992). These reaches, in addition to being habitat for some species, are the essential pathways for downstream dispersal of larvae and upstream migration of post-larvae of the diadromous species. Diversions and channelization may block these pathways, inhibiting the migrations required for successful completion of the life cycle.

Almost three fourths of the known introductions to Hawaiian inland waters have become established (Maciolek, 1984). At least one exotic species has been found in all streams surveyed in Hawaii (Hawaii Stream Assessment, 1991; Timbol and Maciolek, 1978). The number of introduced species in Hawaiian streams is already much larger than the number of native species (Devick, 1991). Introductions of aquatic organisms into Hawaiian streams occurred in essentially four waves (as documented in Eldredge, 1994, 1992; Devick, 1991; Maciolek, 1984; Randall, 1978; Brock, 1960). Prior to 1900, a number of species were introduced, primarily as food, by immigrant Asian workers. Between 1900 and World War II, mosquito control and recreation were the primary focus of introductions. Western mosquitofish (*Gambusia affinis*) and sailfin mollies (*Poecilia latipinna*) were the first fishes introduced into Hawaii as mosquito control agents (Maciolek, 1984). From 1946 to 1961 numerous species were introduced for control of aquatic plants, aquaculture, as baitfish, and for recreational fishing. During these early years, the perception that Hawaiian streams contained a depauperate fauna requiring supplementation led to the stocking of such species as the predatory Tahitian prawn, *Macrobrachium lar*. Most recently, introductions have come mainly from the aquarium trade. Between 1982 and 1990 alone, more than 20 such species have been established (Devick, 1991). For example, guppies (*P. reticulata*), the Mexican molly (*P. mexicana*), swordtails (*Xiphophorus helleri*), and the Cuban limia (*P. vittata*) have become established in diverted and/or channelized streams either "accidentally" (dumped aquarium fishes) or intentionally for mosquito control (Devick, 1991).

In the absence of definitive studies, it is difficult to attribute the success of nonindigenous species to any single factor or group of factors. However, stream alterations appear to have created habitats far more suitable for exotic species such as poeciliids than for the native fishes (Maciolek, 1977). The native species, which require sufficient streamflow to provide clean, cool, fresh water (Timbol and Maciolek, 1978), are unable to maintain populations under degraded conditions. But exotic species, which have broader environmental tolerances, are able to flourish (Norton et al., 1978). In laboratory experiments, native species were less tolerant of high temperatures than exotics, and some native species had upper lethal temperature limits within the range measured in channelized streams (Hathaway, 1978). Concrete-lined channels, especially, appear to favor introduced fishes and to serve as nurseries for exotic poeciliids (Hathaway, 1978). Facultative air breathing by the introduced loricariid catfishes (*Hypostomus* sp. and *Pterygoplichthys multiradiatus*), Chinese catfish (*Clarias fuscus*), and dojo loach (*Misgurnus anguillicaudatus*) may have allowed these species to become established in Oahu streams with low water quality.

The direct effects of the successful invaders of altered streams on native species are unknown but may be extensive. Predation on the post-larvae of native species may be substantial as they attempt to migrate through altered habitat to the upstream reaches of streams or attempt to settle in such habitat (Maciolek, 1984). In addition to whatever competition and predation introduced species directly exert on native fishes in altered streams, introduced species are a source of a number of helminth parasites previously unknown in Hawaii (Font and Tate, 1994). These include the nematode, *Camallanus cotti*, the Asian tapeworm, *Bothriocephalus acheilognathi*, and the leech, *Myzobdella lugubris*.

The obvious success of introduced species in Hawaiian streams suggests that these streams are easily invaded, but most invading species only do well in altered habitats. This is similar to the situation in the Eel River drainage, where few introduced species have done well in the river itself. On the Island of Oahu, native species dominate in relatively unaltered streams, while exotic species dominate altered streams (Timbol and Maciolek, 1978). Exotic species are predominant in the lower reaches of streams, which typically are altered. Upper reaches, even of altered streams, tend to be occupied primarily by native species. Most of the introduced species are poorly adapted for the large fluctuations in streamflow that occur in unaltered Hawaiian streams. For example, restoring the majority of natural streamflow to the previously diverted Waiahole Stream on Oahu reduced populations of introduced poeciliids, presumably because they were displaced downstream by increased water velocities (W. Font, pers. commun.).

As in the Eel River drainage, several introduced species have become established in Hawaiian reservoirs but have not successfully invaded streams. Largemouth bass and other game species such as peacock cichlid (*Cichla ocellaris*) and channel catfish (*Ictalurus punctatus*) have become established in reservoirs, but are unable to become established in the high-velocity Hawaiian streams. Similarly, the freshwater needlefish (*Xenentodon cancila*) is extremely abundant in Wahiawa Reservoir on Oahu (where it preys on introduced largemouth bass) but also does not appear suited to physical conditions in Hawaiian streams (Devick, 1991).

Despite the large number of successful introductions of nonindigenous aquatic species, there are many examples of failed invasions of Hawaiian streams. The attempted introductions of salmonids, including brown trout, brook trout, and chinook salmon, starting in the late 1800s, provide noteworthy examples (Eldredge, 1994; Maciolek, 1984; Brock, 1960; Needham and Welsh, 1953). In the early 1900s, attempts were made to establish rainbow trout in streams on the islands of Kauai and Oahu. Warm stream and ocean temperatures, and the absence of appropriate food (the Hawaiian native stream fauna lacks typical swift-water dwelling species including stoneflies, mayflies and caddisflies) probably made it difficult for these species to survive and reproduce. Interestingly, rainbow trout stocked in a stream near Kokee, on the Island of Kauai, consumed almost exclusively terrestrial insects and other terrestrial invertebrates (Needham and Welsh, 1953). A combination of high stream temperatures and periodic flooding appear to limit spawning success of Kokee rainbow trout. Water temperatures in Kokee area streams, located at an elevation of 3000–4000 ft, are relatively cool, but there is probably not a sufficiently long period of low temperature each winter to permit proper development of the gonads (Needham and Welsh, 1953). In addition, severe floods may displace trout downstream. Rainbow trout are still stocked into a reservoir in the Kokee area.

In addition to the salmonids, a prawn (*M. rosenbergii*) introduced in the early 1960s also failed to invade Hawaiian streams. Eldredge (1994) suggested that the failure of the prawn to become established may be based on its need for slow flowing, large river habitat. Such habitat is rare in the generally small, high-gradient Hawaiian streams.

Only three nonindigenous species have invaded relatively unaltered Hawaiian streams to a significant extent. The Tahitian prawn was introduced in 1962, and has subsequently invaded almost every stream in Hawaii. The amphidromous life history of this species has allowed it (like the native stream macrofauna) to colonize both altered and unaltered streams throughout the state. Tahitian prawn is now common in streams throughout Hawaii. An interesting exception is Hanawi Stream on the Island of Maui, where the prawn is very uncommon. This stream is largely fed by cool spring water, suggesting that Tahitian prawn may be intolerant of low stream temperatures. The widespread abundance of Tahitian prawn is important, because it is a major predator on native species that have few native predators other than the black crowned night heron (*Nycticorax nycticorax*) and the estuarine eleotrid.

The smallmouth bass (*Micropterus dolomieu*) has also been a success invader. Smallmouth bass was originally introduced for recreational fishing to Oahu in 1953 and Kauai in 1955. Unlike salmonids, smallmouth bass are well adapted to Hawaiian stream conditions and have expanded their range within the drainage basins where they were originally introduced (Heacock, pers. commun.). Smallmouth bass are native to warm and cool water streams and rivers in the eastern U.S. which often experience spates due to summertime thunderstorms. Smallmouth bass have been extremely damaging to Hawaiian streams. Smallmouth bass may compete for space with native fishes because their habitat preferences are very similar. Smallmouth bass and gobies may also compete for food. For example, smallmouth bass feed on an introduced caddisfly (Heacock, pers. commun.) that is also a common food for one of the native gobies (Kido et al., 1993). Smallmouth bass also prey on a number of other aquatic organisms including native gobies, shrimp, and damselflies (Heacock, pers. commun.). Future expansion of the range of smallmouth bass is possible because of its dispersal ability within stream systems and the possibility it will be introduced to new streams to enhance recreational fishing.

Another predator that has become established in Hawaiian streams is the blackchin tilapia, *Tilapia melanotheron*. It escaped in 1965 from an aquaculture facility into Honolulu Harbor (Randall, 1978). Apparently, the fish escaped from holding tanks that were drained without proper screening in place (Devick, 1991). The species appears to have been further distributed through human intervention (probably in bait buckets) into a number of reservoirs on Oahu. Tilapia are tolerant of a wide range of environmental conditions, including extreme temperatures, turbidity, and widely fluctuating salinities. *T. melanotheron* has become abundant in the lower reaches of many streams and estuaries on the Island of Oahu. It also appears to be spreading around coral reefs on Oahu, suggesting it is capable of spawning in marine habitats (Devick, 1991). It has recently appeared on the Island of Kauai, apparently escaping from an aquaculture operation.

In summary, the success of most introduced species in Hawaiian streams is closely linked to habitat alteration. The small number of species successful in unaltered habitats have habitat requirements which match well with habitats available in Hawaiian streams. Neither reservoir adapted introduced species nor introduced species inhabiting channelized portions of streams have successfully invaded unaltered stream reaches. However, the larvae and post-larvae of the native species often must migrate through altered stream reaches, usually located in downstream areas, to complete their life cycle. These periods of spatial overlap provide a significant opportunity for biotic interactions among native and introduced species. Also, the smallmouth bass has successfully invaded relatively unaltered stream habitats as did the Tahitian prawn. These species can interact with juveniles and adults of native species. As is often the case with endemics that have evolved in isolation from vigorous competition or predation pressure, the native species in Hawaii appear to be poorly equipped to compete with the hardy exotics in situations where they co-occur (Hathaway, 1978).

CONCLUSIONS

In both the Eel River drainage and Hawaiian stream systems, the success of an invasion is closely linked to adaptations of the invading species to the physical habitat. The more similar the new habitat is to the one in which they evolved, the more likely the invaders will be successful. Baltz and Moyle (1993) provide two general hypotheses to explain invasion resistance by California native stream fish assemblages: (1) the environmental resistance hypothesis, that introduced fishes cannot adapt to the highly seasonal flow regimes of unregulated streams, and (2) the biotic resistance hypothesis, that introduced fishes cannot break into an established assemblage structured by strong biotic interactions. The former hypothesis appears more important in the Eel River and Hawaiian streams.

In the Eel River, the strong seasonal pattern in stream discharge seems to limit successful invaders to species from similar systems in California. Native stream communities in Hawaii tend to be structured by physical factors, particularly the unpredictable and flashy flow regimes, and by ocean conditions that determine the species and number of individuals returning to a particular stream at a particular time. It appears that these same physical factors are the source of resistance to invasion by exotic species. Biotic resistance may occur but appears to be less important. In the Eel River, biotic resistance may be a factor in the failure of speckled dace to invade areas downstream of its present range. In Hawaii, however, there are no known cases of biotic resistance, and native fish likely present few biological barriers to invasion.

Both the Eel River and Hawaiian streams have naturally depauperate fish faunas compared to many other regions of the world. Such depauperate faunas might seem more prone to invasion than more speciose ones; however, Moyle and Light (1996a) summarize evidence indicating that the species diversity of the receiving fauna is unrelated to the success of invasions. Additional evidence from other depauperate areas also indicates that environmental resistance is most important in determining the success of nonindigenous species. Failed introductions and invasions may be even more instructive than successful ones.

Coates (1987) documented the status of three introduced species in the Sepik River drainage, New Guinea. Western mosquitofish was introduced in the 1930s for mosquito control. Tilapia (*Oreochromis mossambicus*) and common carp (*Cyprinus carpio*) were introduced to New Guinea for aquaculture purposes in 1954 and 1960, respectively. Both subsequently escaped into the river. All three species have become locally abundant in the lower river system, and tilapia is now the most important species in the local fishery (Coates, 1985). However, none of the three introduced species has successfully invaded upland streams. Baltz and Moyle (1993) have documented the failure of nonindigenous species to invade Deer Creek, California, despite the existence of both upstream (brown trout) and downstream (green sunfish and smallmouth bass) source populations. They attribute the failures primarily to the inability of the introduced species to adapt to the natural hydrologic regime, although predation by Sacramento squawfish and other native fishes is likely a secondary factor. Cooper (1983) documented a situation in the Pit River, California, where a disturbed portion of the river was dominated by introduced species, but an undisturbed upstream portion of the river was dominated by native species. Bramblett and Fausch (1991) observed the failure of nonindigenous species to invade a reach of a Colorado stream from a more physically benign upstream reach in Colorado. The downstream reach retained the natural hydrologic regime of generally low flow but severe summer flash floods. Meffe (1991) documented a failed invasion of a southeastern stream by bluegill where the species became common in a reservoir but was not able to establish a population in a nearby stream. The failure was attributed to poor adaptation to living in streams, even those with low velocities (7–25 cm/s).

Studies on desert fishes in the North American southwest have shown that in habitats where frequent flooding occurs, native fish, adapted to this type of periodic disturbance, are able to persist, while introduced fish tend to be displaced by the floods (Meffe, 1984). Laboratory studies have confirmed that fish that have evolved with such environmental disturbances are more tolerant of these flow regimes than those that have evolved in lakes and more sluggish rivers (Castleberry and Cech, 1986; Meffe, 1984).

While environmental resistance is important in determining invasion success, many species have ecological characteristics that predapt them to be successful invaders in some situations. For example, green sunfish, one of the native species in a Colorado stream assemblage exhibiting invasion resistance (Bramblett and Fausch, 1991), has successfully invaded many small streams in California (Brown and Moyle, 1993; Moyle, 1976; Moyle and Nichols, 1973). Sacramento squawfish, seen as a key species contributing to invasion resistance in a California stream where it is native, has successfully invaded the Eel River. In California it has been documented that predators and omnivores/detritivores are the most successful invaders of relatively unaltered systems (Moyle and Light, 1996b). Introduced predators are able to take advantage of prey that may be vulnerable to the new behaviors of an invading predator. Omnivores and detritivores utilize resources that are rarely limiting in stream systems. In Hawaii, the Tahitian prawn, smallmouth bass, and blackchin tilapia are invading streams where predation of one stream resident fish on another was a relatively rare event, except in the estuarine portions of streams where the piscivorous eleotrid may be present. In the Eel River, the squawfish invaded a system without a large, resident warmwater predator. The California roach is an omnivore that feeds on whatever algae or invertebrates are available.

Predicting the success or failure of any particular invasion remains largely a guessing game for ecologists. Many intentional introductions have failed, indicating that our judgments about the availability of appropriate habitat require refinement. For example, Rinne (1975) documented a case where a small stream was selected to be a refuge for two endangered cyprinids, the spikedace, *Meda fulgida*, and the loach minnow, *Tiaroga cobitis*. The creek was chemically treated to remove a nonindigenous fish, the longfin dace, *Agosia chrysogaster*. Individuals of two native species, the Gila chub, *Gila intermedia*, and the speckled dace, were captured for restocking along with individuals of the two rare species. The two rare species never became established, but all three of the resident species recovered. Failure to eradicate all of the longfin dace resulted in recovery of the population following the poisoning operation. The same process was repeated again the following year with the same result.

In two recent papers, Moyle and Light (1996a,b) develop a number of predictions or rules concerning the success of fish invasions into California streams and aquatic habitats in general. Aspects of environmental resistance are strongly incorporated in their rules, but they note that the rules are probably too general to allow accurate predictions of the success of an invasion at the local level. They state that at the small scales of interest to ecosystem managers, reasonable predictions can only be made based on past experience with the species and systems of interest. However, our ability to predict the success of intentional introductions and the effects of unintentional ones should improve as our knowledge of the ecological requirements of different species increases and more care is given to understanding the ecological conditions within the receiving systems. For example, three detritivorous fish species were introduced to the Sepik River, New Guinea, as food fish after extensive studies of the native fish fauna indicated that the introductions would pose little threat to them (Coates, 1993). Similarly, the success of Sacramento squawfish and California roach in the Eel River was completely predictable, though the relatively poor success of the speckled dace emphasizes the danger of equating a general assessment of habitat suitability with specific studies of the physical and biotic conditions in a particular stream. Unfortunately, given the present rate of species introduction, there will be many opportunities to improve our predictive ability.

GLOSSARY

Anadromous: Diadromous fishes that spend most of their lives in the sea and migrate to fresh water to spawn.
Amphidromous: Diadromous fishes whose migration from freshwater to the sea or vice versa is not for the purpose of breeding but occurs regularly at some stage of the life cycle. The species in this chapter migrate to sea as larvae and return to fresh water as post-larvae.
Demersal: Associated with the bottom of a body of water.
Diadromous: Those fish that normally, as a routine phase of their life cycle, and for the vast majority of the population, migrate between marine and fresh waters.
Diel: A 24-hour period, usually including a day and the adjoining night.
Lentic: Relating to still waters such as ponds and lakes.
Semelparous: Describes a species of fish of which the adults spawn once and then die.

REFERENCES

Allan, J.D. and A.S. Flecker. Biodiversity Conservation in Running Waters: Identifying the Major Factors that Affect Destruction of Riverine Species and Ecosystems. *Bioscience*. 43, pp. 32–43, 1993.
Bain, M.B., J.T. Finn and H.E. Booke. Streamflow Regulation and Fish Community Structure. *Ecology*. 69, pp. 382–391, 1988.
Baltz, D.M. and P.B. Moyle. Invasion Resistance to Introduced Species by a Native Assemblage of California Stream Fishes. *Ecological Applications*. 3, pp. 246–255, 1993.
Baltz, D.M., P.B. Moyle, and N.J. Knight. Competitive Interactions between Benthic Stream Fishes, Riffle Sculpin, *Cottus gulosus*, and Speckled Dace, *Rhinichthys osculus*. *Canadian Journal of Fisheries and Aquatic Sciences*. 39, pp. 1502–1511, 1982.

Bramblett, R.G. and K.D. Fausch. Fishes, Macroinvertebrates, and Aquatic Habitats of the Pugatoire River in Pinon Canyon, Colorado. *The Southwestern Naturalist*. 36, pp. 281–294, 1991.

Brock, V.E. The Introduction of Aquatic Animals into Hawaiian Waters. *Internationale Revue der gesamten Hydrobiologie and Hydrographie*. 45, pp. 463–480, 1960.

Brown, L.R. and A. Brasher. Effects of Predation by Sacramento Squawfish (*Ptychocheilus grandis*) on Habitat Choice of California Roach (*Lavinia symmetricus*) and Rainbow Trout (*Oncorhynchus mykiss*) in Artificial Streams. *Canadian Journal of Fisheries and Aquatic Sciences*. 52, pp. 1639–1646, 1995.

Brown, L.R. and P.B. Moyle. Invading Species in the Eel River, California: Success, Failures, and Relationships with Resident Species. *Environmental Biology of Fishes*, 49; pp. 271–291, 1997.

Brown, L.R. and P.B. Moyle. Changes in Habitat and Microhabitat Partitioning within an Assemblage of Stream Fishes in Response to Predation by Sacramento Squawfish (*Ptychocheilus grandis*). *Canadian Journal of Fisheries and Aquatic Sciences*. 48, pp. 849–856, 1991.

Brown, L.R. and P.B. Moyle. Distribution, Ecology, and Status of the Fishes of the San Joaquin River Drainage, California. *California Fish and Game*. 79, pp. 96–114, 1993.

Brown, L.R., S.A. Matern and P.B. Moyle. Comparative Ecology of Prickly Sculpin, *Cottus asper*, and Coastrange Sculpin, *Cottus aleuticus*, in the Eel River, California. *Environmental Biology of Fishes*. 42, pp. 329–343, 1995.

Castleberry D.T. and J.J. Cech Jr. Physiological Responses of a Native and Introduced Fish to Environmental Stressors. *Ecology*. 67, pp. 912–918, 1986.

Coates, D. Fish Yield Estimates for the Sepik River, Papua New Guinea, a Large Floodplain System East of 'Wallaces Line.' *Journal of Fish Biology*. 27, pp. 431–443, 1985.

Coates, D. Consideration of Fish Introductions in the Sepik River, Papua New Guinea. *Aquaculture and Fisheries Management*. 18, pp. 231–241, 1987.

Coates, D. Fisheries Ecology and Management of a Large Australasian River Basin, the Sepik-Ramu, New Guinea. *Environmental Biology of Fishes*. 38, pp. 345–368, 1993.

Cooper, J.J. Distributional Ecology of Native and Introduced Fishes of the Pit River System, Northeastern California, with Notes on the Modoc Sucker. *California Fish and Game*. 69, pp. 39–53, 1983.

Cushman, R.M. Review of Ecological Effects of Rapidly Varying Flows Downstream from Hydroelectric Facilities. *North American Journal of Fisheries Management* 5, pp. 330–339, 1985.

Day, J.S. *A Study of Downstream Migration of Fish Past Cape Horn Dam on the Upper Eel River, Mendocino County, as Related to the Pacific Gas and Electric Company's Van Arsdale Diversion*. California Department of Fish and Game, Marine Resources Administrative Report 68–4. 1968.

Devick, W.S. Patterns of Introductions of Aquatic Organisms to Hawaiian Freshwater Habitats, in *New Directions in Research, Management and Conservation of Hawaiian Freshwater Stream Ecosystems: Proceedings of the 1990 Symposium on Freshwater Stream Biology and Fisheries Management*. Division of Aquatic Resources, Department of Land and Natural Resources, Honolulu, Hawaii, pp. 189–237, 1991.

Dill, W.A. and A.J. Cordone. *History and Status of Introduced Fishes in California, 1871-1996*. California Department of Fish and Game, Fish Bulletin 178, 1997.

Dudgeon, D. Endangered Ecosystems: a Review of the Conservation Status of Tropical Asian Rivers. *Hydrobiologia*. 248, pp. 167–191, 1992.

Eldredge, L.G. Unwanted Strangers: an Overview of Animals Introduced into Pacific Islands. *Pacific Science*. 46, pp. 384–386, 1992.

Eldredge, L.G. *Perspectives in Aquatic Species Management in the Pacific Islands*, vol. 1, *Introductions of Commercially Significant Aquatic Organisms to the Pacific Islands*. South Pacific Commission, Noumea, New Caledonia, 1994, p. 127.

Fisk, L.O. and O.E. Pelgren *A Limnological Survey of Lake Pillsbury, Lake County*. California Department of Fish and Game, Inland Fisheries Administration Report No. 57–30. 1955.

Font, W.F. and D.C. Tate. Helminth Parasites of Native Hawaiian Freshwater Fishes: an Example of Extreme Ecological Isolation. *Journal of Parasitology*. 80, pp. 682–688, 1994.

Hathaway, C.B. Jr. *Stream Channel Modification in Hawaii*, Part C: *Tolerance of Native Stream Species to Observed Levels of Environmental Variability*. FWS/OBS-78/18, USFWS National Stream Alteration Team, Columbia, MO. 1978.

Hawaii Stream Assessment. *A Preliminary Appraisal of Hawaii's Stream Resources. Report R84*. Prepared for the Commission on Water Resources Management, Hawaii Cooperative Park Service Unit, Honolulu, HI. 1991.

Kido, M.H., P. Ha and R.A. Kinzie III. Insect Introductions and Diet Changes in an Endemic Hawaiian Amphidromous Goby, *Awaous stamineus* (Pisces: Gobiidae). *Pacific Science.* 47, pp. 43–50, 1993.

Kinzie, R.A. III. Habitat Utilization by Hawaiian Stream Fishes with Reference to Community Structure in Oceanic Island Streams. *Environmental Biology of Fishes.* 22, pp. 179–192, 1988.

Li, H. and P.B. Moyle. Ecological Analysis of Species Introductions into Aquatic Systems. *Transactions of the American Fisheries Society.* 110, pp. 772–782, 1981.

Ligon, F.K., W.E. Dietrich, and W.J. Trush. Downstream Ecological Effects of Dams. *Bioscience.* 45, pp. 183–192, 1995.

Lodge, D.M. Species Invasions and Deletions: Community Effects and Responses to Habitat and Climate Change, in *Biotic Interactions and Global Change*, Kareiva, P.M., Kingsolver, J.G. and Huey, R.B., Eds., Sinauer Associates, Sunderland, 1993a.

Lodge, D.M. 1993b. Biological Invasions: Lessons for Ecology. *Trends in Ecology and Evolution.* 8, pp. 133–137, 1993b.

Maciolek, J.A. Taxonomic Status, Biology, and Distribution of Hawaiian *Lentipes*, a Diadromous Goby. *Pacific Science.* 31, pp. 355–362, 1977.

Maciolek, J.A. Exotic Fishes in Hawaii and Other Islands of Oceania, in *Distribution, Biology and Management of Exotic Fishes*, Courtenay, W.R. and Stauffer, J.R., Jr., Eds., The Johns Hopkins University Press, Baltimore, MD, 1984.

McDowall, R.M. *Diadromy in Fishes.* Timber Press, Portland, OR, p. 308, 1988.

Meffe, G.K. Effects of Abiotic Disturbance on Coexistence of Predator–Prey Fish Species. *Ecology.* 65, pp. 1525–1534, 1984.

Meffe, G.K. Failed Invasion of a Southeastern Blackwater Stream by Bluegills: Implications for Conservation of Native Communities. *Transactions of the American Fisheries Society.* 120, pp. 333–338, 1991.

Moyle, P.B. *Inland Fishes of California.* University of California Press, Berkeley, CA, p. 405, 1976.

Moyle, P.B. and R.A. Leidy. Loss of Biodiversity in Aquatic Ecosystems: Evidence from Fish Faunas, in *Conservation Biology: the Theory and Practice of Nature Conservation, Preservation, and Management*, Fiedler, P.L. and Jain, S.K., Eds., Chapman and Hall, New York, 1992.

Moyle, P.B. and T. Light. Biological Invasions of Fresh Water: Empirical Rules and Assembly Theory. *Biological Conservation.* 78, pp. 149–162, 1996a.

Moyle, P.B. and T. Light. Fish Invasions in California: Do Abiotic Factors Determine Success? *Ecology.* 77, pp. 1666–1670, 1996b.

Moyle, P.B. and G.M. Sato. On the Design of Preserves to Protect Native Fishes, in *Battle Against Extinction: Native Fish Management in the American West*, Minckley, W.L. and Deacon, J.E., Eds., University of Arizona Press, Tucson, AZ, 1991.

Moyle, P.B. and R.M. Yoshiyama. Protection of Aquatic Biodiversity in California: a Five-tiered Approach. *Fisheries.* 19, pp. 6–19, 1994.

Moyle, P.B., H.W. Li and B. Barton. The Frankenstein Effect: Impact of Introduced Fishes on Native Fishes of North America, in *Fish Culture in Fisheries Management*, Stroud, R.H., Ed., American Fisheries Society, Bethesda, MD, 1986.

Needham, P.R. and J.P. Welsh. Rainbow trout (*Salmo gairdneri* Richardson) in the Hawaiian islands. *The Journal of Wildlife Management.* 17, pp. 233–255, 1953.

Norton, S.E., A.S. Timbol and J.D. Parrish. *Stream Channel Modification in Hawaii*, part B: *Effect of Channelization on the Distribution and Abundance of Fauna in Selected Streams.* FWS/OBS-78/17, USFWS National Stream Alteration Team, Columbia, MO. 1978.

Pastene, L.A., K. Numachi, and K. Tsukamoto. Examination of Reproductive Success of Transplanted Stocks in an Amphidromous Fish, *Plecoglossus altivelis* (Temmink et Schlegel) Using Mitochondrial DNA and Isozyme Markers. *Journal of Fish Biology.* 39 (Suppl. A), pp. 93–100, 1991.

Poff, N.L. and J.D. Allan. Functional Organization of Stream Fish Assemblages in Relation to Hydrological Variability. *Ecology.* 76, pp. 606–627, 1995.

Poff, N.L. and J.V. Ward. Implications of Streamflow Variability and Predictability for Lotic Community Structure: a Regional Analysis of Streamflow Patterns. *Canadian Journal of Fisheries and Aquatic Sciences.* 46, pp. 1805–1818, 1989.

Power, M. Effects of Fish in River Food Webs. *Science.* 250, pp. 811-814, 1990.

Power, M. Habitat Heterogeneity and the Functional Significance of Fish in River Food Webs. *Ecology* 73, pp. 1675–1688, 1992.

Radtke, R.L., R.A. Kinzie III, and S.D. Folsom. Age at Recruitment of Hawaiian Freshwater Gobies. *Environmental Biology of Fishes*. 23, pp. 205–213, 1988.

Randall, J.E. Introductions of Marine Fishes to the Hawaiian Islands. *Bulletin of Marine Science*. 41, pp. 490–502, 1978.

Resh, V.H., J.R. Barnes, B. Benis-Steger, and D.A. Craig. Life History Features of Some Macroinvertebrates in a French Polynesian Stream. *Studies on Neotropical Fauna and Environment*. 27, pp. 145–153, 1992.

Rinne, J.N. Changes in Minnow Populations in a Small Desert Stream Resulting from Naturally and Artificially Induced Factors. *The Southwestern Naturalist*. 20, pp. 185–195, 1975.

Saruwatari, T. Temporal Utilization of a Brackish Water Lake, Lake Hinuma, as a Nursery Ground by Amphidromous Ayu, *Plecoglossus altivelis* (Plecoglossidae) Larvae. *Environmental Biology of Fishes*. 43, pp. 371–380, 1995.

Shapovalov, L. *Recommendations for Management of the Fisheries of the Eel River Drainage Basin, California*. California Department of Fish and Game, Bureau of Fish Conservation Administration Report 39-2. 1939.

Smith, J.J. Fishes of the Pajaro River System. *University of California Publications in Zoology*. 115, pp. 85-169, 1982.

Strange, E.M., P.B. Moyle, and T.I. Foin. Interactions Between Stochastic and Deterministic Processes in Stream Fish Community Assembly. *Environmental Biology of Fishes*. 36, pp. 1-15, 1992.

Timbol, A.S. and J.A. Maciolek. *Stream Channel Modification in Hawaii*, Part A: *Statewide Inventory of Streams; Habitat Factors and Associated Biota*. FWS/OBS-78/16, USFWS National Stream Alteration Team, Columbia, MO). 1978.

30 Risk Analysis of Species Introductions: Insights from Qualitative Modeling

Hiram W. Li, Philippe A. Rossignol, Gonzalo Castillo

INTRODUCTION

The infestation of alien species through human influences has been massive. Over 4500 exotic species have become established in the continental U.S., alone within the last 100 years (U.S. Congress 1993). During this period the spread of fish diseases world-wide has been unprecedented (Ganzhorn et al. 1992), and well over 300 endemic fishes in Lake Victoria have become extinct due to the introduced Nile perch (*Lates niloticus*) (Hughes, 1986). Many pests were unfortunate mistakes of purposeful introductions. Introductions are now made with more deliberation, and introductions for management purposes are no longer as much in vogue. It is therefore disturbing to note that new records of alien species are still being reported. Many new introductions are often unauthorized or the result of accidents. Numerous examples of such introductions are documented elsewhere in this publication.

By the time the biologist becomes aware of the presence of an alien species, it has often become well established in its new habitat, and its source and natural history are either unknown or poorly known. There is only anecdotal information to rely on at best. The greatest need is for a management framework to assess the following issues: (1) the potential impact of the alien species upon the receiving community, (2) the ecological interactions for which understanding is most critical, (3) strategies that guide management policy to minimize potential deleterious effects of ecological change. Unfortunately, the theory of biological invasions is relatively undeveloped. The following quotes were printed in the proceedings of a recent symposium on invasion ecology:

> The bottom line of this collection is ambiguous. Optimists will read the papers and claim the field of invasion biology has advanced... . Pessimists will emphasize that the predictions are still scarce, and that the predictions being made are not especially impressive. (Kareiva 1996).

> Unfortunately, pressing questions like "what attributes make some species more invasive?" or "what makes some ecosystems more invasible than others?" appear to lack satisfactory answers. (Rejmánek and Richardson 1996).

We suggest that qualitative analysis and modeling can provide the basis for both advancing the understanding of potential ecological effects of alien species and serve as a practical tool for conservation and natural resource biologists. We will demonstrate that qualitative analysis can help identify (1) important interactions preserving community integrity, (2) keystone species, and (3) "risks" of species introduction. By simulating interactions in model communities, one can address the following questions:

1. Can species introductions be benign? If so, under what conditions? Can species introductions be positive?
2. What are the ecological consequences when an invading species introduces novel trophic interactions among species?
3. What are the ecological consequences when an invading species has similar properties to those of a species rich functional group?
4. What is the outcome when an exotic species is introduced to a community regulated by a keystone species?
5. How do species invasions affect the relative importance of direct vs. indirect interactions within a native community?
6. Are communities more sensitive to exotic predators, omnivores, herbivores, or primary producers?

Our last objective is to illustrate the use of qualitative modeling to assess natural systems. Specifically, we will analyze the impacts that the introduction of *Mysis relicta* and *Lates nilotica* have had on the Flathead Lake (Montana) and Lake Victoria (Africa) ecosystems, respectively.

THE TOOL OF QUALITATIVE ANALYSIS

One difficulty of assessing, predicting, and intervening in the spread of exotic species lies in the complexity of the ecosystems affected. Complexity in science and public policy implies extensive collection of data and therefore extreme expense. Part of this hurdle may be overcome if mathematical tools dealing with complex systems are used to full advantage. Most of these procedures rely, by necessity, on qualitative input, are less expensive and in a sense are more reliable. Puccia and Levins (1985) first indicated that much of the requirements of biology are in generality and realism, while many of our analytical techniques, particularly statistics and simulations, emphasize precision at the expense of one or the other. Engineers (Mason 1953) and economists (Quirk and Ruppert 1965; Maybee and Quirk 1969) developed many of the mathematical techniques for evaluating the stability, equilibrium, and effect of input into complex systems that have generality and broad realistic applications. Given that the systems addressed are well beyond numerical and statistical analyses, precision rarely enters as a concern.

Over the last few decades, these techniques have been introduced into ecology, particularly population dynamics. Thus, Levins (1974 1975) adapted the techniques to ecology and made an important contribution; namely, that the components and determinants of the matrices involved represented intuitive and meaningful biological relationships, which he termed "loops," hence loop analysis (Puccia and Levins 1985). Numerous collaborators subsequently used the technique to make important contributions to community structure, evolution and population dynamics (Lane 1986; Bodini 1991; Giavelli et al. 1990; Pilette et al. 1987; Levins and Vandermeer 1989). Other workers also made important contributions to the approach (Roughgarden 1977; May 1973; Jeffries 1974).

The technique, however, did not draw as much attention as we feel it deserves. One important reason is the tediousness of the algorithms, which have not been accessible to a computerized approach unless one introduced numerical values, in essence defeating the qualitative approach. Recently, the introduction of efficient and affordable symbolic processors, such as MAPLE or MATHEMATICA, have made such analyses fairly trivial. A hand-held calculator, the TI-90, has a symbolic processor and is currently capable of much of this analysis.

The technique thus allows for the determination of the presence of equilibrium and its stability, as well as the effect of input into the system. From a conceptual and experimental point of view, this allows not only for powerful hypothesis testing and experimental design, but also for generation of strong management policies since either parameter or input can be human agencies, for example.

This approach has led to important policy and management recommendations in agriculture (Levins and Vandermeer 1989; Levins 1986) and fisheries (Li and Moyle 1981; Barnhisel and Kerfoot 1994).

QUALITATIVE ANALYSIS WITH A SYMBOLIC PROCESSOR

We shall here present simple procedures for determining stability criteria for complex systems and assessing the effects of input. The procedures are specific for MATHCAD, but are easily modified for any symbolic processor. The examples are from Puccia and Levins (1985), which remains the basis of qualitative analysis in biology but presents rather tedious hand procedures. Readers may wish to compare them in order to appreciate the ease provided by modern PC programs.

To analyze a model, it is easiest to first illustrate it as a signed digraph (directed graph). Here, for example, the graph may represent a plant (1), herbivore (2), and predator (3). The double arrows represent linearizable relationships, such as Lotka–Volterra, while the self-feeding bubble arrows represent self-regulation. Nonlinearizable functions, such as functional responses, may be accommodated, but are not addressed below.

The arrows of the graph are then converted into positive (+1), negative (–1), or zero elements

in the so-called *Jacobian* or *community matrix*. An element $a_{i,j}$ should be read as "to i from j."

$$\begin{bmatrix} a_{1,1} & a_{1,2} & a_{1,3} \\ a_{2,1} & a_{2,2} & a_{2,3} \\ a_{3,1} & a_{3,2} & a_{3,3} \end{bmatrix} \quad \text{which for the model above is} \quad \begin{pmatrix} -1 & -1 & 0 \\ 1 & 0 & -1 \\ 0 & 1 & 0 \end{pmatrix}$$

STABILITY OF EQUILIBRIUM

To assess equilibrium and stability criteria, the so-called *Routh–Hurwitz criteria for local stability* must be met. They simply state that feedback must be negative at all levels and that lower levels must be stronger than higher ones. Enter the model matrix and subtract from every diagonal element the symbol λ, the *eigenvalues* or roots of the equation

$$\begin{bmatrix} (-1-\lambda) & -1 & 0 \\ 1 & (0-\lambda) & -1 \\ 0 & 1 & (0-\lambda) \end{bmatrix}$$

In the "Symbolic" menu, "Load Symbolic Processor," select the matrix and evaluate "Determinant of Matrix" symbolically, which yields, $-\lambda^3-\lambda^2-2\lambda-1$, called the *characteristic polynomial,* and is in the general form,

$$F(\lambda) = p_0^n + p_1\lambda^{n-1} + p_2\lambda^{n-2} + \ldots + p_{n-1}\lambda + p_n$$

where n = total number of variables in model. The Routh-Hurwitz criteria are met provided that

1. All coefficients (p_n) are of the same sign (note that absent coefficients mean zero and the test fails),
2. The *Hurwitz determinant* is >0 (i.e., lower feedback is stronger).

The general formula for the Hurwitz determinant is

$$\begin{vmatrix} p_1 & p_3 & p_5 & \cdots & p_n \\ p_0 & p_2 & p_4 & \cdots & p_{n-1} \\ 0 & p_1 & p_3 & \cdots & p_{n-2} \\ \cdot & \cdot & \cdot & \cdots & \cdot \\ 0 & 0 & 0 & \cdots & \dfrac{p_{n+1}}{2} \end{vmatrix}$$

Note that this formula applies to models with an odd number of variables. For an even number, m, use the formula $m - 1 = n$, that is, one less. The Hurwitz determinant of a model with four variables is thus the same as that with three. Therefore, our particular model's Hurwitz determinant is

$$\begin{vmatrix} -1 & 1 \\ -1 & 2 \end{vmatrix} = 1$$

This technique may fail to detect conditionally stable systems that have symmetrical loops of three or more components, such as caused by omnivory. However, conditional stability requirements of these are simply that the strengths of the interactions be equal in both directions. We will assume throughout this chapter that such systems are "stable," since the condition is easily met.

ASSESSING THE EFFECT OF INPUT INTO A SYSTEM

To obtain a table of predictions of the model, follow these procedures. Input is in terms of changes in or affecting fecundity or mortality, and output is in change in population size. Evaluate the determinants of matrices in which the row where input occurs has been replaced by a row containing the value -1 (for a positive input) in the target parameter and zeroes in the other parameters, and then divide by -1 for an odd number of variables and $+1$ for an even number. Evaluate the results as subscripted matrix elements, using the icon Xi, making sure that you have "ORIGIN = 1" above the evaluations. For the input of A into A, $a_{1,1}$, in the first model above, the first row (-1 -1 0) is replaced by (-1 0 0), thus,

$$a_{1,1} := \frac{\begin{vmatrix} -1 & 0 & 0 \\ 1 & 0 & -1 \\ 0 & 1 & 0 \end{vmatrix}}{1} \quad \text{up to} \quad a_{3,3} := \frac{\begin{vmatrix} -1 & -1 & 0 \\ 1 & 0 & -1 \\ 0 & 0 & -1 \end{vmatrix}}{1}$$

then evaluate matrix 'a'

Alternatively, 'a' can be obtained by multiplying the community matrix by -1, then "Invert" and "Transpose."

$$a = \begin{pmatrix} 1 & 0 & 1 \\ 0 & 0 & 1 \\ 1 & -1 & 1 \end{pmatrix}$$

This is a "table of predictions" or transpose of the inverse matrix. To interpret, read a particular row as the resultant effects of positive input to the particular variable on the system. Thus, for the first row, positive input into (1) will lead to its own increase, no change on (2) and increase in (3); negative effects are simply reversed. The values may not always be unity, but it is the sign that is important for this purpose. Note that although population size may not change, population rage structure may do so.

If you have more than one input, simply add them. A problem arises if the signs are opposite (say you have positive input in parameter 1 and negative input into 2). Then (1 0 1) would be added to (0 0 –1) and result in (1 0 ±). The effect on the third parameter would therefore depend on the strength of inputs at the two other parameters. To construct a model from a series of observations on predictions, take the matrix of the predictions and go through the prediction procedures (i.e., get a table of predictions from the table of predictions). The resultant matrix will be the model.

CONDITIONAL STABILITY CRITERIA

To obtain conditions under which a stable equilibrium occurs, given that the above criteria are not fully met, rewrite your matrix using only signed symbols (but with zeroes kept as such) and evaluate the determinant. Thus for the following model, the so-called "keystone predator," which fails both the above tests,

$$\begin{bmatrix} -\lambda & a12 & a13 \\ -a21 & -a22-\lambda & -a23 \\ -a31 & -a32 & -a33-\lambda \end{bmatrix}$$

$-\lambda^3 -(-a22 - a33)\lambda^2 -(-a21a12 - a22a33 - a31a13 + a23a32)\lambda -(a31a12a23 - a21a12a33 + a21a13a32 - a31a13a22)$, which will again be in the form, $F(\lambda) = p_0\lambda^3 + p_1\lambda^2 + p_2\lambda^1 + p_3$. You can extract p_n with "Polynomial Coefficients" and readily evaluate the criteria conditions, in this case, $a31a12a23 - a21a12a33 + a21a13a32 - a31a13a22$, that overall feedback be negative, $a23a32 - a22a33 < a21a12 + a31a13$, competition between 2 and 3 must be weaker than the predator links. For the second criterion, the determinant can be calculated from the Hurwitz matrix stating the last condition, often nonintuitive,

$$\begin{bmatrix} (-a22 - a33)(a31 \cdot a12 \cdot a23 - a21 \cdot a12 \cdot a33) + a21 \cdot a13 \cdot a32 - a31 \cdot a13 \cdot a22 \\ -1 \qquad (-a21 \cdot a12 - a22 \cdot a33 - a31 \cdot a13 + a23 \cdot a32) \end{bmatrix}$$

$a22a23a32 + a33a23a32 < a22a21a12 + a22^2a33 + a22a33^2 + a33a31a13 + a31a12a23 + a21a13a32$, in essence that the competitive relationships be weaker than the predatory ones.

For a conditionally stable system, the table of predictions may not be totally accurate and they must be analyzed symbolically. Let us take the keystone predator relationship. To evaluate input for a_{11}, go through the procedures for calculating predictions, first numerically and then symbolically, and divide by the determinant of the whole system. Thus, starting with a_{11},

$$\begin{vmatrix} -1 & 0 & 0 \\ -a21 & -a22 & -a23 \\ -a31 & -a32 & -a33 \end{vmatrix}$$ select the expression $\dfrac{(-a22 \cdot a33 + a23 \cdot a32)}{(-a21 \cdot a12 \cdot a33 + a21 \cdot a13 \cdot a32 + a31 \cdot a12 \cdot a23 - a31 \cdot a13 \cdot a22)}$

$$\begin{vmatrix} 0 & a12 & a13 \\ -a21 & -a22 & -a23 \\ -a31 & -a32 & -a33 \end{vmatrix}$$ and Shift F - 9

and so on for the other nine elements of this matrix. Then determine which numerators conflict with the R–H criteria. In this particular case, input at a_{11} yields the only difference which, if Routh–Hurwitz criteria are met, indicates a negative effect, since although the numerator is positive, the denominator, or overall feedback, must be negative.

SYSTEMS OF EXTREME COMPLEXITY

Qualitation analysis is a powerful tool but nevertheless has practical computational limits, typically of 10 parameters. In addition, it assumes that equilibrium levels are fairly constant and that the system can recover rapidly. One can still draw important conclusions as these assumptions are dropped, but a more important restriction faces field ecologists and managers, which is that the system is so opaque, newly observed, or so large that it cannot be understood within a reasonable amount of time or funds. Two techniques permit further insight. The first is to simplify the system through a hierarchical order (O'Neill et al. 1986; Tansky 1978), which then reverts the problem back to qualitative analysis. Beyond that, two further mathematical considerations allow investigators to better understand the introduction of exotics into poorly understood complex systems. The two crucial equations that bear on the problem are, first, the variance of eigenvalues (Levins 1975),

$$\text{var}(\lambda) = \text{var}(a_{ii}) + (n - 1)\frac{a_{ij}a_{ji}}{2}$$

var (eigenvalue) = var (self - regulation) + (no. species) *

arith. mean (predator - prey interactions)

which will be positive as long as eigenvalues all have real solutions, in which case population will recover from a disturbance in a nonoscillatory fashion, characteristic of global stability. A large value for the variance would indicate, however, that recovery will be very different for the various populations, potentially indicating some dissociation. Eigenvalues can also have imaginary components to their solutions, in which case variance may be negative and recovery oscillatory or unstable.

The introduction of a highly predatory exotic could thus lower variance, by increasing the mean of predator–prey interactions which is negative, on first appearance a desirable trait, but it will also bring it closer to being negative and introduce oscillations. The important aspect of this equation is that as the number and or strength of predator–prey interactions increases, then variance decreases, since the components of the system respond similarly but approaches a region of oscillation, that is of negative variance. There can thus be a sudden transition from uniformity to unpredictability.

The other equation relates predator–prey interactions to connectance (Tansky 1978),

$$\frac{\left|a_{ij}a_{ji}\right|}{2} \le \frac{\sqrt{a_{ii}a_{jj}}}{\sqrt{l_i l_j}}$$

$$\text{Absolute value of arith. mean (predator - prey interactions)} < \frac{\text{geom. mean (self - regulation)}}{\text{geom. mean (connectance)}}$$

Thus, if a predatory exotic interacts very strongly or with many native species, connectance in the system must decrease (simplification) or oscillations will increase. A highly omnivorous species introduced in a highly connected system might thus throw a system into chaos before it eventually simplifies greatly.

These two equations allow either to anticipate or to analyze the results of introductions in general. Overall, the equations tie in four variables, namely, number of community components (n), connectance (l), strength of predator–prey interactions ($a_{ij}a_{ji}$), and degree and variability of self-regulation (a_{ii}). We suggest that qualitative understanding of these four components allow managers to interpret or predict effects of introduction into existing systems.

These effects, however, are not comforting, in that one can generally expect a short-term increase in oscillations and unpredictability, but a long-term increase in likelihood of extinction of species and hence eventual simplification. The importance of oscillations in biological populations is that it brings populations closer to the extinction level, increases the likelihood of extinction (Levins 1975) and thus brings about simplification through loss of either diversity or connectance or both.

INTRODUCTION OF A SINGLE SPECIES

Overall, introduction of a single species can be interpreted in five important forms, namely,

(1)

1. As a satellite system, wherein the exotic establishes a predator–prey relationship with a single species, called a principal, in the community. The important consideration is that a satellite will buffer a principal from any input except through the satellite and that no input to the principal will have any effect anywhere except on the satellite. From a management point of view, an exotic satellite splinters a system.

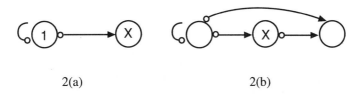

2(a)   2(b)

3. As a component of the community, but with more than one pair of links. In this case, qualitative analysis is required to assess change in the system. Overall there would be less impact if the exotic becomes a link in a trophic chain (2a); note, however, that variance of eigenvalues would change and may lead to unpredictability, as explained below. If the exotic causes the formation of an omnivory loop with the same number of species but

increases connectance, either as prey (2b) or as predator, then conditional equilibrium would occur and most likely a sharp increase in oscillations.

(3)

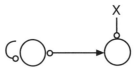

3. As an input or modification to input to another species. In this case of negative input, the exotic is not a direct community participant, but acts as a modifier of birth or death rate to another species or component of the community, such as a nutrient. Again, qualitative analysis will assess its impact. Such indirect effects are often neglected by ecologists.

(4)

4. As introducing a positive self-effect in the community. Loop analysis will again determine the results. The most interesting such modification is through the modification of a self-effect into a positive sign. Human intervention in the form of a quota harvest will frequently result in this effect. If human input into the exotic or native species is modified as such, the effects can be dramatic. We are currently exploring this aspect of fisheries (Mason and Rossignol, in prep.). It should also be noted that many pioneering species have this intrinsically unstable tendency of positive self-effect, as when its presence enhances further success or may exhibit depensation (Myers et al. 1995).

(5)

5. An exotic may have a functional response or may induce such a response somewhere in a system. Mathematically speaking, qualitative analysis requires that relationships can be linearized for the analysis to be applicable to parameter size. In the case of Holling's nonlinearizable functional response II, qualitative analysis results will apply to statistical parameters such as variance, and stability will vary according to population size. Puccia and Levins (1985) discuss this situation. Needless to say, it makes for a highly unpredictable situation.

SIMULATIONS AND MODEL ECOSYSTEMS

The purpose of this section is to examine the proposition that the arrangement of nodes (i.e., species) within a trophic network affects its stability. For this reason, the types of interactions limited to paired, "predator–prey" (i.e., $-a_{ij}a_{ji}$) and single, self-regulating (i.e., $-a_{ii}$) or amensal links and parsimoniously eliminated commensal, interference competition, or one-way interactions from the analysis. Trophic networks of two, three, and four species were examined, and a system was considered stable if criteria described in the previous section were met. The role of self-regulating loops for each trophic network was determined by adding loops to the network in a step-wise fashion ranging from none to one for every species. Different permutations of self-regulating loops were assigned in a systematic fashion so that all combinations were exhausted. We were particularly interested in the number of negative feedback loops it would take for each trophic arrangement to be stable. Those trophic networks requiring the fewest self-regulating loops were considered to be more inherently stable.

The arrangement of the trophic system did matter (Figure 30.1, Table 30.1). The percentage of inherently stable systems declined from 81% to 71% when the number of species in the trophic network increased from three to four. Moreover, strong negative correlations between percentage of connectance and the percentage of stable trophic communities were observed ($r = -1.0$ and $r = -0.84$ for three-species and four-species systems, respectively). These results reflect that as systems become more complex, either through adding species or increasing connectance, the architecture of the system becomes more critical to the stability of the trophic network. Simply put, adding species to the system increases the risk of destabilizing the community structure (May, 1973). From this prospective, the risk of community destabilization from an unplanned or unauthorized introduction can be calculated by $1 - (\%$ stability$)$.

ANALYSIS OF MYSIS SHRIMP INVASION OF FLATHEAD LAKE, MONTANA

The mysid shrimp, *Mysis relicta*, invaded the Flathead Lake from lakes in its upper watershed where it was introduced by fisheries managers. Prior to its invasion, the pelagic zone was largely a community composed of a mixture of alien and native fishes. The trophic structure was as follows: phytoplankton was grazed by cladocerans, primarily *Daphnia* and *Bosmina* (Spencer et al. 1991). The primary predators of the cladocera were the introduced kokanee salmon (*Oncorhynchus nerka*), lake whitefish (*Coregonus clupeaformis*), and cyclopoids. The kokanee and lake whitefish were preyed upon by the introduced lake trout, *Salvelinus namaycush* (Figure 30.1). Assuming that the relationships were of equal strengths, this trophic structure was stable. After the introduction of *Mysis*, the community structure became unstable, and a series of events cascaded from the bottom of the food chain to disrupt the structure of the higher trophic levels, including populations of grizzly bears and bald eagles (Figure 30.1). In essence the reason for the collapse seems to be that *Mysis* competed with kokanee for cladocera, and this interaction set off the following chain of events (Spencer et al. 1991). The kokanee declined, which affected the bears and eagles that preyed upon and scavenged carcasses of kokanee in McDonald Creek, a tributary to Flathead Lake. The mysid introduction did benefit fishes dwelling in the deep water of the pelagic zone: lake whitefish and juvenile lake trout and the growth and survival of the juvenile fishes increased markedly (Richard Hauer, Flathead Biological Station, pers. commun.). The prognosis for the native bull trout (*Salvelinus confluentus*) is not good, as they are receiving increasing competition from the growing population of lake trout. As they are oriented to the shallow waters of the incoming tributaries, they have not received any benefits from the mysid invasion (Richard Hauer, Flathead Biological Station, pers. commun.). Given this information, we predict that the pelagic zone will become stable with the following trophic configuration: kokanee will become rare and lake whitefish will become the dominant forage for lake trout (Figure 30.2).

ANALYSIS OF FISH COMMUNITIES IN AFRICAN RIFT LAKES

Theoretical Framework

The fish communities of the African Rift Lakes are among the most speciose, at least until recent times (Fryer and Iles 1972; Greenwood 1974; Lowe-McConnell 1987). From both historical and experimental evidence, these communities are also very stable (Hori et al. 1993; Kaufman and

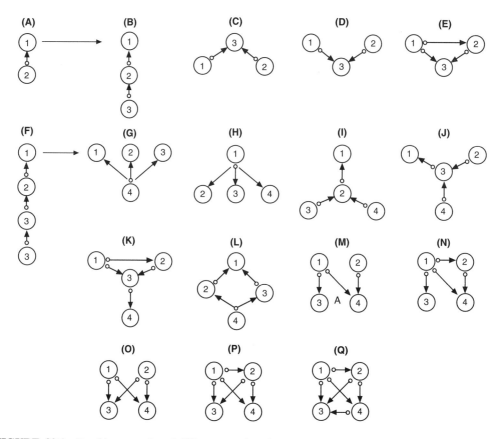

FIGURE 30.1 Trophic networks of different species richness and configurations. The numbered nodes represents species. The arrow with the small circle represents a pair of links, one positive, one negative between two nodes. The head of the arrow denotes positive input in that direction; the small circle denotes negative input in the opposite direction. Note that there are no competitive interactions or self-regulating links presented in these models. See text for details.

Ochumba 1993). This apparently contradicts the finding that adding species to the system can be destabilizing. How do we reconcile this with the real world? Based on qualitative analysis, Levins' variance of eigenvalue, Tansky's connectance formula, the community modeling of de DeRuiter et al. (1995) and Luh and Pimm (1993), and our simulations of model ecosystems, we predict the following characteristics of species-rich communities:

1. The relative strengths of competitive interactions will be weak. This suggests great refinement in niche partitioning (i.e., niche specialization) and a high degree of niche complementarity.
2. The number of interactions are high, but predominantly weak.
3. Interaction strengths will be stronger near the base of the trophic network than at higher trophic levels.
4. The number of alternative, locally stable, trophic networks are few. However, ecosystems may be hierarchically composed of a number of different spatially segregated, communities. Within each of these communities, trophic partitioning occurs by habitat units. Interactions within each habitat trophic unit may be strong, but interactions among units should be weak (Figure 30.3).

TABLE 30.1
Stability Analysis of Communities Varying in Species Richness and Trophic Configuration

Configuration Type	Nodes	C_{max}	C_{stable}	% C_{max}	FB_{min}	% Stability	Stable Combinations at Levels Below FB_{min}
A	2	4	3	75	1	75	NA
B	3	7	5	71	1	88	NA
C	3	7	6	86	2	75	1 or 2 (at level 1)
D	3	7	6	86	2	75	1 or 2 (at level 1)
E	3	7	5	71	1	88	NA
F	4	10	7	70	1	94	NA
G	4	10	9	90	3	50	1,2 or 2,3 or 1,3 (at level 2)
H	4	10	9	90	3	50	1,2 or 2,3 or 1,3 (at level 2)
I	4	10	9	90	3	50	1,3 or 1,4 or 3,4 (at level 2)
J	4	10	9	90	3	50	1,3 or 1,4 or 3,4 (at level 2)
K	4	12	9	75	1	94	NA
L	4	12	11	92	3	56	1,2 or 1,3 or 2,4 or 3,4 (at level 2)
M	4	10	7	70	1	94	NA
N	4	12	9	75	1	94	NA
0	4	12	11	92	3	56	1,4 or 2,4 or 1,3 or 2,4 (at level 2)
P	4	14	13	93	3	75	3 or 4 (at level 1) 1,2 or 1,4 or 2,3 or 1,3 or 2,4 (at level 2)
Q	4	16	13	81	1	94	NA

Notes: Letters in the left-hand column denote different trophic configurations, as shown in Figure 30.1. Nodes = numbers of species, C_{max} = maximum number of links connecting nodes, C_{stable} = the minimum number of pathways needed to stabilize the community network, FB_{min} = the number of self-regulating feedback loops at C_{min}. Stable Combinations identifies node combinations by node code (see Figure 30.1). Stability was determined when both Routh–Hurwitz criteria were met (see details in text).

5. If the ecosystems show strong seasonal patterns, the core trophic structure should be stable, but peripheral elements may change as a phenological phenomenon (Lane 1986).
6. Self-regulation will be strong and common.

Evidence from Rift Lakes

The predictions are in concordance with the findings from the East African Rift Lakes. Niche partitioning, niche complementarity and niche diversification are high (Fryer and Iles 1972; Greenwood 1974; Witte 1981; Lowe-McConnell 1987; Witte and van Oijen 1990; Goldschmidt et al. 1990; Hori et al. 1993). Agonistic behavior is frequent, but territorial behavior, a form of self-regulation, lessens exploitation competition by maintaining spatial segregation. Competitive interaction strength among species is modified and lessened through commensal behavior in feeding groups. Hori et al. (1993) calls this "exploitative mutualism," that is, foraging by one species aids the other because each employs a different strategy, which minimizes escape strategies by the prey.

In one area of Lake Tanganyika, the number of predator–prey interactions are large; but, as judged by percent utilization of various prey, most of the interactions are weak (Table 30.2). More than 80% of the interactions were of 5% utilization or lower. Interestingly, of the 17 prey that the scale-eaters had in common with the large piscivorous fishes, there was only an overlap of 5 species between functional feeding groups. And of course, scale-feeding is more akin to grazing than predation. In accordance with the findings of de DeRuiter et al. (1995), interaction strengths are stronger at lower trophic levels than at the top (Table 30.2).

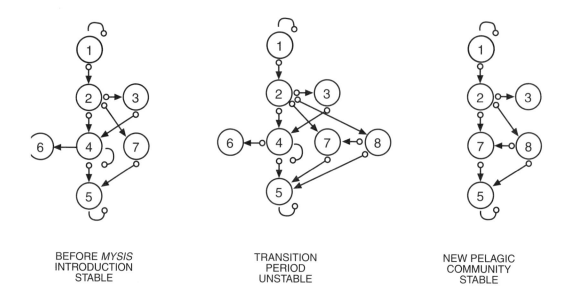

BEFORE *MYSIS*
INTRODUCTION
STABLE

TRANSITION
PERIOD
UNSTABLE

NEW PELAGIC
COMMUNITY
STABLE

FIGURE 30.2 Trophic networks of Flathead Lake before, during, and the predicted community after the invasion of *Mysis relicta* into Flathead Lake, Montana. 1 = algae, 2 = cladocerans, 3 = copepods, 4 = kokanee salmon, 5 = lake trout, 6 = scavengers and predators on spawning salmon, 7 = lake whitefish, 8 = *Mysis*.

Fish communities of the Rift Lakes are arranged in a hierarchy of spatial allotopy. For instance, Lowe-McConnell (1993) suggests that there are three main fish communities: littoral–sublittoral fishes (30–40 m depth), benthic and bathypelagic fishes (40-m oxygen limit), pelagic zone. Littoral and sublittoral communites are much more complex than those in the pelagic zone. The littoral–sublittoral communities can be isolated from each other, each having a unique faunal segment (Hori et al. 1993; Kaufman and Ochumba 1993). In Lake Malawi, for example, islands are isolated by deep water and fauna around each island had its own endemic fishes (Lowe–McConnell 1993). Within each community there are groups of fishes that are associated together in particular habitats stratified horizontally or vertically by trophic pattern. We have analyzed the impacts of the introductions of Nile perch using this hierarchical approach to community structure.

Analysis of the Introduction of Nile Perch

Our depiction of the Lake Victoria trophic network is highly abstract. There are over 15 trophic groups within the haplochromine fishes alone, and there were well over 300 species by various estimates of *Haplochromis*. We modeled the sublittoral–littoral zone where haplochromines were abundant and placed them into three general functional feeding groups associated with different habitat trophic units: zooplanktivores, detrivores, and carnivores. There were certain fishes that had links to all three trophic units. These were integrated as links between them. Before the introduction of Nile perch (*Lates niloticus*), all the habitat trophic units were stable (Figure 30.4).

The community as a whole was examined by using each habitat trophic unit as a node with the exception of three linking species. One of these links was the exotic omnivore, *Oreochromis niloticus*. The system was not stable, but had high inertia because the system conserved input, that is no feedback at the highest level, meaning that the system may change, but is highly resistant to

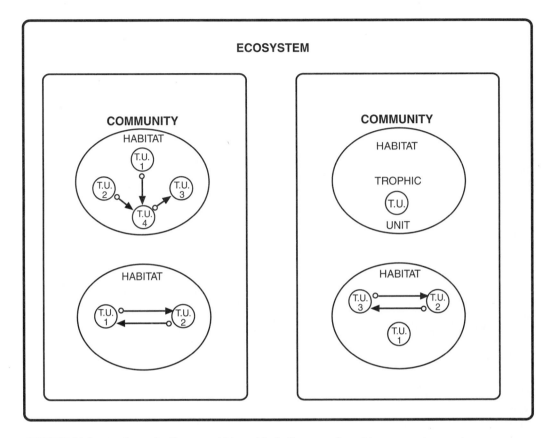

FIGURE 30.3 A schematic diagram of hierarchical allotropy of trophic systems in complex ecosystems. Ecosystems are composed of different communities within which there are different trophic units operation in different habitats. T.U. = trophic unit. The arrow with the small circle represents a pair of links, one positive, one negative, between two nodes. The head of the arrow denotes positive input in that direction; the small circle denotes negative input in the opposite direction.

input, whether from outside or through natural selection (Figure 30.4). Introduction of the Nile perch destabilized the system (Figure 30.4), resulting in a new depauperate community that is stable (Figure 30.4). We suggest that the Nile perch and *Oreochromis nilotica* destabilized the system by being eurytopic; therefore, spatial allotropy, which contributed to the overall stability of the community, ceased to function. Secondly, interaction strength increased. Nile perch ate haplochromines exclusively until haplochromine standing stocks were extremely rare (Goldschmidt et al. 1993). *Rastrineobola argentea,* a zooplanktivorous cyprinid, was not a preferred prey item of Nile perch formerly, and it has now grown to large numbers and replaced the approximately 20 species of haplochromine zooplanktivores (Hughes 1986; Goldschmidt et al. 1993). The native catfish piscivores have been replaced by the Nile perch and have disappeared and the native, microphagous shrimp, *Caradina nilotica,* has also proliferated and has replaced the detritivorous haplochromines, once comprising 31% of the haplochromine community (Goldschmidt et al. 1993). There are other contributing factors such as eutrophication and overfishing, but that is beyond the scope of this chapter. Ultimately, this means that the system will be difficult, if not impossible to restore. Luh and Pimm (1993) concluded from mathematical simulations that reassembling damaged communities may be impossible even if all the species are available for stocking. We suggest that complexity and diversity of a community is a consequence of natural selection and therefore itself as unique as a species.

TABLE 30.2
Strengths of Predator–Prey Interactions of Fishes at One Site in Lake Tanganyika by Trophic Guild

Interaction Strength (% of Utilization)	No. of Interactions	Trophic Interactions
90	9	algivores-algae (3 types)
		larval piscivore-larval fishes
		small fishes-shrimp
60	12	small fishes-algae
		small fishes-shrimp
		small invertivores-small invertebrates
30	10	general algivore-algae
		small fish-larval fishes and invertebrates
		algivore-detrivore-algae and detritus
5	115	large piscivores-small fishes
<5	29	scale predators-large piscivores and small fishes

Source: Hori, M., M.M. Gashagaza, M. Nshombo, and H. Kawanabe, Littoral Fish Communities in Lake Tanganyika: Irreplaceable Diversity Supported by Intricate Interactions among Species. *Conservation Biology,* 7, pp. 657–666, 1993. With permission.

CONCLUSIONS

Qualitative analysis of signed digraphs is a powerful tool to understand ecosystems and communities. Although its predictions lack precision, we can see that it can represent the outcome of perturbations to the community with generality and realism. From examining communities, we argue that qualitative analysis may be more reliable than conventional quantitative modeling approaches. Data for large complex systems are seldom gathered during the same period of time because of the enormity of field or experimental studies; therefore, the various transfer coefficients and initial values of variables may be out of temporal context. Such data sets are also site-specific. The use of qualitative values simplifies the modeling effort.

Qualitative analysis is a particularly useful tool for analyzing problems introduced species may cause. Modeling is a good tactic to use in terms of adaptive management (Walters 1986). However, conventional models require data that are seldom available and gathering that data takes time, time that could be spent managing adaptively to minimize deleterious ecological change. Using qualitative analysis, one can construct hypothetical communities from the information at hand. Several alternative communities might be constructed because the relationships and interactions are unclear. These alternative communities may be weighted in terms of their likelihood or simply given equal weight. The ecological risk may be assessed by determining the percent of the alternative, hypothetical communities are stable. In Bayesian fashion, critical research can be designed by examining model outputs to determine critical interactions (either stabilizing or destabilizing links) and potential keystone species. Management can then be used experimentally, as an adaptive management tool, *sensu* Walters (1986).

In summary, the great universal consequence of introductions may be simplification of communities in the recipient community either through direct or indirect interactions through the trophic

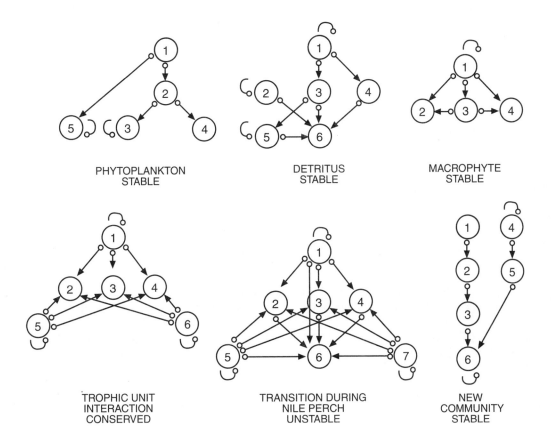

FIGURE 30.4 Trophic networks arranged in terms of hierarchical allotopy in Lake Victoria, before, during and after the introduction of exotic fishes, especially the Nile perch. PHYTOPLANKTON TROPHIC UNIT (1 = phytoplankton, 2 = zooplankton, 3 = zooplanktivorous haplochromine fishes, 4 = *Rastrineobola argentea*, 5 = phytoplanktivorous haplochromine fishes.). DETRITUS TROPHIC UNIT (1 = detritus, 2 = detritivorous haplochromine fishes, 3 = molluscs, 4 = *Caradina nilotica*, 5 = benthic feeding haplochromines, 6 = *Bagrus docmac*). MACROPHYTE TROPHIC UNIT (1 = macrophytes, 2 = *Oreochromis niloticus*, 3 = insects, 4 = *Clarias mossambicus*). TROPHIC UNIT INTERACTION (1 = phtyoplankton trophic unit, 2 = *Oreochromis niloticus*, an alien fish, 3 = *Schilbe mystus*, 4 = *Clarias mossambicus*, 5 = macrophyte trophic unit, 6 = detritus trophic unit). TRANSITION (7 = Nile perch, the rest as in Trophic Unit Interaction). NEW COMMUNITY (1 = algae, 2 = zooplankton, 3 = *Rastrineobola argentea*, 4 = detritus, 5 = *Caradina nilotica*, 6 = Nile perch).

network, disruption of conditional equilibria or increase in variance of self-regulatory properties. For those reasons we suggest that alien species will rarely be benign; however, there are several situations where change will occur glacially because the system shifts from one which is locally stable to one which is conserved (i.e., no feedback at higher levels of integration). Ultimately, one stable system will be replaced by another although at great cost to biodiversity. Species introductions can be beneficial, but only when the ecosystem has already been compromised. For instance, introduction of Pacific salmonids did benefit the communities of the Laurentian Great Lakes by acting as biological control agents on alien fishes which competed with and preyed on eggs of native zooplanktivores (Smith 1968; Crowder 1980). However, the conservation biologist is confronted by the "species unpacking problem": once structural aspects of trophic structure have been lost they may not be reacquired (Tyler et al. 1982; Luh and Pimm 1993).

REFERENCES

Barnhisel, D.R. and W.C. Kerfoot, Modeling Young-of-the-Year Fish Response to an Exotic Invertebrate: Direct and Indirect Interactions. In *Theory and Application in Fish Feeding Ecology*, Eds. D.J. Stouder, K.L. Fresh, R.J. Feller. The Univ. South Carolina Press, 1994.

Bodini, A., What is the role of predation on stability of natural communities? A theoretical investigation. *BioSystems* 26, pp. 21–30, 1991.

Crowder, L.B., Alewife, Rainbow Smelt and Native Fishes in Lake Michigan: Competition or Predation? *Environmental Biology of Fishes,* 5, pp. 225–233, 1980.

de DeRuiter, P.C., Neutel, A.M., and J.C. Moore, Energetics, Patterns of Interaction Strengths and Stability in Real Ecosystems. *Science*, 269, pp. 1257–1260, 1995.

Fryer, G., and T.D. Iles, *The Cichlid Fishes of the Great Lakes of Africa: Their Biology and Evolution,* Oliver and Boyd, Edinburgh, Scotland, 1972.

Ganzhorn, J., J.S. Rohovec, and J.L. Fryer, Dissemination of Microbial Pathogens through Introductions and Transfers of Finfish. In *Dispersal of Living Organisms into Aquatic Ecosystems,* Rosenfeld, A. and R. Mann, Eds., University of Maryland Sea Grant Program, College Park, MD, 1992.

Giavelli, G., O. Rossi, and E. Siri, Stability of Natural Communities. Loop Analysis and Computer Simulation Approach. *Ecological Modeling,* 40, pp. 131–143, 1990.

Goldschmidt, T., F. Witte, and J. de Visser, Ecological Segregation in Zooplanktivorous Haplochromine Species (Pisces: Cichlidae) from Lake Victoria. *Oikos,* 58, pp. 343–355, 1990.

Goldschmidt, T.F., F. Witte, and J. Wanink, Cascading effects of the Introduced Nile Perch on the Detritivorous/Phytoplanktivorous species in the Sublittoral Areas of Lake Victoria. *Conservation Biology,* 7, pp. 686–700, 1993.

Greenwood, P.H., The Cichlid Fishes of Lake Victoria, East Africa: The Biology and Evolution of a Species Flock, *Bulletin of the British Museum of Natural History (Zoology)*, suppl. 6, pp.1–134, 1974.

Hori, M., M.M. Gashagaza, M. Nshombo, and H. Kawanabe, Littoral Fish Communities in Lake Tanganyika: Irreplaceable Diversity Supported by Intricate Interactions among Species, *Conservation Biology,* 7, pp. 657–666, 1993.

Hughes, N.F., Changes in the Feeding Biology of the Nile Perch, *Lates niloticus* (L.) (Pisces: Centropomidae), in Lake Victoria, East Africa Since its Introduction in 1960, and its Impact on the Native Fish Community of the Nyanza Gulf, *Journal of Fish Biology*, 29, pp. 541–548, 1986.

Jeffries, C., Qualitative Stability and Digraphs in Model Ecosystems. *Ecology*, 55, pp. 1415–1419, 1974.

Kareiva, P., Developing a Predictive Ecology for Non-indigenous Species and Ecological Invasions. *Ecology*, 77, pp. 1651–1652, 1996.

Kaufman, L. and P. Ochumba, Evolutinary and Conservation Biology of Cichlid Fishes as Revealed by Faunal Remnants in Northern Lake Victoria. *Conservation Biology*, 7, pp.719–730, 1993.

Lane, P.A., Symmetry, Change, Perturbation, and Observing Mode in Natural Communities, *Ecology*, 67, pp. 223–239, 1986.

Levins, R., The Qualitative Analysis of Partially-specified Systems. *Annals of the New York Academy of Science*, 231, pp. 123–138, 1974.

Levins, R., Evolution of Communities Near Equilibrium. In *Ecology and Evolution of Communities*. Eds. Cody, M.L. and J.M. Diamond. Belknap Press. 1975.

Levins, R., Perspective on IPM: From an Industrial to an Ecological Model. In *Ecological Theory and Integrated Pest Management*. Ed. M. Kogan. Wiley, New York. 1986.

Levins, R. and J.H. Vandermeer, The Agroecosystem Embedded in a Complex Ecological Community. In *Agroecology*, Ed. C.R. Carrol, J.H. Vandermeer and P. Rosset. McGraw-Hill, New York. 1989.

Li, H.W. and P. B. Moyle, Ecological Analysis of Species Introduction into Aquatic Systems. *Tansactions of the American Fisheries Society*, 110, pp. 772–782, 1981.

Lowe-McConnell, R.H., *Ecological Studies in Tropical Fish Communities*, Cambridge University Press, Cambridge, England,1987.

Lowe-McConnell, R.H., Fish Faunas of the African Great Lakes: Origins, Diversity, and Vulnerability, *Conservation Biology,* 7, pp. 634–643, 1993.

Luh, H.K. and S.L. Pimm, The Assembly of Ecological Communities: a Minimalist Approach, *Journal of Animal Ecology,* 63, pp. 749–765, 1993.

Mason, S.J., Feedback Theory — Some Properties of Signal Flow Graphs. *Proceedings of the Institute of Radio Engineering,* 41, pp. 1144–1156, 1953.

May, R.M., Qualitative Stability in Model Ecosystems. *Ecology,* 54, pp. 638–641, 1973.

Maybee, J. and J. Quirk, Qualitative Problems in Matrix Theory. *SIAM Review,* 11, pp. 30–51, 1969.

Myers, R.A., N.J. Barrowman, J.A. Hutchings, and A.A. Rosenberg, Population Dynamics of Exploited Fish Stocks at Low Population Levels, *Science,* 269, pp. 1106–1108, 1995.

O'Neill, R.V., D.L. DeAngelis, J.B. Waide, and T.H.F. Allen, *A Hierarchical Concept of Ecosystems.* Princeton Monographs in Population Biology #23, 1986.

Pilette, R., R. Sigal, and J. Blamire, The Potential for Community Level Evaluation Based on Loop Analysis. *BioSystems,* 21, pp. 25–32, 1987.

Puccia, C.J. and R. Levins, *Qualitative Modeling of Complex Systems. An Introduction to Loop Analysis and Time Averaging.* Harvard Univ. Press. 1985.

Quirk, J., and R. Ruppert, Qualitative economics and the stability of equilibrium. *Review of Economic Studies,* 32, pp. 311–326, 1965.

Rejmánek, M. and D.M. Richardson, What Attributes Make Some Plant Species More Invasive? *Ecology,* 77, pp. 1655–1660, 1996.

Roughgarden, J., Coevolution in Ecological Systems: Results from "Loop Analysis" for Purely Density-dependent Coevolution. In *Measuring Selection in Natural Populations.* Eds. F. B. Christiansen and T. Fenchels. Lecture Notes in Biomathematics, vol. 19. Springer-Verlag, New York, 1977.

Smith, S.H., Species Succession and Fishery Exploitation of the Great Lakes. *Journal of the Fisheries Research Board of Canada,* 25, pp. 667–693, 1968.

Spencer, C.N., B.R. McClelland, and J.A. Stanford, Shrimp Stocking, Salmon Collapse, and Eagle Displacement, Cascading Interaction in the food web of a large aquatic system. *BioScience,* 41, pp.14–21, 1991.

Tansky, M., Stability of Multispecies Predator-prey System. *Memoirs of the Faculty of Science, Kyoto University,* 7, pp. 87–94, 1978.

Tyler, A.V., W.L. Gabriel, and W.J. Overholtz, Adaptive Management Based on Structure of Fish Assemblages of Northern Continental Shelves, *Canadian Special Publication of Fisheries and Aquatic Sciences,* 59, pp 149–156, 1982.

U.S. Congress, Office of Technology Assessment, *Harmful Non-indigenous Species in the United States.* OTA-F-565. U.S. Government Printing Office, Washington, DC, 1993.

Walters, C.J., *Adaptive Management of Renewable Resources,* MacMillan, London, U.K., 1986.

Witte, F., Initial Results of the Ecological Survey of the Haplochromine Cichlid Fishes from the Mwanza Gulf of Lake Victoria: Breeding Patterns, Trophic and Species Dstribution, *Netherlands Journal of Zoology,* 31, pp. 175–202, 1981.

Witte, F. and M.J.P. van Oijen, Taxonomy, Ecology and Fishery of Lake Victoria Haplochromine Trophic Groups. *Zoologische Verhandelingen, Monograph National Naturhistoriche Museum,* Leiden, Germany, 1990.

Appendix 1

GLOSSARY FOR MOLLUSCS

acute	Forming an angle of less than 90°.
angulate	Forming an angle; not rounded.
annulate	Ringed; surrounded by a ring of a different color; formed in ringlike segments.
annulation	Formation of ring-like parts or subdivision of a segment.
annulus	A ring or division.
anterior	In the direction of the head; the opposite of posterior. In bivalve shells, the end with the foot, usually the narrower end.
aperture	Opening, usually referring to opening of snail shell.
apex	The tip or top of a snail shell.
apical	Toward the tip of a structure or appendage.
arch	A curved structure.
asymmetrical	Lacking in symmetry; not capable of being divided into equal or corresponding parts.
basal	At, or pertaining to, the base or point of attachment; nearest the main body.
base	The lowest part of a body or structure; the portion upon which a structure rests; the broadest portion of a cone-shaped structure.
beak	The umbo in bivalves; the apex of the ventral valve of a brachiopod.
benthic	Of, or pertaining to, the bottom of a body of water or the organisms living at the bottom.
bifid	Cleft, or divided into two parts; two-pronged.
bifurcate	Divided partly, or forked into two.
bilamellate	Divided into two lamellae or plates.
bilobed	Having two lobes.
bivalved	Having a shell consisting of two valves or valve-like parts.
blood	In most freshwater molluscs, blood is clear fluid containing hemocyanin.
body whorl	The last, and largest, whorl of a snail shell.
calcareous	Of the nature of, consisting of, or containing calcium carbonate.
carina	An elevated ridge or keel.
cilia	Fringes; series of moderate or thin hairs.
cleft	Split, partly divided longitudinally.
compressed	Having a low width/height ratio; flattened in such a way that the body is deep and narrow, the opposite of depressed.
concave	Having a curved, depressed surface.
concentric	A series of circles or ellipses having a common center.
conical	Cone-shaped.
contiguous	Adjacent to; in contact with; adjoining.
convex	Having a curved rounded exterior, as that of a sphere.
corneous	Hornlike; hardened proteinaceous layer.
crenation	One of the rounded projections on a scalloped edge.
crenulate	With small scallops, evenly rounded and rather deeply curved.
crest	A prominent ridge.

crystalline style	A rod containing digestive enzymes in the stomach of bivalves
deflected	Bent down or turned to one side.
denticle	A small tooth.
denticulate	Set with little teeth or notches.
depressed	Flattened in such a way that the body is low and wide; having a low height/width ratio; the opposite of compressed.
detritus	Loose material; rock fragments and organic material; can be suspended in water column or part of benthos.
dextral	To the right of the median line, or, for snails, coiled to the right.
differentiated	Modified and specialized for the performance of specific functions.
diffuse	Spread out; not localized.
discoidal	Disk-shaped; shaped like a round plate.
dissections	Deeply cut divisions or lobes.
distal	Near or toward the free end of any appendage; the part farthest from the body.
divergent	Spreading out from a common base.
dorsal	Toward the upper surface when the body is in normal walking position; the opposite of ventral.
elliptical	Shaped like an ellipse, with an oval body rounded at both ends.
elongate	Lengthened; slender; longer than wide.
emarginate	Having the margin notched or indented.
expanded	To open outward or spread.
exposed	Open to view; not shielded or protected.
filament	In bivalves, a tube-like structure on the gill. Several filaments make up a gill lamellae.
foot	Single in molluscs. It has creeping, crawling, or burrowing function.
frontal	Of, or pertaining to, the anterior portion of the body or a part.
fusiform	Spindle-shaped; tapered at each end.
gill	Structure especially adapted for the exchange of dissolved gases between a mollusc and a surrounding liquid. In molluscs it is also called a *demi*branch.
globose	Approaching the shape of a sphere.
globule	A small spherical body.
glochidium	A larval unionacean. Most are parasitic on fish.
grooved	Having a narrow channel or depression.
growth lines	In gastropods, raised lines running across the whorls, not to be confused with spiral striae.
hatchet-shaped	Having an ax shape. Bivalve foot is hatchet-shaped.
hinge teeth	Any of various ridges or thickenings on the dorsal side of the interior of a bivalve shell that aid in holding the two halves of the shell together.
incised	Notched or deeply cut into.
incision	The impressed line marking the juncture of two segments.
inclined	To have a bend or slant.
inflated	In a snail shell, refers to swollen, as if the whorls had been pumped full of air. In a bivalve, refers to laterally expanded, as opposed to compressed.
irregular	Not uniform; not conforming to the usual pattern.
keel	An elevated ridge; a carina.
lamella	A sheet or thin plate. In bivalves, each gill is composed of an inner and outer lamella. Plural, lamellae.
lateral	Toward the side of the body; the opposite of mesal.
lentic	Of, or pertaining to, standing water.
limnetic	Pertaining to organisms inhabiting open waters of ponds, lakes and inland seas.

lobate	Composed of, or possessing, lobes or divisions resembling lobes.
lobe	A rounded division or projection of an organ.
longitudinal	Lengthwise; extending along the long axis.
lotic	Of, or pertaining to, flowing water.
macro-	Being large, thick, or exceptionally prominent; of, involving, or intended for use with relatively large quantities or on a large scale.
malleated	Having small flattened areas, as if having been hit with a hammer.
mantle	In molluscs, a thin layer of tissue that secretes the shell and lines the shell's interior.
medial	Lying in the midline of the body.
mesial	Pertaining to the middle; toward the middle; the opposite of lateral.
nacre	The pearly white or colored material covering the inside of a unionid shell.
naiad	Common name for unionid shell.
nuclear whorl	The first whorl on the apex of a snail shell.
obliquely	At or from an angle.
obtuse	Forming an angle exceeding 90°.
operculate	Having an operculum.
operculum	A lid or cover. Present on foot of all prosobranch snails.
organic	Pertaining to, consisting of, or related to organs; pertaining to, originating in, or derived from, living organisms; pertaining to compounds that contain carbon.
ovate	Somewhat oval in shape.
ovoid	See ovate.
patelliform	Limpet-shaped.
paucispiral	Marked by a spiral with few turns, as on operculum of hydrobiid snails.
periostracum	The colored, proteinaceous layer that covers the outside of mollusc shells.
peripheral	Of, toward, or constituting an external surface or boundary; away from a center; toward the surface.
planktonic	Of, or pertaining to, aquatic organisms of fresh, brackish, or sea water that float passively or exhibit limited locomotor activity.
posterior	In the direction away from the head; the opposite of anterior.
process	An elongation of the surface, or a margin, or an appendage; any prominent part of the body not otherwise definable.
proximal	Away from the tip of a structure or appendage; the opposite of distal.
pustule	A large projection lump.
quadrate	Square, or nearly so.
recess	A depression, fossa, cleft, or cavity.
recurved	Bent backward.
reduced	Dimished in size, amount, or number.
respiratory	Of, or pertaining to, the interchange of gases between an organism and its environment.
reticulate	Covered with a network of lines.
retractile	Capable of being drawn in or back.
ridge	A crest or elevation.
robust	Strongly formed or constructed; sturdy.
rudimentary	Of, or pertaining to, an organ or structure that is just beginning to develop or one in which development has been arrested; describing a vestige or the remains of a structure that was functional in an earlier stage of evolutionary development.
secrete	To produce a substance discharged through a duct from a gland.

sedentary	Permanently attached to the substratum; not free-moving; remaining in the same place; not migratory.
segmentation	The state of being divided into segments or parts that are marked off or separated from adjacent parts.
seminal receptacle	A sperm storage structure; spermatheca.
serrate	Sawlike; with notched edges.
serrulate	Finely serrate; bearing minute teeth.
sessile	Attached by the base; without stem or stalk.
seta(e)	Slender, hairlike appendage; hair.
setiferous	Bearing setae.
setose	Furnished or covered with setae or stiff hairs.
shallow	Lacking in depth.
sigmoid	S-shaped; curved in two directions.
simple	Unmodified, not branched.
sinistral	To the left of the median line, or for snails, coiled to the left; opposite to dextral.
sinuous	Curving in and out.
sinus	A cavity, hollow, recess, channel, or space.
siphon	A tube or tubelike structure for drawing in or expelling fluids, as that in bivalve molluscs.
slender	Small or narrow in circumference.
solitary	Existing singly or alone.
sparse	Not thickly grown; scattered.
spatulate	Broad and rounded at the tip, more slender at the base; spoon-shaped.
sphaerical	Having the form of a sphere or globe.
spinous	With many spines.
spinule	A very small spine.
spire	The entire snail shell, excepting the body whorl.
stria	A fine longitudinally impressed line. Plural, striae.
sub-	Prefix; more-or-less.
submersed	Covered with water; growing or adapted to grow under water.
succeeding	Coming next to one another in position; following in order.
suture	A seam or impressed line indicating the division of the distinct parts of the body wall, on a snail shell, separating the whorls.
tapered	Becoming progressively smaller at one end.
terminal	Pertaining to, placed at, or forming an end or extremity; being the last of a series; located at the extreme end of a body or structure.
terminate	To come to, or form, a terminus or end.
tooth	One of a pair of structures on the hinge of a lamellibranch shell that function in locking the valves; a toothlike or pointed process.
transverse	At right angles to the long axis of the body.
truncate	With a shortened end; often squarish.
tubercle	A small bump or pimple-like structure.
tubular	Having the shape of a hollow cylindrical structure; composed of or possessing tubes.
tubule	A small, elongate tubelike structure.
tuft	A small cluster of flexible structures, as hairs, feathers, or blades of grass, closely associated at their bases but with free ends spread apart.
umbilicus	The small opening present in some snail shells behind the base of the aperture.

umbo	The hump on the dorsal margin of a bivalve shell; also called the beak or umbone.
unequal	Of different sizes or proportions.
uniform	Presenting an unvaried appearance of surface, pattern, or colour.
valve	In molluscs, one half of a bivalve shell.
veliger	A larval stage of dreissenid and corbiculid bivalves.
ventral	Toward the lower surface when the body is in normal walking position; the opposite of dorsal.
vertical	Of, or pertaining to, the highest point, summit or apex; upright; perpendicular to the horizontal plane; lengthwise, or in the direction of the long axis.
vestigial	Small or degenerate.
weakly	Not strongly; not overly pronounced.
whorl	One turn of the spiral of a snail shell.

Index

A

Acipenseridae (sturgeon)
introduction in Canada, 405
introduction in Canada and United States, 100
vectors of introduction
    aquaculture, 298
    aquarium/horticulture trade, 130
Adrianichthyidae (ricefish), introduction by
    aquarium/horticulture trade, 131
African Rift lakes, analysis of fish communities in,
    439–444
Agreements
binational, 87–88
international, 88
Alewife, see *Alosa pseudoharengus;* Clupeidae
Alien species, see Nonindigenous aquatic species
Alosa pseudoharengus, 355
Amblyopsidae (cavefish)
introduction in Canada and United States, 103
American bullfrog, see *Rana catesbeiana*
Amiidae (bowfins), introduction in Canada and
    United States, 100
Amphibians, nonindigenous to United States, 5–6
Anabantidae (climbing gouramies), introduction by
    aquaculture, 300
Anguillidae (freshwater eels)
introduction in Canada and United States, 100, 108,
    216
vectors of introduction
    aquaculture, 298
    canals and diversions, 394
Anticipatory policy, 59
Aplexa, see *Stenophysa*
Aplocheilidae (rivulines)
introduction by aquaculture, 299
introduction in Canada and United States, 104
Applesnail, see *Pomacea*
Aquaculture
and baitfish industry, 176
in Canada and United States, 283–287, 297–303
    disease control, 285–286
    fish escape and gene pool, 284–285
    nutrient enrichment by, 286
    sustainability of, 284
in Mexico, 113–123
    ecological impact, 117–123
    risks associated with, 116–117

of molluscs, 305–312
Aquariums, as vectors
of bullfrog introduction, 291–293
of fish introduction, 127–128, 129–133
of mollusc introduction, 135–147, 151–159
Atherinidae (silversides)
introduction as prey species, 73
introduction in Canada and United States, 104, 108
introduction in Mexico, 39–40
Atlantic Coast canals, 361

B

Baitfish industry, 184
aquaculture and, 176
export and import in, 176–177
regulation of, 174–177
Baitfish introduction, 163–177, 181–183, 389
distribution of, 169, 170
ecologic impact of, 170–174, 181–192
evidence for, 164
nontarget organisms and, 183
probability of release of fish from baitbuckets,
    168–170
species of, 165–168
survey of Ontario dealers and users, 168
Ballast, as vector of introduction, 204–208
of fish, 215–216
in Great Lakes, 203–212, 215–216, 273–278
of molluscs, 219–246
solid, 204–205
water, 205–208
Ballast water discharge, see also Ballast
ecological impact of, 210–211
in Great Lakes, 210–211
    microbiological, chemical and physical sur-
    vey of, 273–278
long-term solutions to, 211–212
outstanding issues in control of, 209–210
regulation of, 208–209
Ballast water tank locations, 206
Bass, see Centrarchidae; Moronidae; Percichthyidae
Belonidae (needlefish), introduction by canals and
    diversions, 394
Belontiidae (gouramies)
introduction by aquaculture, 300
introduction by aquarium/horticulture trade, 132